COMPARATIVE SOCIOECOLOGY

Comparative Socioecology

THE BEHAVIOURAL ECOLOGY OF
HUMANS AND OTHER MAMMALS

SPECIAL PUBLICATION NUMBER 8 OF THE
BRITISH ECOLOGICAL SOCIETY

EDITED BY

V. STANDEN
Department of Zoology
University of Durham

R. A. FOLEY
Department of Biological Anthropology
University of Cambridge

BLACKWELL SCIENTIFIC PUBLICATIONS

OXFORD LONDON EDINBURGH

BOSTON MELBOURNE

1989

© 1989 by The British Ecological Society and
published for them by
Blackwell Scientific Publications
Osney Mead, Oxford, OX2 oEL
  (*Orders*: Tel. 0865−240201)
8 John Street, London, WC1N 2ES
23 Ainslie Place, Edinburgh, EH3 6AJ
3 Cambridge Center, Suite 208,
  Cambridge, Massachusetts 02142, USA
107 Barry Street, Carlton
  Victoria 3053, Australia

First published 1989

Set by Setrite Typesetters, Hong Kong,
and printed and bound in Great Britain
at the University Press, Cambridge.

DISTRIBUTORS
USA
  Publishers' Business Services
  PO Box 447
  Brookline Village
  Massachusetts 02147
  (*Orders*: Tel. (617) 524−7678)

Canada
  Oxford University Press
  70 Wynford Drive
  Don Mills
  Ontario M3C 1J9
  (*Orders*: Tel. (416) 441−2941)

Australia
  Blackwell Scientific Publications
  Australia (Pty) Ltd
  107 Barry Street
  Carlton, Victoria 3053
  (*Orders*: Tel. (03) 347 0300)

British Library
Cataloguing in Publication Data
Comparative socioecology. — (Special
  publication of the British Ecological Society;
  no. 8).
  1. Primates. Social behaviour. Ecological
aspects
  I. Standen, V. (Valerie)   II. Foley,
  Robert   III. Series 599.8′0451

ISBN 0−632−02361−9

Library of Congress
Cataloging-in-Publication Data

Comparative socioecology/edited by V. Standen,
  R.A. Foley.
    p.   cm. — (Special publication number 8
  of the British Ecological Society)
    Includes index.
    ISBN 0-632-02361-9
    1. Social behavior in animals.
  2. Sociobiology.   3. Animal ecology.
  4. Human ecology.   I. Standen. V.
  II. Foley, Robert.   III Series: Special
  publication . . . of the British Ecological
  Society: no. 8.
  QL775.C65   1989
  591.5′1   dc19

Contents

Preface

Socioecology is the study of the relationships between social behaviour and natural resources. The idea that the resources available to a population in a particular area will have an influence upon its social organization is an old one in both zoology and anthropology. John Crook's pioneering work with African weaver birds and primates, and the ethnographies reported in Daryll Forde's book, *Habitat, Economy and Society*, are early examples of attempts to understand how birds, mammals and humans are constrained in their behaviour by the environments in which they live. However, the fortune of socioecology in anthropology has been rather different from its fortune in zoology. In biology the study of social behaviour was boosted by the increased interest in evolutionary theory, whereas in anthropology it faltered. Through the work of W.D. Hamilton, R.L. Trivers, G.C. Williams, Richard Dawkins, E.O. Wilson and many others, socioecology became far more explicitly evolutionary, and behaviour was seen as a direct consequence of natural selection operating through the differential reproductive success of individuals. For many anthropologists, however, the growth of sociobiology was a major problem, for much of human social behaviour did not seem to relate closely to an individual's reproductive success, and neither could a close relationship be shown between behaviour and genetic variability. Furthermore, few anthropologists were happy with the commonly made assumption that inter-specific differences among animals were comparable to inter-societal or cultural differences within the human species. Despite these difficulties, however, an interest in socioecology remained both within and outside anthropology.

In April 1987 a conference was held at the University of Durham on the subject of comparative socioecology. The idea behind this meeting was to bring together people working on the socioecology of both humans and other mammals to see whether, when considered in a comparative context, human social and ecological variability follows patterns seen in other species. It was never intended, however, that the conference should stand or fall on how well humans could be made to fit the socioecological framework. Rather, humans were to be just one more species, to be treated alongside others that could often be studied in more detail. Before the problem of humans and socioecology could be dealt with it was necessary for the overall principles and patterns to be discussed. Of particular importance in constructing this comparative framework was the increasing recognition among biologists that behaviour is frequently not 'species-specific' — that is, among many species, not just among humans, behaviour may vary from one population to another according to ecological conditions. This concept of intra-specific variability is central to the task of reconsidering the socioecology of humans. The themes of the conference followed from these concerns, and the sessions were organized to tackle some of the major questions in the subject: Why be social? What determines the size and structure of a social group? What are the principal influences on mating strategy, reproductive rate and parenting behaviour? This book arises from the Durham conference, and most of the papers presented there are included.

What is absent, regrettably, is the often lively discussion that took place, but we hope that while not being comprehensive, this book fulfils two main aims: one, to elucidate the environmental factors that may shape the social strategies of individuals and so the social structures of populations; and second, to see whether humans fit the same patterns we see for other mammals.

The Durham conference was supported by grants from the British Ecological Society, the Royal Anthropological Institute of Great Britain and Ireland, and the Mammal Society, and we wish to thank them for their generous help and encouragement of this inter-disciplinary venture. We would also like to thank the University of Durham for providing the facilities to hold the meeting. Many individuals also contributed to the success of the meeting. An initial steering committee included Robin Dunbar, Geoffrey Harrison and Tim Ingold, and we are grateful to them, and to Monique Borgerhoff Mulder, Phyllis Lee and Nigel Dunstone, for helping determine the academic content of the meeting. We are also grateful to all the people who reviewed the manuscripts so promptly. The smooth running of the sessions was greatly enhanced by those who chaired sessions — Rory Putnam, John Coulson, Hans Kruuk, Gilbert Manley, and Roy Ellen, and by the many people in Durham who worked so hard before and during the conference.

ROBERT FOLEY
VALERIE STANDEN

Introduction:
Socioecological paradigms, evolution and history:
perspectives for the 1990s

J.H. CROOK
Department of Psychology, University of Bristol,
8/10 Berkeley Square, Bristol BS8 1HH, UK

SUMMARY

1 Introduction: dialectical insights.
2 Nineteenth century perspectives on ecology and social behaviour.
3 Classical and social ethology: the rise of a socioecological viewpoint.
4 The sociobiological explosion: from facts to modelling.
5 Ecology, fitness theory and the socioeconomics of human vernacular societies.
6 The inter-generational transmission of strategies for fitness in culture-controlled societies.
7 Socioecology and history.
8 Shifting paradigms — conclusion.

INTRODUCTION: DIALECTICAL INSIGHTS

At any one time a scientific discipline comprises a set of interrelated propositions which have survived a rigorous process of competitive selection: the criteria for viability are a combination of established verifications and plausibility in relation to other hypotheses, together with a conformity to an over-arching theoretical and methodological paradigm (Kuhn 1962). Propositional 'mutations' may, however, invade a paradigm successfully and ultimately lead to its replacement. This occurs when awareness of a new problem creates an intellectual niche within which deviant or revolutionary hypotheses can flourish. The process of intellectual competition is not unlike the natural selection of organismic traits and entails the survival or elimination of the names of peddlers in ideas within a common pool of knowledge: hence the passion with which it is conducted.

The dynamism of Western science as a whole is based in the dialogue between holders of opposed (antinomical) propositions as first described by Immanuel Kant. Propositions, he thought, were of two types: analytic *a priori* (for example, 2 and 2 make 4) which are logically incontrovertible, tautalogical, independent of experience, and, on the other hand, synthetic *a posteriori* (Yesterday was a wet day; Moths are blacker in the Midlands) propositions which are capable of empirical verification or rejection. There are, however, tantalizing propositions which sometimes crop up in the formulation of scientific hypotheses where a suggestion takes the form of an apparently synthetic proposition (for example: The universe exists in time and has a beginning and an ending) which, however, is untestable empirically. Such a synthetic *a priori* proposition may nonetheless form the premise for extensive theoretical exploration and model building. It is then subject to opposition from its negative formulation (The universe does not exist in time — there is no ending or beginning) and the dialectic set up between the two premises develops into a major debate. Recently, for example, the suggestion that the adaptational research project in sociobiology was no more than tautalogical rationalization has been countered by accounts of exactly how sociobiological ideas contribute to evolutionary theory (Wittenberger 1981, p. 14; Dunbar 1982, 1986, 1988; Stephens & Krebs 1986). This dialectical process accounts for the emergence, decay and re-expression of interactively self-modifying paradigms in science which, however, depend ultimately upon empiricism for their validation.

Socioecology, in common with other sciences, is prone to paradigm shifts, and their history has considerable significance for an understanding of contemporary problems. History never repeats itself but incompletely resolved issues tend to spiral through time, re-emerging in new forms through an invasion of discourse of successive generations. This chapter, intended as an introduction to the survey of contemporary projects that comprises this book, provides a brief account of past paradigms in socioecology ending with a focus on the problem which appears to be of crucial interest for the immediate future. Both genetic evolution and historical modifications in tradition appear to track environmental and demographic changes adaptively. A major question therefore arises: how can studies of biological evolution on the one hand and cultural processes on the other engage to create a theory of biohistorical evolution which would encompass the problem of the emergence of mind and hence human history? I believe socioecological research has a major role to play in answering such a question.

NINETEENTH CENTURY PERSPECTIVES ON ECOLOGY AND SOCIAL BEHAVIOUR

The idea that ecological adaptation played an important role in the formation of animal societies was rooted in Darwinian thought but received its first explicit statement in Espinas' book of 1878. Espinas based his stance in the positivist sociological perspective of Auguste Comte, and his view of evolution was closer to that of Herbert Spencer than it was Darwinian. He explored the notion that animal societies were highly structured relationships among individuals of a species. He discovered that such structures varied independently from the taxonomic relationships of species and he therefore interpreted them as expressions of direct adaptation to ecology rather than of phylogenetic descent. A number of passages reveal an early interest in the details of societal adaptation: sea bird colonies are only to be found in food-rich areas, carnivorous birds are territorial but boundaries are relaxed when supplies are abundant, etc.

Espinas' book provided material for use in the vigorous debate within French zoological circles of the nineteenth century concerning the value of field and behavioural studies in biological science. Indeed, we have here the first set of opposed paradigms that influenced the emergence of socioecology. In the 1830s bitter and devisive debates occurred between the laboratory-based anatomist Baron Cuvier and E. Geoffrey Saint-Hilaire who wished to include natural history and behaviour in what was essentially an adaptationist programme for biology. Saint-Hilaire's son, Isidore, in work designed to emphasize his father's perspective, began in 1859 to use the term 'ethology' in more or less its modern sense as the study of living things in their natural environment. His appropriation of the term appeared at about the same time as Haeckel, a German Darwinist, propounded ecology as a science defined in almost the same way. Only gradually did the differential emphasis on behaviour and habitat make these two approaches distinct.

The progress of ethology was slow, however, for P. Flourens, an exceedingly influential proponent of the Cuvier tradition, once more established the pre-eminence of laboratory-based research by creating the term 'comparative psychology' for the study of both human and animal behaviour from the viewpoint of a mechanistic neurology. Flourens, antagonistic to Darwinian theory, used the word comparative without reference to the adaptationist perspective of a phylogenetic analysis and thereby failed to underpin his new science with an evolutionary viewpoint. It was indeed this failure that eventually led to the absorption of this discipline by ethology (Hodos & Campbell 1969; Lockhard 1971).

The subsequent history of nascent socioecological ideas in the early years of the twentieth century has been little explored. Surprisingly perhaps, Thorpe (1979) totally ignores published reference to it and other historians (Klopfer & Hailman 1967; Jaynes 1969; Burckhardt 1981; Durant 1986) likewise seem unaware of the personalities who engaged in debate. While I did a little research on this topic in preparing the introduction to the book *Social Behaviour in Birds and Mammals* (1970; see also Crook & Goss-Custard 1972), a full-fledged doctoral study on the history of these ideas, with adequate reference to continental sources and archives, would be well worthwhile.

The Saint Hilaires' ethological perspective was taken up by Alfred Giard at the Sorbonne around 1900. Giard, however, was a Lamarkian evolutionist and, with the renaissance in Darwinian thought, his focus moved to the back burner. Only among a group of now virtually unrecognized sociologists working in Brussels did his socioecological focus retain some significance. The main trend in behaviour study on the continent become instinct theory centred upon the work of Konrad Lorenz (1950, 1970) and later Niko Tinbergen (1942, 1951) which emphasized the study of fixed action patterns, their causation, function and evolution. The history of this 'classical' ethology has been well documented although even here the historical essays (above) are often incomplete. Sparks' popular account is useful (1982).

In Brussels, the Espinas tradition was taken up by a short-lived school of sociology directed by Emile Waxweiler which came to an end during the upheavals of World War I. Waxweiler (1906) and his colleague Raphael Petrucci (1906) focused on exactly those socioecological themes which were neglected in classical ethology. Espinas had attempted to account for the continuity and durability of animal societies in a set of principles that had a remarkably cybernetic flavour. His work did not interpret social organization simply in terms of interactions between individuals, but rather he tried to show how social systems were related to ecological conditions. Petrucci attempted to explain society in terms of direct determination by ecological factors rather than as an expression of adaptation through natural selection — an issue that reappears today in attempts to understand intra-specific variation in social structure (Caro & Bateson 1986). Waxweiler himself, in a programmatic study, considered sociology itself to be a sub-discipline of a social ethology of the animal kingdom as a whole.

After the First World War it was the young Julian Huxley who, with binoculars and a notebook in hand, undertook simple but critical field studies and began to think of data obtained in this way as fundamental in the construction of a Darwinian natural history in which selectionist principles played the prime role in behavioural interpretation (1923). Huxley's attempts to explain the mating system of birds in terms of ecological adaptation through natural selection were perhaps the first researches in which the principles of contemporary socioecology begin to come into view.

Table 1 summarizes the opposed paradigms in the social ethology of the pre-World War I period. A number of trends are clear:
(a) An interest in the natural history and ecology of a species seen as essential for a complete understanding of its biology is opposed by the traditional comparative anatomy of the museum taxonomists who lacked an evolutionary programme.
(b) An emerging interest in social behaviour and ecological adaptation is confronted by a strong, laboratory-oriented, non-evolutionary programme in 'comparative psychology' and its subsequent development in behaviourism.
(c) The gradual establishment of a Darwinian−Mendelian viewpoint in comparative anatomy is opposed by a Lamarkian perspective stressing the role of the environment in directly shaping animal social structures.
(d) A preference for laboratory research into mechanisms and genetics is matched against the acquisition of field data for a Darwinian programme of research on behaviour and social adaptation to the environment.

Within these oppositions there is a further issue in play. A perspective that focuses on anatomical traits, behaviour characteristics, physiological mechanisms or genes as entities contrasts with whole animal studies carried out in the field (for example Howard 1920) often with a focus on populations or societies as processes within which individuals interact in systemic ways. To this opposition we shall return.

CLASSICAL AND SOCIAL ETHOLOGY: THE RISE OF A SOCIOECOLOGICAL VIEWPOINT

The powerful school of Konrad Lorenz (1950, 1970) and Niko Tinbergen (1942, 1951) dominated the European study of social behaviour in animals from the mid-1930s until the publication of important critiques by Lehrmann (1953) and Hinde

TABLE 1. Opposed themes in interpretations of animal social behaviour, 1830–1914

Laboratory-based anatomical and phylogentic speculation	vs.	Field-oriented natural history
Baron Cuvier	vs.	Geoffroy Saint-Hilaire
P. Flourens	vs.	Isidore Saint-Hilaire
Comparative psychology	vs.	Ethology/ecology
Museum anatomy	vs.	Lamarkian social ethology Espinas Giard Waxweiler; Petrucci
Comparative anatomical and physiological zoology based on laboratory research including Mendelian perspectives	vs.	Comparative social behaviour in field studies; Julian Huxley's fresh Darwinian approach

(1960, 1966), evoked principally by Tinbergen's book *The Study of Instinct* (1951). This classical phase in the history of ethology was preoccupied with analyses of the adaptive significance, evolution, motivational and mechanistic control of innate behavioural patterns. Social behaviour was interpreted in terms of interactions between individual organisms in which the fixed action patterns of one individual elicited reciprocal behaviour in another — as in the courtship movements of the stickleback, for example. The role of social behaviour within groups or within population biology was only rarely considered and was never a focus of serious attention (see, for example, Tinbergen 1953). However, the complexity and multifaceted nature of coadapted aspects of behaviour in relation to the detailed structure of an ecological niche was becoming clear from within Tinbergen's gull studies (1959), in particular through the work of Esther Cullen (1957) and Mike Cullen (1960). In addition, the ecological research of the British Ornithological Union's centenary expedition to Ascension Island came to emphasize the role of contrasts in the seasonality and availability of food resources in the evolution of clutch size, parental behaviour, length of breeding season and other features of reproductive behavioural biology (Ashmole 1963; see also Nelson's later massive study on Sulidae 1978). Skutch's (1935) early work on 'helpers at the nest' had drawn attention to the circumstances under which collaborative behaviour in the rearing of offspring might develop. These themes foreshadowed the later work of MacArthur & Wilson (1967) on r and K selection in the differential evolution of social organizations (Emlen 1978, 1984). David Lack's classic books (1954, 1966) on the 'natural regulation of animal numbers' also included a chapter on the significance of social behaviour in relation to the exploitation of an environment, sociality being interpreted in relation to the patterning of food resource availability (see also Fisher 1954).

My own interest in what is today called behavioural ecology developed out of teenage bird watching while I was an undergraduate in zoology at University College, Southampton. I noticed that the winter population of gulls in Southamptom Water roosted on mud-flats and in spartina marshes in the estuary, but flew up the valleys of the rivers Test and Itchen to forage every day. The patterning and volume of movement in this daily 'dispersal system' was, however, regulated by the state of the tide at dawn. If the tide was low the birds, particularly the black headed gull (*Larus ridibundus*), would spread out over mud flats and pools to feed using a wide range of situation-specific foraging patterns. Only as the tide rose would birds fly inland in numbers. With a high tide at dawn, however, the great majority of all gulls would fly directly inland to feed in the fields and hills using a different range of foraging patterns. Clearly the diurnal rhythmicity of the dispersal system was modulated by the differing rhythmicity of the tide cycle. This evoked varying patterns of environmental exploitation and local movement revealing the adaptability of the birds' behaviour (Crook 1953). Although I was not able to follow up this investigation, my prime interest in the proximate factors relating ecology and behaviour within a system of demographic and social adaptability was established at this time and continues to be a key theme in my thinking.

At Cambridge my research interests took shape around a project initially suggested to me by W.H. Thorpe focusing on the behaviour and ecology of the *Quelea* (Ploceinae), a major pest to developing agricultural schemes in semi-arid areas of Africa. This highly sociable weaver bird lived in enormous numbers in the African savannah and, like the gulls, its demography focused behaviourally on massive diurnal movements from a central roost or breeding colony. It became clear that the interpretation of its behaviour would be enhanced by comparative studies on other weaver birds (Ploceinae) living in contrasting habitats. My research thus developed in two directions, comparative studies on the adaptive radiation of social behaviour within this taxonomic family and detailed field and experimental studies attempting to understand the proximate factors controlling the 'social system'. As time passed, Peter Ward undertook the ecological work fundamental to the economic significance of this species (1965) while I, based by this time in Bristol, tried to dissect some of the proximate factors through behavioural and endocrinological experimentation with caged *Quelea* in the laboratory.

The 'social system' of weaver birds was conceived in terms of three interrelated aspects within each of which intra-specific competition to varying degrees was seen as a prime mover in evolution:

(a) The dispersion of the population in relation to habitat features such as food supply and nest site availability, also in relation to individual behaviour reducing the probability of predation.

(b) The mating system, monogamous or polygamous, interpreted in terms of food resources which either necessitated the involvement of both parents in raising young or produced a super-abundance permitting maximization of male mating efforts with little involvement in rearing.

(c) The communication system of displays, flocking behaviour, etc. functional in relation to habitat characteristics.

The form of any one behavioural characteristic within the system was seen to be 'the result of the interaction of several selective pressures some of which may impose contradictory demands on the organism; the resulting character is thus a compromise of optimum survival value in the particular circumstance prevailing' (Crook 1964, p. 3). Also, 'a full evolutionary interpretation demands therefore an adequate coverage in terms of both Lorenz−Tinbergen−Hinde ethology and the approach to population structure of Lack (1954) and Slobodkin (1961)'. Population dispersion was treated as a function of effective food exploitation in relation to food type (insectivore/granivore), dispersion and seasonality. The mode of dispersion in flocks was, however, also influenced by adaptations reducing risks from predation. Mating systems correlated strongly with the ecological variables determining population dispersion and the details of behaviour patterns in nest building and displays were also interpreted in terms of functional adaptations to breeding niches. The details of this study are well known and will not be repeated here.

In retrospect, this method of correlational comparative study has proved to be instrumental in creating a flexible cybernetic viewpoint very different from the fixed interpretative approach to social interaction of classical ethology. Working with Steve Gartlan, I also found it to be valuable in attempting a preliminary overview of the social systems of primates in 1966. Similar approaches to social organization were also being formulated by others; I was aware of Immelmann's important work on estrildine finches (1967) but not of Winn's study (1958) of social structure in fish. Hinde's (1955/56) remarks on the evolution of behaviour in finches and conversations with Peter Marler in Cambridge on the adaptive radiation of bird songs were important influences, but the work of David Lack was my main inspiration outside the data themselves. Jarman's important study of African ungulates later emphasized the value of this type of approach (1974) and particularly stressed the importance for mammals of female distribution in relation to food resources upon which male distribution was contingent, a point which is now of particular significance in Wrangham's (1980) recent reinterpretation of the evolutionary ecology of primates.

My interest in the proximate causation of behaviour within a social system focused on the *Quelea*, then easily obtainable in Bristol as an imported cage bird. Competition, an evolutionary principle stressed by Lack, took the form of encounters leading to differential success in the acquisition of food, nest sites and mates. Ward (1965) had demonstrated that during the non-breeding dry season in West Africa, males lost weight and endured a mortality less extreme than that of females. This resulted in a shortage of females available for mating in the breeding season even though the short duration of the breeding season apparently called for both genders to be involved in parental care and hence a monogamous pair bond. The mechanisms producing this paradox appeared to be endocrinal. A series of laboratory experiments led to a view strongly implicating luteinizing hormone as an important factor determining male dominance in space-mediated food competition in flocks. Intact (non-ovariectomized) females were subordinate to males but operated birds competed well. While oestrogen in intact birds appeared to reduce the aggressive response in females so that they lost out in food competition in the dry season, it nonetheless played an important role in producing passive courtship patterns making the female acceptable to highly aggressive males attempting to breed in crowded colonies during the rains (Crook & Ward 1968; Crook & Butterfield 1970; Lazarus & Crook 1973; Dunbar & Crook 1975). Comparative endocrinology suggests at least some of the mechanisms responsible for differing gender relations in social systems in birds. More detailed research by Lehrmann (1964) and Hinde & Steel (1966) on other species pointed in the same direction.

Peter Ward's work with *Quelea* (1965) led him into a collaboration with Amotz Zahavi who had demonstrated experimentally that flocking and the holding of individual territories by wagtails wintering in Israel were alternative environmentally driven responses to differing patterns of resource distribution in the habitat (Zahavi 1971; see further research by Davies & Houston 1981). Ward & Zahavi (1973) proposed that the dispersal system of flocks based upon roosts or colonies enabled individuals to assess the whereabouts of foraging areas from observing the state of birds returning to the central locus of the population movement. Working in the Bristol laboratory, Peter de Groot (1980) demonstrated experimentally that hungry caged birds did indeed follow companions known to the experimenter to be knowledgeable not only in the finding of food or water, but also with respect to food resources of two differing qualities. He also showed that hungry birds would choose to join well fed individuals rather than hungry ones. (For field studies see Waltz 1987; Weatherhead 1987.)

Following a different line of thought John Lazarus (1979a, b) demonstrated that individual *Quelea* in larger groups were less anxiously vigilant than birds in smaller ones or alone. This work and similar projects on other species (Siegfried & Underhill 1975; Powell 1974) helped to explain Murton Isaacson & Westwood's finding (1971) that foraging wood pigeons would prefer to attempt to obtain food in flocks even when they would have done better to feed alone, the implication being that flock formation reduced the risk to individuals of predation. These studies in the field supported theoretical treatments of the topic by Hamilton (1971) and Vine (1971). The dispersion of individuals in populations needed therefore to be examined in terms of the evolution of behavioural strategies that optimized the sometimes conflicting needs for both resource management and predation protection. Since flocking could be interpreted in several ways more precise modelling of such relations were clearly required.

In the late 1960s, largely as a result of the death of Professor K.R.L. Hall at Bristol University in 1965 and my assumption of the research direction of the group of avian and primate workers in the Psychology Department, I prepared several programmatic discussions of research in social ethology (Crook 1970a, b; Crook & Goss-Custard 1972). The perspective focused on the significance of social structures of species populations not only as adaptive but also as adaptable systems. The social structure was interpreted as a biotic one in which change through time consisted of several laminated processes with different rates of operation. Lack had viewed dispersion as the means whereby individuals so spaced themselves as to maximize their effective genetic contribution to suc-

ceeding generations. Socioecology in this Darwinian perspective was the study
of social adaptations to the environment and the mode of action of natural
selection producing social traits (Goss-Custard, Dunbar & Aldrich-Blake 1972).
Underpinning this endeavour was research on the proximate mechanisms responsible for the 'social dynamics' which maintained the socioecological system. These
dynamics involved competitive interactions leading, for example, to spacing patterns, territoriality, etc., integrative dynamics such as social facilitation, flocking
behaviour, and bonding dynamics maintaining relationships both between individuals and between and within groups.

Socioecology in the late 1960s thus comprised a range of loosely integrated
perspectives based on a marked 'systems' approach focusing on the biotic environment within which the evolutionary process operated. A book produced by a
group of colleagues at Bristol (Crook 1970c) presented a range of topics within
this theme. The result of this approach was to open up the perspectives on social
evolution and social function formerly constrained by classical ethology and to
relate them to problems of their proximate control in nature.

Other authors played a prominent role in the development of this approach.
In particular, Brown's (1964, 1969) studies on territoriality and Orians' (see 1980)
field and theoretical studies of the Icterids, in many ways paralleling in North
America the social evolution of weavers in Africa, provided important impetus
for the development of socioecology in America. Of paramount importance was
the new modelling approach to ecology initiated by Robert MacArthur. These
themes are well known to workers in this field and will not be treated further in
this account.

This focus on the proximate determination of aspects of socioecological systems
(see, for example, Crook, Ellis & Goss-Custard 1976) left, however, to one side
the question of how the clear adaptedness to environment could be interpreted in
terms of population genetics. In the 1960s the main approach to function was
Tinbergen's field experimentation demonstrating the survival value of behaviour
patterns (see, for example, Tinbergen, Impekoven & Franck 1967). These elegant
studies, however, were focused neither on the role of the behaviour within a
social system nor on the population genetics of the selective process itself.

THE SOCIOBIOLOGICAL EXPLOSION: FROM FACTS TO MODELLING

In 1962, V.C. Wynne-Edwards drew attention to the possibility that many behaviours determining social organization might often be the result of group rather
than individual selection. He argued that group selection was responsible for a
very wide range of social attributes including flocking and mating systems. While
both David Lack (1966) and I (Crook 1965) rejected this interpretation, pointing
out that alternative explanations based on individual selection were both more
plausible in relation to Darwinian theory, and better supported by detailed field
study, points also stressed in G.C. Williams' important book (1966); a number of

issues (the role of territorial behaviour in limiting the numbers of breeding birds, the complexity of cooperative behaviour in the social insects and in the tropical birds studied by Skutch, etc.) remained unresolved. Wynne-Edwards had in fact refocused attention on the complexity of cooperative behaviour in animals, an issue neglected since Kropotkin's (1902) general account early in the century, and on the problems of interpreting it in terms of the selfish maximization of individual reproductive advantages.

In retrospect, it seems strange that British researchers failed to see the great significance of Willliam Hamilton's critical examination and solution of this problem in the context of the social organization of bees (1964). Content with the functional demonstrations of survival value resulting from field study and the inferences to selection coming from studies of mortality factors naturally regulating demographic systems, the need for deeper theory based in Fisher's studies of population genetics (1930) was not perceived. The significance of Hamilton's work on inclusive fitness first received full recognition across the Atlantic in Harvard. In particular, the entomologist Edward O. Wilson, following in the footsteps of the eminent Morton Wheeler, had greatly extended our understanding of the complexity of social insect societies. He quickly saw that Hamilton's work provided the basis for a neo-Darwinian account of biological altruism that had ramifying implications for the animal kingdom as a whole. His important book *Sociobiology — the New Synthesis* (Wilson 1975) soon followed. Of even greater theoretical significance had been the work of Robert Trivers on parental investment (1972, 1974) and reciprocal altruism (1971). These various theoretical contributions put forward a fresh integrative vision which encompassed the field and laboratory findings of the 1960s and early 1970s, and raised the whole *problematique* to a level of sophistication demanding new approaches and methodology.

Hamilton's key realization, which had been foreshadowed by less worked out theory by R.A. Fisher (1930) and J.B.S. Haldane (1953), was that genetic inheritance is a matter involving all relatives not only parents and children. Any action by an adult that increases the fitness of an indirect relative such as a brother's child will improve the probability that genes shared in common between them will be passed to the next generation. This probability increases with the degree of relationship. It is possible to calculate the degree of relatedness (the coefficient of relatedness, r) and hence to predict relationships within which assistance from one relative to another will be beneficial in terms of 'inclusive' fitness to the donor — that is to say its fitness depends not only on its own survival and reproductive success but also on that of its kin to which it may contribute by fitness — enhancing acts of 'altruism' less the costs to its personal fitness. This is the principle of kin selection.

Trivers saw that the investment which a parent may make in improving the chances of survival and reproduction of its offspring is at the cost of its ability to invest in a further one. Since a parent is assumed to have a tendency to maximize its fitness there will be a conflict between it and its offspring with respect to the amount of care provided. Field research does indeed show that parents actively promote the independence of offspring as soon as possible. There will also be a conflict of interests between parents when one of them can maximize reproductive success, say, by investing more in mating than in rearing. Since, due to the differing physiological economics of egg and sperm production, this option arises usually for males, a range of differing scenarios can arise when ecology presents reproductives with contrasting settings — for example, one in which food resource availability requires maximum parental care from both sexes and others where varying degrees of reduction in male care are possible. The linkage between ecology and mating system can therefore be modelled theoretically and checked against sufficiently detailed field studies of a species' socioecology. Comparative studies between species of a taxonomic family or genus which are adapted differentially to a range of ecological features are often highly instructive here.

Trivers also realized that reciprocal altruism, independent of kin selection, could arise in circumstances where individuals lived in closed groups with ample opportunity for mutual recognition and exchange of benefits. Such behaviour, however, would be accompanied by high risks of the beneficiary cheating by failing to provide a return of benefit. Field work by Packer (1977) on baboons provides confirmation of Trivers's ideas but the precise circumstances for the appearance of reciprocal altruism have not otherwise been clearly identified in field studies nor the frequency of their occurrence known. The importance of cheating has been further discussed by Crook (1980, p. 244), Krebs & Dawkins (1984), Trivers (1985) and Byrne & Whiten (1988).

Trivers's work on mating systems gave rise to John Maynard-Smith's (1972) important application of games theory in the analysis of reproductive strategies. He realized that whether parental care should be given or not depended upon the strategy of the mate. A male that fails to care can only do so if his mate is doing so adequately. By comparing all possibilities of male and female options he sought to establish the evolutionarily stable strategy (ESS) for a given circumstance. Such a stable strategy develops when, in terms of fitness, it does not pay for one sex to change his/her strategy so long as the partner continues to play his/hers. By examining the effectiveness of parental caring, whether by one or both parents, the likelihood of a deserting partner finding another mate and the extent to which a female's provision of care reduces her probability of producing further offspring, it is possible to predict whether monogamy, polygyny, polyandry or promiscuity should be favoured.

The ESS approach has now been applied to many socioecological problems where alternative strategies are potentially available and there is a need to calculate which one of a set would be stable — in genetic terms — that gene set which has the best fitness pay-off when 'played' against mutants determining alternative strategies. (For a helpful exposition see Parker 1984.) Since ESS theory is not, moreover, directly concerned with the genetic basis but rather with the fitness outcome of behaviours, culturally controlled strategies can be examined in the same way. While the application of games theory has led to a great deal of abstract speculation with little assessment of the real value of strategies in actual life histories of species as examined in field study, it has contributed greatly to the logic of evolutionary theorizing and the use of extensive computer modelling as an important methodology associated with empirical investigation.

The importance of the relationship between the availability of resources and the demography of a population in determining the evolution of social systems, and in particular the occurrence of selfish or altruistic strategies, was greatly stressed as a result of MacArthur & Wilson's (1967) analysis of r and K selection. In brief, individuals may maximize r (the intrinsic rate of natural increase) when ample or supra-abundant resources are available during the breeding period. Selfish strategies of maximizing individual fitness are then functionally effective. Under conditions where the carrying capacity (K) of the environment is such that the addition of individuals to the population is difficult for reproductives, altruistic strategies of several sorts, including for example, delayed reproduction combined with the giving of aid to parents in caring for subsequent brothers and sisters, may be an optimum strategy. Emlen (1978, 1984) reviews field studies and theory showing when and under what circumstances helpers would do better to stop helping and attempt to breed themselves.

The emphasis on modes of environmental exploitation has also led to the great development of research and thought in 'foraging theory' (Stephens & Krebs 1986). The approach resembles the games theory approach used in the analysis of strategies, in that both depend on the determination of which of several plausibly adaptive behaviours optimizes reward. For example, in searching for prey should a predator take all prey encountered or only large ones? In exploiting a patch in the company of others how long should an individual stay before moving on in search of another? Such issues can now be subjected to analytically precise modelling in a mathematically sophisticated way, and a considerable literature is accumulating.

The precision and analytical elegance of optimization modelling, decision and games theoretic approaches to understanding behavioural adaptation has combined with inclusive fitness theory to yield an extremely powerful paradigm under the general flag of Wilson's 'sociobiology'. There is no question but that these approaches are an essential basis for future work but a number of problems are appearing, many of which focus on the need for more effective empirical demonstration of the utility of what is often very fine-spun theory. Modelling has gone beyond the 'facts' and the resurgence of interest in socioecological field work as shown by the papers in this volume is motivated in large measure by the need to redress the balance. But some of the difficulties go deeper than this.

(a) A convenient assumption in providing basic propositions or premises for modelling is to take a social behavioural category — reciprocal altruism say, and conduct an analysis as if it were the only process undergoing social selection in a given population. This is unlikely as Wrangham's (1982) discussion of mutualism has made especially clear.

(b) While theoretical discussions of the adaptiveness of traits can be conducted without reference to whether the proximate mechanisms involved are at the genetic or some other level, models have often been constructed on the crude or unspoken assumption that a 'trait' is controlled by a 'gene'. One may then speak of a gene for altruism. Convenient as this may be, both for basic adaptationist theory and for modelling, the premise is of course a gross parody of the actual complexity of genetic determination. The 'selfish gene' is no more than a gross metaphor and as such has allowed the growth of a curious metabiology lacking anchorage in known genetic facts. While simplistic theorizing of this sort has been of great value in clarifying a basic logic for a programme analysing the principles of fitness it becomes subject to subsidence as soon as a researcher requires detailed knowledge on the proximate determination of traits. Recent examinations of the interwoven complexities of the historical processes producing adaptation re-emphasize the need for a less simplistic basis for evolutionary argument (Alexander & Borgia 1978; Caro & Bateson 1986) and this is especially important where social behaviours are largely the expression of learned traditions, direct effects of local environments on dispositions allowing behaviour flexibility, effects of changing demographic pressures, and/or complex relationships within proto-cultures, rather than 'hard-wired' mechanisms. Such a consideration applies to all adaptationist programmes focusing on 'higher' animals.

(c) The use of key terms, in particular 'inclusive fitness', in constructing theory has often been remarkably lax revealing both a willingness to use paradigmatic assumptions uncritically and an unwillingness to penetrate the mysteries of the founding text (Hamilton 1964). The result has been the publication of a number of metabiological culs-de-sac which, because of their ease of exposition, have become texts used in teaching students. The appropriate usage of the 'inclusive fitness' concept has been much clarified by Alan Grafen (1984) who, following Hamilton, points out that whenever field data are concerned the assessment of fitness is often best carried out through the direct comparison of reproductive success between categories. This is because reproductive success is the sum of personal fitness plus success due to help from relatives. The actual calculation of the extent to which an altruist gains or loses in cooperative relationships is a statistic uncommonly difficult to determine in the field. It is clear that the relation between theory and data needs a much more careful statement. Dunbar's (1984, Chapter 7) careful calculations of the relative contributions of the components of inclusive fitness arising from the coalitionary behaviour of female geladas is exemplary and shows the way forward.

(d) The extent of difficulty is applying ESS techniques to field research in cases of all but the simplest social processes remains unknown but the problem is already recognized (Davies 1982; Parker 1984). It is often difficult to identify what propositions about strategies made in modelling are biologically or behaviourally plausible. If a population is actually performing an ESS the examination of pay-offs from possible alternatives would be theoretically impossible to determine. Furthermore, where directional selection is occurring and the habitat changing, any ESS could only be of limited duration. The time-scale of the inference to ESS is rarely discussed. The problem might be better stated in terms of competing strategies in relation to rates of environmental change in time rather than in terms of ESS which seem to require periods of stabilizing selection. Again the testing of models from field data is problematic. The extent to which ESS theory has developed an abstract jargon of its own, making inference to field results difficult, is suggested by the following summarizing passage from Parker's informative review (1984). He writes 'Where a game has more than one ESS, the only way to deduce which ESS is likely to achieve fixation is from a knowledge of the frequencies of strategies at the start of the game. Ideally, we need to examine each step in the possible sequences of changes in games, mutation by mutation, until the present state of strategic complexity is reached'. Whether one can meaningfully talk about mutations in the same breath as discussing the metaphysics of games and strategies is the critical philosophical point at issue here.

(e) Optimality theory has attracted a number of criticisms. Some of these form part of a critique of the adaptationist programme which differs little from the anti-Darwinian theses of former times. They have been effectively answered by Mayr (1977), Dunbar (1982) and Stephens & Krebs (1986) among others. Other criticisms, however, still pose problems. Optimization models usually focus on a small set of design features and ignore other factors that may act to influence these features. In neglecting these wider aspects of overall design the models seem to imply that different aspects of design are effectively independent instead of interactive. Recent models dealing with the trade-off between complementary activities (e.g. foraging risk and predation, time spent feeding versus territorial defence, etc.) are helpful but remain a partial remedy. The advantage of a piecemeal approach is that it handles manageable problems while the attempt to cover all interrelationships demands holistic modelling — and holistic models, apart from difficulties in gaining adequate field data with which to test them, do not encourage experimentation pitting alternative hypotheses against one another. Detailed comparative studies (e.g. Clutton-Brock 1974) can, however, be very useful here.

Optimality models have had a fair degree of success in predicting levels of performance in field situations but the precision of these predictions is often not robust. The issue here again focuses on when an animal can be expected to show optimum behaviour in an environment which may be changing as fast as or faster than the animal's ability to adapt. This problem is the same issue as that raised above in questioning the method of ESS games theory. Yet, as Stephens & Krebs (1986) argue, the important feature of this approach is perhaps not the hypotheses of optimality itself but the technique 'used to work out the testable implications of the specific hypothesis about design and constraint'. I suspect that ways of doing this in the context of a more holistic picture of the quantitative socioecology of populations will be important if the claims of the method are to be further supported.

The emergence of these problems are signs that the original 'sociobiology' paradigm may be losing its hold on behavioural ecology. The achievements of the sociobiological synthesis have been impressive. Hamilton's initial insight insured that evolutionary theory applied to social behaviour cannot advance without detailed reference to population genetics. Together with the work of Trivers and others this discovery provided the first satisfactory account of 'altruistic' behaviour in the animal kingdom and thereby integrated a whole range of problematic issues within a strengthened neo-Darwinian vision. Combined with quantitative ecological theory developed originally by MacArthur the role of ecology in determining the extent of altruism or non-altruism in social systems has been greatly clarified. The precision of Maynard-Smith's games theoretic approach to analysing the fitness of alternative strategies and the development of optimality theory have produced immense gains particularly for theoretic studies. The problems arising within the synthesis centre around the difficulty of encompassing the interdependence of many factors, including many not easily detected, within a modelling approach which becomes progressively inadequate as the actual complexity of socioecological systems becomes clearer in field study.

Perhaps the most glaring defect, however, has been the prevalent use of an over-simplified genetic metaphor (point (b) above). In advanced animals the phenotype is a consequence of the interaction of several layers of interdependent factors acting ontogenetically in a behavioural development framed by the relationships and patterned interactions comprising the structure of the social environment within which selection itself occurs. This laminated developmental process is the expression furthermore of an equally complex set of hierarchically inter-related historical processes that have operated through successive generations to produce a contemporary social system (Crook 1970a, b; Crook & Goss-Custard 1972; Hinde 1983, 1988; Caro & Bateson 1986; Dunbar, this volume). Until the modelling approach develops the ability to depict the complexities of this more holistic picture there is a danger that evolutionary sociobiology may tend towards abstract disjointed theorizing that fails to convince owing to its naïveté in developmental studies; theorizing, furthermore, which sometimes appears to admire itself because of its technical sophistication rather than for its biological insight. Clearly the major role of socioecology now is to provide detailed field studies that allow the testing of hypotheses in relation to adequately demanding sets of data and which thereby pushes the subject towards a more satisfactory empirical basis.

ECOLOGY, FITNESS THEORY AND SOCIOECONOMICS OF HUMAN VERNACULAR SOCIETIES

Social anthropologists have been aware of the importance of ecology in the determination of human social systems for many years (Mauss 1950; Maitland-Bradfield 1973) but only recently have 'etic' arguments (Harris 1968) incorporating the idea that human culture-controlled strategies may be optimizing the fitness of individual participants been seriously considered. The recent application of comparative studies to human communities is opening a whole new field of investigation here. (Alexander 1980; Chagnon & Irons 1979; Reynolds & Blurton Jones, 1978; Flinn & Low 1986 and others in a rapidly expanding literature; see for example, Borgerhoff Mulder 1987a, b; Betzig, Borgerhoff Mulder & Turke 1988 and papers in this volume).

I have found Mildred Dickemann's (1979a, b) sociobiologically based interpretation of polygynous marriage in stratified vernacular societies of Indian and Chinese traditional (i.e. pre-industrial) cultures especially provocative. Accepting the interpretations of animal mating systems of Orians (1969) and Trivers (1971) as applicable in principle to non-genetically controlled social processes, she argued that women should seek to control fertilization by choosing for marriage males of households in which female reproductive success and that of their offspring could be at a predictable optimum. Such a choice would also sustain the fitness of a woman's parents so that parental involvement in a daughter's marriage is, in this argument, to be expected. Correspondingly, males would be expected to compete in order to establish or maintain households (and lineages) which attract women in this way, and thereby to maximize their fertilization rate and reproductive success. Male power to maximize this fitness thus links with the choice by (or for) women of powerful marital partners who provide reproductive security. Dickemann shows that under ecologically and sociologically insecure conditions household lineages owning rich and productive agricultural estates with control over labour offer a buffer against poverty and famine when conditions are poor and great wealth and comfort when they are good. The wealth differential in her survey was found to associate with hypergyny and the payment of dowry. At lower levels in these societies women were in short supply as a result of hypergyny and men (or the families of men) had to pay bride-price in order to secure a bride. As many men did not obtain wives a large population of unmarried men developed. At the top of society a small number of men inheriting estates through primogeniture established harems of women, the female surplus occasioning female infanticide; and primogeniture occasioning a fall-out of male relatives either into collatoral positions in government, the military or a priesthood where their activities sustained the influence of the lineage, or into crime, brigandage or revolution. Hartung (1982) has also demonstrated that, in theory at least, an individual could maximize the 'fitness enhancing potential of his wealth if he (*sic*) transfers it to descendants who have inherited the highest concentration of his genes' and that where descendants are of equal relationship to an ancestor the latter will favour a system of inheritance which transfers wealth to those whose reproductive success is most dependent upon it (see also Borgerhoff Mulder, this volume). The widespread occurrence of the medieval pattern of social stratification, hypergyny and male primogeniture is thus interpretable in terms of fitness theory even though the mechanisms of inter-generational transmission are cultural.

Dickemann's work and Goldstein's pioneering field studies (1971) stimulated my own research on Tibetan fraternal polyandry in the high altitude Himalayan villages of Ladakh (Crook 1980; p. 215; Crook & Shakya 1983; Osmaston 1985; Crook & Crook 1988, 1989; Crook & Osmaston 1989). There is only space to provide a brief survey here. In our study area of Zangskar the montane desert can only be cultivated where streams of snow-melt flow down from above, the largest stemming from the snouts of considerable glaciers. These streams generate spreading alluvial fans into which the water usually disappears before reaching the main river at the valley bottom. The small villages on these alluvia consist of farming estates that are inherited within partrilineages by way of primogeniture to an eldest son who may, however, include other brothers in a joint marriage with his wife. On each estate the so-called 'big house' (*khang.chen*) holds the polyandrous marital group and their children. On the early marriage of the sons, parents and uncles move into 'small houses' (*khang.chung*) elsewhere on the estate, but in the traditional society few of these individuals are reproductive and many males and some females adopt celibate religious statuses. In each generation, therefore, a mono-marital principle (Goldstein 1971) operates on each estate.

Computer simulations suggest that a human population settling on an alluvial fan would increase in numbers of estate-owning families until estate size decreased below that from which adequate resources for maintaining a family could be realized. Beyond this point primogeniture with polyandry allows: (i) the estate to be inherited intact down generations, (ii) the numbers of children to be kept low, yet (iii) the labour force remains sizeable, while (iv) the total population of the community on the alluvium is controlled. In ecological terms, the population has adjusted to a limited carrying capacity which, as in some animals, entails high levels of intra-familial cooperation with reproductive altruism. Furthermore, this polyandrous way of life is only found among land-owning farmers. Among landless peasants and town-dwelling merchants monogamy is the rule and the wealth of the parents is split on their death between children. Polyandry is thus, we claim, a functional adaptation to a very precise set of socioecological conditions. Its form has also doubtless been reinforced by known periodic Tibetan government taxation and land reforms which tend to enhance the value of polyandry for land-owning farmers.

We were most fortunate that in 1938 Prince Peter of Greece had constructed the genealogies of a number of polyandrous families near Leh, Ladakh. In 1981 Tsering Shakya and I contacted six of these families and brought their genealogies up to date. Armed with data on family sizes in different types of household my son and I were able to examine to what degree the polyandrous strategy maintained fitness for individual men and women. The completed family sizes of women were significantly larger in the *khang.chen* than in the *khang.chung* as was also the case for polyandrous rather than monogamous marriages. Furthermore, the contribution to grandmaternal reproductive success of polyandrous daughters' marriages was greater than from monogamous daughters' marriages. Clearly if a woman is interested in her personal reproductive fitness she should marry into a *khang.chen* and accept polyandry.

Polyandrous husbands sharing their one wife equally would have a lowered fitness than any one of them marrying monogamously. We worked out a formula expressing how much larger a polyandrous family would have to be for two brothers in polyandry to be as fit as either of them marrying alone as the sole reproductive of his generation. The data showed that the degree to which polyandrous families were larger than monogamous ones was such to suggest that brothers would lose little fitness if any through co-marrying so long as each had equal access to the wife. In fact, however, this is unlikely and the tendency for males to leave polyandrous households where possible is probably largely due to the reproductive pre-eminence of the elder brother.

Between 1938 and 1981 the six lineages showed great reductions in polyandry in parallel with other changes made possible by the availability of income through employment outside the traditional subsistence farming life. At first sight, one might suppose these changes to be owing to individual's choices when faced with new opportunities. To some extent this may be true, but the family as an extended lineage remains important and estates are still not usually divided. In Zangskar, the patrilineage of several households forms a collaborative family aid society called a *pha.sPun*. Its main function is choice of marriage partner and aid at funerals. The strategies whereby individuals are allocated to households and into monogamy or polyandry are thus at least as much patrilineage decisions as they are individual ones. In the Indus valley, near the capital of Leh, these patrilines seem not to have been sustained — perhaps because of greater geographical and economic mobility reducing contacts between separated kin units. Instead the *pha.sPun* has become a club of not-necessarily-related neighbours who help one another. In such a context individual choice is then likely to have greater weight than lineage decisions.

We can see here then that a traditional society showing a marked adaptation demographically and structurally in relation to the carrying capacity of an estate on limited farmland, together with individual life trajectories that are effective in the maintenance of fitness, becomes modified upon ecological release resulting from monetization. The release into monetization allows relaxation of lineage control over individual lives, greater choice of livelihood and marriage and yields an increasing population. However, the traditional reciprocity and personal constraint is also being replaced by a more competitive and less kin supported way of individualized life. From a socioecological perspective a K adaptation entailing great reproductive variance between individuals of each sex, kin alliances with non-reproductive and nepotistic altruists of both sexes and complexity in institutions of family and patrilineal management is being replaced by a broadly r adaptation to freshly available financial opportunities entailing decreased reproductive variance for both sexes, less nepotistic altruism, and a simplification of familial institutions in the direction of the competitive nuclear family. In each set of circumstances individuals can be seen to be seeking to maximize their economic and reproductive success but in contrasting ways in relation to differing forms of socioeconomic constraint. Taken with Mildred Dickmann's work and other studies these results suggest that etic analyses of human social systems using socioecological premises are valuable new sources for relating biological and social anthropological research.

THE INTER-GENERATIONAL TRANSMISSION OF STRATEGIES FOR FITNESS IN CULTURE-CONTROLLED SOCIETIES

Haldane (1956), followed by Irons (1979), argued that the behavioural differences shown by contrasting human groups are environmentally determined variations in the expression of basically similar genotypes. The prime source of human adaptation to contrasting ecologies is through changes in the structure of social relations based on the ability of individuals to vary their behaviour.

The central prediction in the Darwinian perspective is of course that humans act either consciously or unconsciously to optimize their reproductive success. It is, therefore, as we have seen above, a matter of field research to demonstrate whether or not individuals marrying in differing contexts do so in ways that do promote their genetic fitness. It is furthermore of increasing importance that the means whereby these culture-based strategies are established and maintained from generation to generation be researched and understood. The context-related precision with which behaviours in human marital relations are reproductively successful is often remarkable. 'It is this strategic precision that calls forth comparisons with the behavioural adaptions of non-cultural animals and the suggestion that common ground rules, albeit manifested through very different mechanisms, are often at work' (Crook & Crook 1988). I submit that a renewed socioecological paradigm of the 1990s must unravel in detail the nature of these mechanisms. This will require a focus on a socioecologically referred social psychology of personal behaviour and development. Yet such a perspective needs to be framed within a broader theory of evolution than the current sociobiological paradigm provides and one which transcends the limitations discussed on pp. 9–15 above.

In recent years a number of authors have attempted to provide a theoretical frame which takes into account the several levels of change that make up the adaptive evolutionary process in cultural or protocultural animals.

I first explored this problem in three propositions put forward in a survey of socioecology in 1970.

'(a) Social structure is a dynamic system expressing the interactions of a number of factors within both the ecological and the social milieux that influence the spatial dispersion and grouping tendencies of populations within a range of lability allowed by the behavioural tolerance of the species.

(b) Historical change in a social structure consists of several laminated and interacting processes with different rates of operation. Thus while the direct effect of environment may mould a social structure quickly, the indirect effects of this on learned traditions of social interaction come about more slowly and genetic selection within the society even more slowly still.

(c) Because a major requirement for biological success is for the individual to adapt to the social norms of the group in which it will survive and reproduce, it follows that a major source of genetic selection will be social, individuals maladapted to the group structure being rapidly eliminated. Social selection is thus a major source of biological modification.'

Durham (1978) argued that cultural evolution, like biological evolution, favours behaviour that maintains the ability of individual humans to survive and reproduce in their natural and social environments. The two processes remain, however, based on separate mechanisms, the cultural having evolved from the biological level. The possibility that cultural activities affect genetic selection and that genes provide templates or predispositions that favour the canalization of acquired behaviours into flexible yet constrained routines leads to the idea that mutual feedback between genetic and cultural change may be the basis for advanced protocultural and cultural evolution. The essential notion is that genetic and cultural adaptation is a continuing and mutual process and that the latter does not supercede or replace biological determination but rather adds to the biological level a flexibility in choosing alternative behavioural scenarios that renders socio-environmental tracking more effective.

Cavalli-Sforza & Feldman (1981), Lumsden & Wilson (1981) and Boyd & Richerson (1985) all attempt extensive mathematical modelling of such a coevolutionary process. In order to do this some unit of cultural change has had to be defined. A number of terms for such a unit have been proposed ('cultural instruction': Cloak 1975; 'meme': Dawkins 1976; 'cultural trait': Cavelli-Sforza & Feldman 1981) and these can refer both to categorizations of an artifact type (arrowhead, hand-axe, armour, cars, computers) or of ideas that influence human behaviour in specific ways, such as religious propositions or injunctions in law. Lumsden & Wilson (1981, p. 27) coin the term 'culturgen' as 'a relatively homogenous set of artefacts, behaviours, or mentifacts ... that either share without exception one or more attribute states selected for their functional importance or at least share a consistently recurrent range of such attribute states within a polythetic set'. A culturgen may be identified as a unit through a cluster analysis determining the core features of an attribute set. The concept is a catch-all term referring to whole sets of ideationally created 'objects'. The 'culturgen' is used as the semantic and syntactic equivalent at the cultural level of the 'gene' at the biological level. This definitional move allows neat mathematical modelling to proceed.

Lumsden & Wilson (1981) argue that during the socialization of an individual an array of behaviours and culturgens is processed through a sequence of epigenetic rules. This forms a 'heterarchy' of filters. At the basic level epigenetic developmental rules are extended to include physiological and physical features of perceptual and cognitive mechanisms that act as constraints on learning and cognition generally. At a secondary level cognitive evaluation of the contents of long-term memory leads to differential selection according to cognitive rules, traits and capacities. Some of these features impose an epigenetic control over possible behavioural variance while other behaviours are allowed wide variance within a very open programme. Both Cavalli-Sforza & Feldman on the one hand and Lumsden & Wilson on the other attempt to examine the process whereby alternative culturgens are evaluated and selected for behavioural expression. Taking a life cycle as their time unit, Lumsden & Wilson model situations in which offspring resulting from a large randomly mating population are socialized by peers, parents and older relatives. Alternative culturgens are acquired and selected in the contexts of exploration, play and the observation of others both in clarifying the same age cohort and in older ones. The preferred culturgens are employed in prereproductive life in the obtaining of resources which may mean actual commodities or social positions that control access to commodities. The range of culturgenic response results from (i) the variance in the biological epigenetic basis, (ii) from differential sensitivity to companions and elders, and (iii) variance in skill whereby resource control is converted into genetic fitness with varying success.

The use of mathematical modelling here has required the establishment of elementary premises. A modeller may 'start with units that are clearly definable, establish them as paradigms and then proceed to more complex phenomena entailing less easily definable units' (Lumsden & Wilson 1981). That which is 'clearly definable' may, however, consist solely in ideas or definitions that allow the key premises to be handled by a modelling method. The vagueness and width of meaning inherent in the term 'culturgen' as a unit in analysis may be tolerable within the discourse provided by those authors but the structure of their overall paradigm rests heavily on basically metaphysical assumptions. These highly intelligent books rely only weakly on empirical findings, and while the more detailed intuitions mathematically modelled within them may yet turn out to be heuristic, they remain currently little more than brilliant advocacy.

The weakness in these theoretical approaches results from a failure to identify adequately the prime mover(s) of cultural change. It must be made clear that, whatever the process may be, it differs markedly from the differential selection of random copying-errors of replicating genes which is central to change at the purely biological level. Cultural innovation occurs within the setting of cultural transmission, the passage of acquired behaviour from one generation to another. Since acquisition depends on learning, innovation will arise when the behaviour or concept in learning theory terms no longer gives rise to positive 'reinforcement' during its expression. It will then tend to drift into a divergent form until rewards required by the individual in the context are re-established. If this fails to happen the 'culturgen' in question will 'extinguish'. Novelty may also arise as a result of deliberately focused trial and error behaviour strongly based in teleological motivation. The key process here is the motivational system underlying the behavioural performance and the system of reinforcement that leads to satisfaction. Cultural behaviour is primarily acquired through the differential satisfaction of motivational states which are often highly intentional in nature.

The developing viewpoint of Plotkin & Odling-Smee (1981) has an embrace that seems sufficiently wide to meet at least the initial requirements for a psychological perspective in a coevolutionary paradigm. They have sought to provide a model that allows room for both genotypic evolution (phylogenesis) through the natural selection of genes and for phenotypic adaptability based in the acquisition of knowledgeable behaviour during the lifetime of individuals. In addition, their model aims to allow for both the range of learning processes found widely in the animal kingdom and for the specific diversity of learning applications due to species-specific learning constraints based in genetic constitutions. The model thus attempts to go beyond the instinct–culture and nature–nurture dichotomies which have been so resistant to theory and such barriers to advance.

The Plotkin & Odling-Smee model proposes a multiple-level interaction between processes which differ in the manner in which information about the environment, on which adaptation is based, is stored, Fig. 1. In any specific case of adaptation the integration of information on which it is based may be due to 'retrieval' from more than one storage base. The four levels of information storage essential to a complete account of socioecological adaptation are (i) gene pool, (ii) unlearnt non-traditional phenotypes dependent upon information accrued within open programmes in the genetic constitution, (iii) individual learning, and (iv) cultural pool. The relations between these four (see Fig. 1) comprises a nesting of levels within one another. Since the capacity for variable epigenesis responsive to ecological contrasts within the species habitat (level 2) is dependent upon the genetic constitution which fixes the degree of openness in development, it is necessarily constrained by the gene pool. The third level (capacity for learning) is likewise constrained by the first, even though it is itself an autonomous information gaining and storing system. Canalization of learning into particular areas of information and skill acquisition arises through the effect of genetic constraints (Hinde & Stevenson-Hinde 1973). The fourth level acts both as a constraint upon the third, which thus rests within it, and is itself, like the third, constrained by levels 1 and 2. The cultural pool (Durham 1978) comprises evolutionary flexible strategies (Crook 1980) made up of acquired components (or 'culturgens' as Lumsden & Wilson call them) which are culturally transmissible between individuals.

Important to Plotkin & Odling-Smee's (1981) approach is the emphasis on adaptation as a process of acquisition of knowledge from the physical and social environments in the context of both inter- and intra-generational time scales. Where rates and frequency of environmental changes are high within a generation it is vital for individuals to evolve mechanisms that are responsive to changes occurring within a lifetime. On each level a heuristic involving a match–mismatch test for dissonance is the common denominator. The g(enerate)–t(est)– r(egenerate) heuristic operates recursively on each level but within contrasting time scales and with differing mechanisms. The inter-relationship between these processes constitutes the biological and cognitive responsiveness to change that gives rise to environmental tracking, often with highly successful effects in terms of optimizing reproductive success. In principle, the model should allow us to go beyond tests of static adaptations to questionably stable environmental conditions to the consideration of adaptation as a dynamic process of tracking in which rates of change in both environment and response mechanisms are examined together.

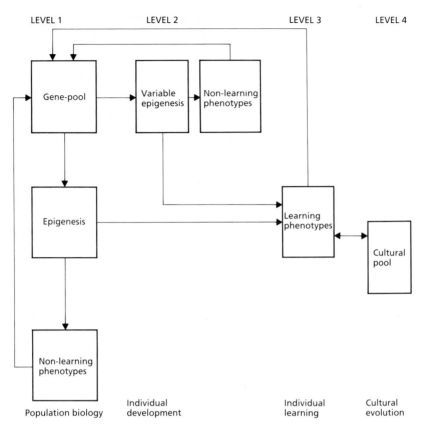

LEVEL 1   LEVEL 2   LEVEL 3   LEVEL 4

Fig. 1. Levels of evolution. (From Plotkin & Odling-Smee 1981.)

Clearly a new generation of general and specific systems modelling will be called for in which the calculus of rates of change will come to the fore as a prime consideration. The whole problem of the developmental process responsible for intraspecific variation in social behaviour and the range of life trajectories or 'routes' triggered in the course of individual experience acquires major significance here. Caro & Bateson (1986) have opened a discussion on the theory required in the study of these 'alternative tactics'.

One of the ways in which the human species differs in degree if not entirely in kind from other culture-creating animals, for example chimpanzees (de Waal 1982; Goodall 1986), is in the enhanced capacity for information transfer between self-aware individuals within and between groups and between succeeding generations. In particular, the societal conventions of one generation become the basis for the construction of culture and personal identity in individuals of the succeeding one.

Humphrey (1975, 1983) and Crook (1980) provide accounts of the plausible course of evolutionary events whereby social selection for enhanced abilities at interpreting the motives of others in groups of mutually dependent individuals gave rise phylogenetically to the characteristic form of the human person, in particular the awareness and conceptualization of self. Trewarthen (1979) argues that human infants show evidence for inborn capacities to evoke inter-individual responses with caregivers which produce interactions that go far beyond the mere provision of basic needs and which prefigure and provide templates for the intersubjective reciprocality of adults with language. Humans show evidence therefore of innate predispositions for an intentional reciprocality that results in the formation of individual identity (Mead 1934; Luckmann 1979) with its dependent processes of self-esteem. It is, however, clear that the particular quality of a person is subject to environmentally induced shaping of an adaptive nature in response to successive stages in the historical development of human society and civilization (Snell 1960; Jaynes 1976; Luckmann 1979; Crook 1980). We are not today the same 'persons' as ancient Egyptians or Homeric Greeks, nor are Western tourists personally constituted in the same way as a Ladahki villager. The depth and significance of this aspect of cross-cultural psychology is so far little studied.

Some social scientists in reaction to 'sociobiology' have argued that: 'We know of no relevant constraints placed on social processes by human biology. There is no evidence . . . that would enable us to circumscribe the limits of possible human social organization' (Allen *et al.* 1976). While deploring these authors' ignorance of the professional biological literature and their over-sensitivity to E.O. Wilson's (1975) rather blunt reduction of their science to genetics, this view is a common one outside biological circles. Yet, if it were true, it would suggest that a culturally instantiated idea of how a society might constitute itself could take practical expression without constraint from factors rooted in human biology and ecology.

An examination of ideologically driven events in human history suggests that social change imposed in this way often breaks rules set by the cybernetic relationship between personal constitution, social frame and the economic/ecologic base in ways that rapidly become maladaptive for the individuals involved (Corning 1976; Rappaport 1982). We need here an understanding of the mechanism whereby information accruing in an open programme is constrained so that an adaptive rather than a maladaptive behavioural stance is sustained. The answer must be in terms of a process whereby actions tending to become maladaptive are unrewarded by comparison with those that are adaptive. Since actions in the short term may be difficult to evaluate in terms of either future well-being or reproductive success, it seems likely that personal actions will be accompanied by experiential clues that act indexically as premonitions of future consequences. One aspect of the question concerns what constitutes the 'reinforcement' of cultural behaviour and its instantiation as tradition. It also concerns the means whereby an animal engaged in making decisions proceeds beyond a choice point and commences one line of action rather than another. The question concerns the nature of the 'test' in the generate−test−regenerate heuristic of levels 2, 3 and 4 of Plotkin & Odling-Smee's general theory in relation to the goal(s) of individual motivation.

At the level of non-human primates the studies of Hans Kummer (1975) and Robin and Patsy Dunbar (1975) on the gelada provide very important information which may provide some purchase on this issue. Kummer convened dyads of geladas by placing unrelated individuals together in pens. The animals engaged in interactions forming a stereotyped series that led to relationships in which individuals maximized their compatibilities. If a third individual was added to the dyad a predictable sequence of events emerged based on the contrasting compatibilities of same-sex and cross-sex partners, the agonistic and sexual repertoire and specific patterns of social interaction and inhibition Group relations were derived from the behaviours of individuals, each attempting to maximize their own sum of compatibilities with others while minimizing that sum in their social competitors. The structure of the groups so formed closely resembled those studied in the wild by Dunbar, although, in nature, the existence of long-term kinship lineages and historically determined patterns of relationships greatly modified the picture.

Dunbar's extensive research on Gelada baboons, suggests that individuals of both sexes behave in a way that seeks to maximize their reproductive output over a lifetime. This entails taking decisions about which course of action to adopt; actions which in given environmental, demographic, social and personal circumstances will lead to an optimum outcome. Females, for example, take decisions leading them to adopt coalition partners either in a mother−daughter pair, a sister−sister pair or a female−male-of-unit pair. Dunbar demonstrates that contrasts in coalitionary behaviour are associated with differences in the dominance of individuals which predict their reproductive success. The optimum relationship predicting dominance is an alliance with a socially powerful mother or daughter. Coalitions with sisters are less effective in that it will be in the sister's interest to desert an alliance as soon as her own daughter matures. Since males only remain a short time as 'owners' of a harem the effectiveness of a coalition with one of them is limited. The choice of partner depends upon an individual's circumstance, her birth order, mother's rank, group size and so on in the group in which she is born, and appear often to be optimal in relation to the possibilities available. These conclusions and comparable arguments on male tactics are well supported by Dunbar's analysis of his meticulously collected field data.

The psychological nature of the decision-taking process remains the point at issue here. While the language of assessment and decision does not necessarily imply consciousness in the sense of a metacognitive awareness of attributed categories such as 'self' and 'other', 'male' or 'female', neither does it imply a knowledge of the genetic consequences that may result from the behaviour. The animals do seem to be responding according to 'rules of thumb' which must be more than merely 'arithmetic' calculations. It is becoming theoretically helpful to consider the likelihood that the animals are responding to feelings excited by situations in which they fiind themselves, such feelings acting as indices of positive or negative outcomes within relationships which may or may not be related with high/low reproductive success. Performances that are broadly in accordance with an ESS are likely to 'feel' right and thereby reinforced while those that do not may be aborted. In this way protocultural behaviour could become canalized along lines normative for a species in its given socioecological situation. Such experiential states may include evaluative responses to such variables as proximity to certain relatives, grooming frequencies and types, learnt consequences of associations in terms of rewards (food items, access to grooming partners, etc.), mating frequencies and results of agonistic encounters.

I have previously argued that 'rules' akin to those discovered by Kummer may be significant in the determination of the more universal patterns of relationship shown in social groupings of human beings (Crook 1980, p. 236). Quite plausibly genetically controlled dispositions to respond with negative or positive affect to certain social interactions or intersubjective relations promote a canalization of interactional outcomes, so that characteristic societal configurations arise whenever human groups are convened. Cultural structures in vernacular societies would then comprise 'tactical' modifications of those configurations to suit the contrasting necessities of environmental exploitation in differing habitats. The question remains — what exactly produces these environmentally shaped modifications which seem so often not only to optimize the acquisition of resources but also to sustain fitness in diverse ways in contrasting classes within a society (see Ecology, fitness theory and the socioeconomics of human vernacular societies, above).

Two inter-relating approaches to this issue can be put forward as candidate theories in a very preliminary form. Csikzentmihalyi (1975) has shown that human action is optimally rewarding to an individual when the relationship between project and skill is neither so discrepant that either acute anxiety (skill low: task difficult) or boredom (skill high: task easy) are induced but where the balance is such (skill faced with appropriately challenging task) that the performance of action is experienced with an involved excitement described as 'flow'. Flow is a condition of heightened awareness in which the reduction in tension about performance allows a positively evaluated enjoyment of personal experience. Csikzentmihalyi & Massimini (1985) argue further that individuals make choices between courses of action based on a psychologically controlled selection favouring performances associated with flow. Individuals, they argue, tend to pay attention to and act upon cultural instructions that facilitate flow experiences. The viewpoint of these authors, moreover, is based not merely on a supposition that people follow up on themes which make them 'feel good' but upon an extensive experiential and observational research programme. Experiences of flow associate with the repetition of actions that preserve and restore personal well-being and facilitate relationships of high compatibility with others. In the performance of culturally determined 'games' in which the pay-off may be in terms of provisioning, security, self esteem, family welfare or in-group success, the survival of any particular game depends, they argue, on how many people experience flow within the context of the game and hence promote it. In the context of relatively simple subsistence economies such psychological selection could quite plausibly result in behaviour that would sustain or enhance 'inclusive fitness'. Csikzentmihalyi & Massimini are thus proposing a psychological mechanism that acts to favour preferences between social options that enhance a sense of well-being and which functions as a reliable guide or index of actions tending towards a maintenance of fitness in both the proximate sense of health and well-being and hence also in the sense of reproductive success.

This idea, I suggest, may also be of value in the context of a second approach. In vernacular societies personal identity is constituted through a socialization process embedded in the most important of the systems sustaining social structure — the network of kin relations. These relationships are, as Luckmann (1979) argues 'familiar, stable, highly individualized and systematically connected'. They are in fact the root of those societally instituted behaviour systems that maintain family welfare (pp. 15–18, above) and hence also reproductive success. Personal identity is profoundly influenced by emotional evaluations by the self of the self — which, when negative, can produce highly depressive personal states culminating in social withdrawal. In many vernacular societies assessments promoting self esteem are rooted in the credit mainly attributed by kin to an individual resulting from his positive actions in sustaining his/her estate and/or kin relationships in an extended family lineage. When credit is given or implied an individual is likely to experience satisfaction, his/her merit both in his/her own, and in others' eyes, will be high and his/her action in life likely to be sustained and effective.

Self-identity is the means whereby the human subject evaluates and guides his/her transactions with others (Ashworth 1979). Its quality is subject to depressive subsidence if self esteem falls but to enhanced pleasure in existence when self esteem dependent on the fulfilment of basic familial and social commitments is sustained. The underlying mechanisms may well consist in innate predispositions that canalize individual behaviour towards positive affect, sustaining patterns of relationship which constitute environments effective for child rearing. If this were so, one would expect that whenever a social system under the influence of an extreme ideology becomes maladaptive ecologically, economically or socially, constraints acting initially within the psychological systems of individuals will soon come into play. Much more work is required on this topic but at least some research (e.g. Bischof 1975) in Israeli kibbutzim, suggests this line of thought to be fruitful.

Hopefully the arguments put forward in this section may go some way towards bridging the gap which Ingold (this volume) creates between behavioural studies of human and animal subjects, for there is a need to link his valuable discussion of human sentience to the less psychologically oriented studies of Dunbar, Voland, Layton and Foley in this volume.

SOCIOECOLOGY AND HISTORY

Robin Dunbar argues cogently that social ethologists should not underestimate the structural complexity of non-human mammalian societies which arises largely through the formation of coalitions. This argument links well to the current emphasis on social complexity as the initiating matrix for the evolution of advanced cognition, the capacity for self-referencing identity and consciousness itself (Humphrey 1975, 1983; Crook 1980; Hinde 1983; Byrne & Whiten 1988).

One of the consequences of coalitionary behaviour is the development of networks of relationships that form the basis for grouping patterns. Geladas convene themselves into reproductive and all-male groups which associate together into congregations of a herd-like nature. Individuals in such a band associate together more frequently than they do with members of other bands which may range over the same area. Such bands may, however, share a number of units and thus together make up a wider community. Comparable complexity is reported from the hamadryas, the vervet monkey, chimpanzees, elephants, mongoose and other social species including birds. The intriguing point about these nested hierarchies of groups is that at each congregational level the association concerned functions to facilitate the cooperative (mutualistic) solution of the problems of individual members. Thus, the bands of gelada probably have an anti-predator or anti-harassment function that can be teased apart from the functions of the constituent reproductive and all-male groups. One of Dunbar's main points is that the development of a more inclusive structure may impose stresses upon the included structures that sets up a demographic process that intensifies as population increases. In particular, the fission–fusion process, that develops as gelada reproductive groups grow and eventually fragment, yields a society that tends to alternate between states (many small groups *or* fewer larger less stable groups) even when the demography itself has undergone a relatively minor change. Demographically led reconstructions of societal groupings are thus triggered by threshold reactions of individuals to patterns of changing stress arising within the social environment and the successive fissioning of units leads to the formation of lineages with distinctive family trees.

These nested hierarchies are emergent consequences of the historical relationships between groups of animals undergoing interactive changes in succeeding generations. If a population is expanding each generation re-establishes a pattern of nuclear reproductive units distributed in an expanding network of distant kin. Alliances at higher levels in such a structure are usually between groups which are mutually closer in kinship (i.e. sharing a common lineage) than they are close to other units in the population. Lineages with strong internal alliances may emerge as socially dominant kin groups or clans. The potential advantages to be gained from alliances in structures as complex as this may be very great — and mistakes by individuals or groups can be as costly. Contrasting societal configurations arising within differing ecological or demographic conditions mean that individuals and/or units may pursue differing tactical 'routes' (Caro & Bateson 1986), the success of which will be highly contingent on a continuous evaluation of a range of variables.

Viewing animal societies as potentially expanding networks of coalitions in extended lineages is a major advance from earlier views that treated social organizations as species-typical or only subject to minor ecologically responsive variations. It allows us to perceive a continuity between the longitudinal study of animal lineages and research on the history of human institutions. The comparison is particularly clear perhaps in the study of historical sequences in tribal societies.

An example of human societal radiation in response to both demographic and ecological factors is provided by Bonte's (1982) review of non-stratified social formations in the nilo-hamitic pastoral nomads of East Africa (of whom Dyson Hudson's Turkhana, this volume, are one group). Bonte argues that as certain of these peoples became increasingly dependent on pure pastoralism rather than on agropastoralism in areas ecologically less permissive of farming, so there were progressive developments in social organization that comprised adaptations to demographically driven social tensions arising within the population. In this sequence of increasing pastoralism individuals show ecological and demographic adaptability of the greatest interest. Pastoralists expand their grazing areas when demographically successful but the dispersed nature of their life encourages smaller familial units. The emergent age—set differentiation means that young men are often responsible for the peripheral expansion that may be resisted by other peoples. In the case of Masai, expansion reached a limit and the resulting internal stresses continued to promote societal change. One can see in this series of transformations that an increasing complexity in social behaviour and societal organization results from the intrasocietal stresses generated as the population became constricted due to ecological and demographic pressures. Ecodemographic processes thus play major roles in the historical emergence of social complexity within social structures of both non-human mammals and our own species.

Roles and relationships in vernacular human societies shift in response to changing ecodemographic conditions. Symbolic extensions of kinship roles to non-kin may serve to broaden the range of social ties and to adjust groupings, division of labour and gender roles to pressures of demography and ecology (Keesing 1975). In relatively simple cultures these adjustments are responses to changes made collectively by the individuals concerned. The result is the fine attunement to ecology of the life ways of subsistence cultures. Maitland-Bradfield's (1973) review of the Shoshonean Amerindian cultural radiation and Feil's (1987) comparative survey of the remarkable societies of the eastern and western highlands of Papua provide excellent examples from a growing literature.

In modern times, such vernacular systems have often become invaded by politically driven monetization and 'development'. The result has only too often been an unhappy breakdown in the ecological adjustment of the people leading to over-exploitation of limited resources, over-correction using imported technologically produced chemicals, erosion, social disruption, culture loss, emigration and even famine and poverty. Rappaport (1982) vividly describes such maladaptive change which is commonly the result of governmental failure to comprehend the nature of the indigenous system of subtle checks and balances. Such ideological enthusiasm for social change may eventually become subject to constraints that might reinstate systems of relations that allow for a restoration of family welfare sustaining the reproductive success of individuals. If this does not happen social, ecological and reproductive disaster quickly follows. E.O. Wilson's sensitive account (1984) of his response to ecological tragedy in South America must be essential reading for all concerned with environmental and cultural conservation.

SHIFTING PARADIGMS — CONCLUSION

Changes in perspective in the biological sciences are rarely either total replacements or reversals. Paradigms (Kuhn 1962) here do not collapse but rather undeveloped themes emerge or, more usually re-emerge, as a prime focus of interest (Mayr 1977; Ruse 1985; Kitcher 1986.) In behaviour study in particular there tends to be an oscillation between emphases on genetic and environmental themes in the process of evolution. The focus of classical ethology on innate factors in the determination of behaviour formed a counterpoint to behaviouristic environmentalism while the emergence of socioecology led to a renewed interest in social complexity and social relations as biotic systems responsive to ecological factors. The development of sociobiology once more emphasized the importance of genetic determination and stressed the analysis of behavioural adaptation as strategic. Today the pendulum is once more swinging because, as sociobiology encounters problems which its sophisticated but theory-laden approaches fail to resolve, there is a realization that the way forward must lie in a more extensive development of socioecological and ecocultural field study. While this constitutes the actual testing of sociobiological premises it also will require, as we have argued, a more holistic model of adaptation which relates environmental, societal and cultural processes to those of genetic selection.

The emerging 'coevolutionary' models force us to consider the psychological processes by which societies based on culture appear to copy the fitness maintaining strategies of non-cultural animals. This necessitates an approach which involves work on the cognitive competences of protocultural animals, motivational factors underlying the reinforcement or elimination of cultural themes and the emergence of metacognition in the human species. The material surveyed above show that ways of thinking that go beyond the simplicisms of early sociobiology are now well in the making. In particular, the need to analyse behavioural decisions and tactical routes of individual animals in relation to their ecological, social and demographic contexts is likely to emerge as a key focus. A similar focus on individual and collective practice in human cultures will also become of increasing interest. The analysis of the hierarchical structuring of social groups, the functions of groups at different levels of structure and the processes whereby complexity of structure emerges as a result of social and demographic stressors arising within particular environmental constraints, is likely to become especially important as it forms a bridge to the study of human history.

The themes we have been tracing also reflect shifts in ideas on a wider stage. The difficulties encountered within the sociobiological paradigm reflect the shortcomings of the Cartesian methodology underlying much contemporary experimentation and theory. As in psychology, it is becoming clear that where a behavioural phenomenon is not only complex but the result of the interaction of several different modes of determination a more holistic, even 'Hegelian', perspective becomes essential if science is not to stagnate in prejudicial antinomies (Markhova 1982). Socioecology is a field with wide implications for neighbouring disciplines. E.O. Wilson (1975) predicted the absorption of the social sciences by his version of sociobiology. A more cautious assessment of the relations between these disciplines reveals a prospect with a rather different promise. The emergence of civilization has resulted from technological solutions to major problems of ecological exploitation in agriculture, animal husbandry, irrigation and transportation, all associated with elaborating cultural patterns of alliance and exchange. The patterns of socioeconomic determination that have produced the rise and fall of cultures have rarely been separable from their socioecological base. The prime determinants of major changes in human history are ultimately related via technology to socioecology and we may doubt whether the purely biological layers of determination (levels 1 and 2 in Plotkin & Odling-Smee 1981) have provided more than the essential basis for these changes. As individual and collective behaviour tracks economic stress the old ground-rules of human biogrammar may do little more than decide the limits of particular trends in historical cycles (Modelski 1987). Nonetheless a continuing exploration of the complex inter-relations between genetic, neurological and endocrine systems on the one hand and individual expression on the levels of interaction, relationship and societal organization on the other remains the way forward to a more inclusive understanding (Hinde 1988). In exploring the social psychological and anthropological aspects of protocultural and cultural adaptation to environment socioecological research has an assured and vital place.

REFERENCES

Alexander, R.D. (1980). *Darwininism and Human Affairs*. Pitman, London.
Alexander, R.D. & Borgia, G. (1978). Group selection, altruism and the levels of organisation of life. *Annual Review of Ecology and Systemics*, 9, 449–74.
Allen, L. *et al.* (1976). Sociobiology — another biological determinism. *Bio Science*, 26, 182–6.
Ashmole, N.P. (1963). The regulation of numbers of tropical oceanic birds. *Ibis*, 103G, 458–73.
Ashworth, P.D. (1979). *Social Interaction and Consciousness*. Wiley, New York.
Betzig, L., Borgerhoff Mulder, M. & Turke, P. (Eds) (1988). *Human Reproductive Behaviour*. Cambridge University Press, Cambridge.
Bischof, N. (1975). Comparative ethology of incest avoidance. *Biosocial Anthropology* (Ed. by R. Fox). Malaby, London.
Bonte, P. (1982). Non-stratified social formations among pastoral nomads. *The Evolution of Social Systems* (Ed. by J. Friedman & M.J. Rowlands), pp. 37–68. Duckworth, London.
Borgerhoff Mulder, M.B. (1981a). Adaptation and evolutionary approaches to Anthropology. *Man (N.S.)*, 22, 25–41.
Borgerhoff Mulder, M.B. (1981b). Progress in human sociobiology. *Anthropology Today*, 3, 5–8.
Boyd, R. & Richerson, P.J. (1985). *Culture and the Evolutionary Process*. University of Chicago Press, Chicago, Illinois.
Brown, J.L. (1964). The evolution of diversity in avian territorial systems. *Wilson Bulletin*, 76, 160–7.
Brown, J.L. (1969). Territorial behaviour and population regulation in birds. *Wilson Bulletin*, 81, 293–329.

Burckhardt, R.W. (1981). On the emergence of ethology as a scientific discipline. *Conspectus History*, 1, 62–81.

Byrne, R. & Whiten, A. (1988). *Machiavellian Intelligence: Social Expertise and the Evolution of Intellect.* Oxford University Press, Oxford.

Caro T. & Bateson, P. (1986). Organisation and ontogeny of alternative tactics. *Animal Behaviour*, **34(5)**, 1483–1500.

Cavelli-Sforza, L.L. & Feldman, M.W. (1981). *Cultural Transmission and Evolution: a Quantitative Approach.* Princeton University Press, Princeton, New Jersey.

Chagnon, N. & Irons, W. (Eds) **(1979).** *Evolutionary Biology and Human Social Behaviour: an Anthropological Perspective.* Duxbury, North Scituate, Massachussets.

Cloak, F. (1975). Is a cultural ethology possible? *Human Ecology*, **3**, 161–82.

Corning, A. (1976). Toward a survival oriented policy science. *Biology and Politics* (Ed. by A. Somit), pp. 127–54. Mouton, The Hague.

Clutton-Brock, T. (1974). Primate social organisation and ecology. *Nature (London)*, **250**, 539–42.

Crook, J.H. (1953). An observational study of the gulls of Southampton Water. *British Birds*, **46**, 386–97.

Crook, J.H. (1964). *The Evolution of Social Organisation and Visual Communication in the Weaver Birds (Ploceinae).* Behaviour Monograph 10. Brill, Leiden.

Crook, J.H. (1965). *The adaptive significance of Avian social organisations The Social Organization of Animal Communities* (Ed. by P.E. Ellis). Symposia of the Zoological Society of London, 14, pp. 181–218. Academic Press, London.

Crook, J.H. (1970a). Social organisation and the environment: aspects of contemporary social ethology. *Animal Behaviour*, **18**, 197–209.

Crook, J.H. (1970b). Introduction — social behaviour and ethology. *Social Behaviour in Birds and Mammals* (Ed. by J.H. Crook) pp. xxi–xL. Academic Press, London.

Crook, J.H. (Ed.) **(1970c.)** *Social Behaviour in Birds and Mammals.* Academic Press, London.

Crook, J.H. (1980). *The Evolution of Human Consciousness.* Oxford University Press, Oxford.

Crook, J.H. & Butterfield, P. (1970). Gender role in the social system of *Quelea. Social Behaviour in Birds and Mammals* (Ed. by J.H. Crook), pp. 211–48. Academic Press, London.

Crook, J.H. & Crook, S.J. (1988). Tibetan polyandry: problems of adaptation and fitness. *Human Reproductive Behaviour.* (Ed. by L. Betzig, M. Borgerhoff-Mulder, & P. Turke), pp. 97–114. Cambridge University Press, Cambridge.

Crook, J.H. & Crook, S.J. (1989). Explaining Tibetan polyandry, socio-cultural, demographic and biological perspectives. *Himalayan Buddhist Villages* (Ed. by J.H. Crook & H. Omaston). Aris & Phillips, Warminster.

Crook, J.H., Ellis, J.E. & Goss-Custard, J.D. (1976). Mammalian social systems: structure and function. *Animal Behaviour*, **24**, 261–74.

Crook, J.H. & Gartlan, J.S. (1966). Evolution of primate societies. *Nature*, **210**, 1200–3.

Crook, J.H. & Goss-Custard, J.D. (1972). Social ethology. *Annual Review of Psychology*, **23**, 277–312.

Crook J.H. & Osmaston H. (Eds) **(1989).** *Himalayan Buddhist Villages.* Aris & Phillips, Warminster.

Crook, J.H. & Shakya, T. (1983). Six families of Leh. *Recent Research on Ladakh* (Ed. by D. Kantowsky & R. Sander) Shriftenreihe Internationales Asien Forum. Weltforum Verlag 1, Munchen.

Crook, J.H. & Ward, P. (1968). The Quelea problem in Africa. *The Problem of Birds as Pests* (Ed. by R.K. Murton & E.N. Wright), pp. 211–29. Symposia of the Institute of Biology, 17, Academic Press, London.

Cullen, E. (1957). Adaptations in the kittiwake to cliff nesting. *Ibis*, **99**, 275–302.

Cullen, J.M. (1960). Some adaptions in the nesting behaviour of terns. *Proceedings 12th International Ornithological Congress*, **12**, 153–7.

Czikszentmihalyi, M. (1975). *Beyond Boredom and Anxiety.* Jossey-Bass, San Francisco.

Czikszentmihalyi, M. & Massimini, F. (1985). On the psychological selection of bio-cultural information. *New Ideas in Psychology*, **3(2)**, 115–38.

Davies, N.B. (1982). Behaviour and competition for scarce resources. *Current Problems in Sociobiology* (Ed. by King's College Sociobiology Group), pp. 363–80. Cambridge University Press, Cambridge.

Davies, N.B. & Houston, A.I. (1981). Owners and satellites: the economics of territory defence in the Pied Wagtail, *Motacilla alba. Journal of Animal Ecology*, **50**, 157–80.

Dawkins, R. (1976). *The Selfish Gene*. Oxford University Press, Oxford.

Dickemann, M. (1979a). The ecology of mating systems in hypergynous dowry societies. *Social Science Information*, **18**, 163–95.

Dickemann, M. (1979b). Female infanticide, reproductive strategies and social stratification. *Evolutionary Biology and Human Social Behaviour* (Ed. by N.A. Chagnon & W. Irons). Duxbury, North Scituate, Massachusetts.

Dunbar, R.I.M. (1982). Adaptation, fitness and evolutionary tautology. *Current Problems in Sociobiology* (Kings College Sociobiology Group). Cambridge University Press, Cambridge.

Dunbar, R.I.M. (1984). *Reproductive Decisions: an Economic Analysis of Gelada Baboon Social Strategies*. Princeton University Press, Princeton, New Jersey.

Dunbar, R.I.M. (1986). Sociobiological explanations and the evolution of ethnocentrism. *The Sociobiology of Ethnocentrism* (Ed. by V. Reynolds, V. Falger & I. Vine), pp. 48–59. Croom Helm, London.

Dunbar, R.I.M. (1988). Darwinising Man: a commentary. *Human Reproductive Behaviour* (Ed. by L. Betzig, M. Borgerhoff-Mulder & P. Turke), pp. 161–9. Cambridge University Press, Cambridge.

Dunbar, R.I.M. & Crook, J.H. (1975). Aggression and dominance in the Weaver Bird, *Quelea quelea*. *Animal Behaviour*, **23**, 450–9.

Dunbar, R.I.M. & Dunbar, P. (1975). *Social Dynamics of Gelada Baboons*. Karger, Basel.

Durant, J.R. (1986). The making of ethology: the Association for the Study of Animal Behaviour 1936–1986. *Animal Behaviour*, **34**, 1601–16.

Durham, W.H. (1978). The co-evolution of human biology and culture. *Human Behaviour and Adaptation*. (Ed. by V. Reynolds & N. Blurton-Jones), pp. 11–32. Symposium for the Society for the Study of Human Biology, 18. Taylor & Francis, London.

Emlen, S.T. (1978). The evolution of co-operative breeding in birds. *Behavioural Ecology: an Evolutionary Approach* (Ed. by J.R. Krebs & N.B. Davies). Blackwell Scientific Publications, Oxford.

Emlen, S.T. (1984). Co-operative breeding in birds and mammals. *Behavioural Ecology: an Evolutionary Approach* 2nd edn. (Ed. by J.R. Krebs & N.B. Davies), pp. 305–339 Blackwell Scientific Publications, Oxford.

Espinas, A. (1878). *Des Sociétés Animales*. Bailliére, Paris.

Feil, D.K. (1987). *The Evolution of Highland Papua New Guinea Societies*. Cambridge University Press, Cambridge.

Fisher, J. (1954). Evolution and Bird Sociality. *Evolution as a Process* (Ed. by J. Huxley, A.C. Hardy & E.B. Ford). Allen & Unwin, London.

Fisher, R.A. (1930). *The Genetic Theory of Natural Selection*. Clarendon Press, Oxford.

Flinn, M.V. & Low, B.S. (1986). Resource distribution, social competition and mating patterns in human societies. *Ecological Aspects of Social Evolution* (Ed. by D.I. Rubenstein & R.W. Wrangham). Princeton University Press, Princeton, New Jersey.

Goldstein, M.C. (1971). Stratification, polyandry and family structure in Central Tibet. *Southwestern Journal of Anthropology*, **27**, 64–74.

Goodall, J. (1986). *The Chimpanzees of Gombe: Patterns of Behaviour*. Belknap/Harvard, Cambridge, Massachusetts.

Goss-Custard, J.D., Dunbar, R.I.M. & Aldrich-Blake, P.G. (1972). Survival, mating and rearing strategies in the evolution of primate social structure. *Folia Primatologica*, **17**, 1–19.

Grafen, A. (1984). Natural Selection, kin selection and group selection. *Behavioural Ecology: an Evolutionary Approach*, 2nd edn., (Ed. by J.R. Krebs & N.B. Davies), Blackwell Scientific Publications, Oxford.

Groot, P. de (1980). Information transfer in a socially roosting weaver bird (*Quelea quelea*; Ploceinae): an experimental study. *Animal Behaviour*, **128**, 1249–59.

Haldane, J.B.S. (1953). Animal populations and their regulation. *Penguin Modern Biology*, **15**, 9–24.

Haldane, J.B.S. (1956). The argument from animals to man — an examination of its validity for anthropology. *Journal of the Royal Anthropological Institute*, **86**, 1–14.

Hamilton, W.D. (1964). The genetical evolution of social behaviour, I, II. *Journal of Theoretical Biology*, **7**, 1–52.

Hamilton, W.D. (1971). Geometry for the selfish bird. *Journal of Theoretical Biology*, **31**, 295–311.

Harris, M. (1968). *The Rise of Anthropological Theory*. Crowell, New York.

Hartung, J. (1982). Polygyny and the inheritance of wealth. *Current Anthropology*, 23, 1–12.

Hinde, R.A. (1955/6). A comparative study of the courtship of certain finches (Fringillidae). *Ibis*, 97, 706–45; 98, 1–23.

Hinde, R.A. (1960). Energy models of motivation. Symposia of the Society for Experimental Biology, 14, pp. 199–213. Cambridge University Press, Cambridge.

Hinde, R.A. (1966). *Animal Behaviour. A Synthesis of Ethology and Comparative Psychology*. McGraw-Hill, New York.

Hinde, R.A. (Ed.) (1983). *Primate Social Relationships: an Integrated Approach*. Blackwell Scientific Publications, Oxford.

Hinde, R.A. (1988). *Individuals, Relationships and Culture*. Cambridge University Press, Cambridge.

Hinde, R.A. & Steel, E. (1966). Integration of the reproductive behaviour of female canaries. *Nervous and Hormonal Mechanisms of Integration*, Symposia of the Society for Experimental Biology, 20, pp. 401–26. Cambridge University Press, Cambridge.

Hinde, R.A. & Stevenson-Hinde, J. (Eds) (1973). *Constraints on Learning: Limitations and Predisposition*. Academic Press, New York.

Hodos, W. & Campbell, C.B.E. (1969). Scala naturae: why there is no theory in comparative psychology. *Psychology Review*, 76(4), 337–50.

Howard, H.E. (1920). *Territory in Bird Life*. Murray, London.

Humphrey, N.K. (1975). The social function of intellect. *Growing Points in Ethology* (Ed. by R.P.G. Bateson & R.A. Hinde) Cambridge University Press, Cambridge.

Humphrey, N.K. (1983). *Consciousness Regained*. Oxford University Press, Oxford.

Huxley, J. (1923). Courtship activities in the red-throated diver together with a discussion of the evolution of courtship in birds. *Journal of the Linnean Society*, 35, 253–92.

Immelmann, K. (1967). Verhaltensökologische Studien an afrikanischen und australischen Estrildiden. *Zoologische Jahrbuch Systematik*, 94, 1–67.

Irons, W. (1979). Natural selection, adaptation and human social behaviour. *Evolutionary Biology and Human Behaviour* (Ed. by N. Chagnon & W. Irons). Duxbury, North Scituate, Massachusetts.

Jarman, P.J. (1974). The social organisation of Antelope in relation to the ecology. *Behaviour*, 48, 215–67.

Jaynes, J. (1969). The historical origins of "Ethology" and "Comparative Psychology". *Animal Behaviour*, 4, 601–6.

Jaynes, J. (1976). *The Origins of Consciousness in the Breakdown of the Bicameral Mind*. Mifflin, New York.

Keesing, R.M. (1975). *Kin groups and Social Structure*. Holt, Rinehart & Winston, New York.

Kitcher, P. (1985). *Vaulting ambition: Sociobiology and the Quest for Human Nature*. M.I.T. Press, Cambridge, Massachusetts.

Klopfer, P.H. & Hailman, J.P. (1967). *An Introduction to Animal Behaviour. Ethology's First Century*. Prentice-Hall, New Jersey.

Krebs, J.R. & Dawkins, R. (1984). Animal signals, mind reading and manipulation. *Behavioural Ecology: an Evolutionary Approach*, 2nd edn. (Ed. by J.R. Krebs & N.B. Davies), pp. 380–402. Blackwell Scientific Publications, Oxford.

Kropotkin, P. Prince (1902). *Mutual Aid — a Factor of Evolution*. Heinemann, London.

Kuhn, T.S. (1962). *The Structure of Scientific Revolutions*. University of Chicago Press, Chicago, Illinois.

Kummer, H. (1975). *Rules of Dyad and Group Formation among Captive Gelada Baboons (Theropithecus gelada)*. Proceedings of the Symposium of the 5th Congress of the International Primatological Society, Nagoya, pp. 129–160. Japan Science Press, Tokyo.

Lack, D. (1954). *The Natural Regulation of Animal Numbers*. Clarendon Press, Oxford.

Lack, D. (1966). *Population Studies of Birds*. Clarendon Press, Oxford.

Lazarus, J. (1979a). Flock size and behaviour in captive red-billed weaver birds (*Quelea quelea*): implications for social facilitation and the functions of flocking. *Behaviour*, 71, 1–2; 127–45.

Lazarus, J. (1979b). The early warning function of flocking in birds: an experimental study with captive *Quelea*. *Animal Behaviour*, 27, 855–68.

Lazarus, J. & Crook, J.H. (1973). The effects of luteinizing hormone, oestrogen and ovariectomy on the agonistic behaviour of female *Quelea quelea*. *Animal Behaviour*, 21, 49–60.

Lehrmann, D.S. (1953). A critique of Konrad Lorenz's theory of instinctive behaviour. *Annual Review of Biology*, **28**, 337–63.

Lehrmann, D.S. (1964). The reproductive behaviour of ring doves. *Scientific American*, **211**, 48–54.

Lockhard, R.B. (1971). Reflections on the fall of comparative psychology: Is there a message for us all? *American Psychologist*, **26(2)**, 14 pp.

Lorenz, K. (1950). The comparative method in studying innate behaviour patterns. *Physiological Mechanisms in Animal Behaviour*. Symposia of the Society for Experimental Biology, 4, pp. 221–68. Cambridge University Press, Cambridge.

Lorenz, K. (1970) *Studies in Animal and Human Behaviour*. (Translated by R. Martin) *Vol. 1*. Methuen, New York.

Luckmann, T. (1979). Personal identity as an evolutionary and historical problem. *Human Ethology: Claims and Limits of a New Discipline*. (Ed. by M. von Cranach, K. Foppa., W. Pepenies & D. Ploog, pp. 56–74. Cambridge University Press, Cambridge.

Lumsden, C.T. & Wilson, E.O. (1981). *Genes, Mind and Culture: the Co-evolutionary Process*. Harvard University Press, Cambridge, Massachusetts.

Macarthur, R.H. & Wilson, E.O. (1967). *The Theory of Island Biogeography*. Monographs in population biology 1. Princeton University Press, Princeton, New Jersey.

Maitland-Bradfield, R. (1973). *A Natural History of Associations: a Study in the Meaning of Community*. Duckworth, London.

Markhova, I. (1982). *Paradigms, Thought and Language*. Wiley, New York.

Mauss, M. (1950). *Sociologie et Anthropologie*. Paris.

Maynard-Smith, J. (1972). *On Evolution*. Edinburgh University Press, Edinburgh.

Mayr, E. (1977). Concepts in the study of Animal Behaviour. *Reproductive Behaviour and Evolution* (Ed. by J.S. Rosenblatt & B.R. Komisaruk), pp. 1–16. Plenum, New York.

Mead, G.H. (1934). *Mind, Self and Society*. University of Chicago Press, Chicago, Illinois.

Modelski, G. (1987). *Long Cycles in World Politics*. MacMillan, London.

Murton, R.K., Isaacson, A. & Westwood, N.J. (1971). The significance of gregarious feeding behaviour and adrenal stress in a population of Woodpigeons, *Columba palumbus*. *Behaviour*, **40**, 10–42.

Nelson, J.B. (1978). *The Sulidae: Gannets and Boobies*. Oxford University Press, Oxford.

Orians, G.H. (1969). On the evolution of mating systems in birds and mammals. *American Naturalist*, **103**, 589–603.

Orians, G.H. (1980). *Adaptations of Marsh-nesting Blackbirds*. Princeton University Press, Princeton, New Jersey.

Osmaston, H. (1985). The productivity of the agricultural and pastoral system in Zangskar (N.W. Himalaya). *Acta Biologica Montana*, **5**, 75–89.

Packer, C. (1977). Reciprocal altruism in *Papio anubis*. *Nature (London)*, **265**, 441–3.

Parker, G.A. (1984). Evolutionarily stable strategies. *Behavioural Ecology: an Evolutionary Approach* 2nd edn. (Ed. by J.R. Krebs & N.B. Davies), pp. 30–61. Blackwell Scientific Publications, Oxford.

Peter, Prince of Greece and Denmark (1963). *A Study of Polyandry*. Mouton, La Hague.

Petrucci, R. (1906). *Origine Polyphylétique, Homotypie et Noncomparabilité Directe des Sociétés Animales*. Institut Solvay, Travaux de l'Institut de Sociologie, Notes et Memoires 3. Misch et Thon., Brussells.

Plotkin, H.C. & Odling-Smee, F.J. (1981). A multiple-level model of evolution and its implications for sociobiology. *Behavioural and Brain Sciences*, **4(2)**, 225–68.

Powell, G.V.N. (1974). Experimental analysis of the social value of flocking by Starlings (*Sturnus valgaris*) in relation to predation and foraging. *Animal Behaviour*, **22**, 501–5.

Rappaport, R.A. (1982). Maladaption in social systems. *The Evolution of Social Systems* (Ed. by J. Friedman & M.J. Rowlands), pp. 49–71. Duckworth, London.

Reynolds, V. & Blurton-Jones, N. (Eds) (1978). *Human Biology and Adaptation*. Symposium of the Society for Human Biology. Taylor & Francis, London.

Ruse, M. (1979). *Sociobiology: Sense or Nonsense?* Reidel, Dordrecht.

Seigfried, W. & Underhill, L.G. (1975). Flocking as an anti-predation strategy in Doves. *Animal Behaviour*, **23**, 504–8.

Skutch, A.F. (1935). Helpers at the nest. *Auk*, **52**, 257–73.

Slobodkin, L.B. (1961). *Growth and Regulation of Animal Populations.* Holt, Rinehart & Winston, New York.

Snell, B. (1960). *The Discovery of the Mind: the Greek Origins of European thought.* Harper, New York.

Sparks, J. (1982). *The Discovery of Animal Behaviour.* Collins, London.

Stephens, D.W. & Krebs, T.R. (1986). *Foraging Theory.* Princeton University Press, Princeton, New Jersey.

Thorpe, W.H. (1979). *The Origins and Rise of Ethology.* Praeger, London.

Tinbergen, N. (1942). An objectivistic study of the innate behaviour of animals. *Bibliotheca Biotheoretica,* **1**, 39–98.

Tinbergen, N. (1951). *The Study of Instinct.* Clarendon Press, Oxford.

Tinbergen, N. (1953). *Social Behaviour in Animals.* Methuen, London.

Tinbergen, N. (1959). Comparative studies of the behaviour of gulls (*Laridae*): a progress report. *Behaviour,* **15**, 1–70.

Tinbergen, N., Impekovin, M. & Franke, D. (1967). An experiment on spacing out as a defence against predation. *Behaviour,* **28**, 307–21.

Trewarthan, C. (1979). Instincts for human understanding and for cultural cooperation: their development in infancy. *Human Ethology: Claims and Limits of a New Discipline.* (Ed. by M. von. Cranach, K. Foppa., W. Lapenies & D. Ploog), pp. 530–71. Cambridge University Press, Cambridge.

Trivers, R. (1971). The evolution of reciprocal altruism. *Quarterly Review of Biology,* **46**, 35–57.

Trivers, R. (1972). Parental investment and sexual selection. *Sexual Selection and the Descent of Man* (Ed. by B. Campbell). Aldine, Chicago, Illinois.

Trivers, R. (1974). Parent-offspring conflict. *American Zoologist,* **14**, 249–65.

Trivers, R. (1985). *Social Evolution.* Cummings, Menlo Park, California.

Vine, I. (1971). The risk of visual detection and pursuit by a predator and the selective advantage of flocking behaviour. *Journal of Theoretical Biology,* **30**, 405–22.

Waal, F.B.M. de (1982). *Chimpanzee Politics.* Jonathan Cape, London.

Waltz, E.C. (1987). A test of the information-centre hypothesis in two colonies of common term *Sterna hirundo. Animal Behaviour,* **35(1)**, 48–59.

Ward, P. (1965). Feeding ecology of the black-faced dioch *Quelea quelea* in Nigeria. *Ibis,* **107**, 173–214, 326–49.

Ward, P. & Zahavi, A. (1973). The importance of certain assemblages of birds as "information centres" for food finding. *Ibis,* **119**, 517–34.

Waxweiler, E. (1906). *Esquisse d'une Sociologie.* Instituts Solvay. Travaux de l'Institut de Sociologie. Notes et mémoires 2. Misch et Thon, Bruxelles.

Weatherhead, P. (1987). Field tests of information transfer in communally roosting birds. *Animal Behaviour,* **35(2)**, 614.

Williams, G.C. (1966). *Adaptation and Natural Selection: a Critique of some Current Evolutionary Thought.* Princeton University Press, Princeton, New Jersey.

Wilson, E.O. (1975). *Sociobiology — the New Synthesis.* Belknap/Harvard, Cambridge, Massachusetts.

Wilson, E.O. (1984). *Biophilia.* Harvard University Press, Cambridge, Massachusetts.

Winn, H.E. (1958). The comparative ecology and reproductive behaviour of 14 species of Darter (Percidae). *Ecological Monographs,* **28**, 155–91.

Wittenberger, J.F. (1981). *Animal Social Behaviour.* Duxbury, North Scituate, Massachusetts.

Wrangham, R.W. (1980). An ecological model of female-bonded primate groups. *Behaviour,* **75**, 262–300.

Wrangham, R.W. (1982). Mutualism, kinship and social evolution. *Current Problems in Sociobiology* (Ed. by King's College Sociobiology Group). Cambridge University Press, Cambridge.

Wynne-Edwards, V.C. (1962). *Animal Dispersion.* Oliver & Boyd, Edinburgh.

Zahavi, A. (1971). The social behaviour of the white wagtail (*Moticilla alba alba*) wintering in Israel. *Ibis,* **113**, 203–11.

Section I
The Causes of Sociality

Almost by definition socioecologists are interested in animals that live in social groups, or at least in some social context. One of the principal areas of interest is the pattern and determinants of different types of social organization. However, preceding this question is perhaps an even more fundamental one — why be social at all?

The simple answer, in terms of evolutionary ecology, is that an animal will be social when the benefits of doing so exceed the costs. This in turn prompts the question — what *are* the costs and benefits of living in groups? We should perhaps start to answer this question by considering what is the natural state for most animals (social or asocial), for this may well form the best starting assumption. Certainly it would appear that most animals are asocial, or have only limited sociality, confined to courtship/copulation and to a limited extent to the raising of young. If this is correct, then the best way of phrasing questions about the causes of sociality lie in determining when the benefits of sociality exceed the costs, rather than the other way around. This seems fairly straightforward, but is important to stress because the advantages of sociality must be considered at least in its initial stages *in the context of asocial living*. There may be many reasons why once established sociality may be beneficial, but these cannot account for its original (and no doubt repeated) evolution.

Various factors that prevent animals living in groups have been recognized (see Krebs & Davies 1987 for a discussion). Among these would be the loss of access to food, or increased costs of acquiring food, caused by interference and competitive utilization by conspecifics, the potential for increased vulnerability to disease and pathogens, and the competitive threat that conspecifics may pose to an individual's offspring. In several species unrelated animals may actively attempt to kill the young within a group (e.g. lions: Bertram 1975; langurs: Hrdy 1977; chimpanzees: Goodall 1977). Only when costs such as these are overcome can animals be expected to live in social groups.

The benefits that may result from group living are varied. The most frequent advantage is predator avoidance. Animals living in groups may well increase their total vigilance, while decreasing individual vigilance, and therefore reduce their vulnerability to the approach of a predator. Furthermore, these groups may be able to act in concert in some way that either confuses the predator or else actively resists and deters it. For example, many birds may 'mob' a predator, using numbers and erratic movement to overcome their disadvantages of size.

Other factors that may promote sociality include the advantages that may occur in foraging — either actively through cooperative hunting behaviour as is proposed for lions, hyaenas and hunting dogs (Kruuk 1972; Caraco & Wolf 1975; Packer 1986) — or because of the increased access to food provided by a better potential for defence of resources or a greater chance of locating them. The benefits of sociality will also be enhanced by demographic and genetic factors — where related individuals are concerned increased fitness may accrue from helping relatives either in acquiring mates or in bringing up offspring.

Overall, we may expect sociality to occur when conditions favour these factors, and furthermore, the intensity of the level of sociality may also be expected to be enhanced by these same conditions. The causes of sociality may therefore be said to be the occurrence of these conditions.

This is the problem directly addressed by Caro, with respect to felids. The felids offer a particularly apt illustration of the problems of sociality, for by and large they are asocial. Among the Felidae as a whole, only the lion is consistently social and the cheetah rather more variably so. Caro argues that the evolution of sociality among the felids has been limited by the availability of prey species of appropriate size — the costs of 'sharing' resources limit sociality, and so it should only occur when the size of prey, relative to the size of the predator, is sufficiently large *and* abundant for more than one animal (particularly females) to live together. Under circumstances when this is not the case — for example, when the prey are too small, or larger prey too rare to be depended upon, the costs in terms of hunting time or the probability of success would be too high. Caro analyses the data available on the prey size and density for the principal species of felid and shows that the conditions necessary for sociality, according to his model, occur only occasionally. This approach throws interesting light on the differences in behaviour in a group of closely related animals that are on the margins of sociality, and therefore may slip across the boundary conditions relatively (in evolutionary terms) frequently. Furthermore, the model raises some interesting questions about the co-evolution of body size in predators and prey — a long running game that may have had consequences for the frequency of sociality at different periods — and hence the pattern of evolution above the level of the species.

Social animals, by definition, live together. They consequently share space, so that spatial factors are closely linked to the development of sociality. In some respects sociality can be treated as the inverse of one of the central concepts of ethology, territoriality. To be territorial is to exclude other animals from the area in which an animal lives; to be social is to accept the presence of others. Macdonald (1983) has argued that sociality may exist as a result of the tolerance that may occur when the smallest economically denfensible unit may also accommodate additional animals, and therefore the costs of being social are reduced. This has been referred to as the resource dispersion hypothesis as the conditions under which this is most likely to occur are when resources are dispersed. The logic of this argument implies that sociality is linked to territoriality through the character of the resources on which an animal depends and may or may not defend.

In their paper (this volume) Macdonald and Carr make a further point — that the factors influencing sociality fall into two categories: ecological and socio-logical. They show that for individuals living in the same resource environment the social or demographic environment may vary considerably (see also Section 3, especially Datta, this volume), and consequently so too does the potential fitness of each individual animal. Their model shows that because of the uniqueness of each individual's environment the costs and benefits of staying (being social) or dispersing are variable.

Models such as these take us a long way from the simple 'species-specific' behaviour of classical ethology. When socioecologists refer to the costs and benefits of an environment, each individual's unique, often socially determined, environment is critical and will determine the potential fitness derived from pursuing any particular behavioural strategy.

The papers by Caro and Macdonald & Carr bring the focus for explaining the causes of sociality on to the nature of the resources utilized by an animal. For Caro it was package size, for Macdonald & Carr the extent to which resources were dispersed. Powell continues this theme by looking at another resource characteristic, the variance in its availability through time and space, and argues that the type of mating strategy — polygyny or monogamy — is sensitive not just to the total amount of food available, but to how reliably available it is.

Powell tests his mathematical model against a variety of data drawn from the literature, including some on human populations. To some extent, given that primates are among the most social of all animals, the question of why humans should be social is an uninteresting one. However, the issues involved in the causes of sociality are important for understanding the intensity and the complexity of social behaviour, for the factors that are likely to result in social interactions may also determine their extent. Thus for humans, who are probably the most socially complex species in existence, it may well be that all the factors that promote sociality are particularly important for them. For example, several authors (Isaac 1978; Potts 1984) have suggested that human sharing is related to a higher level of meat consumption than that found in other extant primates. When this hypothesis is considered in the light of Caro's model of prey size and sociality — that animals will forage together when size and density of prey is sufficient — and of Macdonald's idea that tolerance will be enhanced when resources are clumped, then we have a potential insight into the underlying reasons for human social intensification during the course of their evolution.

At the other end of the social spectrum, a comparative approach may throw light on the breakdown of human sociality — those occasions when humans do not tolerate the presence of others. According to socioecological theory, this should occur only when the costs exceed the benefits. In a species with a very long-established tendency towards a pattern of sociality these costs would have to be very high. However, it is certainly true to say that the intensity and extent of human social relationships vary considerably, but their complete breakdown is extremely rare. One case has been reported ethnographically — that of the Ik, a group of hunter-gatherers in north-east Uganda, who became, according to

Turnbull (1973), largely asocial under extreme conditions of starvation and economic stress following the collapse of their traditional subsistence strategies. Members of this population were reported to go to great lengths to avoid social interactions, particularly involving food. It would be naïve to suggest that this alone was the process going on, but it could be argued that for the Ik the ecological costs of sociality were too great, and solitary foraging was a more apposite behaviour under these extreme conditions. Whether this is the case or not, it focuses on an important aspect of socioecology — that of the within-species variability in social intensity and complexity.

REFERENCES

Bertram, B.C.R. (1975). Social factors influencing reproduction in wild lions. *Journal of the Zoological Society of London*, **177**, 463–82.

Caraco, T. & Wolf, L. (1975). Ecological determinants of group sizes of foraging lions. *American Naturalist*, **109**, 343–52.

Goodall, J. (1977). Infant killing and cannibalism in free-living chimpanzees in the Gombe National Park, Tanzania. *Zeirschrift für Tierpsychologie*, **61**, 1–60.

Hrdy, S. (1977). *The Langurs of Abu: Female and Male Strategies of Reproduction*. Harvard University Press, Cambridge, Massachusetts.

Isaac, G.Ll. (1978). Food sharing and human evolution: archaeological evidence from the Plio-Pleistocene of East Africa. *Journal of Anthropological Research*, **34**, 311–25.

Krebs, J.R. & Davies, N.B. (1987). *An Introduction to Behavioural Ecology*, 2nd edn. Blackwell Scientific Publications, Oxford.

Kruuk, H. (1972). *The Spotted Hyaena*. University of Chicago Press, Chicago, Illinois.

MacDonald, D.W. (1983). The ecology of carnivore social behaviour. *Nature*, **301**, 379–84.

Packer, C. (1986). The ecology of sociality in felids. *Ecological Aspects of Social Evolution* (Ed. by R.W. Wrangam & D.I. Rubenstein), pp. 429–51. Princeton University Press, Princeton, New Jersey.

Potts, R. (1984). Hominid hunters? Identifying the earliest hunter-gatherers. *Hominid Evolution and Community Ecology* (Ed. by R. Foley), pp. 129–66. Academic Press, London.

Turnbull, C.M. (1973). *The Mountain People*. Jonathan Cape, London.

Determinants of asociality in felids

T.M. CARO

*Evolution and Human Behavior Program, Rackham Building, University of Michigan,
Ann Arbor, Michigan 48109–1070, USA*

SUMMARY

1 Despite a number of benefits that could accrue from living in groups, adult members of most felid species live alone, which suggests that there are considerable costs to living together for members of this family.

2 The reasons why male felids of most species live alone, but why male cheetahs and lions live in groups are first discussed. I then address the more problematic question of why the great majority of adult female cats do not live together.

3 The idea that females actually live alone for most of their lives is dismissed. Using data from free-living cheetahs, it is shown that their companions (dependent cubs) consume a large share of the food that females acquire, but help their mothers little in catching prey.

4 Data are presented on the time that cheetah mothers spend hunting with litters of different sizes. They show that the amount of extra time necessary to sustain members of a social group would be prohibitively high.

5 For cheetahs, hunting larger prey in order to feed group members would be no more profitable than hunting smaller prey because of the difficulties in capturing large prey items. Then data from males are presented to show that increasing group size does not result in significant increases in hunting success.

6 These results suggest that in becoming social, cheetah mothers would have to spend too much time hunting unless there were sufficient numbers of large or vulnerable prey in the habitat.

7 The implications of these findings are then extended to other species of felid. Prey larger than that normally taken by females, necessary to sustain two families living together, is relatively scarce in nearly all the habitats that have been studied. The review suggests that conditions necessary for sociality to evolve are absent for virtually all extant felids.

INTRODUCTION

In most of the thirty-seven species of felids, males and females are asocial. Only in cheetahs (*Acinonyx jubatus*) (Frame & Frame 1981) and lions (*Panthera leo*) (Schaller 1972) do males live in groups, and females live communally only in the latter species (van Orsdol, Hanby & Bygott 1985). A number of hypotheses have been advanced to explain why carnivores, including lions (see, for example, Caraco & Wolf 1975; Clark 1987), live in groups. For example, increased group size might benefit individuals of both sexes because it might increase the chances of detecting predators (Rood 1986), it could reduce the risk of losing carcasses to other predators (Lamprecht 1978, 1981), or it might lower risks of injury during prey capture (see, for example, Gashwiler & Robinette 1957). For females in particular, the probability of infanticide by males is likely to be reduced if they defend their cubs together (Packer & Pusey 1983), while for males, group living might enhance access to females through advantages in contests with single males (Bygott, Bertram & Hanby 1979; Caro & Collins 1986). However, advantages that predators might gain through hunting cooperatively, previously thought to be an important evolutionary cause of sociality in many carnivores (Kruuk 1972, 1975; Curio 1976), have now been convincingly dismissed across a wide range of species, including lions (Packer 1986), especially when large or single prey times are hunted and when individual hunting success is reasonably high (Packer, in press; Packer & Ruttan 1988).

In the Felidae, both sexes could, in theory, benefit from group living in a number of different ways in so far as small cats suffer from predation (for example, servals *Leptailurus serval* by leopards *Panthera pardus*: Schaller 1972), felids lose carcasses to other predators (such as servals and cheetahs to spotted hyaenas, *Crocuta crocuta*: Geertsema 1985; Schaller 1972 respectively), cubs of several species are subject to infanticide by males (tigers *Panthera tigris*: Schaller 1967; cougars, *Felis concolor*: Seidensticker *et al*. 1973; possibly cheetahs: Burney 1980; see also Packer & Pusey 1984), and all male Felidae compete over access to females (for example, lions: Owens & Owens 1984; caracals, *Felis caracal*: Pringle & Pringle 1979). Indeed, in situations where domestic cats *Felis catus* are provisioned on farms and live socially, several females appear to guard and defend kittens from visiting males (MacDonald & Apps 1978). Therefore, considering the theoretical benefits of group living, it seems probable that there must also be considerable costs to explain why most members of this family are not social, yet to date these costs have not been examined.

This chapter focuses particularly on the costs of sociality rather than discussing its possible benefits. Packer (1986) also addresses the topic of felid sociality arguing that in those species where female felids live at high population densities, in open habitats, and which capture large prey, it would pay females to share their visible carcasses with female relatives rather than inevitably relinquish parts of them to unrelated conspecifics (see also Waser & Waser 1985). This reasoning may explain why female lions living in savannah regions could minimize foraging costs by living amongst groups of female relatives that do not disperse from their natal home range.

Three sets of observations question the generality of this explanation. First, lionesses should be solitary, in situations where prey carcasses would not be found by conspecifics, for example, in habitats where visibility is poor, such as thick woodlands, or where lionesses live at low densities. At present, there are insufficient data to test this proposition critically, but it is known that pride size does remain relatively constant across a wide range of lion populations living in different habitats (Eloff 1973; van Orsdol 1981). Moreover, lionesses living at reduced densities in many areas where human hunting pressure is high should now live alone, but there is no strong evidence in support of this.

Second, in palaearctic and subarctic regions both predators and the carcasses of their prey can sometimes be seen from long distances, especially if snow is lying. Moreover, carcasses in these regions often last longer than in the tropics because of freezing temperatures. In some of these areas, felids such as lynxes, *Lynx canadensis*, and bobcats, *Lynx rufus*, live at high densities and often feed on large prey (see below), so females of these species would also be expected to be social in some regions if Packer's suggestion is applicable to species other than lions.

Third, it is known that nearly all felids scavenge (see, for example, Konecny 1987), including the morphologically specialized ones (Skinner 1979; Caro 1982), and there is growing evidence as studies accumulate that solitary, and presumably unrelated, individuals form temporary aggregations at kills in species other than lions (tigers: Schaller 1967; cougars: Seidensticker *et al.* 1973; cheetahs: personal observations). Thus, females in these species would also be expected to reduce foraging costs by sharing their kills with related adult females, as argued for lions.

Most troubling perhaps, are the tigresses living in the dry deciduous habitat of Ranthambore National Park, where visibility is good, at population densities higher ($10/100$ km^2) than savannah lions ($7·9/100$ km^2). Adult females sometimes take large sambar, *Cervus unicolor*, are attracted to each other's kills by vultures, and have been seen to compete over carcasses (Thapar 1986). According to the hypothesis, these tigresses should live in groups but they do not.

In short, this combination of ecological factors may allow lions to be social in savannah regions, but it seems somewhat surprising that more felids would not be following Packer's suggestion in other habitats, unless there were considerable costs to group living. Here, the time costs incurred by cheetah mothers feeding litters of different size and the difficulty they experience in capturing prey of different weight are examined, in order to estimate the costs of group living. There then follows a discussion on the extent to which conditions necessary for sociality are absent in the other felids, but I start with a brief review of sociality in males.

MALES

As a consequence of internal fertilization and gestation, parental investment by female mammals is usually considerably greater than that of males, and reproductive success of females is restricted by access to resources (Trivers 1972). Male reproductive success is usually limited by access to females. Thus the distribution of females will be of major importance in determining male social organization in most species (Bradbury & Vehrencamp 1977; Emlen & Oring 1977; Wrangham 1979). In all but one of the felid species, females are characterized as living alone or with their dependent cubs, and either have exclusive home ranges which they defend against other females (for example, bobcats: Bailey 1974; European wild-cats, *Felis sylvestris*: Corbett 1979; tigers; Sunquist 1981; leopards: Bertram 1982) (see Fig. 1), or have home ranges with a considerable degree of overlap (e.g. cougars: Seidensticker *et al.* 1973; jaguars, *Panthera onca*: Schaller & Crawshaw 1980; North American lynxes: Carbyn & Patriquin 1983; ocelots *Felis pardalis*: Ludlow & Sunquist 1987). In both cases solitary males defend ranges that overlap those of several females.

In cheetahs, females have extremely large home ranges, 800 km^2 in the Serengeti (Frame 1984), compared to only 16 km^2 for similar-sized leopards living in the same ecosystem (Bertram 1978) (Fig. 1), because female cheetahs annually follow the migratory movements of their principal prey species, Thomson's gazelles *Gazella thomsoni* (Frame 1984; Durant *et al.* in press). Females do not defend their ranges against other females, and even one female range would probably be too large to be defended by a male. However, female cheetahs appear to collect in localities affording localized cover when Thomson's gazelles move into the area (Caro & Collins 1987a). By monopolizing such an area, only a fraction of a female's home range, a male cheetah may encounter many transient females. Males that join others have a much better chance of obtaining and maintaining exclusive access to these areas than do single males because they can oust residents and repel intruders of smaller coalition size (Caro & Collins 1986, 1987b).

In lions, females live in stable prides of two to eighteen related females with their dependent offspring (Schaller 1972; Bertram 1975). Male lions accrue reproductive benefits from living in groups because they have enhanced competitive ability to take over prides, they have extended pride tenure and they can occupy more than one pride; indeed per capita reproductive success is higher in larger

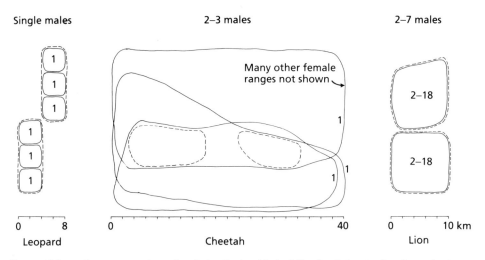

FIG. 1. Schematic representation of male territories (dashed lines) relative to female territories or home ranges (solid lines) in three species of felid. Numerals denote number of females that would typically occupy the ranges. Scale in kilometres shown below.

coalitions (Bygott, Bertram & Hanby 1979; Packer & Pusey 1982; Packer *et al.* 1988 (see Fig. 1). Thus, for different reasons, male cheetahs and male lions gain access to more females by living as coalitions because the costs of sharing matings are probably outweighed by the ability of coalitions to monopolize large numbers of females.

In the other Felidae where females' ranges overlap far less and females do not collect in small areas, or in species where female home ranges are exclusive, a pair of males would have to range over more than twice as many female ranges as would a single male in order to gain sufficient reproductive benefits to outweigh the costs of shared matings (Fig. 2). Assuming that, in these species, single males are currently occupying the maximum size of range they can defend successfully, it seems improbable that a pair of males could range over and defend effectively an area that was nearly twice this size. Only if they split up could they cover such an enlarged circuit, but in so doing, would lose their advantage in fights with single males. Hence, under current female distributions, the most successful reproductive option for males may be to remain single and attempt to limit range incursions by all other males.

In summary, males will form groups only if they can increase individual access to females. Larger coalitions of male lions can exclude smaller groups of males from prides of females, while coalitions of male cheetahs can monopolize areas where females collect. In other felid species, where females are more dispersed, a male's reproductive interests will be served best by excluding all other males from the greatest number of female ranges he can encompass.

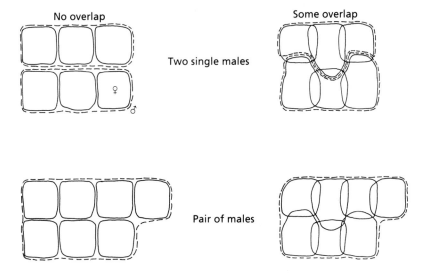

Fig. 2. Schematic representation of singleton male territories (dashed lines) relative to female territories or home ranges (solid lines) where no overlap occurs (left) or where some occurs (right). Diagrams below show the male territory size necessary to allow a hypothetical pair of males to gain reproductive advantages over a singleton.

FEMALES

Female life histories

Although adult female felids usually live apart and avoid contact with each other (for example African wildcats, *Felis lybica*: Smithers 1971), they spend a considerable proportion of their lives with dependent cubs. Table 1 shows that the percentage of time adult females have cubs in their charge is normally over 80% for the ten species for which data are available. Moreover, adult females in each species had weaned cubs accompanying them for an average of nearly 60% of their lives, assuming an unbroken reproductive career. In practice, loss of cubs at birth and post-natal mortality caused by factors such as disease (for example cheetahs: O'Brien *et al.* 1985) or infanticide (lions: Packer & Pusey 1983) would reduce the total amount of time that adult females had cubs accompanying them. However, an example from the first 4 years of my field study of cheetahs shows that mothers were usually accompanied by offspring. Of the 173, 298, 254 and 208 sightings of independent female cheetahs seen in each year, 54·3%, 65·4%, 64·2% and 58·7% of these sightings respectively were of females with cubs accompanying them.

During the period of post-weaning dependence, cubs consume large quantities of food. In cheetahs, cubs leave the den in which they have been hidden by their mother at about 6 weeks and accompany her on hunting expeditions, remaining with her for between 13 and 20 months (Frame 1980). Thus mothers have to share a proportion of their food with offspring for between 46 and 74 weeks, or 71–80% of the time between pregnancies (Frame 1984). By 8 months, cheetah cubs' jaw size has reached that of adult females' (personal observations) suggesting they consume as much food as their mothers per unit time. Thus mothers with only a single cub may have to share carcasses approximately equally for 31–52% of their reproductive careers. Given that two or three cubs (the most common

TABLE 1. Reproductive parameters of ten species of felid (calculated from Gittleman 1986)

	% of life female has cubs*	% of life female has attendant cubs[†]
Acinonyx jubatus	83·6	64·0
Panthera pardus	86·0	66·0
Panthera leo	91·1	78·4
Felis caracal	83·2	55·2
Panthera tigris	84·6	60·1
Lynx lynx	78·0	41·3
Lynx rufus	85·3	71·3
Puma concolor	82·4	—
Felis lybica	71·1	—
Felis sylvestris	67·6	27·1

* Age at independence/(length of gestation + age at independence).
[†] (Age at independence−age at weaning)/(length of gestation + age at independence).

litter size at this older age in the Serengeti) are feeding for an average of 73% of the time that all family members are feeding (Table 2), adult female cheetahs may relinquish two-thirds of the meat they catch to other individuals for half their lives.

Cubs are not only a drain on the food that mothers capture, they are of little help in acquiring food. The percentage of the family's successful hunts made by cheetah cubs alone was very low compared to those made by their mother (Fig. 3). Indeed, of the 178 successful hunts that fifty-four cheetah families were seen to make during 2773 hours of observation, only eleven were actually made by cubs alone, and these were usually of easy-to-catch hares.

Although cubs began to initiate hunts from 5½ months of age onwards, they would usually hang back after their mother started to hunt, and rarely contacted prey until their mother had captured it. Mothers' hunting success rates did not increase with cub age despite cubs' participation in hunts (Fig. 4), nor was there any indication that mothers attempted to capture large prey when they had older cubs (Table 3). These results indicate that in the one felid species where quantitative data are available on hunting behaviour of both offspring and mothers, cubs are essentially parasitic on adult females for food.

Time costs to sociality

If adult female felids did live together (were social) and bred seasonally, then on average, the minimum size of a social group would reach two adult females plus twice the mean litter size. Even if breeding lacked seasonality, mothers might nevertheless live in groups of this size during some periods of their lives. If two females lived together, their individual rates of food intake would decrease unless

TABLE 2. Total numbers of minutes mothers and individual cubs from different litters ate from carcasses during the period they were observed

Cub age in months	Number of days observed	Mother	Time spent eating by: First cub	Second cub	Third cub	Total % of time cubs ate
8	7	238·2	221·3	253·1		66·6
8	6	95·4	146·3	128·2		74·2
8	7	189·5	285·1	256·5		74·1
8	3	0	7·9	3·8		100·0
8	6	76·4	87·0	107·1		71·8
8	4	28·9	52·8	27·7	15·5	76·9
10	6	97·1	68·3	90·4		62·0
10	5	36·7	49·1	55·5	71·8	82·8
≥12	1	37·7	33·7	37·3		65·3
≥12	1	11·0	9·5	6·9		59·9
≥12	6	105·6	75·7	92·3		61·4
≥12	7	79·9	86·7	79·0		67·5
≥12	5	46·8	70·6	79·4	72·2	82·6

Felid asociality

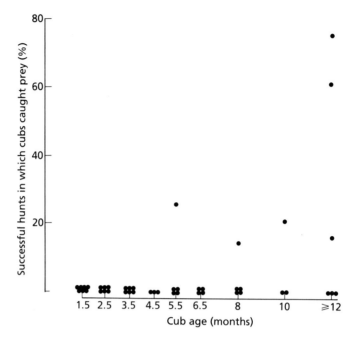

FIG. 3. Percentage of families' successful hunts that were made by dependent cubs without the aid of their mothers, separated by cub age ($n=45$, $r_s=0.440$, $P<0.002$).

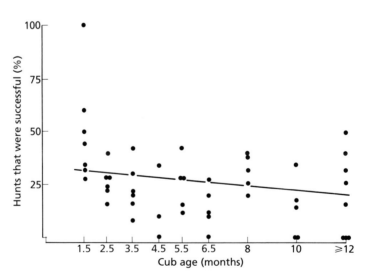

FIG. 4. Mothers' successful hunts expressed as a percentage of all mothers' hunts, separated by cub age ($n=51$ mothers, $r=-0.207$, N.S.; $y=0.333-0.011x$).

TABLE 3. Attempts on known prey larger than an adult male Thomson's gazelle by cheetah mothers with cubs of different age. Prey were subadult male, adult female and adult male Grant's gazelles; neonate, yearling and 2 year old wildebeests, an adult male topi, neonate zebra, a subadult and adult female reedbuck, and a hartebeest of unknown age or sex.

	Cub age in months								
	$1\frac{1}{2}$	$2\frac{1}{2}$	$3\frac{1}{2}$	$4\frac{1}{2}$	$5\frac{1}{2}$	$6\frac{1}{2}$	8	10	$\geqslant 12$
Grant's gazelle	3	1	2*	2	7	1	3	4	0
Wildebeest	3*	1	2	0	0	0	1	0	4
Other	0	1	0	0	1	0	2*	4	1
Total	6	3	4	2	8	1	6	8	5

* Denotes that one of the attempts was successful.

compensation could be made for loss of food. Additional food could be provided effectively in only two ways: either by hunting more often per group, or by capturing larger prey per hunt, or some combination of the two.

A third way of increasing food intake without increasing hunting rates or size of prey might be for the two mothers to take alternate turns in hunting. However, in this situation all other non-hunting group members (cubs) would have to double the amount of time and energy spent in accompanying the huntresses so as to be able to consume prey every time it was caught by each of the two mothers. It is argued below that the pooled dependent cubs would suffer prohibitive costs under this new regime.

A fourth possibility might be for mothers to bring food back to a common den where cubs spent a large proportion of their time, in order to free cubs of travel costs spent in accompanying them on hunts. This sometimes occurs in spotted hyaenas (Kruuk 1972). In this situation, however, food intake provided by a mother per unit time would be reduced unless mothers increased their hunting rates, because they would have difficulty in carrying all but the smallest prey back to the den, especially since felids have relatively small heads (Van Valkenburgh & Ruff 1987) lacking the musculature required to carry large pieces of meat. Moreover, cubs that spent a large proportion of their time in a den might have little opportunity to learn hunting skills before independence.

Only in one scenario would hunting habits not need to change, that is if females lived together but if they and their own cubs travelled, hunted and ate separately. Yet in these circumstances it is unlikely they could capitalize on the benefits of sociality derived from early detection of predators, cooperative defence of carcasses, or protection of cubs.

Could mothers then hunt twice as often as they do when they are asocial, in order to feed a social group as argued above? The mean number of minutes that cheetah mothers hunted per hour (defined as stalking, trotting towards, crouching at, rushing or chasing a group of prey animals; see Caro 1987a) decreased with age of their cubs from 2·47 min/h when cubs were young ($1\frac{1}{2}$–$3\frac{1}{2}$ months old) to

1·24 min/h when cubs were 8–18 months old ($F_{2,51}=2\cdot895$, $P=0\cdot064$; see Caro 1987b for methods). This measure is an underestimate of the time spent hunting, however, because many of the prey caught by mothers were neonate Thomson's gazelles, and hares *Lepus crayshawi* and *Lepus capensis* (mean percentage representation in prey killed by forty-three mothers: 30·4% S.D.=33·0; 14·9%, S.D.=25·2 respectively). These prey items were not caught using a concealed approach, but were located and briefly pursued only when they had been disturbed by mothers walking through vegetation in which they were hiding. Maternal effort to acquire prey is therefore better measured as the percentage of time mothers hunted, moved and sat up observing their surroundings (searched for prey) during the course of the day; the last behaviour was strongly associated with prey capture (Caro 1987b).

Figure 5 shows the percentage of time that mothers searched for prey when cubs were young (when they spent most time hunting), separated by litter size. Mothers spent an average of 41·9% of the time searching for prey when they had three young cubs. By fitting a regression to these data, one can estimate the percentage of time mothers would have to search for prey to support increasing numbers of cubs of this age, assuming search time increases linearly with litter size. With litters of six young cubs, which have been seen in the Serengeti (G. Frame, personal communication; personal observations), mothers would have to search for prey for 57·0% of daylight hours. For groups of seven or eight (two litters of the normal three or four young cubs, Schaller 1972), mothers would have to search for between 63·4 and 70·0% of the day.

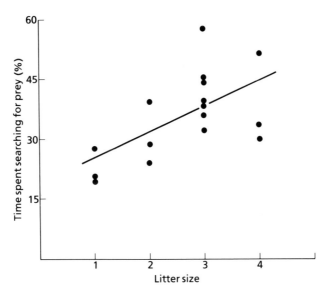

FIG. 5. Percentage of instantaneous scans taken every 15 minutes that mothers with young cubs (1½–3½ months old) searched for prey (hunted, moved or sat up observing), separated by litter size ($n=16$, $r=0\cdot612$, $P=0\cdot012$; $y=18\cdot583+6\cdot409x$).

Given that mothers nurse litters of one to four young cubs for an average of 11·5% ($n=19$, S.D.$=11·0$) of the day (18·9% in 1½ month old cubs) and that they spend an average of 5·8% ($n=14$, S.D.$=2·8$) of the day eating during this period, only 12·7% of the day would be available, on average, for other activities if mothers were with eight group members. Mothers only caught prey on an average of a quarter of all the hunts they embarked on; moreover, there was considerable variability in hunting success ($n=51$, $\overline{X}=27\%$, S.D.$=21\%$; see Fig. 4), some mothers having runs of up to ten hunts without success. Thus, having only approximately 7·5 minutes 'free' in each hour might be risky in terms of securing prey, energetic costs aside. If pooled cubs continuously accompanied two mothers that hunted alternately, as argued in the third scenario, they would spend 80–90% of the day involved in searching for prey (double that shown for three or four cubs in Fig. 5).

Prey size, ease of procurement and sociality

If female cheetahs were social they might be able to feed group members by regularly catching large prey. Serengeti cheetah mothers were less successful at capturing prey, however, as the carcass weight of the prey they attempted increased in size (Fig. 6a), and there was no indication that success rates levelled off as prey size increased. Also, the total time spent in hunting necessary to capture different sorts of prey items did not level off as prey size increased (Fig. 6b). In effect, cheetah mothers could not improve on their food intake per unit time by concentrating on large prey because these took too long to hunt successfully. Only by focusing on very small prey (hares) could they marginally increase food intake. In fact, hungry cheetahs tried to minimize the risk of starvation by walking long distances searching for hares and neonate Thomson's gazelles in vegetation, and not by hunting medium or large-sized prey (personal observations).

Moreover, the larger the prey item, the more likely it is to be an adult member of a species rather than a neonate or juvenile which is usually less vigilant and less competent at escaping predators. Indeed, mothers attempted to capture the youngest age classes of Thomson's gazelles disproportionately relative to their abundance (relative abundance of adult males was 27·9%, adult females 58·2%, subadult males and females 9·3%, and halfgrowns and fawns 4·7% ($n=$ 30642 counted, Borner *et al.* 1987), whereas these classes were attempted in respectively 33·2%, 20·4%, 6·8% and 39·6% of the 265 hunts where the age sex class of this quarry was known ($\chi^2=754·07$, df$=3$, $P<0·001$).

It is also possible to estimate the amount of prey that a female cheetah living socially might need to capture. Figure 7 shows the number of kilograms of edible flesh that mothers made available to cubs per hour separated according to litter size. Mothers made an average of 0·14 kg of flesh available to single cub litters per hour, 0·22 kg/h to two cubs, 0·29 kg/h to three cubs, and 0·24 kg/h to four cubs ($F_{3,50}=1·132$, N.S.). Being conservative, extrapolation suggests mothers with seven or eight cubs would have to procure approximately 0·30 kg of flesh/h, but assuming

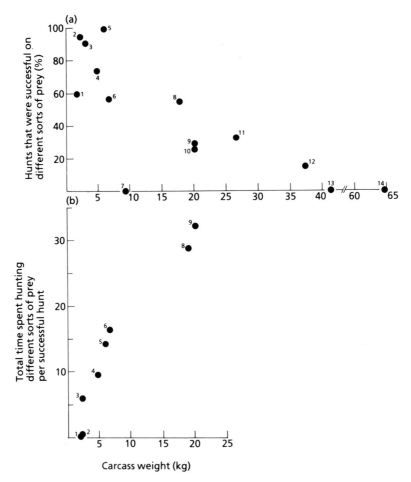

FIG. 6. (a) Hunting success rates of cheetah mothers on prey of different body weights for prey that were hunted more than once, calculated by dividing the total number of successful hunts made by mothers of fifty-five litters by the number of hunts attempted on each sort of prey item. $n=14$, $r_s=-0.822$, $P<0.002$. (b) Time expended by cheetah mothers in capturing prey of different body size that were hunted more than once, calculated by dividing the total time spent in successful and unsuccessful hunts by the number of successful hunts made by mothers of fifty-five litters on each sort of prey item. $n=8$, $r_s=1.000$, $P<0.002$. 1. Subadult hare; 2 Adult hare; 3. Neonate (N) Thomson's gazelle; 4. Halfgrown Thomson's gazelle; 5. N Grant's gazelle; 6. Subadult female (SF) Thomson's gazelle; 7. Subadult male (SM) Thomson's gazelle; 8. Adult female (AF) Thomson's gazelle; 9. Adult male (AM) Thomson's gazelle; 10. N wildebeest; 11. SM Grant's gazelle; 12. AF reedbuck; 13. AF Grant's gazelle; 14. AM Grant's gazelle.

a linear increase, they might have to capture 0.44−0.49 kg/h for groups of this size.

If, for the sake of argument, mothers exclusively fed on neonate Thomson's gazelles (edible flesh weight 1·7 kg: Blumenschine & Caro 1986), those with one cub would need to kill once every 10 daylight hours, but make one hunting

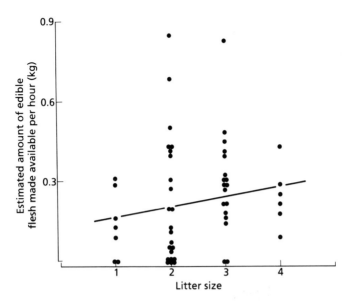

FIG. 7. Estimated number of kilograms of edible flesh that mothers of fifty-four litters made available to their cubs per hour, separated by litter size ($r=0.189$, N.S.; $y=0.129+0.044x$).

attempt every 9 hours when failed hunts are taken into account (see Fig. 6a). Mothers with eight group members would need to attempt this prey every $3.2-5.1$ hours, depending on whether they needed to capture 0.30 or 0.49 kg/h. If however, mothers sought larger prey such as 20 kg adult male Thomson's gazelles then mothers with eight mouths to feed would have to procure the 10 kg of edible flesh once every $20.4-33.3$ daylight hours. But to realize this intake, they would need to attempt to capture adult males every $5.8-9.5$ hours, nearly twice a day, using hunting success rates from Fig. 6a. If mothers switched to larger prey still, they would need to hunt almost continuously to have a chance of securing prey because success rate is extremely low on prey of large body size (Fig. 6a). These arguments hold whether the burden of hunting fell on one, or on two mothers, because the pooled cubs would still suffer the same failure rates and have to hunt for these lengths of time in order to be present when each kill was made.

Two caveats

Although these arguments suggest that social cheetah mothers would face considerable difficulties in finding either enough time to catch prey, or in being able to catch sufficient large prey items necessary to feed dependent cubs, they make two assumptions: that the amount of food cubs consume does not decline appreciably with litter size, and that hunting efficiency is not appreciably greater with increasing numbers of adult females.

Amount of food each cub ate per unit time was estimated by dividing the proportion of time a cub fed from each carcass compared to other family members and multiplying by the estimated weight of edible flesh consumed on that carcass; these amounts were then summed across carcasses and divided by the total time a family was observed. When cubs were young ($1\frac{1}{2}-3\frac{1}{2}$ months) and had small mouths, I estimated mothers ate three times as much meat per unit time as each cub; when $4\frac{1}{2}-6\frac{1}{2}$ months old, twice as much; but when 8 months or older, equal amounts per unit time as each cub. Of those families that were seen eating, the estimated average amount of food cubs actually ate per hour, rather than that made available to them, was unaffected by litter size ($n=44$, $\overline{X}=0\cdot05$ kg/h, S.D.$=0\cdot05$, $F_{3,39}=0\cdot542$, N.S.). Furthermore, when litters were separated by age group, food ingested per cub was not significantly affected by litter size (one and two vs. three and four cubs respectively, $1\frac{1}{2}-3\frac{1}{2}$ months, $\overline{X}$s$=0\cdot05$, $0\cdot04$ kg/h, $t_{6,13}=1\cdot12$, N.S.; $4\frac{1}{2}-6\frac{1}{2}$ months, $\overline{X}$s$=0\cdot04$, $0\cdot05$, $t_{6,6}=-0\cdot59$, N.S.; $8-18$ months, $\overline{X}$s$=0\cdot09$, $0\cdot06$, $t_{10,3}=0\cdot68$, N.S.). Thus, cubs in my sample of large litters did not consume markedly less flesh than those in small litters because mothers made somewhat more kilograms available to them per unit time (see Fig. 7).

The second assumption, that groups of adult females would not hunt more effectively than singletons, cannot be tested directly in cheetahs because females are solitary. However, some male cheetahs do live in groups, providing a comparison for examining the effects of group size on hunting success. Differences in food intake in different sized male groups could not be used as an analogy for females because males are up to 25% heavier (Smithers 1983) and, in the Serengeti, attempt to capture different sorts of prey. Whereas only forty-three out of 641 hunts by mothers were attempts to capture prey larger than an adult male Thomson's gazelle (20 kg), single males attempted such prey items on a total of thirty-two out of 163 occasions ($X^2=24\cdot16$, df$=1$, $P<0\cdot001$). These differences notwithstanding, Table 4 shows that there were no significant differences in the number of hunts made per hour, hunting success per group, or number of successful hunts per hour in male groups of differing size.

In sum, cheetah cubs did not get significantly less food in large litters, and hunting success of male cheetahs did not change appreciably with increasing number of males (see also Eaton 1974; Packer & Ruttan, in press), suggesting female hunting success would not increase appreciably if females lived in groups. Indeed, it is difficult to imagine how felid hunting methods, a concealed approach followed by a short chase, could benefit substantially by the participation of several individuals.

Conclusions

Female felids do not live alone because mothers live with cubs of varying sizes for much of their lives. Cubs in all felid species appear to require extensive practice to hunt efficiently (see also Caro 1980) and they help their mothers little in catching food, but mothers have to share the food they capture with their cubs.

TABLE 4. Effects of male group size on measures of hunting in male cheetahs. Data refer to group hunts because coalition members nearly always hunted together; analysis is restricted to groups observed for 12 or more hours. Mean values, and standard deviations in brackets

Male group size

Measure	1	2	3	F-test	P value
Hours watched	491·1	464·4	305·5		
Number of hunts per hour*	0·15 (0·11)	0·22 (0·14)	0·15 (0·05)	$F_{2,24}=1·039$	N.S.
% hunting success[†]	22·6 (26·6)	24·8 (20·9)	47·1 (33·0)	$F_{2,22}=1·659$	N.S.
Number of successful hunts per hour*	0·04 (0·04)	0·06 (0·08)	0·07 (0·05)	$F_{2,24}=0·500$	N.S.

* $n=11$, 11 and 5 groups respectively for male group sizes of 1, 2 and 3.
[†] $n=10$, 10 and 5 groups respectively.

Thus if females were to live with other females and benefit from some of the advantages that sociality is known to bring in other carnivores, mothers would either have to hunt more often or catch larger prey to feed an increased number of group members.

Cheetah mothers spent a large proportion of each day searching and hunting for food and appeared to be near the upper limit of the time that they were able to spend in this activity. By extrapolation, it would be difficult for mothers to feed additional group members by hunting more often. If cheetah mothers were to live in groups, mothers would also find it difficult to feed group members by capturing larger prey because hunting success decreased and time spent hunting increased with prey size attempted. To realize an increased food intake through taking large prey, the family would again be forced to be almost continuously involved in hunting. Female cheetahs in groups would be unlikely to enjoy enhanced hunting success; hunting success in males did not increase significantly with group size.

In short, if female felids were to live socially, they would need to catch large prey that were easy to capture because most felids are likely to be at the upper limit of the time they can spend hunting with normal litter sizes. However, more information is needed on the proportion of time that mothers of different felid species allocate to hunting and food acquisition.

COMPARATIVE EVIDENCE

These data have shown that the only real means by which mothers could support a group of cheetah cubs would be to concentrate on large prey that were easy to catch. The working hypothesis is that this would also hold for other species of felid because it is assumed that each species gives birth to the maximum number of cubs it can usually raise successfully. Some prey species are more vulnerable than others because they are less vigilant, live in smaller groups, or run slowly, but unfortunately very little is known about species' differences in anti-predator behaviour (but see Johnsingh 1983) and the problems predators must overcome in capturing different sorts of prey (Hornocker 1970). At present, a first broad approximation of how easily prey can be caught is the relative frequency with which they might be encountered (but see Bertram 1973). If prey species relatively large to a female's body weight were numerous in the habitat then, other things being equal, the opportunity to increase food intake necessary to sustain sociality would be open. Here I examine different-sized prey available to free-living felids in the habitats where these data have been collected.

Cheetahs

For cheetahs, modal prey size across different studies is approximately 30 kg or 80% of an adult female's body weight (38 kg, Caro *et al.* 1987) (Table 5). In order to take large prey that were easy to catch, as measured by abundance (necessary to sustain a group of animals as argued above), female cheetahs in the Serengeti would have to feed regularly on subadult and adult wildebeests, *Connochaetes taurinus*, (32, 39, 40 in Fig. 8), but these animals are three to seven times the size of prey cheetahs normally take, or 200–400% of their own body weight. The most numerous prey closest to their own body weight are small Thomson's gazelles (11, 12) and neonate wildebeests (13) which may effectively preclude living in permanent groups in this habitat. Similarly, Fig. 9 shows that in Kruger National Park, cheetahs would have to switch from impalas, *Aepyceros melampus* (12, 20, 24), about their own body weight, to female greater kudus, *Tragelaphus strepsiceros* (36), zebras, *Equus burchelli* (35, 47, 48) and wildebeests (37, 39), up to ten times their modal prey size, to take advantage of large prey that were abundant and so be able to feed as a group; there are relatively few prey in the 60–120 kg range.

Presence of abundant domestic livestock, usually poor in detecting predators, means that felids can have an increased opportunity to catch large prey easily, if livestock are not well protected by people. In south-west Africa, where farms carry high concentrations of domestic sheep and goats, McVittie (1979) reported that twenty-three out of 123 sightings of adult female cheetahs consisted of two females associating together, with or without cubs. In Serengeti, only one out of 390 female sightings was of this sort (Frame & Frame 1976) $\chi^2=67\cdot25$, df=1, $P<0\cdot001$). McVittie's data came from information provided by farmers, some of whom caught cheetahs in box traps, so it is possible that different individuals that had never met before were caught together over a period of time using this method (D. Morsbach, personal communication). Nevertheless, the data do suggest that under certain conditions where prey are easy to catch, cheetahs can be social. Further information on cheetahs' prey capture and feeding rates is clearly needed from this region.

TABLE 5. Identity of the most frequently eaten prey species by some of the felids described in the text (some studies reported two species); data could not be separated by sex of felid. Prey weights (kg) are given as 75% female body weight because the proportion of adults of both sexes and juveniles eaten was unknown. K=kills, S=scats, St=stomach contents

Prey species	Weight	K	S	St	Location	Source
Cheetah						
Impala	31·6	X			Nairobi N.P.	1
Puku	45·9	X			Kafue N.P.	2
Thomson's gazelle	13·3	X			Serengeti N.P.	3, 4
Impala	31·6	X			Kruger N.P.	5
Leopard						
Reedbuck	28·1	X			Kafue N.P.	2
Thomson's gazelle	13·3	X			Serengeti N.P.	3
Impala	31·6	X			Kruger N.P.	5
Impala	31·6	X			Serengeti N.P.	6
Thomson's gazelle	13·3	X			Serengeti N.P.	6
Impala	31·6		X		Tsavo N.P.	7
Procavidae	2·3		X		Matopos N.P.	8, 9
Lion						
Buffalo	562·5	X			Kafue N.P.	2
Wildebeest	122·3	X			Serengeti N.P.	3, 4
Blue wildebeest	135·0	X			Kruger N.P.	5
Zebra	226·7	X			Serengeti N.P.	6
Hartebeest	94·5	X			Nairobi N.P.	10
Zebra	226·7	X			Nairobi N.P.	11
Wildebeest	122·3	X			Nairobi N.P.	11
Tiger						
Chital	42·6	X	X		Kahna N.P.	12
Chital	42·6		X		Bandipur	13
Sambar	122·7	X			Chitawan N.P.	14
Chital	42·6		X		Chitawan N.P.	14
Wapiti	152·3		X		Primorje	15
Wapiti	152·3		X		Lazorski reserve	16
Wild pig	42·0		X		Sikhote Alin reserve	17
Wapiti	152·3		X		Sikhote Alin reserve	18
Cougar						
Elk calves	109·3	X			Idaho Primitive area	19
Mule deer	48·8			X	Western USA	20
Varying hare	1·1		X		Washington State	20
Mule deer	48·8			X	Utah & Nevada	21
Mule deer	48·8			X	British Columbia	22
European hare	2·8		X		Torres del Paine N.P.	23
Agouti	4·0		X		Cocha Cashu N.P.	24

1. Eaton 1974; 2. Mitchell, Shenton & Uys 1965; 3. Kruuk & Turner 1967; 4. Schaller 1972; 5. Pienaar 1969; 6. Bertram 1982; 7. Hamilton 1976; 8. Grobler & Wilson 1972; 9. Smith 1978; 10. Rudnai 1973; 11. Foster & Kearney 1967; 12. Schaller 1967; 13. Jonhsingh 1983; 14. Sunquist 1981; 15. Abramov 1962; 16. Matjushkin, Zhivotchenko & Smirnov 1980; 17. Yudakov 1973; 18. Gromov & Matjushkin 1974; 19. Hornocker 1970; 20. Young & Goldman 1946; 21. Robinette, Gashwiler & Morris 1959; 22. Spalding & Lesowski 1971; 23. Yanez et al. 1986; 24. Emmons 1987a.

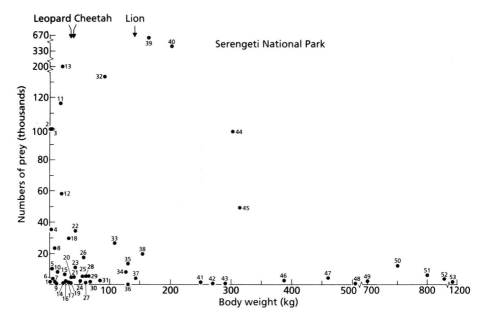

FIG. 8. Estimated abundance (in 1000s) and body weights (kg) of prey in the Serengeti National Park, Tanzania (after several sources). Also shown are weights of an adult female leopard, female cheetah and lioness. 1. Neonate (N) warthog; 2. Subadult (S) hare; 3. Adult hare; 4. N Thomson's gazelle; 5. N impala; 6. N Grant's gazelle; 7. N 'other' (average of roan, oryx, reedbuck, bushbuck, etc., from Schaller 1972); 8. S Thomson's gazelle; 9. N waterbuck; 10. N topi; 11. Adult female (AF) Thomson's gazelle; 12. Adult male (AM) Thomson's gazelle; 13. N wildebeest; 14. S 'other'; 15. S impala; 16. S Grant's gazelle; 17. N eland; 18. N zebra; 19. S warthog; 20. N buffalo; 21. AF 'other'; 22. AF impala; 23. AF Grant's gazelle; 24. AM 'other'; 25. AF warthog; 26. AM impala; 27. N giraffe; 28. S topi; 29. AM Grant's gazelle; 30. S hartebeest; 31. AM warthog; 32. S wildebeest; 33. AF topi; 34. AF hartebeest; 35. AM topi; 36. S waterbuck; 37. AM hartebeest; 38. S zebra; 39. AF wildebeest; 40. AM wildebeest; 41. AF waterbuck; 42. AM waterbuck; 43. S eland; 44. AF zebra; 45. AM zebra; 46. S buffalo; 47. AF eland; 48. S giraffe; 49. AM eland; 50. AF buffalo; 51. AM buffalo; 52. AF giraffe; 53. AM giraffe. Weights from Schaller (1972), Smithers (1983), Georgiadis (1985), Blumenschine & Caro (1986), while subadult weights were estimated as half mean adult male and female weights. A few age classes of uncommon species were omitted from the figure where weights were unobtainable. Abundance for each species calculated from Schaller (1972), Frame & Wagner (1981), Borner *et al.* (1987), K. Campbell, personal communication; age classes were calculated as AM 25%, AF 50%, S 10% and N 15% of each population.

Leopards

Leopards (female body weight, 34 kg: Wilson 1968) usually take prey of similar size to cheetahs (about 30 kg: Table 5), although a leopard is capable of taking prey several times its own body weight (e.g. sable antelope, *Hippotragus niger*: Pienaar 1969). In Serengeti and Kruger National Parks (Figs 8 and 9), abundant prey five to nine times an adult female's body weight are present, but these may be too large to catch on a regular basis. Prey two to four times female body weight are not particularly common in either of these habitats suggesting regular

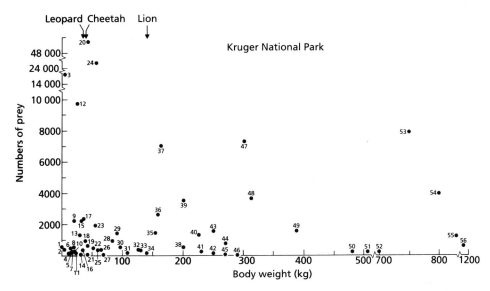

FIG. 9. Estimated abundance and body weights (kg) of prey in the Kruger National Park, South Africa (after Pienaar 1969). Also shown are weights of an adult female leopard, female cheetah and lioness. 1. Neonate (N) warthog; 2. N bushbuck; 3. N impala; 4. Subadult (S) 'other' (average of baboon, bushpig, klipspringer, steenbok, grysbok, suni, red duiker, grey duiker); 5. N sable antelope; 6. N waterbuck; 7. S bushbuck; 8. Adult female (AF) 'other'; 9. N wildebeest; 10. AM 'other'; 11. S reedbuck; 12. S impala; 13. AF bushbuck; 14. N eland; 15. N zebra; 16. S warthog; 17. N buffalo; 18. AF reedbuck; 19. AM bushbuck; 20. AF impala; 21. S nyala; 22. AM reedbuck; 23. AF warthog; 24. AM impala; 25. N giraffe; 26. AF nyala; 27. S tsessebe; 28. AM warthog; 29. S wildebeest; 30. S kudu; 31. AM nyala; 32. AF tsessebe; 33. S waterbuck; 34. AM tsessebe; 35. S zebra; 36. AF kudu; 37. AF wildebeest; 38. AF sable antelope; 39. AM wildebeest; 40. AM kudu; 41. AM sable antelope; 42. AF roan antelope; 43. AF waterbuck; 44. AM waterbuck; 45. AM roan antelope; 46. S eland; 47. AF zebra; 48. AM zebra; 49. N buffalo; 50. AF eland; 51. S giraffe; 52. AM eland; 53. AF buffalo; 54. AM buffalo; 55. AF giraffe; 56. AM giraffe. Weights from Sachs (1967), Eisenberg (1981), Smithers (1983), Georgiadis (1985), while subadult weights were estimated as described in legend, Fig. 8. A few uncommon age classes were omitted from the figure where weights were unobtainable; see Fig. 8 for calculation of age−class abundance.

acquisition of large prey items in these areas might also be difficult for this species as it is for cheetahs.

Lions

Currently there are no quantitative data specifically on the hunting of single lionesses (weight 141 kg: Wilson 1968) although single females are known to have killed an adult zebra, an adult female topi, *Damaliscus korrigum*, an adult warthog, *Phacochoerus aethiopicus* and a warthog piglet on separate occasions (C. Packer & D. Scheel, personal communication); the data in Table 5 are mostly derived from kills made by groups of lions. Although cases of solitary lions attacking prey the size of an adult buffalo, *Syncerus caffer* do exist (Schaller 1972;

C. Packer, personal communication) usually several lions do so simultaneously (Schaller 1972); thus modal prey size for single lions, at least, is probably 155 kg, or just above a female's body weight. In both Serengeti and Kruger National Parks there is abundant prey of this body size and upwards to 300 kg (Figs 8 and 9). Thus the opportunities for lionesses and their cubs to live together in groups are far greater than they are for cheetahs and leopards because relatively large prey, greater and up to double the weight of a lioness, are numerous, will be encountered often and hence will be reasonably easy to capture, other things being equal. In both these ecosystems therefore, lionesses are the only large female felid that could afford to capitalize on any benefits that might result from group living. Similar arguments can be applied to the numerous buffalo in Lake Manyara National Park which constitute 62% of lions' diet there (Schaller 1972).

Servals

There are few quantitative data on prey availability for other species of African felid. Servals (females 9·7 kg: Smithers 1983) can take prey as large as impala fawns (*c.* 5 kg: Pienaar 1969) but these are likely to be far less numerous than the rodents on which servals specialize. Geertsema (1985) argues that servals' hunting technique makes it difficult for them either to chase hares or catch bounding springhares, *Pedetes capensis*, which are 10–15 times the body weight of *Otomys*, *Mus* and *Arvicanthis*, the most common items in the diet. Thus servals may find it difficult to catch larger prey not only because they are less numerous than rodents (see Senzota 1978; Frame & Wagner 1981) but because the medium-sized species are difficult to capture.

Caracals

Caracals (females 11·5 kg: Smithers 1983) take a wide range of prey from small birds to juvenile impalas but Pienaar (1969) and Pringle & Pringle (1979) report hyraxes, *Procavia capensis*, as the most frequent prey item. Certain individuals, however, specialize on sheep and goats in some areas (Pringle & Pringle 1979) and in the Mountain Zebra National Park, South Africa, fifteen out of twenty-one caracal kills were of mountain reedbuck, *Redunca fulvorufula*, representing 19·6% of items found in scats (Grobler 1981). Given that a tame subadult female caracal ate approximately 1 kg of meat per day, a pair of females each with two cubs (see Smithers 1971 for litter sizes) would require only 6 kg/day. The estimated 26.1 kg available on a mountain reedbuck carcass (Grobler 1981) would theoretically support a caracal group for 3 days providing the meat could be defended (Stoddart 1979) and would not putrify.

   Although probable encounter rates of caracal with mountain reedbucks are unknown, it seems possible that female caracals could live socially in this area. In support of this, Grobler reported that over a 2-year period eleven out of fifty-seven sightings were of pairs of adults (which he was careful to separate from five sightings of females with kittens). The possibility that these were pairs of females and not consorting male–female pairs clearly requires further investigation.

Tigers

Tigers (females 147 kg, Sunquist 1981) eat comparatively small prey for their body weight in lower latitudes but larger prey in the USSR (see Table 5). In Kahna National Park, they feed primarily on chital *Axis axis*, the most abundant wild ungulate (Schaller 1967) (Fig. 10). Sambar, barasingha, *Cervus duvauceli* and gaur, *Bos gaurus*, are also eaten to a lesser degree; analysis of kills rather than scats suggests gaurs are eaten infrequently. Large prey are not particularly abundant: there are only one-third as many female sambar (22, Fig. 10) as female chital (13), and less than a fifth as many gaur (27, 28), which probably makes it difficult for tigresses to locate large prey sufficiently often to support a group. Tigers living outside Schaller's central study area would have encountered large herds of domestic cattle and buffalo (>200 kg, not shown in Fig. 10) which were probably easy to catch and might have allowed peripheral tigresses to live together with attendant cubs. Schaller observed tigresses occasionally feeding together at the same kills.

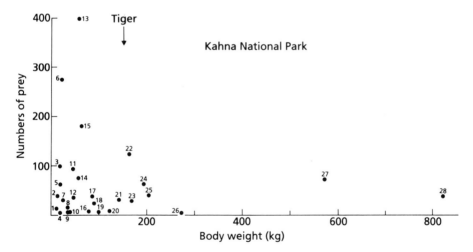

FIG. 10. Estimated abundance and body weights (kg) of prey in the Kahna National Park, India (after Schaller 1967). Also shown is weight of an adult female tiger. 1. Subadult (S) barking deer; 2. Neonate (N) sambar; 3. 4-horned antelope (average of all age classes); 4. S blackbuck; 5. Adult female (AF) barking deer; 6. N chital; 7. Adult male (AM) barking deer; 8. S wild pig; 9. AF blackbuck; 10. AM blackbuck; 11. S chital; 12. N gaur; 13. AF chital; 14. AF wild pig; 15. AM chital; 16. S barasingha; 17. AM wild pig; 18. S sambar; 19. S nilgai; 20. AF nilgai; 21. AF barasinga; 22. AF sambar; 23. AM barasinga; 24. AM sambar; 25. S gaur; 26. AM nilgai; 27. AF gaur; 28. AM gaur. Weights from Schaller (1967), Eisenberg (1981), Sunquist (1981), while subadult classes were estimated (see legend, Fig. 8). A few uncommon age classes were omitted from the figure where weights were unobtainable. Abundance calculated as in Fig. 8.

In the Royal Chitawan National Park, absolute prey numbers are greater than in Kahna (Seidensticker 1976; Tamang 1979) (Fig. 11). Here chital are taken most frequently as measured by scat analysis but sambar are as measured from kills (Sunquist 1981) (Table 5). Sunquist suggests sambar (13, 14, 15, Fig. 11) are the preferred prey or are most vulnerable, although hogdeer, *Axis porcinus* (4, 7, 8) and chital (5, 9, 11) are more numerous (Sunquist & Sunquist, in press). There seems little reason why tigresses could not be social in this habitat given that adult sambar body weights are greater than that of tigresses, and that they occur at a moderately high density (2·3−2·6/km^2), although they are not as common as other species. Larger prey still, adult and juvenile rhinoceroses, *Rhinoceros unicornis* (not shown in Fig. 11) are probably too dangerous to capture on a regular basis (Sunquist 1981).

Snow leopards

The scant data on snow leopards, *Panther uncia* (females 31·7 kg, Schaller 1977) suggests they feed primarily on bharals, *Pseudois nayaur* but also on much smaller marmots, *Marmota caudata* as determined by scat analysis (Schaller 1977; Schaller et al. 1987). Although bharals are large (males 60 kg, females 39 kg), they are probably sufficiently scarce to preclude regular capture required of group living: Schaller suggests snow leopards must move large distances to locate and catch wild ungulates. Thus, feeding almost exclusively on livestock without harassment probably represents the only possible way to sociality in this species.

European lynxes

European lynxes, *Lynx lynx* (females 16·8 kg, Haglund 1966) normally feed on hares, *Lepus timidus* in Sweden, but they take an increasing proportion of roe deer, *Capreolus capreolus* and reindeer, *Rangifer tarandus* as winter progresses (Haglund 1966). In some areas 50% of their kills consisted of these species as measured from stomach contents and these prey appear easy to capture: forty-five out of sixty-six hunts on reindeer and twenty-three out of thirty-five hunts on roe deer were successful. Similarly, lynxes reintroduced into the Swiss alps took large prey: of eighty-eight prey items found, forty-eight were of roe deer and thirty were of chamois, *Rupicapra rupicapra* (Breitenmoser & Haller 1987). Given that lynxes eat approximately 2 kg per meal and that Haglund saw three lynxes feeding on a medium-sized roe deer for 2 days, these large ungulates could have provided sufficient food for lynxes to live socially at least in the Renomraden and Radjuvsonraden districts of Sweden in the early 1960s. One situation in which two adults killed and fed together (on a hare) is described but, in general, female lynxes living in northern latitudes are thought to be solitary.

Spanish lynxes, *Lynx pardina* in the Coto Donana take only a small proportion of deer in their diet, and the annual availability of fallow deer fawns *Dama dama* there is probably too low to make it worthwhile switching from rabbits, *Oryctolagus cuniculus* to larger prey (Beltran et al. 1985).

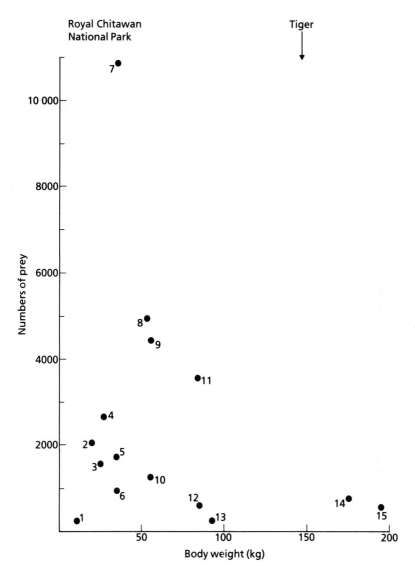

Fig. 11. Estimated abundance and body weights (kg) of prey in the Royal Chitawan National Park, Nepal (after Sunquist 1981). Also shown is weight of an adult female tiger. 1. Young (Y) barking deer; 2. Adult female (AF) barking deer; 3. Adult male (AM) barking deer; 4. Y hogdeer; 5 Y chital; 6. Y wild pig; 7. AF hogdeer; 8. AM hogdeer; 9. AF chital; 10. AF wild pig; 11. AM chital; 12. AM wild pig; 13. Y sambar; 14. AF sambar; 15. AM sambar. Weights from Schaller (1967), Eisenberg (1981), Sunquist (1981), abundance calculated as a mean of data from Seidensticker (1976) and Tamang (1979).

North American lynxes

North American lynxes, *Lynx canadensis* (females, 18·9 kg: Saunders 1964) consume snowshoe hares, *Lepus americanus* more than any other prey item in

each of the habitats they have been studied whether their populations are increasing
or in decline (see, for example, Saunders 1963; van Zyll de Jong 1966; Nellis &
Keith 1968; Brand & Keith 1979). In one study of lynxes on Cape Breton Island,
Newfoundland, 160 hares and 155 red-backed voles, *Clethrionomys gapperi*, the
most common small mammal, were caught on 1000 trapnights over the whole
study area; grouse numbers were not estimated but were reported as common
(Parker *et al.* 1983). Numbers of larger animals, white-tailed deer, *Odocoileus
virginianus*, were said to be common when the ground was free of snow, but only
ten to twelve moose, *Alces alces* were present. Lynxes can kill large ungulates,
caribou, *Rangifer caribou* (*c.* 100 kg) (Saunders 1963) or mule deer, *Odocoileus
hemionus* (57 kg) (Sheppard 1960), and also scavenge them (moose: Saunders
1963). However, snowshoe hares are likely to be far more numerous than these
species in nearly every habitat, although quantitative data are difficult to find. As
a lynx requires 0·6 kg of meat per day (Nellis *et al.* 1972) and hares weigh 1·4 kg,
mothers with three or four kits would already have to kill hares more than once a
day, and would thus be obliged to switch to larger but much less numerous
ungulate prey to sustain group living.

Bobcats

Bobcat diet is extremely variable as determined by the numerous analyses of
bobcat stomach contents (see, for example, Blum & Escherich 1979). In general,
representation of large prey, white-tailed deer, in the diet increases in the north-
east of the USA accounting for 16–35% occurrence in this region (summarized in
Maehr & Brady 1986); unfortunately data on comparable densities of prey species
available to bobcats are difficult to locate (Benson & Moore 1977). Although
strong representation of deer in bobcat diet in high latitudes suggests females
could live in groups, surveys that separate diet by sex show that males (body
weight 12·3 kg) eat considerably greater proportions of deer than do females
(body weight 7·2 kg: Litvaitis, Stevens & Mautz 1984): occurrence of deer in
Arkansas bobcat stomachs was 10% in males, but 3% in females (Fritz & Selander
1978), and in Maine was 19% ($n=72$) and 7% ($n=95$) respectively ($\chi^2=5·26$,
$P<0·05$: Litvaitis, Clark & Hunt 1986). Also, larger bobcats (>12 kg) eat more
deer than smaller individuals (Litvaitis, Clark & Hunt 1986). It therefore seems
probable that the majority of bobcat deer kills (up to 55 kg: Marston 1942; see
also Pollack 1951; Petraborg & Gunvalson 1962) are made by males, although
females might scavenge them. If the smaller females can only take deer fawns or
juveniles, and those in low numbers, then the opportunities for sociality in this
species appear limited.

Cougars

Cougars (females 42 kg: Young & Goldman 1946) have been studied most
extensively in the northern part of their range where they capture large prey such

as elk *Cervus canadensis* and mule deer (Table 5). Cougars concentrated on these species because they were forced into terrain offering ideal hunting conditions during winter (Hornocker 1970). Abundance of smaller prey such as snowshoe hares, or even beavers or porcupines, known to be taken by cougars (Robinette, Gashwiler & Morris 1959) are not available for comparison with ungulate densities in this area (Fig. 12). Superficially, it appears that exploiting prey this size (even

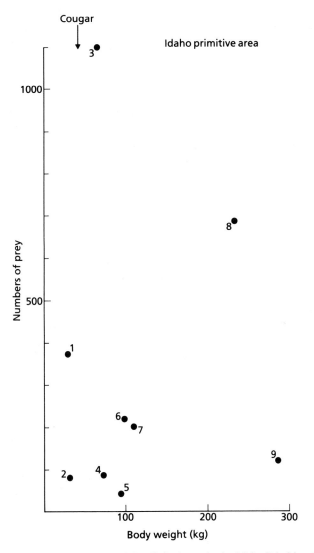

Fig. 12. Estimated abundance and body weights (kg) of prey in the Idaho Primitive Area, USA (after Hornocker 1970). Also shown is weight of a North American adult female cougar. 1. Mule deer calf; 2. Big horn sheep lamb; 3. Adult female (AF) mule deer; 4. AF big horn sheep; 5. Adult male (AM) big horn sheep; 6. AM mule deer; 7. Elk calf; 8. AF elk; 9. AM elk. Weights from Hornocker (1970), Schaller (1977).

though many mule deer were fawns) is sufficient to sustain a pair of female cougars with a total of four or five pooled cubs (see Robinette, Gashwiler & Morris 1961 for litter sizes), and temporary associations of cougars at large kills were seen (Seidensticker *et al.* 1973). Hornocker calculated that adults require 1·8−2·7 kg of meat per day, thus a group would need 10·8−18·9 kg. He also estimated that elk calves (7, Fig. 12), and adult male mule deer (6), 28% and 16% respectively of ungulate prey taken by Idaho cougars, would have 77 kg and 69 kg of edible flesh on them. This would mean that social mothers would have to kill large prey about once every 5 days. A mother with three 32 kg cubs did kill four deer in 18 days, so this estimated killing rate is within the range found in this habitat during this season. The key variable is whether these size categories of ungulates are encountered sufficiently often to sustain such a killing rate throughout the year. Densities of these species (elk: 0·2/km^2, mule deer: 0·3/km^2) suggest it might be difficult. Moreover, large prey disperse during summer resulting in much larger female cougar ranges (Seidensticker *et al.* 1973) which implies that catching such prey becomes problematic during this period.

In tropical regions, cougars feed on much smaller prey (Table 5). Although cougars are smaller in these regions (for example, females 23·6 kg: Sanborn 1954; 29 kg: Emmons 1987a) and might thus prefer smaller prey species, there are few large species on which to prey. In Cocha Cashu National Park, Peru, collared peccaries, *Tayassu tajacu* (7, Fig. 13), red brocket deer, *Mazama americana* (8) and capybara, *Hydrochaeris hydrochaeris* (9) were present in low numbers and were not found in the small number of cougar scats sampled by Emmons (1987a). Predation by jaguars on these species may have reduced the rate at which cougars could encounter these larger prey.

Finally, cougars living in Torres del Paine National Park, Chile, ate introduced hares, *Lepus capensis* most frequently but also caught guanacos, *Lama guanicoe* (males up to 130 kg: Yanez *et al.* 1986). Although 540 guanacos are estimated to live in the park, Wilson (1984) considers adult and subadult males, 39% of the population, to be relatively invulnerable to attack because they live in large groups. Nevertheless, adult female and young guanacos might be sufficiently numerous to enable cougars to live in groups.

Ocelots

At Cocha Cashu, 31% of ocelot scats contained *Proechimys* sp. (1, Fig. 13), and 21% *Oryzomys* sp. (2) while a smaller number contained oppossum and bird remains (Emmons 1987a). Ocelots (9·3 kg) fulfilled their food requirements of 0·6−0·8 kg/day by capturing prey usually weighing less than 5% of their body weight, which agrees with studies in Belize (Konecny, in press) and Venezuela (Ludlow & Sunquist 1982). Although ocelots did take prey between 1 kg and 8 kg (4·5% representation in scats), these prey were far less numerous than small rodents (Fig. 13) and may have been difficult to find sufficiently often. Feeding on rodents 'required spending many hours foraging to catch several prey each day'

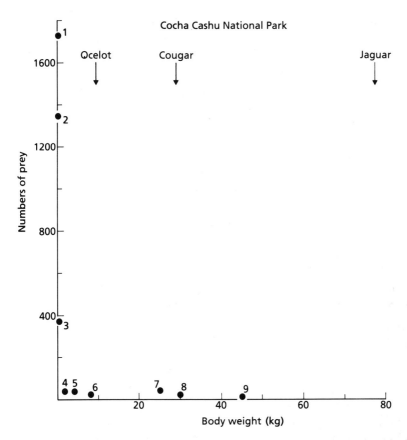

FIG. 13. Estimated abundance and body weights (kg) of prey in the Cocha Cashu National Park, Peru (after Emmons 1987a). Also shown are weights of an adult female ocelot, female cougar and female jaguar in this habitat. 1. *Proechimys* sp.; 2. *Oryzomys* spp.; 3. Opposum sp.; 4. *Myoprocta pratti*; 5. *Dasyprocta variegata*; 6. *Agouti paca*; 7. *Tayassu tajacu*; 8. *Mazama americana*; 9. *Hydrochaeris hydrochaeris*. Weights from Emmons (1987a).

(Emmons 1987a, see also Emmons 1987b), thus it appeared difficult for a hypothetical group-living female to increase food intake through taking small prey because of time limitations, and through taking large prey because of lack of availability.

Jaguars

The bulk of jaguar scats in Cocha Cashu contained medium-sized mammals, collared peccaries (7, Fig. 13), agoutis, *Dasyprocta variegata* (5), pacas, *Agouti paca* (6) and brocket deer (8), as well as tortoises, *Geochelone denticulata* (7 kg). These represented the largest prey that were reasonably abundant in the park; tapirs *Tapirus terrestris* (150 kg), are reportedly difficult to bring down (Emmons

1987a) and were rarely seen in the study area (M. Symington, personal communication. This suggested that female jaguars (78 kg: Schaller & Vasconcelos 1978) would have little opportunity to exploit prey larger than 60% of their own body weight in this habitat.

In Belize, prey most commonly found in scats was of similar body weight, nine-banded armadillos *Dasypus novemcinctus* (3 kg), paca (8 kg) and red brocket deer (30 kg) (Rabinowitz & Nottingham 1986); larger domestic livestock were not taken (Rabinowitz 1986). Nevertheless, in Acurizal, Brazil, jaguars readily take 250 kg cattle, although they prefer to take calves (Schaller 1983). Such observations suggest jaguars, and even cougars, could become social in Acurizal. However, calves are poorly represented in the area owing to bad management, and cattle reportedly have a tendency to bunch at nights (Schaller 1983), so in practice there might be limited opportunity for exploiting large prey in this habitat.

CONCLUSIONS

A review of felids living in twenty-one situations suggests females of most species have little opportunity to be social because large prey are not usually abundant. However, further data are required on the abundance, age structure and sizes of prey available to felids in many of these and other habitats. Only in a few situations might there be opportunities for sociality. Tigers in the Royal Chitawan National Park, European lynxes in northern Sweden, cougars in the Torres del Paine National Park, and jaguars on the Acurizal ranch are all reported to live in areas with fairly high densities of large prey. Given the flexibility of carnivore social organization in this (MacDonald & Apps 1978) and other families (Kruuk 1972; MacDonald 1983), felids might be expected to respond to these favourable conditions and have taken advantage of some of the benefits of sociality. That they have not, suggests that prey in these regions are still difficult to catch despite their abundance, perhaps because of grouping patterns, vigilance, or difficulty in subduing them. Further work on anti-predator behaviour of prey in response to felid predation is needed in these areas. Alternatively, there may be no benefits to sociality in these habitats; however, known cases of infanticide by cougar males in other areas, for example, suggests females could repel males if they lived together, as argued for lions (Packer & Pusey 1983). In one species, caracals in the Mountain Zebra National Park, some females may now live in groups as a consequence of catching large prey.

Lionesses have abundant prey available to them that is slightly larger than themselves, which enables them to catch relatively big prey frequently and without much difficulty. The costs of feeding additional group members are thereby greatly reduced compared to other felids, allowing lionesses to capitalize on advantages that group living provides. This explanation emphasizes that availability of abundant prey, somewhat larger than a lioness, is the factor that allows lions to feed additional group members, whereas Packer (1986) argues that open habitat, high female density and the relatively large size of their observed kills forces lions to share carcasses with relatives.

In summary, this review shows that in most situations where adequate data are available, prey larger than that normally taken by females and up to double that of an adult female's body weight, necessary to sustain two families living together, are relatively scarce. At present then, the hypothesis that sociality in felids is limited by the costs rather than by the benefits needs serious consideration.

ACKNOWLEDGEMENTS

I thank the Government of Tanzania and Tanzania National Parks for permission to conduct research, the Serengeti Wildlife Research Institute for facilities, and the Royal Society and National Geographic Society for support. I am very grateful to Anthony Collins and Clare FitzGibbon for collecting additional data on male cheetahs, to Sarah Durant and Monique Borgerhoff Mulder for fruitful discussions, and to the latter for rescuing the manuscript from the word processor. I thank Monique Borgerhoff Mulder, Hans Kruuk, James Malcolm, Craig Packer, David Scheel, Valerie Standen and Richard Wrangham for commenting on the manuscript; Paul Harvey pointed out that cats don't eat logs.

REFERENCES

Abramov, V.K. (1962). K. biologii amurskogo tigra (On the biology of the Amur tiger). *Panthera tigris longipilis* Fitzinger 1868. *Acta Zoologia Societie Bohemoslovenicae*, **26**(2), 189–202.

Bailey, T.N. (1974). Social organization in a bobcat population. *Journal of Wildlife Management*, **38**, 435–46.

Beltran, J.F., Sanjose, C., Delibes, M. & Braza, F. (1985). An analysis of the Iberian lynx predation upon fallow deer in the Coto Donana, SW Spain. *XVIIth Congress of the International Union of Game Biologists*, pp. 961–7. Brussels, September 17–21.

Benson, S.L. & Moore, R.A. (1977). Bobcat food habit response to a change in prey abundance. *Southwest Naturalist*, **21**, 451–7.

Bertram, B.C.R. (1973). Lion population regulation. *East African Wildlife Journal*, **11**, 215–25.

Bertram, B.C.R. (1975). Social factors influencing reproduction in wild lions. *Journal of Zoology*, **177**, 463–82.

Bertram, B.C.R. (1978). *Pride of Lions*. Dent, London.

Bertram, B.C.R. (1982). Leopard ecology as studied by radio tracking. *Telemetric Studies of Vertebrates* (Ed. by C.L. Cheeseman). Symposium of the Zoological Society of London, 49, pp. 341–52. Academic Press, London.

Blum, L.G. & Escherich, P.C. (1979). *Bobcat Research Conference Proceedings: Current Research on Biology and Management of Lynx rufus*. Smithsonian Institution, Front Royal, Virginia.

Blumenschine, R.J. & Caro, T.M. (1986). Unit flesh weights of some East African bovids. *African Journal of Ecology*, **24**, 273–86.

Borner, M., FitzGibbon, C.D., Borner, Mo., Caro, T.M., Lindsay, W.K., Collins, D.A. & Holt, M.E. (1987). The decline in the Serengeti Thomson's gazelle population. *Oecologia*, **73**, 32–40..

Bradbury, J.W. & Vehrencamp, S.L. (1977). Social organization and foraging in emballonurid bats. III. Mating systems. *Behavioral Ecology and Sociobiology*, **2**, 1–17.

Brand, C.J. & Keith, L.B. (1979). Lynx demography during a snowshoe hare decline in Alberta. *Journal of Wildlife Management*, **43**, 827–49.

Breitenmoser, von U. & Haller, H. (1987). Zur Nährungsökologie des Luchses *Lynx lynx* in den schweizerischen Nordalpen. *Zeitschrift für Saugetierkunde*, **52** (special issue), 168–91.

Burney, D.A. (1980). *The effects of human activities on cheetahs (Acinonyx jubatus Schr.) in the Mara Region of Kenya.* M.Sc. thesis, University of Nairobi.

Bygott, J.D., Bertram, B.C.R. & Hanby, J.P. (1979). Male lions in large coalitions gain reproductive advantages. *Nature*, **282**, 839–41.

Caraco, T. & Wolf, L.L. (1975). Ecological determinants of group sizes of foraging lions. *American Naturalist*, **109**, 343–52.

Carbyn, L.N. & Patriquin, D. (1983). Observations on home range sizes, movements and social organization of lynx, *Lynx canadensis*, in Riding Mountain National Park, Manitoba. *Canadian Field Naturalist*, **97**, 262–7.

Caro, T.M. (1980). The effects of experience on the predatory patterns of cats. *Behavioral and Neural Biology*, **23**, 1–28.

Caro, T.M. (1982). A record of cheetah scavenging in the Serengeti. *African Journal of Ecology*, **20**, 213–14.

Caro, T.M. (1987a). Indirect costs of play: cheetah cubs reduce maternal hunting success. *Animal Behaviour*, **35**, 295–7.

Caro, T.M. (1987b). Cheetah mothers' vigilance: looking out for prey or for predators? *Behavioral Ecology and Sociobiology*, **20**, 351–61.

Caro, T.M. & Collins, D.A. (1986). Male cheetahs of the Serengeti. *National Geographic Research*, **2**, 75–86.

Caro, T.M. & Collins, D.A. (1987a). Ecological characteristics of territories of male cheetahs (*Acinonyx jubatus*). *Journal of Zoology*, **211**, 89–105.

Caro, T.M. & Collins, D.A. (1987b). Male cheetah social organization and territoriality. *Ethology*, **74**, 52–64.

Caro, T.M., Holt, M.E., FitzGibbon, C.D., Bush, M., Hawkey, C.M. & Kock, R.A. (1987). Health of adult free-living cheetahs. *Journal of Zoology*, **212**, 573–84.

Clark, C.W. (1987). The lazy, adaptable lions: a Markovian model of group foraging. *Animal Behaviour*, **35**, 361–8.

Corbett, L.K. (1979). *Feeding ecology and social organization of wildcats (Felis sylvestris) and domestic cats (Felis catus) in Scotland.* Ph.D. thesis, Aberdeen University.

Curio, E.B. (1976). *The Ethology of Predation.* Springer-Verlag, Berlin.

Durant, S.M., Caro, T.M., Collins, D.A., Alawi, R.M. & FitzGibbon, C.D. (in press). Migration patterns of Thomson's gazelles and cheetahs on the Serengeti Plains. *African Journal of Ecology*.

Eaton, R.L. (1974). *The Cheetah: The Biology, Ecology, and Behavior of an Endangered Species.* Van Nostrand Reinhold, New York.

Eisenberg, J.F. (1981). *The Mammalian Radiations.* Athlone Press, London.

Eloff, F.C. (1973). Ecology and behavior of the Kalahari lion. *The World's Cats, Vol 1* (Ed. by R.L. Eaton), pp. 90–126. World Wildlife Safari, Winston, Oregon.

Emlen, S.T. & Oring, L.W. (1977). Ecology, sexual selection and the evolution of mating systems. *Science*, **197**, 215–23.

Emmons, L.H. (1987a). Comparative feeding ecology of felids in a neotropical rainforest. *Behavioral Ecology and Sociobiology*, **20**, 271–83.

Emmons, L.H. (1987b). Jungle cruisers. *Animal Kingdom*, **90(1)**, 22–30.

Foster, J. & Kearney, D. (1967). Nairobi National Park game census, 1966. *East African Wildlife Journal*, **5**, 112–20.

Frame, G. & Frame, L. (1976). Interim cheetah report for the Serengeti Research Institute, annual report, mimeographed.

Frame, G. & Frame, L. (1981). *Swift and Enduring: Cheetahs and Wild Dogs of the Serengeti.* E.P. Dutton, New York.

Frame, G.W. (1980). *Cheetah Social Organisation in the Serengeti Ecosystem of Tanzania.* Paper presented at the Animal Behavior Society, Fort Collins, Colorado.

Frame, G.W. (1984). Cheetah. *The Encyclopedia of Mammals, Vol 1.* (Ed. by D.W. MacDonald), pp. 40–3. Allen & Unwin, London.

Frame, G.W. & Wagner, F.H. (1981). Hares on the Serengeti Plains, Tanzania. *Proceedings of the World Lagomorph Conference* (Ed. K. Myers & C.D. MacInnnes), pp. 790–802. IUCN, Morges, Switzerland.

Fritz, S.H. & Selander, J.A. (1978). Diets of bobcats in Arkansas with special reference to age and sex differences. *Journal of Wildlife Management*, 42, 533−9.

Gashwiler, J.S. & Robinette, W.L. (1957). Accidental fatalities of the Utah cougar. *Journal of Mammalogy*, 38, 123−6.

Geertsema, A.A. (1985). Aspects of the ecology of the serval *Leptailurus serval* in the Ngorongoro crater, Tanzania. *Netherlands Journal of Zoology*, 35, 527−610.

Georgiadis, N. (1985). Growth patterns, sexual dimorphism and reproduction in African ruminants. *African Journal of Ecology*, 23, 75−87.

Gittleman, J.L. (1986). Carnivore life history patterns: allometric, phylogenetic, and ecological associations. *American Naturalist*, 127, 744−71.

Grobler, J.H. (1981). Feeding behaviour of the caracal *Felis caracal* Schreber 1776 in the Mountain Zebra National Park. *South African Journal of Zoology*, 16, 259−62.

Grobler, J.H. & Wilson, V.J. (1972). Food of the leopard *Panthera pardus* (Linn.) in the Rhodes Matopos National Park, Rhodesia, as determined by faecal analysis. *Arnoldia*, 35(5) 1−10.

Gromov, E.N. & Matjuschkin, E.N. (1974). K analizu konkurentnih otnoshenii tigra i volka v Sikhote-Aline (Analysis of the competing relations of the tiger and wolf in Sikhote-Alin). *"Biologisch. Nauki"*, N2, 20−5.

Haglund, B. (1966). Winter habits of the lynx (*Lynx lynx* L.) and wolverine (*Gulo gulo* L.) as revealed by tracking in the snow. Summary in English. *Viltrevy*, 4, 245−83.

Hamilton, P.H. (1976). *The movements of leopards in Tsavo National Park, Kenya, as determined by radio-tracking*. M.Sc. thesis, University of Nairobi.

Hornocker, M.G. (1970). An analysis of mountain lion predation upon mule deer and elk in the Idaho Primitive Area. *Wildlife Monographs*, 21, 1−39.

Johnsingh, A.J.T. (1983). Large mammalian prey − predators in Bandipur. *Journal of the Bombay Natural History Society*, 80, 1−57.

Konecny, M.J. (1987). Food habits and energetics of feral house cats in the Galapagos Islands. *Oikos*, 50, 24−32.

Konecny, M.J. (in press). Movement patterns and food habits of four sympatric carnivore species in Belize, Central America. *Mammals of the Americas: Essays in Honor of Ralph M. Wetzel* (Ed. by J.F. Eisenberg).

Kruuk, H. (1972). *The Spotted Hyaena: A Study of Predation and Social Behavior*. University of Chicago Press, Chicago, Illinois.

Kruuk, H. (1975). Functional aspects of social hunting in carnivores. *Function and Evolution in Behaviour* (Ed. by G. Baerends, C. Beer & A. Manning), pp. 119−41. Oxford University Press, Oxford.

Kruuk, H. & Turner, M. (1967). Comparative notes on predation by lion, leopard, cheetah and wild dog in the Serengeti area, East Africa. *Mammalia*, 31, 1−27.

Lamprecht, J. (1978). The relationship between food competition and foraging group size in some larger carnivores. A hypothesis. *Zeitschrift für Tierpsychologie*, 46, 337−43.

Lamprecht, J. (1981). The function of social hunting in larger terrestrial carnivores. *Mammal Review*, 11, 169−79.

Litvaitis, J.A., Clark, A.G. & Hunt, J.H. (1986), Prey selection and fat deposits of bobcats (*Felis rufus*) during autumn and winter in Maine. *Journal of Mammalogy*, 67, 389−92.

Livaitis, J.A., Stevens, C.L. & Mautz, W.W. (1984). Age, sex, and weight of bobcats in relation to winter diet. *Journal of Wildlife Management*, 48, 632−5.

Ludlow, M.E. & Sunquist, M.E. (1987). Ecology and behavior of ocelots in Venezuela. *National Geographic Research* 3, 444−61.

MacDonald, D.W. (1983). The ecology of carnivore social behaviour. *Nature*, 301, 379−84.

MacDonald, D.W. & Apps, P.J. (1978). The behaviour of a group of semi-dependent farm cats, *Felis catus*: a progress report. *Carnivore Genetics Newsletter*, 3, 256−86.

Maehr, D.S. & Brady, J.R. (1986). Food habits of bobcats in Florida. *Journal of Mammalogy*, 67, 133−8.

Marston, M.A. (1942).Winter relations of bobcats to white-tailed deer in Maine. *Journal of Wildlife Management*, 6, 328−37.

Matjushkin, E.N., Zhivotchenko, V.I. & Smirnov, E.N. (1980). *The Amur Tiger in the USSR*. IUCN, Gland, Switzerland.

McVittie, R. (1979). Changes in the social behaviour of South West African cheetah. *Madoqua*, **11**, 171–84.

Mitchell, B.L., Shenton, J.B. & Uys, C.M. (1965). Predation on large mammals in the Kafue National Park, Zambia. *Zoologica Africana*, **1**, 297–318.

Nellis, C.H. & Keith, L.B. (1968). Hunting activities and success of lynxes in Alberta. *Journal of Wildlife Management*, **32**, 718–22.

Nellis, C.N., Wetmore, S.P. & Keith, L.B. (1972). Lynx–prey interactions in central Alberta. *Journal of Wildlife Management*, **36**, 320–9.

O'Brien, S.J., Roelke, M.E., Marker, L., Newman, A., Winkler, C.A., Meltzer, D., Colly, L., Evermann, J.F., Bush, M., & Wildt, D.E. (1985). Genetic basis for species vulnerability in the cheetah. *Science*, **227**, 1428–34.

Orsdol, K.G. van, (1981). *Lion predation in Rwenzori National Park, Uganda.* Ph.D. thesis, University of Cambridge.

Orsdol, K.G. van, Hanby, J.P. & Bygott, J.D. (1985). Ecological correlates of lion social organisation (*Panthera leo*). *Journal of Zoology*, **206**, 97–112.

Owens, M. & Owens, D. (1984). *Cry of the Kalahari.* Houghton Mufflin Co., Boston.

Packer, C. (1986). The ecology of sociality in felids. *Ecological Aspects of Social Evolution: Birds and Mammals* (Ed. by D.I. Rubenstein & R.W. Wrangham), pp. 429–51. Princeton University Press, Princeton, New Jersey.

Packer, C. (1988). Constraints on the evolution of reciprocity: lessons from cooperative hunting. *Ethology and Sociobiology*, **9**, 137–47.

Packer, C., Herbst, L., Pusey, A.E., Bygott, D., Hanby, J.P., Cairns, S.J. & Borgerhoff Mulder, M. (1988). Reproductive success of lions. *Reproductive Success: Studies of Individual Variation in Contrasting Breeding Systems* (Ed. by T.H. Clutton-Brock), pp. 363–83. University of Chicago Press, Chicago, Illinois.

Packer, C. & Pusey, A.E. (1982). Cooperation and competition within coalitions of male lions: kin selection or game theory? *Nature*, **296**, 740–2.

Packer, C. & Pusey, A.E. (1983). Adaptations of female lions to infanticide by incoming males. *American Naturalist*, **121**, 716–28.

Packer, C. & Pusey, A.E. (1984). Infanticide in carnivores. *Infanticide: Comparative and Evolutionary Perspectives* (Ed. by G. Hausfater & S.B. Hrdy), pp. 31–42. Aldine, New York.

Packer, C. & Ruttan, L. (in press). The evolution of cooperative hunting. *American Naturalist*.

Parker, G.R., Maxwell, J.W., Morton, L.D. & Smith, G.E.J. (1983). The ecology of the lynx (*Lynx canadensis*) on Cape Breton Island. *Canadian Journal of Zoology*, **61**, 770–86.

Petraborg, W.H. & Gunvalson, V.E. (1962). Observations on bobcat mortality and bobcat predation on deer. *Journal of Mammalogy*, **43**, 430–1.

Pienaar, U. de (1969). Predator–prey relations amongst the larger mammals of Kruger National Park. *Koedoe*, **12**, 108–76.

Pollack, M.E. (1951). Food habits of the bobcat in the New England states. *Journal of Wildlife Management*, **15**, 209–13.

Pringle, J.A. & Pringle, V.L. (1979). Observations on the lynx *Felis caracal* in the Bedford District. *South African Journal of Zoology*, **14**, 1–4.

Rabinowitz, A.R. (1986). Jaguar predation on domestic livestock in Belize. *Wildlife Society Bulletin*, **14**, 170–4.

Rabinowitz, A.R. & Nottingham, B.G. (1986). Ecology and behaviour of the jaguar (*Panthera onca*) in Belize, Central America. *Journal of Zoology*, **210**, 149–59.

Robinette, W.L., Gashwiler, J.S. & Morris, O.W. (1959). Food habits of the cougar in Utah and Nevada. *Journal of Wildlife Management*, **23**, 261–73.

Robinette, W.L., Gashwiler, J.S. & Morris, O.W. (1961). Notes on cougar productivity and life history. *Journal of Mammalogy*, **42**, 204–17.

Rood, J.P. (1986). Ecology and social evolution in the mongooses. *Ecological Aspects of Social Evolution: Birds and Mammals.* (Ed. by D.I. Rubenstein & R.W. Wrangham), pp. 131–52. Princeton University Press, Princeton, New Jersey.

Rudnai, J.A. (1973) *The Social Life of the Lion.* Medical and Technical Publishing Co., Lancaster.

Sachs, R. (1967). Live weights and body measurements of Serengeti game animals. *East African Wildlife Journal*, **5**, 24–36.

Sanborn, C.C. (1954). Weights, measurements, and color of the Chilean forest puma. *Journal of Mammalogy*, **35**, 126–8.

Saunders, J.K. Jr. (1963). Food habits of the lynx in Newfoundland. *Journal of Wildlife Management*, **27**, 384–90.

Saunders, J.K. Jr. (1964). Physical characteristics of the Newfoundland lynx. *Journal of Mammalogy*, **45**, 36–47.

Schaller, G.B. (1967). *The Deer and the Tiger: A Study of Wildlife in India.* University of Chicago Press, Chicago, Illinois.

Schaller, G.B. (1972). *The Serengeti Lion: A Study of Predator–Prey Relations.* University of Chicago Press, Chicago, Illinois.

Schaller, G.B. (1977). *Mountain Monarchs: Wild Sheep and Goats of the Himalaya.* University of Chicago Press, Chicago, Illinois.

Schaller, G.B. (1983). Mammals and their biomass on a Brazilian Ranch. *Arquivos de Zoologia*, **31(1)**, 1–36.

Schaller, G.B. & Crawshaw, P.G. Jr. (1980). Movement patterns of jaguar. *Biotropica*, **12**, 161–8.

Schaller, G.B., Talipu, L.H., Hua, L., Junrang, R., Mingjiang, Q. & Haibin, W. (1987). Status of large mammals in the Taxkorgan Reserve, Xinjiang, China. *Biological Conservation*, **42**, 53–71.

Schaller, G.B. & Vasconcelos, J.M.C. (1978). Jaguar predation on capybara. *Zeitschrift für Saugetierkunde*, **43**, 296–301.

Seidensticker, J. (1976). Ungulate populations in the Chitawan valley, Nepal. *Biological Conservation*, **10**, 183–210.

Seidensticker, J., Hornocker, M.G., Wiles, W.V. & Messick, J.P. (1973). Mountain lion social organization in the Idaho Primitive Area. *Wildlife Monographs*, **35**, 1–60.

Senzota, R.B.M. (1978). *Some aspects of the ecology of two dominant rodents in the Serengeti Ecosystem.* M.Sc. thesis, University of Dar es Salaam.

Sheppard, D.H. (1960). *The ecology of the mule deer of the Sheep River region.* M.Sc. thesis, University of Alberta.

Skinner, J.D. (1979). Feeding behaviour in caracal (*Felis caracal*). *Journal of Zoology*, **189**, 523–55.

Smith, R.M. (1978). Movement patterns and feeding behavior of the leopard in the Rhodes Matopos National Park, Rhodesia. *Carnivore*, **1(3)**, 58–69.

Smithers, R.H.N. (1971). *The Mammals of Botswana.* Museum memoir No. 4, National Museums of Rhodesia, Salisbury.

Smithers, R.H.N. (1983). *The Mammals of the Southern African Subregion.* University of Pretoria, Pretoria, South Africa.

Spalding, D.J. & Lesowski, J. (1971). Winter food of the cougar in south-central British Columbia. *Journal of Wildlife Management*, **35**, 378–81.

Stoddart, D.M. (1979). Feeding behaviour in caracal *Felis caracal*. *Journal of Zoology*, **189**, 523–57.

Sunquist, M.E. (1981). The social organization of tigers (*Panthera tigris*) in Royal Chitawan National Park. *Smithsonian Contributions to Zoology*, **336**, 1–98.

Sunquist, M.E. & Sunquist, F.C. (in press). Ecological constraints on predation in large felids. *Carnivore Behavior, Ecology and Evolution* (Ed. by J.L. Gittleman). Cornell University Press, Ithaca, New York.

Tamang, K.M. (1979). Population characteristics of the tiger and its prey. Unpublished paper presented at the International Symposium on the Tiger. New Delhi, India.

Thapar, V. (1986). *Tiger: Portrait of a Predator.* Fact on File Publications, New York.

Trivers, R.L. (1972). **Parental investment and sexual selection.** *Sexual Selection and the Descent of Man, 1871–1971.* (Ed. by B. Campbell), pp. 136–79. Aldine Press, Chicago, Illinois.

Valkenburgh, B. van & Ruff, C.B. (1987). Canine tooth strength and killing behaviour in large carnivores. *Journal of Zoology*, **212**, 379–97.

Waser, P.M. & Waser, M.S. (1985). *Ichneumia albicanda* and the evolution of viverrid gregariousness. *Zeitschrift für Tierpsychologie*, **68**, 137–51.

Wilson, P. (1984). Puma predation on guanacos in Torres del Paine National Park, Chile. *Mammalia*, **48**, 515–22.

Wilson, V.J. (1968). Weights of some mammals from eastern Zambia. *Arnoldia*, **32**(3) 1−20.

Wrangham, R.W. (1979). On the evolution of ape social systems. *Social Science Information*, **18**, 335−68.

Yanez, J.L., Cardenas, J.C., Gezelle, P. & Jaksic, F.M. (1986). Food habits of the southernmost mountain lions (*Felis concolor*) in South America: natural versus livestocked ranges. *Journal of Mammalogy*, **67**, 604−6.

Young, S.P. & Goldman, E.A. (1946). *The Puma: Mysterious American Cat.* American Wildlife Institute, Washington, D.C.

Yudakov, A.G. (1973). O vliganii tigra na chislennost kopitnih (On the influence of the tiger upon the ungulate population). *Redkie vidi mlekopitajuschih fauni SSR i ih ohrana*, Sbornik materialov, str. 93−94. Izd. 'Nauka', Moskva.

Zyll de Jong, C.G. van (1966). Food habits of the lynx in Alberta and the Macenzie District, N.W.T. *Canadian Field-Naturalist*, **80**, 18−23.

Food security and the rewards of tolerance

D.W. MACDONALD AND G.M. CARR
Department of Zoology, University of Oxford, South Parks Road,
Oxford OX1 3PS, UK

*Nos numerus sumus et fruges consumere nati**
Horace, Epistles ii, 14

SUMMARY

1 Group territory sizes vary between populations, and even between neighbours, for many territorial species. With especial reference to the Carnivora, we review some ecological and sociological factors which may help to explain these variations.

2 The resource dispersion hypothesis (RDH) proposes that groups may develop in an environment where resources are dispersed such that the smallest economically defensible territory for a pair can also sustain additional animals. Such patterns of resource dispersion are seen as facilitating the evolution of sociality, in that they reduce the cost of tolerating additional occupants in minimum territories.

3 The food security of a territory is defined as the probability that it will satisfy the nutritional requirements of its occupants for a given feeding period. Applying this concept, we use RDH to show how the frequency distribution of resources available per unit time within a territory may permit the formation of groups, even in the absence of any functional advantage to any individual from the presence of any other.

4 Amongst conditions facilitating group formation are (i) heterogeneity in the pattern of resource availability, and (ii) a difference in the food security demanded by primary and secondary members of the group.

5 Territorial occupants can increase their food security if their borders are expanded to accommodate additional group members.

6 We present a model which shows how variation in the survival probabilities of breeding primary occupants (P_α), non-breeding secondary occupants (P_i) and emigrants (P_e) interact with the marginal contributions (δf_n) made by non-breeders to the primary occupants' reproductive success, to determine whether it is advantageous for a given individual to join its parent's group or to disperse. This model differs from some of its predecessors (e.g. Brown 1978; Emlen 1982a, b) in estimating, through binomial expansion, the likely gains involved when more than one 'helper' is present, and in making explicit (through their marginal values) the

---

* Translation: We are just statistics, born to consume resources.

75

consequences of diminishing returns on the investment of helpers as their numbers increase.

7 Simulations using this model illustrate the consequences for the fitnesses of both mother and daughter of the daughter dispersing or acting as a helper in groups of varying sizes. These show how small changes in circumstances can greatly affect the outcome for both parties, determining whether the interests of both mother and daughter may be in agreement or discord, and whether betas tolerate or attempt to expel one another, or disperse voluntarily.

8 Both the ecological and sociological models presented here utilize variables that can be measured in the field, and therefore their predictions can be tested.

INTRODUCTION

Many animals maintain territories and some live as members of social groups. The sizes of such territories and groups vary greatly, between species, populations and even neighbours. These two variables, territory area and group size, are fundamental elements of social organization, and identifying the selective pressures that affect them is a prerequisite to understanding the adaptive significance of social systems for their members. We argue that it is often helpful to distinguish two categories of selective pressure exerted on social animals. The first category involves ecological factors, such as the availability, predictability and dispersion of resources in the environment. These set the framework within which the second category, sociological factors, may operate. We present two simple models that illustrate the ways in which ecological and sociological factors may operate as selective pressures affecting territory and group size.

Intra-specific variation in group and territory size is well illustrated by studies of the Canidae. In this family, the basic breeding unit is the monogamous pair, and spatial organization is fundamentally territorial. However, some canid species, although they have been found living in such territorial pairs, also live in groups. Wolves, *Canis lupus*, form packs of up to fifteen animals (home range 60—350 km²: Oosenberg & Carbyn 1982; Scott & Shackleton 1982), golden jackals, *Canis aureus*, live in groups of up to twenty (home range from 10 to at least 300 ha: van Lawick 1973; Macdonald 1979a), silver-backed jackals, *C. mesomelas*, form groups of up to five (home range 1·5—5·0 km²: Moehlman (1983), coyotes, *C. latrans*, live in large transient groups of up to twenty-two (home range 14—128 km²: Ozoga & Harger 1966; Berg & Chesness 1978; Camenzind 1978), red foxes, *Vulpes vulpes*, form groups of up to six (home range 10—3000 ha: Macdonald 1981) and Arctic foxes, *Alopex lagopus*, live in groups of up to three (home range 290—6000 ha: Kaikusalo 1971; Hersteinsson & Macdonald 1982. There is also great variation among those species that seem to be obligate pack-dwellers: Dholes, *Cuon alpinus*, have been recorded in packs varying in size between five and twenty (home range up to 40 km²; Johnsingh 1984, and wild dogs, *Lycaon pictus*, live in packs of five to twenty-five animals (home range 500—2000 km²; Malcolm 1979; Frame *et al* 1980; Riche 1981).

RESOURCE DISPERSION AS A SOCIAL PROMOTER

There is evidence that the variations in group and territory size can be explained in part by the pattern of resource availability (another factor likely to affect group and territory size is the pattern of mortality). The pattern of resource availability has been widely recognized as affecting social organization in various animal taxa, (see, for example, Crook 1964; Jarman 1974; Kruuk 1975) and in general models (see, for example, Orians 1961; Waser 1981; Brown 1982, 1987; Lindström 1986). One such family of ideas, developed by authors such as Bradbury & Vehrencamp (1976), Macdonald (1981), Kruuk & Parish (1982), and Mills (1982) has been called the Resource Dispersion Hypothesis (RDH). It can be summarized as follows: groups may develop in an environment where resources are dispersed such that, under certain circumstances, the smallest economically defensible territory for a pair can also sustain additional animals (Macdonald 1983). According to this hypothesis, ecological circumstances, by creating the potential to accommodate additional animals into a pair's territory, may reduce the costs of group formation to such an extent that they are outweighed by the sociological benefits. For example, Kruuk & Macdonald (1985) suggested that the minimum size of a spotted hyaena, *Crocuta crocuta*, territory might be constrained by the distance over which it must chase prey; territories of this size might encompass widely different numbers of prey and thereby have the potential to support groups of different sizes.

Another, perhaps more general, case may exist when food is dispersed in discrete patches. According to the RDH, territory size will be determined by the dispersion of such patches, whereas group size will be determined by their richness. For example, Kruuk & Parish (1982) suggested that the distance between patches where earthworms may be caught determines the size of a badger's territory, whereas the abundance of available worms in the patches determines the number of badgers occupying the territory.

The model we present here is a general version of the RDH, which shows how the frequency distribution of resources available per unit time within a territory may permit the formation of groups, even in the absence of any functional advantage to one individual from the presence of another. A more detailed account of our model can be found in Carr & Macdonald (1986). Recent theoretical work by Bacon *et al.* (in preparation) has led to a mathematical analysis that yields a more comprehensive general model which confirms the qualitative principles outlined below in our more limited (but less mathematical) version, namely that RDH describes conditions that can indeed lead to group formation.

Economic principles (see, for example, Brown 1964; Davies 1978) suggest that animals will defend a territory of the minimum size necessary to satisfy their requirements for the resource being defended. The animals' problem, and this is the keystone of our argument, is that such resources may vary in availability over time, in both the short and the long term. This temporal variation in resource availability led us to the concept of 'food security' upon which our model rests.

We define the food security of a territory as the probability that it will satisfy the nutritional requirements of its occupants for a given feeding period. Each animal will have a certain gross requirement for food (or other potentially limiting resource) during each feeding period (e.g. one night of foraging). Occasionally, most individuals will fail to satisfy their requirements due to a bad night's hunting, for example. However, an individual must realize its food requirements for a certain proportion of these feeding periods, otherwise it risks reproductive failure and ultimately starvation. For a territory to be viable, it must offer a guarantee of sufficient resources to satisfy an occupant's gross requirements for a certain proportion of feeding periods. That is, for any given feeding period, the individual must have a particular probability of achieving its requirements. We have called this the critical probability (Cp). If the food security of a territory falls below this critical probability, it will not sustain its occupants.

For most species, the smallest viable social unit for reproduction is the pair, although some species require a larger group, e.g. dwarf mongooses, *Helogale parvula* (Rasa 1985). In a territorial species, such reproductive units (referred to as the 'primary occupants' of territories) must acquire a territory with sufficient resources to provide adequate food security. The burden of our argument is that the pattern in which food becomes available has far-reaching implications not only for the resource richness of the territory the occupants use, but also for their social system.

In the following model we show how the frequency distribution of resources available per unit unit time within a territory may permit the formation of groups even in the absence of any functional advantage to any individual from the presence of another, and without the necessity of them feeding or travelling together. Consider first an environment where resources are distributed such that the frequency distribution over which they become available in a territory approaches a smooth curve, i.e. food patches are very small and numerous. In the ideal case, this distribution will be normal (the assumptions underlying this model are detailed in Carr & Macdonald 1986).

A primary pair will have certain gross food requirements (R_α units of resources per animal) and seek a territory where food security matches their own critical probability ($Cp = Cp_\alpha$). The relationship between these variables and the average value of available resources ($\bar{R}$) is illustrated in Fig. 1 for territories in two different environments. The variation in resource availability with time is small in the first environment; in the second, it is greater. As the variance in the resources available per feeding period (R) differs between these two habitats, the average resource richness of the minimum territory necessary to guarantee a yield of $2R_\alpha$ units of resources to the primary pair at the required food security (Cp_α) also varies. It is possible to envisage two habitats whose resources available per unit area per unit time are the same, but which differ in the variance of their resources per unit area — the habitats in the two territories depicted in Fig. 1 have these qualities (S.D. = 1 and 3). We will use the term heterogeneity, H, in quantifying environmental variability, where H is the standard deviation of the distribution of resources available per feeding period in a territory.

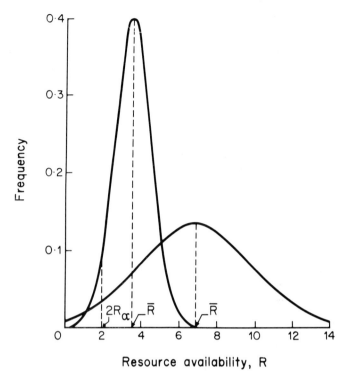

Fig. 1. Frequency distributions of resources available per feeding period in minimum territories, showing the difference between more and less variable circumstances. In both circumstances, the primary pairs can rely upon at least $2R_\alpha$ units of resources at a food security of 95% ($Cp_\alpha=0\cdot95$). The two territories differ in that environmental heterogeneity (i.e. the S.D.) is $1\cdot0$ in the less variable territory, and $3\cdot0$ in the more variable one; hence, the average value of available resources is less ($\bar R=3\cdot6$) in the former than in the latter ($\bar R=6\cdot9$).

Figure 1 illustrates that increased heterogeneity requires an increase in the mean richness ($\bar R$) of a territory to ensure that $R>2R_\alpha$ for Cp_α. This increase in mean richness greatly affects the prospects of a secondary animal wishing to join the primary occupants in their minimum territory. Suppose that the food requirement of an additional group member is R_β, (where R_β is less than or equal to R_α) then if Cp_β is less than Cp_α, the primary and additional occupants may be able to coexist without enlarging the territory beyond the size required by the primary occupants alone (see Fig. 2). The food requirements of such secondary animals may be lower than those of the primaries because additional group members, perhaps through lower status, might not reproduce, and they may also accept a lower than optimum nutrition rate if the alternative were emigration (and increased risks of starvation and death).

Our model shows that coexistence within the primary territory is possible even if the primary pair prevent the secondary from feeding until they have satisfied their own requirements. If the secondary did interfere with the primaries' feeding

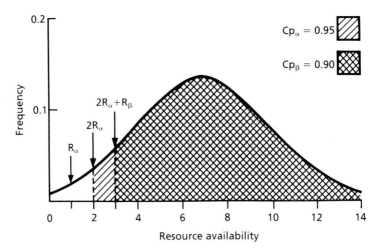

Fɪɢ. 2. The pattern of resource availability in the territory in the more variable of the two environments shown in Fig. 1. This territory provides 95% food security of $2R_\alpha$ for the primary occupants, together with 90% food security of R_β for the secondary. The hatched areas under the graph correspond to the proportions of time when sufficient food is available for the primaries and for the secondary given that the primaries have fed first.

then the principle would be unaffected, but any rewards to the primaries of tolerating the secondaries would be diminished accordingly. The opportunity for coexistence rests upon environmental heterogeneity: high values of H facilitate the accommodation of additional group members, low values of H preclude it, unless the occupants undertake territorial expansion (see Kruuk & Macdonald 1985). The significance of environmental heterogeneity in facilitating group formation is emphasized by further examination of Fig. 1, in which $H=1$ and $H=3$ in the less and the more variable environments, respectively ($Cp_\alpha=0.95$ in both). The territory in the more stable environment offers a secondary group member a food security of 72%, whereas a secondary living in the territory in the more variable environment would enjoy a food security of 90%. The model shows, therefore, how a group could come into existence with no advantages to members other than that of supporting the secondary.

   The extent to which additional group members can bear a food security lower than that of the primaries will depend on the species involved: for small insectivores, a difference of only a few per cent in food security might be critical, whereas for larger generalist species it might be inconsequential.

   The illustrations in Figs 1 and 2 apply to circumstances in which resources become available according to a continuous normal distribution. In reality, many resources are patchily distributed. The same principles apply, *mutatis mutandis*, to patchy distributions. In environments where food patches become available according to a binomial schedule, the primary occupants seeking adequate food security from such resources face a problem analogous to the familiar one of a gambler throwing dice. Imagine that each die represents a food patch and one throw decides that patch's performance during each feeding period. A patch yields food only if the die lands as a '6', and one such 6−patch is sufficient to satisfy the gross food requirements of the animal. Under these circumstances, with only one food patch (one die), the primary occupant's (player's) territory offers a food security of only 0·167. If the player were prudent he would not gamble his fortunes (or his inclusive fitness) on such poor odds. The player might refrain from gambling unless there was an 80% chance of throwing at least one '6' (Cp_α=0·8), in which case the security demanded would require nine dice (nine food patches in the territory). This is because a six-sided die carries a 5/6th risk of failure, and nine throws are needed to reduce this risk to less than the required 20%, i.e. $(1-(5/6)^9)$=0·806. The greater the player's determination to win, the more dice are required. Similarly, the greater the food security the primary occupants require from their territory, the more food patches they will need; the greater the number of patches (or dice), the greater the chance of there being additional food (surplus 6s). It is this surplus that allows additional members to join the group. Figure 3 illustrates some biological consequences of the dice game. It shows how the number of patches per territory, their richness and the probability of each being fruitful during a given feeding period affect the food security provided by that territory to the primary and secondary occupants.

The general conclusions to emerge from this consideration of resource dispersion is that the minimum territory needed by primary occupants may sometimes support extra group members (we suggest that these conditions have facilitated the evolution of group living). Such secondary animals are more easily accommodated into a territory (i) the more heterogeneous is the pattern of resource availability and (ii) the greater is the difference in food security demanded by the primary occupants relative to that demanded by the secondary.

MORTALITY AND COOPERATION AS PROMOTERS OF SOCIALITY

Variance in the pattern of resource availability appears to create ecological conditions conducive to the formation of groups. So far, we have attributed neither cost nor benefit to the primary occupants from accommodating additional group members in their territory. It is likely there will be costs to the primary occupants. For example, the presence of secondary animals may cause the primaries to lose feeding opportunities occasionally, or to defend resources. On the other hand, the admission of secondaries could bring benefits. For example, additional group members might cooperate in the defence of the territory. The various advantages enjoyed by animals that live in groups are reviewed by Bertram (1978). The question for the primary is whether the costs of tolerating a secondary exceed those of excluding it, and, if they do, whether they are outweighed by any benefits. Davies & Houston (1981) illustrate how the answer to this question varies with circumstances for pied wagtails, *Motacilla alba*, amongst which territorial birds tolerate secondaries when the benefits of doing so outweigh the costs.

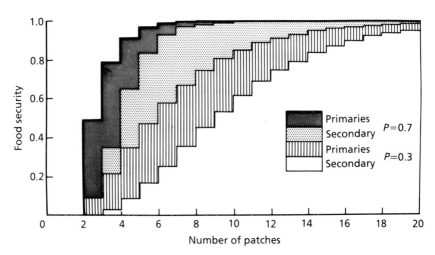

FIG. 3. Food security for primary occupants and a secondary in circumstances under which resources become available in patches of unit size and according to a binomial schedule. Two habitats are illustrated, one in which patches have a higher probability (0·7) of being fruitful during a feeding period, and one in which this probability is lower (0·3). In both cases, food security is greater in territories containing more patches, and primary and secondary values converge at higher levels of food security.

In forming groups larger than the minimum reproductive unit, territorial animals can follow one of two strategies: contractionism or expansionism (Kruuk & Macdonald 1985). Contractionists increase their group size only up to the number of animals that can be accommodated within the minimum territory (indicating that the cost of further expansion of the territory outweighs any benefits that additional group members might bring). Expansionists live in groups larger than the size that can be sustained by the minimum territory (indicating that the benefits of larger groups outweigh the costs of territorial expansion). All else being equal (which is unlikely in field studies), and assuming resource patch dispersion and richness vary independently of each other, a positive correlation between group size and territory area would be expected among populations of expansionists, but not among populations of contractionists (examples of both cases are given in Macdonald 1983). For expansionists, the advantages of cooperative territorial defence could have significant consequences for food security. Consider the case of the less variable territory illustrated in Fig. 1. If the secondary's presence led to a 50% increase in the territory's resources (from $\bar{R}=3.6$ to 5·4), not only would the primaries' food security rise close to unity, but also the secondary's food security would increase from 0·72 to 0·99 (see Carr & Macdonald 1986).

Clearly, there are diverse advantages and disadvantages to sociality, and the net resolution of these can vary between species and populations, age and sex classes, and even individual members of one group. The following examples illustrate the diversity of selective pressures thought to favour group formation: Packs of wild dogs, *Lycaon pictus*, hunt cooperatively (Frame *et al.* 1980). Larger packs of coyotes, *Canis latrans*, are better able to defend the carcasses of their prey than smaller packs (Bowen 1981). Non-breeding female red foxes help feed the cubs of their group-mates and will adopt orphans (Macdonald 1979b, von Schantz 1984). Male lions, *Panthera leo*, form coalitions to secure and defend prides of lionesses (Bygott *et al.* 1979, Packer & Pusey 1982), and female farm cats, *Felis catus*, form coalitions to nurse each other's kittens (Macdonald *et al.* 1987). Dwarf mongooses, *Helogale parvula*, take care of their sick (Rasa 1984), and banded mongooses, *Mungos mungos*, rescue their companions from raptors (Rood 1983).

It is true, but unhelpfully vague, to conclude that whether or not a given individual joins a group depends on the balance of a complicated series of costs and benefits. As a step towards testable predictions it is possible to build models based on the sum of the overall effects of these costs and benefits, without having to enquire too closely into their exact nature.

Our model applies to the type of social system exhibited by the red fox, as mentioned above. The group includes an individual who is a breeder (an alpha individual) and a number of non-breeding individuals (beta individuals) of the same sex as the alpha. For mammalian species, these group members are likely to be a mother and her daughters, and we shall use this case for illustrative purposes. In our model, any beta in the group has a choice between two strategies for the following breeding season: either staying in the natal territory or dispersing. It is then possible to calculate, within our chosen parameters, the consequences for the inclusive fitness accruing to the beta and the alpha individuals in that season of the beta pursuing each of these strategies.

We assume that if the daughter joins her mother's group she foregoes the opportunity to breed, but she may contribute to her own and her mother's fitnesses in other ways (e.g. by becoming a helper). The value of this contribution will be affected by the number of her sisters also choosing to stay. We have also assumed that daughters which do not disperse have the possibility of inheriting the territory on their mother's death. Daughters which disperse may acquire territories and mates, and thus become primary occupants in their own right.

The daughter's decision on whether to disperse, and the mother's on whether she should tolerate her daughter's presence, depend on several factors. In the model the following are considered:

P_α, P_i, P_e

The probabilities of an individual surviving until the following breeding season if it is, respectively, an alpha, a beta choosing to remain in the natal territory and a beta choosing to emigrate. 'Survival', in this context, means being in a position to reproduce, or help another to reproduce so, for the emigrant, it incorporates physical survival, mate acquisition and territory acquisition.

N, n

N is the total number of betas opting to stay in the natal territory; n is the number, other than the individual under consideration, surviving to the following breeding season (i.e. the maximum value of n is $(N - 1)$).

$L, \delta f_n$

L is the average weaned litter size of an unassisted mother. For simplicity, this is assumed to be the same for either an alpha or a successfully dispersing beta. δf_n is the marginal contribution made by a helping beta to an alpha's litter (as a fraction of L) in the presence of n other betas, i.e. the loss to the mother of potential offspring due to the beta withdrawing her assistance by dispersing.

$r_{\alpha y}$, $r_{\alpha\beta y}$, etc.

The coefficients of relationship between an alpha and its young, an alpha and a beta's young, and so on.

The model

When considering similar factors, Emlen (1982a,b; see also Brown 1978) proposed that helper behaviour will be favoured when, in our terminology,

$$(L_{(\alpha+\beta)} - L_\alpha) \cdot r_{\beta\alpha y} > P_e \cdot L_\beta \cdot r_{\beta y},$$

where $(L_{(\alpha+\beta)} - L_\alpha)$ is the increase in the number of young successfully produced by the breeding female as a result of the activities of the beta as a helper, L_β is the number of young produced by a beta which emigrates and breeds independently, and P_e is its chance of doing so successfully. The terms $r_{\beta\alpha y}$ and $r_{\beta y}$ refer to the coefficients of relatedness of the beta to the alpha's young and to its own young. Emlen suggested that this condition will rarely be met, because $r_{\beta y}$ will be greater than $r_{\beta\alpha y}$ in most cases, and because by securing the parental aid of its own mate the emigrant benefits from the efforts of an additional individual (see Charnov 1981). Low external survivorship would seem to be the principal countervailing force.

Our model incorporates the possibility of inheriting the territory from the alpha, as an additional incentive to stay, and the effects of several additional group members. The marginal contribution made by a daughter to the mother's production of offspring may arise from one or more of the various advantages of sociality (e.g. the improved food security mentioned above, greater vigilance for predators, etc). However, for the purposes of this model, it is simpler to consider the beta as a helper at the den, making a direct contribution to the well-being of the alpha's young through alloparental behaviour.

Equations

According to Emlen (1982a) the pay-off associated with dispersal is given by:

$$P_e \cdot L \cdot r_1,$$

where $r_1 = r_{\alpha\beta y}$ when the inclusive fitness of the alpha is under consideration, and where $r_1 = r_{\beta y}$ when the inclusive fitness of the beta is under consideration.

To decide whether dispersal is the better strategy, this value must be compared with the pay-off associated with remaining to help. This can be expressed as:

$$P_i \left(P_\alpha F_1 + (1 - P_\alpha) F_2 \right).$$

As is the case with dispersal, a beta remaining to help must survive (probability = P_i) to the next breeding season to be able to make any contribution at all. Assuming that she does survive, there are two possible outcomes: the alpha also survives (P_α) and the beta acts as a helper with pay-off F_1; the alpha dies ($1 - P_\alpha$) and the beta has a chance to become the new alpha or, if she fails, to help whichever of her sisters inherited the alpha role, with a combined pay-off F_2. Thus, emigration is worthwhile if:

$$P_e \, L_\beta \, r_{\beta y} > P_i \left(P_\alpha \, F_1 + (1 - P_\alpha) F_2 \right).$$

The probabilities, and pay-offs, of each of these outcomes can be estimated from the survival probabilities of the alpha and the non-dispersing beta (P_α, P_i) and the marginal contributions of other daughters which remain and survive to the following breeding season. These other helpers are important because, in general, the more helpers that are available, the less a given individual contributes. Thus, the greater number of sisters a female has, the more likely it is that either she or her mother will favour her dispersal because her contribution in staying will be small. (An issue to which we will return is whether a female may benefit by evicting her sister.) In making the calculation, however, it is insufficient to subtract $P_i \cdot \delta f_n$ (i.e: the chance the helper will survive times her contribution if she does) from the mother's productivity the following season, because the mother and all or some of the other helpers may also die in the meantime, and these risks must also be taken into account. Consequently, it is necessary to consider the binomial expansions of the various within-group mortality probabilities (P_α and P_i). Once again, the problem is analogous to throwing dice. Imagine a dice game in which players lose (group members perish) if they throw either a 5 or a 6 (a 0·33 probability of losing or survival probabilities, P_α and P_i, of 0·66). In this case, the probability of a mother benefiting from a third daughter's presence would be the product of her own chance of survival to the next breeding season and that of all three of her daughters doing so, i.e. ($P_\alpha (P_i)^3$). This would be equivalent to the probability of avoiding 5s and 6s in four successive throws of the die (i.e. $0·66(0·66)^3 = 0·19$). If the mother does benefit from the presence of a third daughter, she does so to the extent of its marginal contribution, therefore the actual benefit accruing to her fitness (the probable rewards on which the gamblers should base their bets) is 0·19 times this marginal value (δf_n), i.e. the chance of it happening times the value if it does. To calculate the total likely credit accruing to the mother if she embarks upon the approach to the next breeding season with a group of three daughters (a throw of three dice), the products of the probabilities and the pay-offs for each should be summed (i.e. $P_0 \cdot \delta f_0$, $P_1 \cdot \delta f_1$, etc., for the outcomes: none, one, etc., daughters surviving (zero, one, etc. dice avoiding the deadly 5s and 6s).

The general expression for calculating the chance of exactly m daughters surviving from an initial group of n is given by:

$$\frac{n!}{m!(n-m)!}\, p^m q^{(n-m)},$$

where p is the probability of a daughter surviving and q the probability of her dying. Where a weighted summation is carried out, the notation

$$\sum_{m=0}^{n} \begin{vmatrix} n \\ m \end{vmatrix} p^m q^{(n-m)} \cdot \text{(variable dependent on value of } r)$$

indicates that each term is calculated from the formula above multiplied by the relevant value of the dependent variable, and all terms in the series are then added. These formulae can be recognized, suitably reorganized, in the equations that follow. Readers unfamiliar with the binomial theorem will find a readable introduction in Saunders *et al.* (1985).

When the beta stays, and survives to help her mother, her expected average contribution is given by the summed products of the probabilities (P) of given numbers of betas surviving and their marginal contributions (δf_n) (appropriately devalued by the coefficient of relationship of the individuals involved):

$$(P_0 \cdot \delta f_0 + P_1 \cdot \delta f_1 \ \ldots \ P_{(N-1)} \cdot \delta f_{(N-1)})\, r_2,$$

i.e.

$$F1 = \sum_{n=0}^{N-1} \begin{vmatrix} N-1 \\ n \end{vmatrix} P_i^n\,(1 - P_i)^{(N-1)-n} \cdot \delta f_n \cdot r_2.$$

As defined earlier, the value of n is the number of other sisters which survive. Thus, the term $P_0 \cdot \delta f_0 \cdot r_2$ describes the pay-off when none of her sisters survives, and she is the only helper, whereas the term $P_{(N-1)} \cdot \delta f_{(N-1)} \cdot r_2$ describes the pay-off when all her sisters survive.

If the alpha dies, the probability distribution is the same, but the pay-off associated with each term is composed of the chance of inheritance and its benefits, and the chance of helping another beta and its benefit:

$$F2 = \sum_{n=0}^{N-1} \begin{vmatrix} N-1 \\ n \end{vmatrix} P_i^n (1 - P_i)^{(N-1)-n} \cdot [(n/(n+1)) \cdot \delta f_n \cdot r_3 + (1/(n+1)) \cdot L \cdot r_4].$$

For the calculation of these terms, $r_2=r_{\alpha\gamma}$, $r_3=r_4=r_{\alpha\beta\gamma}$ when the alpha's fitness is considered, and $r_2=r_{\beta\alpha\gamma}$, $r_3=r_{\beta\beta\gamma}$, $r_4=r_{\beta\gamma}$ when considering the beta's fitness.

If the pay-off arising from a daughter's dispersal is referred to as Bd and the pay-off from her remaining at home is given by Br, the net benefit of the beta dispersing can be calculated as $B_\alpha=Bd_\alpha - Br_\alpha$ for the alpha and $B_\beta=Bd_\beta - Br_\beta$ for the beta. If B_β is positive, the beta's fitness will be greater if she disperses from the natal territory; if it is negative, her fitness will be greater if she stays. The same argument applies, *mutatis mutandis*, to the alpha.

If B_α and B_β are plotted against one another (see Fig. 4), the resulting 'dispersal space' describes all four combinations of the alpha and the beta individuals' best strategies. In two cases, there is conflict between the animals, and in two cases accord.

The potential for conflict is not restricted to mother versus daughter. Sisters may also be in conflict with one another. So far, we have considered only two beta strategies, those of remaining or dispersing. There is a third potential strategy, that of eviction. The pay-off to a beta which evicts one of her n sisters from the group is given by:

$$Be=Br_{(n-1)} + (P_e \cdot L \cdot r_{\beta\beta\gamma} - Br_n),$$

i.e. her own expected gain in a group in which she now has $(n-1)$ fellow helpers, plus her effect via the expelled sister, which is composed of a reduction in the alpha's output but a gain in nephews and nieces.

If $Be>Br$ or Bd, then a beta will attempt to expel another and be prepared to accept a cost, in terms of inclusive fitness, up to the value of the difference between the strategies.

SIMULATIONS

Using this model, it is possible to simulate the parameters that govern the behaviour of particular species. In some cases, field data are available for these simulations; in others, reasonable estimates can be made from what is known about the animals. In these examples, we shall examine the effects of changing the survival probabilities (P_α, P_i, P_e) and the value of the helper's marginal contribution, δf_n. As perfect outbreeding and fidelity have been assumed, there are two levels of relationship: mother:daughter and sister:sister; and grandparent: grandoffspring and aunt:nephew/niece (i.e. $r_{\alpha\gamma}=r_{\beta\gamma}=0\cdot5$; $r_{\alpha\beta\gamma}=r_{\beta\beta\gamma} = 0\cdot25$).

One purpose of the simulations that follow is to emphasize that minor changes in circumstances (i.e. in the values of P_α, P_i, P_e and δf_n) can cause shifts between the possible outcomes for an alpha and the nth beta of a group, and, for any given set of circumstances, a beta's choice of whether to bide or to disperse depends on the marginal contribution it makes to the alpha's productivity by staying (δf_n). To illustrate these general points, we will consider two classes of hypothetical groups.

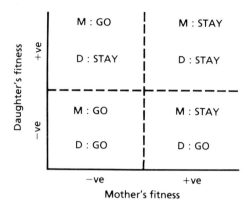

Fig. 4. Four possible outcomes in terms of the fitness of a mother and of an average daughter in a group with *N* betas, each outcome arising from the probable average pay-offs to mother and daughter of that daughter dispersing or staying as a non-breeder in the natal territory. Thus if the mother's fitness is positive, it is worthwhile to the mother if the daughter stays. Conversely if the mother's fitness is negative, the mother does better if the daughter leaves. The upper left quadrant corresponds to Emlen's (1982b) Type I conflict; the lower right one to his Type II conflict. For full explanation see text.

Case 1: a spatial group

We have shown how ecological circumstances, in the form of the pattern of resource dispersion, may facilitate the formation of a group whose members gain no fundamental benefit from their association. In this case, the betas, or secondary group members, would be 'squatters' in the primary occupants' territory. As such groups are based primarily on spatial rather than cooperative relationships, Macdonald (1983) termed them spatial groups. A possible case in point is the Eurasian badger, *Meles meles*, which lives in clans of about five adults, whose members have never been seen to cooperate (Neal 1977). It is possible that they may benefit from living as a group either through collective defence of the territory or because of the excavations of others — most members of a clan share a communal sett. However, as badgers appear to be contractionists (i.e. maintain the smallest economically defensible territories which encompass sufficient resources for reproduction), any such benefit does not seem to outweigh the advantages of occupying the smallest possible territory (Kruuk & Macdonald, 1985). Given that these cooperative benefits of sociality are based on conjecture, for illustrative purposes we will consider badger clans as spatial groups devoid of cooperation. Female badgers in a group are probably kin and, although opinions differ as to whether more than one female may breed in a clan, it is generally accepted that not all adult females in a clan do breed (Kruuk 1978, Evans, Macdonald & Cheeseman in press). In some populations, badgers are long lived, their groups are rather stable, and female dispersal is uncommon (Cheeseman *et al.* in press).

Figure 5 depicts the outcomes for a hypothetical badger sow and up to five daughters under conditions in which the survivorship of territory residents is high ($P_\alpha = P_i = 0.7$) and that of emigrants considerably lower ($P_e = 0.05 - 0.125$). As we are assuming that there is no direct advantage to sociality among these hypothetical badgers, the daughter's marginal contributions (δf_n) are zero. Therefore, the outcomes rest solely on the survival probabilities weighted by the coefficients of relatedness.

If an emigrant's chance of success is at the lower end of the range (it increases from Fig. 5a–d), betas in groups of up to four will maximize their fitness by remaining in the natal territory because their prospects of survival as dispersers are less than their prospects of breeding by inheriting the alpha position in the natal territory. The smaller the group, the greater this benefit (due to the greater probability of inheriting dominant status if the alpha should die). The mother's fitness will also benefit if up to four daughters remain, because they are more likely to produce grandoffspring for her by inheriting her territory than by emigrating to establish their own. However, both mother and daughters would favour the fifth daughter's dispersal in a group containing five betas.

Comparison of Figs 5a and 5b shows that, as the survival prospects of dispersers improve, so the number of daughters for which it is advantageous to stay at home diminishes. Four will remain when $P_e = 0.05$, two when $P_e = 0.075$, one when $P_e = 0.1$, and none when $P_e = 0.125$. Figure 5f shows that such dispersal would be voluntary. Even if the cost of evicting a sister is zero, it is nonetheless a better strategy for a beta to disperse than to force another to do so. However, if we assume that groups with more than two betas are directly disadvantageous to the alpha (perhaps through competition for food), the outcomes are changed as shown in Fig. 5e. Although dispersal is still an advantageous strategy, eviction is even more profitable and conflict may be expected to occur among the betas.

Case 2: a cooperative group

We will consider the red fox, *Vulpes vulpes*, which forms groups composed of a male and up to five related females. In some populations, reproduction is generally confined to the dominant vixen, and some non-breeding vixens tend and provision the cubs of the dominant (often their mother) (Macdonald 1979b). Figure 6(a–d) depicts the consequences of varying the probabilities and patterns of survival, the marginal benefits of cooperation and group size.

Number of offspring (N)

The graphs presented in Fig. 6(a–d) show that the strategies of both mother and daughters with respect to dispersal are affected by group size. For example, where $P_\alpha = 0.8$, $P_i = 0.6$, $P_e = 0.4$ and the law of diminishing returns applies to the marginal contributions of helpers, the outcome is markedly different for each case as the number of betas increases from 1 to 5 (Fig. 6a). In a group with one beta, both

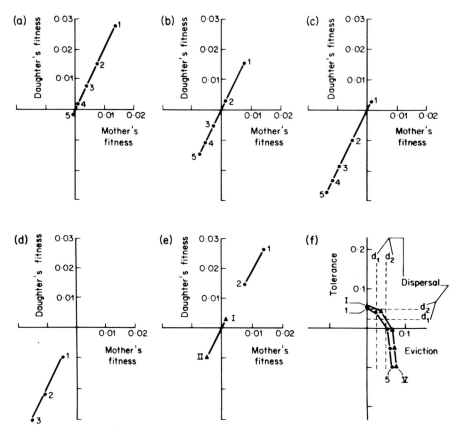

FIG. 5. Graphs (following the design of Fig. 4) of the outcomes in terms of mother's and her average daughter's fitness in groups of different sizes, depending on whether an nth, average daughter disperses or stays in the natal territory under various circumstances for a badger-like society in which offspring do not cooperate ($\delta f_n \leq 0$) with their parent. (a–d) illustrate cases where survivorship $P_\alpha = P_i = 0.7$ and Pe increases from (a) 0.05, (b) 0.075, (c) 0.1, and (d) 0.125, when $\delta f_n = 0$ throughout. In (e) the consequences of making larger groups directly disadvantageous to the primary occupants are illustrated (i.e. marginal contributions of third to fifth non-breeders are increasingly negative at −0.25, −0.5, −0.75). The outcomes for groups with two daughters are shown when $P = 0.05$ (●) and 0.1 (▲) Figure 5f (which has different axes to Figs 5a–e) illustrates the effect on the betas' fitness of following each of three strategies: dispersing (shown as $d1$ when $P_e = 0.05$, and $d2$ when $P_e = 0.1$) as opposed to either tolerating or evicting another beta (shown on the axes). Marginal values are the same as for Fig. 5e. For both values of P_e ($P_e = 0.05$ (●) and 0.1 (▲)) a single beta will do better by remaining at home, whereas two betas in a group maximize their fitness by dispersing if $P_e = 0.1$, but do better by tolerating each other if $P_e = 0.05$, whereas those in larger groups do best by evicting another sister, assuming the cost of doing so does not outweigh the potential benefit to be gained.

mother and daughter would benefit if the daughter remained; in a group with two betas, the benefit to either daughter of biding is very small, although the benefit to her mother would be substantial. With three betas, there is a conflict of interest because if one daughter disperses it is advantageous for her but the mother will benefit if all three remain. When there are four betas, a dispersing daughter

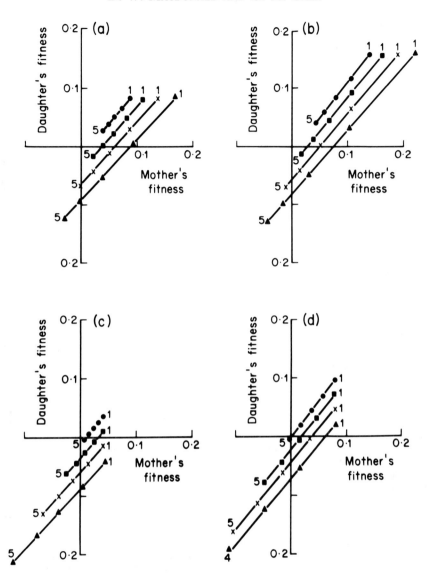

FIG. 6. Graphs (following the design of Fig. 4) of a fox-like system in which all or some daughters act as helpers. (a) illustrates the effects of changing mortality pressures under circumstances in which the survival of alpha occupants (P_α) is higher than that of the betas, and where the presence of up to five betas brings marginal benefits to the alpha's fitness, according to the law of diminishing returns ($\delta f_e - \delta f_4 = 1\cdot0, 0\cdot5, 0\cdot25, 0\cdot125, 0\cdot0625$). The survival probabilities, P_α, P_i, and P_e, are (i) $0\cdot5, 0\cdot3, 0\cdot1$ (●), (ii) $0\cdot6, 0\cdot4, 0\cdot2$ (■), (iii) $0\cdot7, 0\cdot5, 0\cdot3$ (x), (iv) $0\cdot8, 0\cdot6, 0\cdot4$ (▲). The situation is reversed in (b), in so far as survival of the alpha is less than that of betas which stay, the values of P_α, P_i, and P_e being (i) $0\cdot3, 0\cdot5, 0\cdot1$, (●), (ii) $0\cdot4, 0\cdot6, 0\cdot2$ (■), (iii) $0\cdot5, 0\cdot7, 0\cdot3$, (x), (iv) $0\cdot6, 0\cdot8, 0\cdot4$ (▲) whereas in (c) and (d) the schedule of marginal contributions by betas, i.e. 'helpers', is such that alphas with four or five daughters are directly disadvantaged by the presence of these daughters ($\delta f_o - \delta f_4 = 0\cdot5, 0\cdot25, 0\cdot0,$ $-0\cdot25, -0.5$). The mortality probabilities for Fig. 6c equal those of Fig. 6a, whereas those of Fig. 6d equal those of Fig. 6b.

would benefit greatly, whereas the mother is little affected if she stays or goes. Finally, in a group with five betas, the mother and the dispersing daughter will benefit from that daughter's dispersal.

The marginal contribution (δf_n)

Figure 6(a−d) illustrates the consequences of two different schedules of marginal contributions. In Fig. 6a and b owing to the law of diminishing returns, each successive beta is half as useful as the previous one. In Fig. 6c and d, only the first two betas are directly helpful; successive betas bring progressively greater disbenefits to the alpha. For any given pattern of mortality, the result of introducing these disbenefits is to reduce the advantages to both alpha and beta of a beta biding. For example, in the case where $P_\alpha=0.6$, $P_i=0.4$ and $P_e=0.2$, up to four betas benefit from remaining in the natal territory (Fig. 6a) and the alpha will also benefit. In contrast, when the less rewarding schedule of marginal contributions is considered for the same survival probabilities, it is advantageous only for one beta to remain (Fig. 6e). Unlike the hypothetical spatial group of badgers, however, the betas will not attempt to evict one another, even when larger groups result in negative marginal contributions.

The pattern of mortality (P_α, P_i, P_e)

There are three relevant measures of mortality, which are expressed as the survival probabilities of the matriarch (P_α), the helper (P_i) and the disperser (P_e). The values of all three, and their ratios, affect the outcomes. Figure 6a illustrates four mortality schedules, starting with low chances of survival ($P_\alpha=0.5$, $P_i=0.3$, $P_e=0.1$) and which improve by increments of 0.1 to each probability. The results of these systematic changes in all three survival probabilities differ for the five betas. In groups with one beta, the beta's situation is little affected by altering the mortality risks, whereas it becomes progressively more advantageous to the alpha for the beta to remain as values of P_α, P_i and P_e increase. In contrast, in a group with three betas, increasing the mortality makes little difference to the mother's view, but causes marked changes in the strategy of her daughters because they benefit most by staying when mortality is high, but by dispersing when it is lower.

In Fig. 6a and c, the matriarch's survival is higher than that of other group members. This is plausible in so far as her high status indicates greater strength, prowess, and access to resources. In Fig. 6b and d, the situation is reversed. This too is plausible, in so far that the matriarch endures the physiological strain of reproduction and is likely to be older; she may also be subject to more severe predation (e.g. in some areas, gamekeepers spoon powder emitting hydrocyanic gas into the dens of breeding foxes). A considerable change in outcome is wrought by this reversal in mortality pressure. For example, the differences in outcome between successive betas are minimized, and the advantage to the first daughter of remaining in the natal territory is greatly increased by the reversal of

P_α and P_i (Fig. 6a and b). Indeed, in cases where it was previously disadvantageous for any of the daughters to remain in the natal territory, it becomes advantageous for at least one of them to do so (Fig. 6c vs. d) because the helpers have a greater chance of outliving their mother and thereby inheriting the matriarchy.

The potential predictive value of simulations such as these can be illustrated by a real example. The silver-backed jackal, *Canis mesomelas*, forms groups of up to five adults in which only the matriarch breeds and non-breeders of both sexes help to rear the young (we will consider the sexes as one, see Discussion). Moehlman (1983) has shown a positive correlation between the number of helpers and the survival of pups. Figure 7 summarizes her data, from which we have read the approximate observed values of the marginal contributions of one to three helpers. We also know from Moehlman's studies that the survival of members of territorial groups is rather high (she tells us that it is reasonable to estimate $P_\alpha=0\cdot7$, $P_i=0\cdot6$). However, little is known of the mortality of dispersers (P_e), or the marginal values contributed by additional non-breeders beyond three. We have simulated the effects of varying P_e between 0·1 and 0·5 under four schedules of marginal contributions by betas. In all four schedules, the contributions of the first three helpers are read from Fig. 7. In every case, the fitness of both mother and daughter benefits when up to three offspring remain to act as helpers. In the four schedules, we make the hypothetical fourth and fifth non-breeders progress-ively less useful: (i) fourth and fifth marginal contributions estimated by extrapol-ation from Fig. 7; (ii) fourth and fifth marginal contributions zero; (iii) the fourth non-breeder contributes nothing, whereas the fifth is directly disadvantageous; (iv) both fourth and fifth non-breeders are directly disadvantageous. Table 1 summarizes the results of these simulations, which illustrate how the outcomes vary in terms of whether mother and/or daughter benefit by the daughter staying (S) or going (G). In groups with four non-breeders, even at high external survival (P_e) and irrespective of the schedule of marginal contributions, it is invariably to the mother's advantage for all her offspring to stay, and it is generally to the offspring's advantage too. On the other hand, in groups with five non-breeders of which the fourth and fifth are unhelpful, then it is to the advantage of both mother and offspring for the fifth non-breeder to disperse.

DISCUSSION

Why do some animals live in groups? Because it increases their fitness to do so. As with many seemingly straightforward answers to simple questions, this one glosses over a tangled web of factors whose interactions are poorly understood and rarely quantified. In this paper, we have explored two types of answer to this question. One type concerns ecological factors, the second sociological ones. Although we have addressed them separately, our point in presenting them side by side is to emphasize that neither set of factors needs stand alone. Although the particular, simple, models we have presented are probably poor shadows, or even distortions, of reality, they serve to make the following points: complete answers

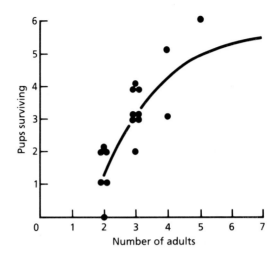

FIG. 7. Pup survival versus adult group size for silver-backed jackals (data from Moehlman 1983, curve fitted by eye).

to questions about the functions of societies require both ecological and behavioural information, and, for any given society, the answers are likely to differ between individuals.

Many societies involve cooperation between members, but our version of the resource dispersion hypothesis shows that cooperation is not a prerequisite for the evolution of groups. On the contrary, heterogeneity in the pattern of resource availability, often in the form of patchiness, can create conditions that promote group formation in the absence of any direct benefit from sociality *per se*. Indeed, such conditions, especially if augmented by the effects of kin selection, form a plausible template for the evolution of more complicated, cooperative societies. In this context Hamilton's Rule clarifies how natural selection can favour individuals promoting the well-being of others (Hamilton 1964, reviewed by Grafen 1984).

Reviewers such as Bertram (1978), Macdonald (1983) and Emlen (1984) list some of the benefits that may be associated with cooperation. First, it can bring access to the advantages of territorial life, in which survival might be expected to be higher. Gaston (1978) argues that seemingly altruistic behaviour can enable ingratiating individuals to 'buy' access to these advantages. (This might explain the assiduous babysitting by immigrants into groups of dwarf mongoose; Rood 1980.) Such advantages must be balanced against the risks involved in dispersing and the difficulties involved in acquiring a territory and a mate. Second, there is the possibility of inheriting the natal territory. Third, there are the advantages of reciprocal altruism (Axelrod & Hamilton 1981) and mutualism, the latter being well illustrated by the wolf pack whose individual members would not be able to tackle prey as large as an adult moose alone. Finally, for helpers at the den, there is the possibility of learning the skills of raising offspring through practice (Lawton & Guindon 1981, Macdonald & Moehlman 1982).

TABLE 1. The outcomes for mother and daughter silver-backed jackals, in terms of the best strategy for maximizing their fitness, of altering the schedule of marginal contributions by helpers (i.e. δf_n), and the probabilities of dispersing offspring surviving to become breeders elsewhere (i.e. P_e). Results are shown only for fourth and fifth offspring, because for all combinations of the values shown it is invariably mutually advantageous to both mother and offspring for up to three offspring to remain as non-breeding helpers in the natal territory. The values for marginal contributions made by successive helpers are extracted from data in Moehlman (1983). '?' indicates minimal difference between strategies. See text for full explanation.

δf_n	1·85; 1·0; 0·77; ·385; ·15		1·85; 1·0; 0·77; 0, 0		1·85; 1·0, 0·77; 0; −0·77		1·85; 1·0; 0·77; −0·77; −1·0	
P_e	4th	5th	4th	5th	4th	5th	4th	5th
0·1	S/S	S/S	S/S	S/S	S/S	S/S	S/S	?/G
0·2	S/S	S/S	S/S	S/S	S/S	S/S	S/S	G/G
0·3	S/S	S/S	S/S	S/G	S/S	S/S	S/S	G/G
0·4	S/S	S/G	S/?	S/G	S/S	?/G	S/G	G/G
0·5	S/G	S/G	S/G	?/G	S/G	G/G	S/G	G/G

There are also potential disbenefits to both cooperating partners. For example, with regard to non-breeding helpers, a mother must balance the increase in her own reproductive output against the decrease in that of a related helper, because she loses fitness in the loss of grandoffspring. The helper should act on the basis of a similar calculation, offsetting the extent to which her activities generate an increased number of sibs or nieces and nephews against the loss of her own potential offspring. It is relevant to both parties how many other helpers are staying because the law of diminishing returns is likely to apply. Our second model incorporates these factors and shows how the answer as to whether to join a group depends not only on the species' adaptation to sociality and cooperation, but also on individual circumstances. This point is illustrated by our model, despite its many simplifications — of which we will now discuss three.

First, we have considered only one sex of parent and offspring. In reality, members of each sex may make different marginal contributions and be subject to different mortality pressures, thereby complicating the estimation of fitness even within the parameters of our model. The consequences of this simplification for the particular simulations we chose may not be too great, because almost all subadult male foxes disperse irrespective of their circumstances, and there is no obvious sex-linked difference in marginal contribution, survival or tendency to become a helper among silver-backed jackals.

Second, in some societies there is undoubtedly positive feedback between group size and the survival chances of group members. For example, in a wolf pack where hunting efficiency increases (at least initially) with group size, P_α and P_i may also increase with group size. Such details could be readily accommodated within our model and, in the example of the wolf pack, would alter the outcomes in favour of group membership.

Third, and most important, our consideration of all betas as equal serves to highlight how, in reality, differences in social status would have a major effect on the outcome. Furthermore, some betas may be better helpers and some better dispersers. The consequences of a hierarchy among the betas would emerge when it came to deciding which of several of them was to assume alpha status. We have considered the average daughter's prospects in this regard, but an individual's assessment of its own position in the hierarchy would affect the best strategy for it to follow. Similarly, a hierarchy might come into play in deciding which beta might evict another. In this respect, our model has shown that changes in circumstances can affect the relative advantages of the three strategies open to betas — dispersal, tolerance or eviction. If our model describes reality adequately then, in some circumstances betas will be struggling to stay in a group, whereas in others they will be racing to leave.

ACKNOWLEDGEMENTS

We are grateful to Drs P.D. Moehlman and C.L. Cheeseman for estimating plausible values for our hypothetical jackal and badger groups, and to them and

Drs F. Ball, J.L. Brown, N.B. Davies, S. Emlen, and R.M. Sibly and Messrs M. Berdoy, P. Blackwell and J. da Silva for critical comments on the manuscript. We are especially grateful to our colleague Dr P. Hersteinsson for remarks that triggered our exploration of the sociological model, to Dr A. Grafan for checking our algebra, and to J. da Silva for steering us around the pitfalls associated with calculating fitness. DWM gratefully acknowledges the support of the Nuffield Foundation.

REFERENCES

Axelrod, R. & Hamilton, W.D. (1981). The evolution of cooperation. *Science*, **211**, 1390–6.

Bertram, B.C.R. (1978). Living in groups: predators and prey. *Behavioural Ecology: an Evolutionary Approach* (Ed. by J.R. Krebs & N.B. Davies), pp. 64–96. Blackwell Scientific Publications, Oxford.

Berg, W.E. & Chesness, R.A. (1978). Ecology of coyotes in northern Minnesota. *Coyotes: Biology, Behavior and Management* (Ed. by M. Bekoff), pp. 229–47. Academic Press, New York.

Bowen, W.D. (1981). Variation in coyote social organisation: the influence of prey size. *Canadian Journal of Zoology*, **59**, 639–52.

Bradbury, J.W. & Vehrencamp, S.L. (1976). Social organisation and foraging in Emballonurid bats. 11. A model for the determination of group size. *Behavioural Ecology and Sociobiology*, **1**, 383–404.

Brown, J.L. (1964). The evolution of diversity in avian territorial systems. *Wilson Bulletin*, **76**, 160–9.

Brown, J.L. (1978). Avian communal breeding systems. *Annual Review of Ecology and Systematics*, **9**, 123–55.

Brown, J.L. (1982). Optimal group size in territorial animals. *Journal of Theoretical Biology*, **95**, 793–810.

Brown, J.L. (1987). Helping and communal breeding in birds: ecology and evolution. Princeton University Press, Princeton, New Jersey.

Bygott, J.D., Bertram, B.C.R. & Hanby, J.P. (1979). Male lions in large coalitions gain reproductive advantage. *Nature*, **282**, 839–41.

Camenzind, F.J. (1978). Behavioral ecology of coyotes on the National Elk Refuge, Jackson, Wyoming. *Coyotes: Biology, Behavior and Management* (Ed. by M. Bekoff), pp. 267–94. Academic Press, New York.

Carr, G.M. & Macdonald, D.W. (1986). The sociality of solitary foragers: a model based on resource dispersion. *Animal Behaviour*, **34**, 1540–9.

Charnov, E.L. (1981). Kin selection and helpers at the nest: effects of paternity and biparental care. *Animal Behaviour*, **29**, 631–2.

Cheeseman, C.L., Cresswell, W.J., Harris, S. & Mallinson, P.J. (in press). A comparison of dispersal and other movements in two badger (*Meles meles*) populations. *Mammal Review*.

Crook, J.H. (1964). The evolution of social organization and visual communication in weaverbirds (Ploceinae). *Behaviour (Suppl.)*, **10**, 1–178.

Davies, N.B. (1978). Ecological questions about territorial behaviour. *Behavioural Ecology: an Evolutionary Approach* (Ed. by J.R. Krebs & N.B. Davies), pp. 317–50 Blackwell Scientific Publications, Oxford.

Davies, N.B. & Houston, A.J. (1981). Owners and satellites: the economics of territory defence in the pied wagtail, *Motacilla alba*. *Journal of Animal Ecology*, **119**, 29–39.

Emlen, S.T. (1982a). The evolution of helping, I. An ecological restraints model. *American Naturalist*, **119**, 29–39.

Emlen, S.T. (1982b). The evolution of helping, II. The role of behavioural conflict. *American Naturalist*, **119**, 40–53.

Emlen, S.T. (1984). Cooperative breeding in birds and mammals. *Behavioural Ecology: An Evolutionary Approach*, 2nd edn. (Ed. by J.R. Krebs & N.B. Davies), pp. 305–39. Blackwell Scientific Publications, Oxford.

Evans, P.G.H., Macdonald, D.W. & Cheeseman, C.L. (in press). The social organization of Eurasian badgers: genetic evidence. *Journal of Zoology (London)*.

Frame, L.H., Malcolm, J.L., Frame, G.W., & Lawick, H., van (1980). Social organisation of African wild dogs (*Lycaon pictus*) on the Serengeti Plains, Tanzania 1967–1978. *Zeitschrift für Tierpsychologie*, 50, 225–49.

Gaston, A.J. (1978). The evolution of group territorial behavior and cooperative breeding. *American Naturalist*, 112, 1091–1100.

Grafen, A. (1982). How not to measure inclusive fitness. *Nature*, 298, 425–6.

Grafen, A. (1984). Natural selection, kin selection and group selection. *Behavioural Ecology: an Evolutionary Approach* (2nd edn.) (Ed. By J.R. Krebs & N.B. Davies), pp. 62–84. Blackwell Scientific Publications, Oxford.

Hamilton, W.D. (1964). The genetical evolution of social behaviour, I, II. *Journal of Theoretical Biology*, 7, 1–52.

Hersteinsson, P. & Macdonald, D.W. (1982). Some comparisons between red and Arctic foxes, *Vulpes vulpes* and *Alopex lagopus*, as revealed by radio tracking. *Proceedings of the Symposia of the Zoological Society of London*, 49, 259–88.

Jarman, P.J. (1974). The social organisation of antelope in relation to their ecology. *Behaviour*, 48, 215–67.

Johnsing, A.J.T. (1984). Dholes. *The Encyclopaedia of Mammals, I* (Ed. by D.W. Macdonald), pp. 80–1. Allen & Unwin, London.

Kaikusalo, A. (1971). Naalin pesimisesta Luoteis — Enontekiossa. *Suomen Riista*, 23, 7–16.

Kruuk, H. (1975). Functional aspects of social hunting by carnivores. *Function and Evolution in Behaviour* (Ed. by G. Baerends, C. Beer & A. Manning), pp. 119–41. Clarendon Press, Oxford.

Kruuk, H. (1978). Spatial organization and territorial behaviour of the European badger, *Meles meles*. *Journal of Zoology (London)*, 184, 1–19.

Kruuk, H. & Macdonald, D.W. (1985). Group territories of carnivores: empires and enclaves. *Behavioural Ecology: Ecological Consequences of Adaptive Behaviour* (Ed. by R.M. Sibly & R.H. Smith), pp. 521–36. Blackwell Scientific Publications, Oxford.

Kruuk, H. & Parish, T. (1982). Factors affecting population density, group size and territory size of the European badger, *Meles meles*. *Journal of Zoology (London)*, 196, 31–9.

Lawick, H. van (1973). *Innocent killers*. Collins, London.

Lawton, M.F. & Guindon, C.F. (1981). Flock composition, breeding success and learning in the brown jay. *Condor*, 83, 27–33.

Lindström, E. (1986). Territory inheritance in the evolution of group-living in carnivores. *Animal Behaviour*, 34, 1540–9.

Macdonald, D.W. (1979a). The flexible social system of the golden jackal, *Canis aureus*. *Behavioural Ecology and Sociobiology*, 5, 17–38.

Macdonald, D.W. (1979b). Helpers in fox society. *Nature*, 282, 69–71.

Macdonald, D.W. (1981). Resource dispersion and the social organisation of the red fox, *Vulpes vulpes*. Proceedings of the Worldwide Furbearer Conference, Vol. 2. (Ed. by J.A. Chapman & D. Pursley), pp. 918–49. University of Maryland Press, Maryland.

Macdonald, D.W. (1983). The ecology of carnivore social behaviour. *Nature*, 301, 379–84.

Macdonald, D.W., Apps, P.J., Carr, G.M., & Kerby, G. (1987). Social dynamics, nursing coalitions and infanticide among farm cats, *Felis catus*. *Advances in Ethology*, 28, 1–66.

Macdonald, D.W. & Moehlman, P.D. (1982). Cooperation, altruism and restraint in the reproduction of carnivores. *Perspectives in Ethology*, 5 (Ed. by P.P.G. Bateson & P. Klopfer), pp. 433–67. Plenum Press, New York.

Malcolm, J.R. (1979). *Social organization and communal rearing of pups in African wild dogs (Lycaon pictus)*. Ph.D. thesis, Harvard University, Massachusetts.

Mills, M.G.M. (1982). Factors affecting group size and territory size in the Brown Hyaena, *Hyaena brunnea*, in the southern Kalahari. *Journal of Zoology (London)*, 198, 39–51.

Moehlman, P.D. (1979). Jackal helpers and pup survival. *Nature (London)*, 277, 382–3.

Moehlman, P.D. (1983). Socioecology of the silverbacked and golden jackals (*Canis mesomelas* and *Canis aureus*). *Advances in the study of Mammalian Behaviour* (Ed. by J.F. Eisenberg & D.G. Leiman), pp. 423–53. American Society of Mammalogists, Pittsburgh, Pennsylvania.

Neal, E.G. (1977). *Badgers.* Blandford Press, Poole, Dorset.

Oosenberg, S.M. & Carbyn, L.N. (1982). Winter predation on bison and activity patterns of a wolf pack in Wood Buffalo National Park. *Wolves of the World: Perspectives of Ecology, Behavior and Conservation* (Ed. by F.H. Harrington & P.C. Paquet), pp. 43–53. Noyes, New Jersey.

Orians, G.H. (1961). The ecology of blackbird (*Agelaius*) social systems. *Ecological Monographs,* **31,** 285–312.

Orsdol, K. van (1981). *Lion predation in Rwenzori National Park, Uganda.* Ph.D thesis, University of Cambridge.

Ozoga, J.J. & Harger, E.M. (1966). Winter activities and feeding habits of northern Michigan coyotes. *Journal of Wildlife Management,* **30,** 809–18.

Packer, C. & Pusey, A.E. (1982). Cooperation and competition within coalitions of male lions: kin selection or game theory? *Nature,* **296,** 740–2.

Rasa, O.A.E. (1984). A case of invalid care in wild dwarf mongooses. *Zeitschrift für Tierpsychologie,* **62,** 181–268.

Rasa, O.A.E. (1985). *Mongoose watch.* John Murray, London.

Riche, A. (1981). *Behavior and Ecology of African Wild Dogs, Lycaon pictus, in Kruger National Park.* Ph.d thesis, Yale University, Connecticut.

Rood, J.P. (1980). Mating relationships and breeding suppression in the dwarf mongoose. *Animal Behaviour,* **28,** 143–50.

Rood, J.P. (1983). Banded mongoose rescues pack member from eagle. *Animal Behaviour,* **31,** 1261–2.

Saunders, D.H., Eng, R.J. & Murph, A.F. (1985). *Statistics: a Fresh Approach,* 3rd edn. McGraw Hill, New York.

Schantz, T. von (1984). 'Non-breeders' in the red fox, *Vulpes vulpes:* a case of resource surplus. *Oikos,* **42,** 59–65.

Scott, B.M.V. & Shackleton, D.M. (1982). A preliminary study of the social organization of the Vancouver Island wolf. In: *Wolves: Perspectives of Behavior, Ecology and Conservation.* (Ed. by F.H. Harrington & P.C. Paquet), pp. 12–25. Noyes, New Jersey.

Waser, P.M. (1981). Sociality or territorial defence? The influence of resource renewal. *Behavioural Ecology and Sociobiology,* **8,** 231–7.

Effects of resource productivity, patchiness and predictability on mating and dispersal strategies

R.A. POWELL

Department of Zoology, North Carolina State University, Raleigh, North Carolina 27695–7617, USA

SUMMARY

1 Animals are assumed to adopt mating strategies that maximize expected lifetime reproductive output, which can be modelled by dividing life histories into three stages: dependency on parents, dispersal and reproduction.

2 Resource productivity affects reproductive output but animals also respond to variances in resource productivity. Because reproductive output responds in a diminishing returns fashion to increasing resource productivity, resource variance can be modelled as a discount on mean resource productivity.

3 The resource productivity−variance model is developed for mammals, incorporating dependence of reproductive output on resource productivity and on variance in that productivity in space (resource patchiness) and in time (resource predictability). For given resource characteristics, the model can be used to predict optimal mating and dispersal strategies. Examples are given for the eight extreme cases of the model.

4 The model correctly predicts the mating systems for several canids, several herbivores and an aborginal human culture.

5 For species with two limiting resources, the model is more complex but still applicable. It is less precise for animals in which primary parental care need not be the responsibility of the mother. The model is applicable at both the individual and species levels and can predict alternate mating strategies. However, the model is appropriate only for members of species for whom resources place the major limits on population size and social organization. The model does not discount the impact of variance by its distance in the future.

INTRODUCTION

In this paper I present a model for lifetime reproductive output of animals and use this model to predict optimal mating and dispersal strategies under different ecological conditions. This predictive approach has the strength of being applicable to a wide variety of conditions and it is falsifiable. I shall first give background to the approach and the model, then present the model with predictions for different conditions of resource productivity, patchiness and predictability, then present examples of mammalian mating systems that appear consistent with predictions of the model.

I shall not review mating strategies and correlate strategies of different mammals with different sets of ecological conditions. Emlen & Oring (1977) have derived a taxonomy of mating systems and described the ecological conditions under which each has been documented to occur. They graphed the probabilities for polygynous mating systems under different qualitative resource conditions but they did not state how to quantify those conditions nor did they present a model capable of making more than limited and general predictions. Other reviews of mating systems (Wittenberger & Tilson 1980; Wittenberger 1981; Rubenstein & Wrangham 1986) have also largely taken descriptive approaches that have generated intriguing hypotheses. Many models have been used to explain specific mating strategies (e.g. Orians 1969; Wittenberger 1980). However, I hope that an approach more structured than Emlen's and Oring's but still general will allow more precise predictions and more general applications and will generate testable hypotheses, all coming from a central structure.

Animals are assumed to adopt mating strategies that maximize expected lifetime reproductive output. Woolfenden & Fitzpatrick (1984) developed a model conceptualizing the ecological constraint hypothesis for cooperative breeding systems (Emlen 1982). They compared predicted lifetime reproduction for Florida scrub jays (*Aphelocoma coerulescens*) under different mating strategies and ecological conditions to find the conditions under which cooperative breeding should occur. Generalizing from this work provides a robust approach to understanding mating and dispersal strategies in animals in general, and in mammals in particular.

Mammal life histories can be modelled as having three life stages with major effects on mortality and reproduction: the periods of dependence on parental care, dispersal, and reproduction (including bringing offspring to age of weaning).

The period of parental care after weaning is quite short for some mammals (for example, some mustelids: Powell 1982) but for others it is quite long (wolves, *Canis lupus*: Mech 1970; pine voles, *Microtus pinetorum*: Smolen 1981: yellow baboons, *Papio cynocephalus*: Altmann 1980).

The transition from parental care to becoming a breeder, the period of dispersal, comes at various relative ages in mammals because duration of parental care varies. The duration of dispersal varies among mammals as well. For many rodents, reproductive life begins promptly after leaving the maternal nest (prairie voles, *Microtus ochrogaster*: Carter *et al.* 1980). For other rodents (marmots, *Marmota* spp.: Armitage 1981, 1986: Armitage & Downhower 1974) and for many large mammals (black bears, *Ursus americanus*: Rogers 1987), there may be years between the termination of parental care and becoming an established breeder. For some mammals this period may be difficult to define because reproductive life begins as parental care wanes (red deer, *Cervus elaphus*: Clutton-Brock, Guinness & Albon 1982).

The final important period, that of reproductive life, is as variable as the other life stages. For mammals, this period is most clearly defined by first parturition because first oestrus is not unequivically independent of other life history stages. In some deer, parental care can extend beyond first oestrus (red deer: Clutton-Brock, Guinness & Albon 1982), and female stoats (*Mustela erminea*) have their first oestrus when only 6 weeks old, which may be before they open their eyes (DonCarlos, Peterson & Tilson 1986).

Adopting terminology modified from that used by Woolfenden & Fitzpatrick (1984), a newborn individual's expected lifetime reproductive output ('fitness' in an imprecise sense), W, can be expressed as

$$W = L_i \cdot D_d \cdot R, \tag{1}$$

where L_i is the probability of living to dispersal age i, D_d is the probability of living through the dispersal period to age d and becoming an established breeder, and R is the expected production of weaned offspring once the mammal has become an established breeder at age d. R is not R_o, the term for the expected lifetime reproductive output in demography. R_o sums over $l_x \cdot m_x$ for all ages x, where l_x is the probability of living to age x and m_x is the expected number of female offspring a female will produce when age x. R sums over $(l_x/l_d) \cdot M_x$, which assumes that an individual has already lived to breeding age d and has become an established breeder ($x > d$: M_x is the expected number of female offspring an established breeding female will produce when age x).

For a stable population, this approach is identical to using Fisher's (1930) reproductive value (v_x) to analyse expected reproductive output. Fisher's v_x can be written

$$v_x = \sum_x \frac{l_a}{l_x} \cdot m_x \cdot \frac{e^{-rx}}{e^{-rd}},$$

where r is the population's intrinsic rate of increase and e is the root to natural logarithms. For $x=0$, age of dispersal i, and age of first reproduction equal to d this can be rearranged to

$$v_o = l_i \cdot \frac{l_d}{l_i} \cdot \sum_{a=d}^{\infty} \frac{l_a}{l_d} \cdot m_d \cdot \frac{e^{-rd}}{e^{-rd}}.$$

This equation differs from equation (1) by including the expected reproduction of all females at any age (m_x) and not just those who become established breeders (M_x) and by including a term for contribution to future population growth through timing of reproduction (the cost of waiting, e^{-ra}/e^{-rd}). For a stable population, $r=0$ and the cost of waiting becomes 1 and therefore has no effect on v_o. Thus for individuals in a stable population v_o differs from equation (1) only by using m_x instead of M_x. For individuals in a population that is not stable, using a reproductive value approach requires knowledge of r, which is usually impossible to estimate. The approach used here in equation (1) is not hampered with this problem but therefore ignores the possible importance of delayed reproduction on future genetic contributions (the cost of waiting).

RESOURCE PRODUCTIVITY AND VARIANCE

Productivity of resources affects reproduction indirectly through animals' abilities to acquire requisite resources for survival and directly through animals' abilities to acquire additional resources for reproduction. Resource productivity also affects mating and dispersal strategies, either directly or indirectly (e.g. Shank 1986). Recent theoretical and empirical work shows that foraging animals should and do respond to variance in resource productivity (Real 1980; Real, Ott & Silverfine 1982; Lima, Valone & Caraco 1985). Variance in resource productivity should also have affected the evolution of mating and dispersal strategies because that variance affects survivorship and reproduction, which are important in determining these strategies.

Variance can be incorporated into models used to investigate mating strategies by realizing that a function maximized by natural selection, such as survivorship or reproductive output, can be approximated by a Taylor series around its mean value (Real 1980). Thus if $F(X)$ is such a function of available productivity, X, of a resource, expanding it around its mean, μ, yields

$$F(X) = F(\mu) + F'(\mu)(X-\mu) + (1/2)F''(\mu)(X-\mu)^2 + \ldots$$

Applying expected value operators to this expression yields

$$E\{F(X)\} = F(\mu) + F'(\mu)E(X-\mu) + (1/2)F''(\mu)E([X-\mu]^2) + \ldots$$

Since $E(X-\mu)=0$, the first order term disappears. $E([X-\mu]^2)$ is by definition the variance of $F(X)$. And with reasonable assumptions the function's origin can be translated so that $F(\mu)=\mu$. Our expression now becomes

$$E\{F(X)\} = \mu + (1/2)F''(\mu)\sigma^2 + \ldots$$

Empirical work shows that animals tend to avoid variance in resources (Caraco 1981, 1982; Real 1981; Real, Ott & Silverfine 1982). In addition, because reproductive output is not unlimited, reproductive output must respond to increased resource productivity in a diminishing returns fashion. Mammal species have maximum litter sizes and characteristic numbers of teats, which limit reproduction. Thus the second order term is expected to be negative and can be expressed as a constant, $-A$. Finally, behaviour of higher order derivatives are unknown and diminish in contribution to the expression, therefore the expression may be truncated after the second order term (Real 1980). Our expression is now

$$E\{F(X)\} = \mu - A\sigma^2.$$

This means that animals should respond to variance in resources as though variance discounts the mean of resource availability.

Resources can vary in space and in time. There has been considerable theoretical and empirical research on the effects of habitat patchiness (variance in space) on foraging (Pyke, Pulliam & Charnov 1977; Pyke 1984). This patchiness can be fine-grained or coarse-grained. I assume that variance among good habitat patches is small, and that breeding habitat for an individual is a habitat patch within which there is little or no spatial variance (or, if in a fine grained environment, breeding habitat is an area with many good, small patches). The impact of spatial variance comes from variance due to good and bad patches, not from variance within or among good patches. Seen another way, there is no variance in space for breeding animals because they stay within known good patches.

Resource availability also varies on daily, seasonal and annual bases. How predictable resource availability is on each of these scales can be measured as variance in time, and this affects how animals can anticipate resources. Assuming that variance is resource availability in space (patchiness) and time (predictability) do not covary, then variance can be split into its two additive parts, σ_s^2 and σ_t^2, variance in space and time. Now our expression becomes

$$E\{F(X)\} = \mu - A_s\sigma_s^2 - A_t\sigma_t^2.$$

The expected value of our function is its mean discounted by its variances.

This means that in the sense of 'evolutionary tactics', an animal is unable to 'anticipate' gaining the full effect of mean resource productivity. Therefore the animal must 'anticipate' less. We can use the above expression to model the effects of resource productivity, patchiness and predictability on expected fitness.

MODELLING FITNESS AND RESOURCE PRODUCTIVITY AND VARIANCE

Assume that an animal's survivorship and reproduction are limited by a single resource that has mean productivity μ and variances in space and time σ_s^2 and σ_t^2. (Relaxing this assumption of a single limiting resource does not affect the following approach but makes the equations more complex.) Productivity and variance of this resource, once normalized to the rate at which the members of a species can consume or otherwise use up the resource, can be related to all three of the major periods of the animal's life through equation (1)

$$W = L_i \cdot D_d \cdot R.$$

Survival to dispersal

L_i, now an individual's anticipated probability of living to age of dispersal, i, should be a function of mean resource productivity within its parents' breeding patch, μ_i, the variances in that productivity and age i:

$$L_i = f_l(\mu_i, \sigma_s^2, \sigma_t^2, i).$$

A logical form for this function is to discount the demographic l_i through resource mean and variances:

$$L_i = \frac{l_i c_i (\mu_i - s_i \sigma_s^2 - t_i \sigma_t^2)}{l_i + c_i \mu},$$

where c_i is a coefficient of proportionality that converts productivity into an effect on survivorship, and where s_i and t_i are coefficients that convert patchiness (spatial variance, σ_s^2) and predictability (temporal variance, σ_t^2) into their effects through productivity on survivorship. In this equation, l_i is survivorship to age i given no limitation by the limiting resource; it is thus an underlying survivorship as affected by other factors (predation, weather, etc.). l_i will thus have its own variance but animals should respond negatively to that variance as they do to variance in resource productivity. Thus the variance of l_i will not be considered further. When σ_s^2 and σ_t^2 equal 0, this function approaches l_i for very large μ_i and approaches 0 for very small μ_i. Thus when resource productivity is very large, survivorship to age i is as large as it can be, given other factors. When resource productivity is very low, L_i is far below l_i and may approach 0. Variances in resource productivity decrease anticipated survivorship below that expected for any given level of productivity.

σ_s^2 should have no effect on L_i because a mammal that has not yet dispersed is still within the patch used by its parents. σ_t^2, however, should affect survival to age i when variation in resource productivity varies from seasonal means. The important scale for this variance will depend on a mammal species' life history and life span. Daily to monthly variation will be more important to mice than to elephants. Thus

$$L_i = \frac{l_i c_i (\mu_i - t_i \sigma_t^2)}{l_i + c_i \mu}. \tag{2}$$

Dispersal and becoming a breeder

D_d, now an individual's anticipated probability of becoming a breeder at age d once having dispersed at age i, is the product of independent anticipated probabilities D, of living from age i to age d, and B, of becoming a breeder at age d. D should be a function of resource mean productivity (μ) within dispersal range, variances in that productivity, and ages i and d:

$$D = f_d(\mu, \sigma_s^2, \sigma_t^2, i, d).$$

The anticipated probability of surviving from dispersal age i through dispersal to age d should logically take a similar form to that for L_i:

$$D = \frac{(l_d/l_i) c_d (\mu - s_d \sigma_s^2 - t_d \sigma_t^2}{(l_d/l_i) + c_d^\mu}. \tag{3}$$

The amount of patchiness (σ_s^2) an animal experiences will vary with its dispersal strategy. For those animals that stay at home and wait for a nearby breeding space to be vacant or wait to take over a parent's breeding position, patchiness experienced will be 0 and $\mu = \mu_i$ (however, for these animals age i will likely be old). Those animals that disperse over large distances will experience much of the patchiness that exists and may have difficulty finding a patch in which to become an established breeder. σ_t^2 should affect both delayed and early dispersers because variance of resource productivity over time will affect survival probability.

The anticipated probability of becoming a breeder, B, will depend on the number of and turnover rate of breeding spaces. The number of breeding spaces should be positively affected by resource productivity but discounted by variance in space. Turnover rate, however, should be positively affected by variance in time, as this will negatively affect established breeders, increasing turnover rate. Thus, for an underlying probability, b, of finding a breeding space

$$B = \frac{b \cdot c_b(\mu - s_b\sigma_s^2 + t_b\sigma_t^2)}{b + c_b\mu}. \tag{4}$$

By multiplying equations (3) and (4) we get $D_d = D \cdot B$.

Reproduction

R, the anticipated lifetime production of weaned offspring once an animal has become an established breeder, should be different for males and females and should thus be a function of mean resource productivity of the breeding patch chosen, μ_b, variances in that productivity, age d and older ages x, and whether a male or helpers help raise young:

$$R = f_r(\mu_b, \sigma_s^2, \sigma_t^2, sex, d, x, \delta, h),$$

where δ is the number of adult males helping a female and h is the number of older offspring helpers. For each sex we can start with the underlying expression for anticipated reproductive output from age d through maximum age θ:

$$R = \sum_{x=d}^{\theta} S_x \cdot R_x, \tag{5}$$

where S_x is anticipated survivorship and R_x is anticipated reproduction.

Anticipated survivorship for a female (S_{xf}) will be affected by mean resource productivity and by variance in resource productivity in time but not in space. Survivorship can be written:

$$S_{xf} = \frac{(l_x/l_d)c_x(\mu_b - t_t\sigma_t^2)}{(l_x/l_d) + c_x\mu}. \tag{6a}$$

Anticipated reproductive output of a female at age x will depend on the productivity of her chosen breeding patch, μ_b. It should not depend on resource variance in space, as for survivorship, but should depend on resource variance in time. Female reproductive output should also depend on numbers of helpers. Thus female reproductive output can be written:

$$R_{xf} = \frac{M_x c_m(\mu_b - t_m\sigma_t^2 + r_m\delta_x + r_h h_x)}{M_x + c_m\mu}, \tag{7a}$$

where r_m and r_h are coefficients relating numbers of adult male, δ_x, or older offspring helpers, h_x, to increase in a female's reproductive output when she is age x.

Anticipated survivorship of a male will be affected by resource productivity and by variance of that productivity in time. Whether S_{xm} is affected by variance in space will depend on whether he is monogamous ($\mu=\mu_b$, $\sigma_s^2=0$) or polygynous ($\mu=\mu$, $\sigma_s^2>0$). S_{xm} will be the same as S_{xf} for monogamous males. For polygynous males

$$S_{xm} = \frac{(l_x/l_d)c_x(\mu - s_t\sigma_s^2 - t_t\sigma_t^2)}{(l_x/l_d) + c_x\mu}. \qquad (6b)$$

Anticipated reproductive output of a male at age x will depend on the effective numbers of females with whom he breeds at age x, N_x. This will be the number of females with whom he breeds times his probability of fathering offspring from those females. Thus if a male breeds with many females but so do many other males, N_x may be very low. Male reproductive output can be written

$$R_{xm} = R_{xf} \cdot N_x. \qquad (7b)$$

Anticipated reproductive outputs of a female (R_f) and of a male (R_m) are obtained by multiplying equations (6a) by (7a) or (6b) by (7b).

Now the complete equations for anticipated W for a male or a female can be written by inserting the appropriate equations for L_i, D_d and R into equation (1):

$$W = L_i \cdot D_d \cdot R.$$

Note that this resource productivity−variance model for mating and dispersal strategies allows interaction between strategy decisions and demographic variables. Decisions to be monogamous or to delay dispersal, for example, are represented in the model by eliminating or introducing variances or by changing ages for actions. Thus the model can deal with interactions between survival probabilities and decisions on mating and dispersal strategies.

To predict mating strategies, equations for R_f and R_m (equation (5) for each sex) must be inspected to determine for each sex which strategies yield highest reproductive output under different ecological conditions (Fig. 1). When help from one or more adult males and from one or more weaned offspring will increase a female's ability to raise offspring to weaning ($r_m>0$ and $r_h>0$), there should be selection on females to accept such help. Whether an adult male (or more than one male) is willing to help a female with whom he has bred will depend on whether his reproductive output will be greater if he helps the female or if he does not and breeds with other females. To decide this, R_m must be compared for $\mu=\mu_b$, $\delta=1$ and $N=1$ with R_m for $\mu=\mu$, $\delta=0$ and $N>1$. If R_f is so low with $\delta=0$ that no matter how great N, R_m is greater for $\delta=1$ and $N=1$, then there will be selection for a male to help his mate raise her young, monogamy. If R_f is not too small for $\delta=0$ and if the male has higher anticipated reproductive output by breeding effectively with many females, then there will be selection for polygyny, even if the females would individually have higher anticipated reproductive outputs when monogamous. If a female has higher fitness by sharing with

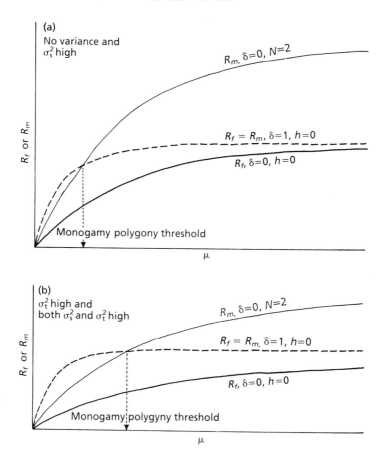

FIG. 1. Expected lifetime reproductive output of breeding adults that have acquired breeding spaces. The curve labelled $R_f = R_m$ represents monogamy with the male helping the female raise the young. The curves labelled R_f represents a female who must raise young on her own under the same conditions. The curve labelled R_m represents a male who effectively breeds with two females. (a) Expected reproductive output when there is no variance in resource productivity or only variance in space. (b) Expected reproductive output when there is variance in time only or variance in both time and space. Curves were generated from equation 5.

another female a breeding space defended by a male than by settling in a poor quality site, then there will be polygyny.

After the strategies of parents have been determined, dispersal strategies of offspring can be determined from the completed equation (1) (Fig. 2). The major offspring decision is whether to disperse when young (i small) or to delay dispersal (i large) and possibly help parents raise younger siblings ($h \geqslant 1$). If the probability of surviving through dispersal is (l_d/l_i large) and the probability of becoming established as a breeder (b) is large, then there is selection for early dispersal. However, if the probability of surviving dispersal is small (l_d/l_i small) but survival

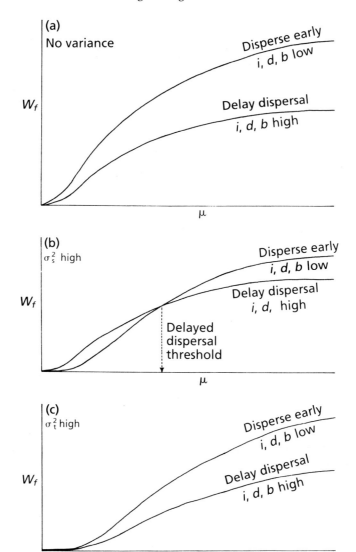

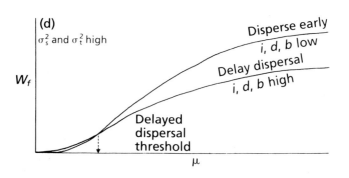

within the parents' breeding patch is high (l_i large even for large i), and if the probability of becoming an established breeder (b) is low (because of habitat patchiness), then there is selection for delaying dispersal to wait for a vacant breeding patch nearby. Under these conditions, offspring increase their chances of surviving until a breeding site becomes available and also minimize time spent in dispersal ($d-i$ is small). If delayed dispersal overlaps with the parents' next reproductive effort, there should be selection for cooperative breeding whether offspring contribute to their parents' fitness or not.

Productivity, patchiness and predictability of resources all vary on continua. Examination of the eight extreme cases shows how predicted mating and dispersal strategies vary with these resource characteristics.

PREDICTED MALE AND FEMALE MATING STRATEGIES UNDER DIFFERENT ENVIRONMENTAL CONDITIONS

Case 1 (Figs 1a, 2a)

Productivity low (μ low).
Patchiness low (σ_s^2 low).
Predictability high (σ_t^2 low).
Predicted mating strategy: Monogamy.
Predicted dispersal strategy: Early dispersal.
Breeding pair: This case has no variance and thus leads to the mating strategy that would be predicted for low resource productivity were resource productivity not discounted by its variances. Female anticipated reproductive output, R_f, is low because μ and σ_s^2 are low, therefore μ_b must be low. As long as a monogamous male is able to make a significant contribution to raising his offspring (r_m not small), there should be selection on males for $\delta=1$, monogamy. The monogamy–polygyny threshold (Fig. 1a, Fig. 3) will depend on how many females a male can inseminate and his probabilities of actually fathering offspring from those females. In Fig. 1a, N_x, effective number of females bred, is set at 2.
Offspring: From the entire equation, W is low because μ is low. Because no variables in equation (1) are discounted (no variances in productivity), offspring should not delay dispersal (Fig. 2a, Fig. 4).

FACING PAGE

FIG. 2. Expected lifetime fitness for animals at age of weaning. The curves labelled 'disperse early' represent animals that disperse at a young age; the curves labelled 'delay dispersal' represent animals that delay dispersal and stay in their parents' home ranges, thus being subjected to less variance in space. (a) Expected lifetime fitness when there is no variance in resource productivity. (b) Fitness when resource productivity is patchily distributed, there is variance in space only. (c) Fitness when resource productivity in unpredictable, there is variance in time only. (d) Fitness when resource productivity is both patchily distributed and unpredictable. Curves were generated from equation 1.

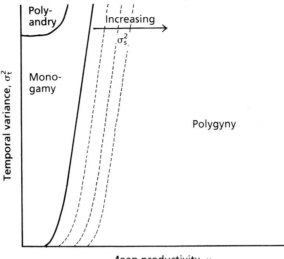

FIG. 3. Male mating strategies as they relate to productivity of the limiting resource and its variance in space (σ_s^2) and time (σ_t^2). At high resource productivity, males should be polygynous. At low resource productivity they should be monogamous, unless temporal variance in that productivity is extremely high, in which case they should be willing to maintain a long-term bond with a polyandrous female. High spatial or temporal variance in resource productivity should cause males to be monogamous at higher levels of productivity than when there is no such variance.

Case 2 (Figs 1a, 2b)

Productivity low (μ low).
Patchiness high (σ_s^2 high).
Predictability high (σ_t^2 low).
Predicted mating strategy: Monogamy, cooperative breeding.
Predicted dispersal strategy: Delayed dispersal (cooperative breeding).
Breeding pair: Predictions are the same as for Case 1 because only σ_s^2 is high and
   R_f is only affected by σ_t^2 (Fig. 1a, Fig. 3). μ_b should be greater than μ because
   σ_s^2 is high but because μ is low, μ_b will still be relatively low.
Offspring: W is low because μ is low and D and B are both discounted by σ_s^2.
   Offspring should delay dispersal in hopes finding nearby breeding spaces (Fig.
   2b, Fig. 4). Selection for delayed dispersal is even stronger if the probability of
   inheriting parents' breeding spaces is greater than that of finding other, patchily
   distributed breeding spaces. Therefore there should be strong selection for
   delayed dispersal and juveniles should be helpers if dispersal is delayed until
   the parents' next breeding attempt (cooperative breeding).

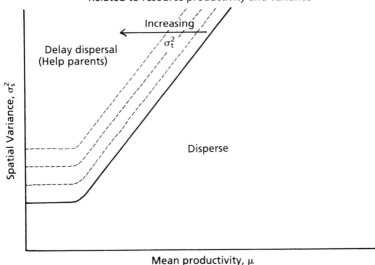

FIG. 4. Dispersal strategies as they relate to productivity of the limiting resource and its variance in space (σ_s^2) and time (σ_t^2). At high resource productivity or at low resource productivity with no resource variance, offspring should disperse early. When there is high spatial variance in resource productivity but limited temporal variance, offspring should delay dispersal. If the delay overlaps the parents' next reproductive effort, the offspring should help their parents raise the next litter (cooperative breeding).

Case 3 (Figs 1b, 2c)

Productivity low (μ low).
Patchiness low (σ_s^2 low).
Predictability low (σ_t^2 high).
Predicted mating strategy: Monogamy, possibly polyandry.
Predicted dispersal strategy: Early dispersal.
Breeding pair: R_f is lower than in Cases 1 and 2 because μ is low and discounted by σ_t^2. In addition, the effect of the male helping the female is most pronounced at low effective μ_b (μ discounted by σ_t^2). These two factors combine to move the monogamy–polygyny threshold to larger μ (Fig. 1b). Thus monogamy will occur over a larger range of μ values than in Cases 1 and 2. If for $\delta=1$, R_f is still very low there could be selection for $\delta>1$, polyandry (Fig. 3.).
Offspring: W is low as in Case 1 and B is increased by high σ_t^2, thus offspring should not delay dispersal (Fig. 2c, Fig. 4).

Case 4 (Figs 1b, 2d)

Productivity low (μ low).
Patchiness high (σ_s^2 high).
Predictability low (σ_t^2 high).
Predicted mating strategy: Monogamy, possibly polyandry.
Predicted dispersal strategy: Probably delayed dispersal (cooperative breeding).
Breeding pair: Predictions are the same as in Case 3 because R_f is only affected
   by σ_t^2 (Fig. 1b, Fig. 3). Polyandry is less likely because weaned offspring are
   more likely to delay dispersal and help parents than in case 3.
Offspring: W is low because μ is low and L_i, D and B are all discounted by σ_s^2.
   However, B is also increased by σ_t^2. Offspring should delay dispersal in hopes
   of finding nearby breeding spaces as in Case 2. However, the delayed dispersal
   threshold (Fig. 2d, Fig. 4) is lower than in Case 2 because σ_t^2 is high, leading
   to higher turnover of breeding spaces. Thus delayed dispersal should occur
   over a smaller range of low μ than in Case 2. If offspring help with future
   parent reproduction, helpers should decrease the strength of selection for
   polyandry in parents.

Case 5 (Fig. 1a, 2a)

Productivity high (μ high).
Patchiness low (σ_s^2 low).
Predictability high (σ_t^2 low).
Predicted mating strategy: Selective promiscuity.
Predicted dispersal strategy: Early dispersal.
Breeding pair: This case has no variance and thus leads to the mating strategy that
   would be predicted for high resource productivity were resource productivity
   not discounted by its variances. R_f is high because μ is high. At high μ, female
   reproductive output without help approaches that with help and R_m male
   reproductive output for N only slightly larger than 1 will exceed R_f. Therefore
   there should be no selection for $\delta > 0$, monogamy. Because both σ_t^2 and σ_s^2 are
   low, breeding females should have many places to breed and have high
   survivorship. Therefore 'good' males should be able to breed with many
   females and there should be selection for $N > 1$, polygyny (μ is above the
   monogamy–polygyny threshold, Fig. 1b, Fig. 3). Females should always select
   to breed with the best male possible but there should be no selection against
   breeding with more than one 'good' male. Thus both males and females
   should breed with more than one member of the opposite sex, leading to
   selective promiscuity. This can not be determined directly from the model,
   however.
Offspring: W is high because μ is high and W is higher for early dispersers delayed
   dispersers under all conditions, so offspring should disperse early (Fig. 2a,
   Fig. 4).

Case 6 (Fig. 1a, 2b)

Productivity high (μ high).
Patchiness high (σ_s^2 high).
Predictability high (σ_t^2 low).
Predicted mating strategy: Polygyny.
Predicted dispersal strategy: Probably early dispersal.
Breeding pair: Since M_x is not discounted by σ_s^2, R_f should be high, N should be greater than one and σ should be 0 because μ is above the monogamy–polygyny threshold (Fig. 1a, Fig. 3), as in case 5. Because σ_s^2 is high, breeding spaces will be patchily distributed and males may be able to defend good patches and attract more than one female, selecting for polygyny and against selective promiscuity. This can not be determined directly from the model, however.
Offspring: W is high because μ is high and higher for early dispersers than for delayed dispersers under most conditions, so offspring should disperse. The switch from delayed dispersal to early dispersal (the delayed dispersal threshold) occurs at an intermediate level of μ under these conditions (Fig. 2b, Fig. 4). If σ_s^2 is very high (D and B both high making finding breeding spaces unlikely), the delayed dispersal threshold moves to higher μ. Under these conditions, there will be selection for delayed dispersal and possibly cooperative breeding.

Case 7 (Fig. 1b, 2c)

Productivity high (μ high).
Patchiness low (σ_s^2 low).
Predictability low (σ_t^2 high).
Predicted mating strategy: Selective promiscuity.
Predicted dispersal strategy: Early dispersal.
Breeding pair: R_f is high because μ is high. N should be greater than 1 and σ should equal 0 because μ is above the monogamy-polygyny threshold (Fig. 1b, Fig. 3), as in Cases 5 and 6. Because σ_s^2 is low, as in Case 5, breeding spaces are not patchily distributed and males should not be able to defend locally productive spaces.
Offspring: W is high because μ is high, though L_i and D are discounted by σ_t^2, and W is higher for dispersers than for nondispersers under all conditions (Fig. 2c, Fig. 4). B is as high as it can be because both μ and σ_t^2 are high and σ_s^2 is low. This means that there will be many breeding spaces distributed evenly in the environment and that the turnover rate on those breeding spaces will be high. Thus, the probability of becoming an established breeder should never be higher. Selection for early dispersal should be very high.

Case 8 (Fig. 1b, 2d)

Productivity high (μ high).
Patchiness high (σ_s^2 high).

Predictability low (σ_t^2 high).

Predicted mating strategy: Polygyny.

Predicted dispersal strategy: Early dispersal.

Breeding pair: R_f should be high, N greater than 1 and δ equal 0 (Fig. 1b, Fig. 3) for the same reasons as in Cases 5, 6 and 7. Because σ_s^2 is high, as in Case 6, breeding spaces are patchily distributed and males may be able to defend good patches and attract more than one female, possibly selecting for polygyny and against selective promiscuity.

Offspring: W is high because μ is high, though L_i, D and B are discounted σ_s^2. B is not as high as in Case 7 because it is both increased by σ_t^2 and discounted by σ_s^2. This means that there will be breeding spaces patchily distributed in the environment and that the turnover rate on those breeding spaces will be high. Though high σ_s^2 moves the delayed dispersal threshold to the right, the high B keeps the threshold in the low μ range. W is thus higher for dispersers than for nondispersers (Fig. 2d, Fig. 4).

Figure 3 summarizes how male mating strategy choices are predicted to vary with mean resource productivity and variance. At high resource productivity, males should always be polygynous. But at low resource productivity, both spatial and temporal variances in resource productivity make monogamy the more productive choice for a male. Temporal variance decreases survivorship (S_{xm}) and decreases female reproductive output (R_f) but monogamy increases female reproductive output. Spatial variance decreases survivorship of polygynous males (S_{xm}) but not that of monogamous males. At lower and lower levels of resource productivity, less spatial or temporal variance in productivity is necessary to make monogamy the more productive choice for a male until at low productivity no variance is necessary. Polyandry is predicted only to occur under conditions of very low resource productivity but great temporal variance in that productivity. Under these conditions, female reproductive output is so small that help from multiple males can make each male's proportion of the female's reproductive output greater than that for monogamous males. Polyandry is not predicted to occur unless there is large temporal variance (unpredictability) in the productivity of the limiting resource.

Figure 4 summarizes how dispersal strategies are predicted to vary with resource productivity and variance. Dispersal is always the best option under conditions of high resource productivity. For lower and lower levels of resource productivity, spatial variance in productivity makes delayed dispersal lead to higher anticipated lifetime reproductive output. This is because high spatial variance in resources decreases an individual's probability of living through dispersal (D) and finding a breeding space (B). However, the amount of spatial variance necessary to make helping the best strategy increases with increasing temporal variance. This is because temporal variance increases the probability of finding a breeding space (B). Delayed dispersal is predicted not to occur unless there is spatial variance (patchiness) in the limiting resource.

RESOURCE PRODUCTIVITY AND VARIANCE AND MAMMAL MATING STRATEGIES

No complete data on mating and dispersal strategies of a mammal are available to test the resource productivity−variance model; the best data I know have been collected on a population of red-cockaded woodpeckers (*Picoides borealis*) (Walters in press; Walters, Doerr & Carter 1988; Walters & Doerr, unpublished data). Walters, Doerr and coworkers are collecting data to test the more specific model without variances developed by Woolfenden & Fitzpatrick (1984) to explain Florida scrub jay cooperative breeding. Walters's and Doerr's data include empirical measures of L_i, D_d, and R and thus some empirical measures of variances. In these woodpeckers, the limiting resource for obtaining breeding spaces is nest hole availability, whereas the limiting resources for survivorship through becoming an established breeder must be food. Data on other birds do support qualitative predictions of the model (for example, Dhondt 1987).

Moehlman (1986) summarized considerable literature on mating strategies of canids and found that small canids tend to be polygynous, medium-sized canids tend to be monogamous, and large canids tend to cooperative breeders. Her summary dealt with species' wide patterns and could not explain variation within species or why some species did not fit the pattern predicted by their weights. Using the resource productivity−variance model to examine mating strategies of individual canids within populations exposed to different resource conditions is insightful.

Macdonald (1981) found that mating systems of red foxes in England ranged from monogamy to polygyny. In rural areas, food availability, μ, was low, so the resource productivity−variance model predicts that foxes in rural England should be monogamous except under the rare case of very high temporal variance in the availability of food production, σ_t^2, when polyandry is predicted. Macdonald found that foxes in areas with limited food were indeed monogamous. In the suburban areas, food resources (located at human residences) were moderately productive (μ moderate) but very patchily distributed (σ_s^2 high). Because some food came from human residences, it was probably quite predictable. The resource productivity−variance model predicts polygyny, perhaps resource defense polygyny, and delayed dispersal under these conditions. This is what Macdonald found.

The mating strategies of golden and silverbacked jackals (*Canis aureus, C. mesomelas*) are also variable (Moehlman 1979, 1983, 1986, 1987). These species prey on small to medium-sized mammals including Thomson's gazelles (*Gazella thomsoni*). Moehlman's discussion of these food supplies indicate that they are usually moderately abundant during the jackals' breeding seasons but not so abundant that parents can easily find adequate food to raise litters successfully (μ moderate to low). These foods also have high spatial variance (σ_s^2 high) but moderate to low temporal variance (σ_t^2 low). Golden jackals' breeding habitat is patchily distributed (σ_s^2 high for breeding habitat, a second resource) and the probability of finding a vacant breeding space is very low (B is low for dispersers). The resource productivity−variance model predicts that individual jackals of both species will have highest reproductive output if monogamous. And because of the high spatial variances in food and breeding spaces (for golden jackals) offspring should often delay dispersal, leading to cooperative breeding (more often in golden jackals). Moehlman has found that both jackals sometimes exhibit cooperative breeding and that golden jackals do so more often than do silverbacked jackals. As predicted, when food is abundant more juvenile silverbacked jackals disperse but this is not the case for golden jackals, because breeding spaces have not increased in abundance.

Gosling's (1986) summary of antelope (Bovidae) mating strategies describes antelope food resources as locally productive (μ high) and usually patchily distributed (σ_s^2 high) or unpredictable (σ_t^2 high) or both. All antelope appear to be polygynous or selectively promiscuous and those that exhibit resource defense polygyny depend on productive resources (μ high) that are patchily distributed (σ_s^2 high) and predictable (σ_t^2 low), as predicted by the resource productivity−variance model.

The eastern grey kangaroos (*Macropus giganteus*) studied by Jarman & Southwell (1986) and the red-necked wallabies (*M. rufogriseus*) studied by Johnson (1986) in the same study area lived in productive habitat with evenly distributed and highly dependable food resources (μ high, σ_s^2 and σ_t^2 low). The resource productivity−variance model predicts that members of these species should be polygynous, probably selectively promiscuous, and should not delay dispersal. This was the case.

As a final example, White (this volume) studied aboriginal social organization in an area exhibiting clinal variation in resource productivity. As predicted by the resource−variance model, the aborigines varied from monogamous to polygynous along this cline.

DISCUSSION

Macdonald's (1981, 1983) resource dispersion hypothesis (see also Macdonald & Carr, this volume) explains mammal mating strategies and social organization through qualitative assessments of resource productivity and dispersion (spatial variance in resource productivity). These concepts are incorporated in the resource productivity−variance model in a specific and quantified manner along with effects of variance in time of resource productivity. Thus if the resource productivity available to individuals and the variances of that productivity are known or can be estimated, the resource productivity−variance model allows quantified testing of the effects of resource dispersion on mammal mating strategies. As much as the predictions of the resource productivity−variance model can be tested, they agree with Macdonald's (1981, 1983) explanations for fox mating strategies using his resource dispersion model.

Expected mating strategies for species or populations or for individual animals can be predicted from the model, depending on whether resource productivities and variances are known for individuals' breeding spaces or only for that population's habitat in general. The 'polygyny threshold' for individuals (Verner & Willson 1966; Orians 1969) follows directly from conditions of high resource productivity and high spatial variance. If resource productivity and variance are known within breeding spaces, comparing a female's expected fitness in a good breeding space shared with another female to her fitness in not so good a breeding space alone yields exactly the predictions of the 'polygyny threshold' model. The resource productivity−variance model correctly predicts that animals that usually breed polygynously in a productive, patchy environment should switch to selective promiscuity in a homogenous environment, as found by Wootton, Bolinger & Hibbard (1986).

The best tests of the predictions outlined for Cases 1 through 8 will be on mammals whose mating systems are not known but whose resource characteristics are at least reasonably well understood. Because μ and σ^2 covary (Lande 1977: Wright 1968), I expect Cases 1 (monogamy) and 8 (polygyny) to be the most common mammalian mating strategies (lower left and upper right corners of Fig. 3). Cooperative breeding and polyandry are predicted only when resource productivity is low but its variances are high (upper right corners of Figs 3 and 4), thus these mating strategies should be relatively rare. I know of no lexicon of mammal mating systems but this hypothesis can be tested through an extensive review of mating strategy literature. It appears outwardly true.

For some mammals, more than one resource may be limiting. Numbers of yellow-bellied marmot colonies are limited by the extremely patchy distribution of tallus slopes, small rock-strewn patches of mountainside that can shelter burrow systems from predators (Armitage 1981). Thus breeding sites have extremely low productivity (dependent on rock slides) and are very patchily distributed but very predictable in time. Productivity of food (much higher than that for breeding sites, less patchy but less predictable) affects individual survivorship and reproductive output (Armitage 1981). Similar relationships exist between pine voles and their tunnel systems (low productivity, very patchy but predictable) and food supplies (Fried 1987; Powell, unpublished data). In the resource productivity−variance model, two limiting resources can be represented by having B dependent on productivity and spatial variance in breeding sites while other variables are dependent on productivity and spatial and temporal variance in food supply.

The resource productivity−variance model predicts that monogamous males should attempt to breed with other females with whom no pair bond is maintained. This male behaviour will simply increase N_x in the model if the other females are pair-bonded to males with equivalent R_ms. If those females are not pair-bonded then a second term is necessary within the summation for matings with unbonded females (those with $\delta=0$). In either of these cases, R_m is increased while R_f stays the same. However, if the male's bonded female also breeds with another male, N_x is decreased. Thus monogamous males should only attempt extra breedings if they can keep $N_x \geq 1$.

Under appropriate conditions, the resource productivity—variance model can predict when alternative mating strategies will be profitable. This requires demographic and life history data for subsets of a population and not just population means. If, for example, for a given population subset, the probability of surviving the expected wait to inherit a breeding space is very low, members of that subpopulation ought to disperse even if not dispersing is the better option for most individuals. Data collected by Gross (1984) appear to fit this approach.

The resource productivity—variance model, as outlined here for mammals, assumes that primary responsibility for parental care falls on the *mother* of the offspring because only females lactate. Thus the only form of polyandry predicted is for a single female to have two or more adult male mates, all of whom contribute parental care. In birds, parental care is more flexible and polyandrous mating systems exist where males have all parental responsibilities (Maxson & Oring 1980; Oring & Lank 1986). The model can be generalized to fit these conditions but the predictions become more general; where the model now predicts polygyny, in a more general form it would not be able to predict the polygamous sex.

The resource productivity—variance model is appropriate for species for which resource limitation is the overriding factor regulating population size and social organization. In species such as mongooses (Waser 1981; Waser & Waser 1985; Rood 1986) and most primates (Terborgh 1983; Andelmann 1986), where the overriding selective pressure for group living has been predation, the resource productivity—variance model of mating strategies will not apply. I suspect that resource productivity and variances will still affect social organization and mating strategies in these species but not in as direct a manner as predicted by the model.

Though many animals are sensitive to variance in resource productivity, equal variance is not always treated equally. Variance in the distant future may not be as important in animal decisions as variance in the present or near future (Kagel, Green & Caraco 1986). Thus, for dispersing animals, the way that resource patchiness or productivity influence dispersal decisions may be more important than the way they may influence reproductive output in the distant future.

ACKNOWLEDGEMENTS

N. Aebischer, Mike Reed, Michael Reiss, Erran Seaman, Jeff Walters and Bill Zielinski read an early draft of this manuscript and gave many constructive criticisms. (Val Standen, Robert Foley and Nigel Dunstone arranged my participation in this symposium.) Consie Powell drafted the figures. This paper is published within the guidelines of Sigma Xi (Jackson & Prados 1983).

REFERENCES

Altmann, J. (1980). *Baboon Mothers and Infants.* Harvard University Press, Cambridge, Massachusetts.

Andelman, S.J. (1986). Ecological and social determinants of cercopithecine mating patterns. *Ecological Aspects of Social Evolution: Birds and Mammals* (Ed. by D.I. Rubenstein & R.W. Wrangham), pp. 201–16. Princeton University Press, Princeton, New Jersey.

Armitage, K.B. (1981). Sociality as a life-history tactic of ground squirrels. *Oecologia*, **48**, 36–49.

Armitage, K.B. (1986). Marmot polygyny revisited: determinants of male and female reproductive strategies. *Ecological Aspects of Social Evolution: Birds and Mammals* (Ed. by D.I. Rubenstein & R.W. Wrangham), pp. 303–31. Princeton University Press, Princeton, New Jersey.

Armitage, K.B. & Downhower, J.F. (1974). Demography of yellow-bellied marmot populations. *Ecology*, **55**, 1233–45.

Caraco, T. (1981). Risk-sensitivity and foraging groups. *Ecology*, **62**, 527–31.

Caraco, T. (1982). Aspects of risk-aversion in foraging white-crowned sparrows. *Animal Behaviour*, **30(3)**, 719–27.

Carter, C.S., Getz, L.L., Gavish, L. & Mcdermott, J.L. (1980). Male-related pheromones and the activation of female reproduction in prairie vole. *Biology of Reproduction*, **23**, 1038–45.

Clutton-Brock, T.H., Guinness, F.E. & Albon, S.D. (1982). *Red Deer: Behavior and Ecology of Two Sexes.* University of Chicago Press, Chicago, Illinois.

Dhondt, A.A. (1987). Polygynous blue tits and monogamous great tits: does the polygyny-threshold model hold? *American Naturalist*, **129**, 213–20.

DonCarlos, M.W., Peterson, J.S. & Tilson, R.L. (1986). Captive biology of an asocial mustelid; *Mustela erminea. Zoo Biology*, **5**, 363–70.

Emlen, S.T. (1982). The evolution of helping, I: An ecological restraints model. *American Naturalist*, **119**, 29–39.

Emlen, S.T. & Oring, L.W. (1977). Ecology, sexual selection, and the evolution of mating systems. *Science*, **197**, 215–23.

Fisher, R.A. (1930). *The Genetical Theory of Natural Selection.* Dover, New York.

Fried, J.J. (1987). *Helping behavior and the evolution of cooperative breeding in pine voles (Microtus pinetorum).* M.Sc Thesis, North Carolina State University.

Gosling, L.M. (1986). The evolution of mating systems in male antelopes. *Ecological Aspects of Social Evolution: Birds and Mammals* (Ed. by D.I. Rubenstein & R.W. Wrangham), pp. 244–81. Princeton University Press, Princeton, New Jersey.

Gross, M.R. (1984). Sunfish, salmon, and the evolution of alternative reproductive strategies and tactic in fishes. *Fish Reproduction: Strategies and Tactics* (Ed. by G. Potts & R. Wootton), pp. 55–75. Academic Press, London.

Jackson, C.I. & Prados, J.W. (1983). Honor in science. *American Scientist*, **71**, 462–4.

Jarman, P.J. & Southwell, C.J. (1986). Grouping, associations, and reproductive strategies in eastern grey kangaroos. *Ecological Aspects of Social Evolution: Birds and Mammals* (Ed. by D.I. Rubenstein & R.W. Wrangham), pp. 399–428. Princeton University Press, Princeton, New Jersey.

Johnson, C.N. (1986). Philpatry, reproductive success of females, and maternal investment in the red-necked wallaby. *Behavioral Ecology and Sociobiology*, **19**, 143–50.

Kagel, J.H., Green, L. & Caraco, T. (1986). When foragers discount the future: constraint or adaptation?. *Animal Behaviour*, **34**, 271–83.

Lande, R. (1977). On comparing coefficients of variation. *Systematic Zoology*, **26**, 214–17.

Lima, S.L., Valone, T.J. & Caraco, T. (1985). Foraging–efficiency — predation–risk trade-off in the grey squirrel. *Animal Behaviour*, **33**, 155–65.

Macdonald, D.W. (1981). Resource dispersion and the social organization of the red fox (*Vulpes vulpes*). *Worldwide Furbearer Conference Proceedings* (Ed. by J.A. Chapman & D. Pursley), pp. 918–49. Worldwide Furbearer Conference, Frostburg, Maryland.

Macdonald, D.W. (1983). The ecology of carnivore social behaviour. *Nature*, **301**, 379–84.

Maxson, S.J. & Oring, L.W. (1980). Breeding season time and energy budgets of the polyandrous spotted sandpiper. *Behaviour*, **74**, 200–63.

Mech, L.D. (1970). *The Wolf: The Ecology and Behavior of an Endangered Species.* Natural History Press, Garden City, New York.

Moehlman, P.D. (1979). Jackal helpers and pup survival. *Nature,* **277,** 382–3.

Moehlman, P.D. (1983). Socioecology of silver-backed and golden jackals (*Canis mesomelas, C. aureus*). *Recent Advances in the Study of Mammalian Behavior* (Ed. by J.F. Eisenberg & D.G. Kleiman), pp. 423–53. American Society of Mammalogists.

Moehlman, P.D. (1986). Ecology of cooperation in canids. *Ecological Aspects of Social Evolution: Birds and Mammals* (Ed. by D.I. Rubenstein & R.W. Wrangham), pp. 64–86. Princeton University Press, Princeton, New Jersey.

Moehlman, P.D. (1987). Social organization of jackals. *American Scientist,* **75,** 366–75.

Orians, G.H. (1969). On the evolution of mating systems in birds and mammals. *American Naturalist,* **103,** 589–603.

Oring, L.W. & Lank, D.B. (1986). Polyandry in spotted sandpipers: the impact of environment and experience. *Ecological Aspects of Social Evolution: Birds and Mammals* (Ed. by D.I. Rubenstein & R.W. Wrangham), pp. 21–42. Princeton University Press, Princeton, New Jersey.

Powell, R.A. (1982). *The Fisher: Life History, Ecology and Behavior.* University of Minnesota Press, Minneapolis.

Pyke, G.H. (1984). Optimal foraging theory: a critical review. *Annual Review of Ecology and Systematics,* **15,** 523–76.

Pyke, G.H., Pulliam, H.R. & Charnov, E.L. (1977). Optimal foraging: a selective review of theory and tests. *Quarterly Review of Biology,* **52,** 137–54.

Real, L.A. (1980). On uncertainty and the law of diminishing returns in evolution and behavior. *Limits to Action* (Ed. by J.E.R. Staddon), pp. 37–64. Academic Press, New York.

Real, L.A. (1981). Uncertainty and pollinator–plant interactions: the foraging behavior of bees and wasps on artificial flowers. *Ecology,* **62,** 20–6.

Real, L.A., Ott, J. & Silverfine, E. (1982). On the tradeoff between the mean and the variance in foraging: effect of spatial distribution and color preference. *Ecology,* **63,** 1617–23.

Rogers, L.L. (1987). Effects of food supply and kinship on social behavior, movements, and population growth of black bears in northeastern Minnesota. *Wildlife Monographs,* **97,** 1–72.

Rood, J.P. (1986). Ecology and social evolution in the mongooses. *Ecological Aspects of Social Evolution: Birds and Mammals* (Ed. by D.I. Rubenstein & R.W. Wrangham), pp. 131–52. Princeton University Press, Princeton, New Jersey.

Rubenstein, D.I. & Wrangham, R.W. (Eds) (1986). *Ecological Aspects of Social Evolution: Birds and Mammals.* Princeton University Press, Princeton, New Jersey.

Shank, C.C. (1986). Territory size, energetics, and breeding strategy in the Corvidae. *American Naturalist,* **128,** 642–52.

Smolen, M.J. (1981). Microtus pinetorum. *Mammalian Species,* **147,** 1–7.

Terborgh, J. (1983). *Five New World Primates.* Princeton University Press, Princeton, New Jersey.

Verner, J. & Willson, M.F. (1966). The influence of habitats on mating systems of North American passerine birds. *Ecology,* **47,** 143–7.

Walters, J.R. (in press). The red-cockaded woodpecker: a primitive cooperative breeder. *Cooperative Breeding in Birds: Long-Term Studies of Ecology and Behavior* (Ed. by P.B. Stacey & W.D. Koenig), publisher under negotiation.

Walters, J.R., Doerr, P.D. & Carter, J.H. III (1988). The cooperative breeding system of the red-cockaded woodpecker. *Ethology,* **78,** 275–305.

Waser, P.M. (1981). Sociality or territorial defense? The influence of resource renewal. *Behavioral Ecology and Sociobiology,* **8,** 231–7.

Waser, P.M. & Waser, M.S. (1985). Ichneumia albicauda and the evolution of viverrid gregariousness. *Zeitschrift für Tierpsychologie,* **68,** 137–51.

Wittenberger, J.F. (1980). Group size and polygamy in social mammals. *American Naturalist,* **115,** 197–222.

Wittenberger, J.F. (1981). *Animal Social Behavior.* Duxbury, Boston, Massachusetts.

Wittenberger, J.F. & Tilson, R.L. (1980). The evolution of monogamy: hypotheses and tests. *Annual Review of Ecology and Systematics,* **11,** 197–232.

Woolfenden, G.E. & Fitzpatrick, J.W. (1984). *The Florida Scrub Jay: Demography of a Cooperative-Breeding Bird*. Princeton University Press, Princeton, New Jersey.

Wootton, J.T., Bollinger, E.K. & Hibbard, C.J. (1986). Mating systems in homogenous habitats: The effects of female uncertainty, knowledge costs, and random settlement. *American Naturalist*, 128, 499–512.

Wright, S. (1968). *Evolution and the Genetics of Populations, I. Genetic and Biometric Foundations*. University of Chicago Press, Chicago, Illinois.

Section 2
Social Organization:
The Influence of Resources

It is apparent from only a brief consideration of the papers in the previous section that the issue of sociality is not simply one of when permanent groupings will occur, but the nature and internal structure of such groups. At the heart of socioecology lies the hypothesis that the type of social organization exhibited reflects the demands or the constraints, or indeed the opportunities, provided by the resources available. This idea has a lengthy history, especially in anthropology, and has often been closely associated with the environmental determinism that was central to early, often evolutionary, anthropology and geography. The two dominant figures of nineteenth century thought, Darwin and Marx, both placed considerable importance on the role of the environment in human affairs, and many geographers such as Elsworth Huntingdon and Friedrich Ratzel saw strong causal relationships between climate and the nature of human society (see Ellen 1982).

At its simplest was a basic expectation that there should be a correlation between habitat or environmental zone and human culture or social system. In this respect, the longest standing project was the attempts of Alfred Kroeber and his students to construct maps of human culture in North America that reflected biogeographic factors (Kroeber 1939).

This work met with relatively little success and the expectations, for example, that high levels of environmental diversity would lead to high levels of cultural diversity, were not realized. The patterns of human variability did not appear to be dependent upon the nature of the environment. Only a few features, principally relating to technology and the broad aspects of subsistence, could be predicted on the basis of habitat distributions and the character of the resources exploited (Whiting 1964). The inventiveness of human sociality seemed to override the determinants of the environment.

The failure of this early ecological approach to human behaviour led to three solutions to the problem being adopted at various times by different anthropologists. One was a withdrawal to a position of environmental 'possibilism' — which is, that the environment did not determine social or economic form, but rather made possible a series of alternatives, and that other, principally cultural and social, factors then played the major part in deciding which developed or was adopted (Ellen 1982). With environmental possibilism the environment was relegated to the role of backdrop to the social drama played out by humans.

125

The second solution, which really developed from the theoretical impoverishment of possibilism, was that of cultural ecology. Cultural ecologists recognized that the environment and ecology could and often did play a significant role in human affairs, but that it was not a simple deterministic one. Rather, that culture was a human system analogous to the ecosystem, within which there were a series of cybernetic interactions. This human system did indeed overlap with the ecosystem, and the interaction between the two formed the basis for cultural ecology. Various aspects of culture were shaped by the constraints of the ecosystem, as indeed various aspects of a human ecosystem may have been modified by cultural activities. What was critical to a cultural ecological approach was that the interactions were regulatory and mutual, rather than deterministic and uni-directional. Perhaps the best known example of this approach was that of Rappaport (1968) describing the cultural ecology of the Tsambaga Maring of New Guinea where the ritual cycle was integrated with the ecological constraints of protein availability, population density and warfare.

The third solution was the abandonment of any interest in the role of the natural environment in human affairs — a sort of environmental minimalism. Many social and cultural anthropologists would argue that the form that a human society takes reflects not external conditions (or at least not gross environmental external conditions), but rather the internal structures of specific cultural, cognitive and political situations. Essentially society is structured by social factors, without significant reference to any external ecological factors. The limited appeal of socioecology to modern anthropology is directly related to the dominance of this paradigm within anthropology.

Given the failure of the attempts by earlier anthropologists to account for social organization and behaviour in terms of characteristics of the environment it might well be asked why these discarded hypotheses should be reconsidered, and why a comparative approach to animal and human behaviour might be interesting. The answer to this question lies not in developments in anthropology, but in the changes that have occurred in behavioural ecology as applied to animals, and as illustrated in the papers in this section.

The first change in perspective, as argued by Wrangham & Rubenstein (1986), is that there are now predictive theories within ecology, suggesting not just how social organization might be affected by resource characteristics, but also specifying what form they should take and what mechanisms are involved. Central to this is the point that resources do not affect societies as a whole, and therefore gross correlations of this type are not sought, but that resource availability, distribution and quality affects the strategies pursued by individuals, and the overall social organization derives from the resolution of conflicting individual needs and tactics (see, for example, Alexander 1974). More specific relations between ecological variables and social strategies such as these are of course much easier to investigate and test. For example, Wrangham (1980) has argued that female strategies are principally constrained by access to resources on account of the higher nutritional costs involved in female reproduction and their more limited variance in reproductive output. Males, on the other hand, are able to pursue more 'risky' strategies in relation to food availability, as the most important factor contributing to their reproductive success is not access to food, but access to females. Out of asymmetries such as this in relation to different resource states comes at least some of the variability in mammalian social organization.

Models of this type represent an enormous advance over earlier attempts to understand social behaviour in an ecological context, not just for humans but for animals as well. The advance over previous ideas of environmental determinism arises from the breakdown of monolithic correlations and their replacement by specific interactions between individuals (or categories of individuals) and attributes of resource variability. Of particular importance here has been the dissection of the term 'environment'. One of the main problems in the past has been the assumption that *habitat type* is critical. Modern ecological theory, however, recognizes that different habitats may often share the same characteristics, while others that are superficially similar (for example, forests or woodlands) may differ markedly in resource structure and food availability. Modern behavioural ecologists have developed a series of attributes that better describe the opportunities and problems that resources pose to a population than broad habitat types. Of particular importance have been the idea of a patch — a unit of resource availability. Patches can vary in size, quality, density, distribution and predictability in time and space. Different resources (for example fruits and insects) may have the same properties as far as their exploitation goes, and this may override the significance of gross habitat classifications. From the point of view of human socioecology, this is a far cry from Kroeber's culture/habitat maps, and represents a series of models that have seldom been tested and are not the same as the old ideas of environmental determinism.

Just as the concept of the 'environment' has been replaced by a more sophisticated, heterogenous set of units or variables reflecting the nature of resources, so too has the concept of social organization and structure been modified. As Hinde (1983) pointed out, social structure is an epiphenomenal concept, and underlying it is a hierarchy of principles that allows a far better articulation of social behaviour and resources. At the base lie the behaviour of individual animals — their actual activities carried out in relation to specific goals (acquiring food, etc.). Many of these activities will impinge directly onto the activities of other animals, and hence constitute social interactions (e.g. supplantation of one individual by another at a food site). Because the situations in which social interactions are likely to occur are repetitive and often predictable, the behaviour during social interactions will become patterned, and hence social relationships will be formed between individuals. The regular patterning of these relationships (for example, females tending to associate together in non-aggressive ways) will form the basis for what we see as a social structure — patterned and differentiated social relationships.

The nature of the social relationships formed will provide the basis for the principles by which the social structure is organized. The outcome will be a series of social units, and it is the nature of these units that will be linked to the structure of the resources available. These units will include kin-based units, foraging parties, dyadic pairs of related individuals, alliances between unrelated individuals and so on. For the socioecologist the critical problem is how these are related to the resources available.

Thus, for both anthropology and zoology, the 'new' socioecology is not concerned with the overall relationship between habitat and society, but with the specifics of particular social units, reflecting social relationships, in the context of the structure of resource availability and distribution. The papers in this section reflect this concern, and also show the power of ecological principles in shaping social structure. Dunbar tackles directly the problem of the complexity of animal societies. While it has long been established that animal social groups are far from formless units, it is nonetheless sometimes assumed that they may have a single structure. What Dunbar shows, however, is that not only does structure exist, but that there are often several structures within a single 'social group'. These structures, Dunbar argues, are understandable when viewed as optimality sets that derive from the outcome of the strategies pursued by a number of individuals. Because both the natural and social environment in which these individuals live are complex, parallel and overlapping, structures will emerge simultaneously: foraging parties, harem breeding groups, larger anti-predator defence herds, and so on. What emerges from this perspective is the conclusion that quite marked social complexity can emerge from relatively simple constituent parts — assuming, of course, that the starting point in the analysis is the reproduction maximizing strategies pursued by individuals.

The way these strategies respond in quite precise ways to the nature of resources is the subject of White's paper on the socioecology of *Pan paniscus*. The bonobo or pygmy chimpanzee is a close relative of the common chimpanzee (*Pan troglodytes*), and yet its social behaviour is markedly different; where male−male relationships are the strongest among adult *P. troglodytes*, no such preference is found among *P. pansiscus*. Instead relationships tend to be equally strong among and between both sexes. This, White argues, is a response to the different distribution of foods. Compared to the common chimpanzee, the patches of food available to pygmy chimps in Lomako forest are larger, and as a result females are more closely associated with each other, and males are then brought closer together in order to maintain proximity to the females. Feeding party size, and hence social behaviour within these groups, is closely associated with the size of the food patches available.

A characteristic of all chimpanzee social organization is that fission and fusion of groups occur. In her long-term study of the Turkana pastoralists, Dyson-Hudson has demonstrated the extent to which the same strategy — manipulating the structure and size of human groups — is an essential component of coping with variation in food availability. The Turkana live in the arid regions of north Kenya, where the principal problem is the patchy spatial and temporal distribution of food, and, in particular, its unpredictability. While the factors leading to changes in group structure through the course of a year may be complex involving a variety of economic, social and political factors, the most striking characteristic is that the fluidity and flexibility of the system is dependent upon the active pursuit of herd management strategies by individuals — bearing out for this particular group of humans the expectation that structure arises from the behavioural strategies of individuals and smaller social units.

Among the Turkana it is the males, as owners of the herds, that seem to play the most significant role in driving the dynamics of the social groups that constitute the overall society. In the comparative perspective adopted here, what is striking is that males control, within the limits of the environment, the distribution of resources. This is made possible by the advantages of having domesticated herds of ungulates, and may have had significant effects on their socioecology; in contrast to most species of non-human primate, Turkana women are tied to men because the latter control resource distribution. The basic principal enunciated above (that females are more resource-bound than males) still holds, but the links have come full circle: females are tied to resources, males are tied to females, but females are tied to males because resources are tied to them!

This is a situation that does not occur in most non-human primate species (and, indeed, in many human societies), and van Schaik explores the variability in primate social structure in terms of the effects that different resource distributions have on female social relationships. According to van Schaik, it is the options available to females that are critical in determining overall social structure; depending on how they respond to competition for food (scramble or contest) groups may be stable or unstable, egalitarian or ranked, kin-based or competitive. None of these systems, though, are simple species-specific attributes, but continuous variables that respond to the nature of the resources available to a population.

REFERENCES

Alexander, R.D. (1974). The evolution of social behaviour. *Annual Review of Ecology and Systematics*, **51**, 325–83.

Ellen, R. (1982). *Environment, Subsistence and System: the Ecology of Small Scale Social Formations*. Cambridge University Press, Cambridge.

Hinde, R. (1983). A conceptual framework. *Primate Social Relationships: an Integrated Approach* (Ed. by R. Hinde), pp. 1–7. Blackwell Scientific Publications, Oxford.

Kroeber, A.L. (1939). *Cultural and Natural Areas of Native North America*. University of California Publications in American Archaeology and Ethnology, 38. University of California Press, Berkeley.

Rappaport, R. (1968). *Pigs for the Ancestors: Ritual in the Ecology of a New Guinea People*. Yale University Press, New Haven, Connecticut.

Whiting, J.W.M. (1964). The effects of climate on certain cultural practices. *Explorations in Cultural Anthropology* (Ed. by W.H. Goodenough). McGraw Hill, New York.

Wrangham, R.W. (1980). An ecological model of female-bonded primate groups. *Behaviour*, **75**, 262–300.

Wrangham, R.W. & Rubenstein, D.I. (1986). Socioecology: origins and trends. *Ecological Aspects of Social Evolution* (Ed. by R.W. Wrangham & D.I. Rubenstein), pp. 3–17. Princeton University Press, Princeton, New Jersey.

Social systems as optimal strategy sets: the costs and benefits of sociality

R.I.M. DUNBAR

Department of Zoology, University of Liverpool, P.O. Box 147, Liverpool L69 3BX, UK*

SUMMARY

1 Two themes are developed in this chapter. One is that animal societies are often very much more complex than we have usually tended to assume. This is illustrated with examples of the multi-level societies of various species of mammals and birds.

2 The second is that the groupings found within these social systems function as coalitions that allow their members to solve critical problems of survival and reproduction more effectively.

3 I argue that, in this respect, animal social systems closely resemble the political structures found in many human societies.

INTRODUCTION

Our conventional view of animal social systems has tended to be rather simplistic in two quite different respects. One is that we have been inclined to focus on a rather limited number of features and so to classify animal societies into just a few basic types (see, for example, Crook 1965; Crook & Gartlan 1966; Wilson 1975; Gray 1985). The most common basis of classification has probably been the mating system, with monogamy, polyandry and polygamy as the primary types (with the latter category sometimes being differentiated further into those based on harem-defence and those based on the defence of individual oestrous females). Such a view seriously underestimates the extent to which the societies of advanced vertebrates are designed to allow animals to solve a diverse range of biological problems, with mating often being one of the least important of these.

The other problem is that our view of animal behaviour is still dominated by a behaviourist tradition. As a direct consequence of this, we have tended to view sociality as being synonymous with group living and social complexity as synonymous with group size (see, for example, Wilson 1975). This view implicitly assumes that animals cannot have formal relationships with individuals whom they do not interact with on a day-to-day basis. Furthermore, it implies that the functions of animal social systems are limited to whatever functions are subserved

*Present address (for correspondence): Department of Anthropology, University College London, Gower Street, London WC1E 6BT.

by the most obviously conspicuous groups that animals happen to spend their time in.

Between them, these two considerations have often limited the range of behaviours which biologists have considered as being relevant to questions about social systems. In part, the emphasis on breeding systems derives from the extensive well-established literature on birds. Birds are often at their most intensely social during the breeding season and this has undoubtedly coloured the literature on avian social systems. Many mammals (and primates in particular) live in more permanent groups based on kinship that integrate several generations into a coherent stable consociation. Both Kummer (1971) and Nagel (1979), for example, have stressed that the social systems of primates are essentially group solutions to ecological problems. Thus, what characterizes a group is not only how it functions as a mating system but also the way in which it allows its members to deal collectively with the more mundane problems of day-to-day survival.

Kummer (1982) has also commented that we ignore an animal's intentions at our peril when we come to study social behaviour in species such as primates. Much of what an animal does is, in fact, a compromise between what it would like to do and what its social and ecological context allows it to do (see also Seyfarth 1980; Dunbar 1984, 1988). By merely describing what it does, we cannot explain why an animal behaves in a particular way. We can only arrive at an understanding of why it does so by asking what problem it is trying to solve and how its ideal solution to that problem is being constrained by the context in which it lives. More generally, perhaps, we need to take a broader view of what a social existence means to an animal, to look beyond the immediate confines of the individual(s) with whom it happens to be interacting in order to ask what kinds of relationships that animal has with the other members of the population as a whole.

My aims here are two-fold. I want to suggest (i) that animal societies are very much more complex than we have tended to give them credit for and (ii) that this complexity can best be interpreted as an attempt to generate cooperative solutions to the many problems of survival and reproduction that face an animal if it is to be able to contribute genes to future generations. In functional terms, cooperative solutions of this kind are little different to the alliances we find in human societies. I do not wish to suggest that such alliances among animals are in all cases based on cognitive strategic planning in the sense that we might use that term of human beings. But I do want to suggest that it might be the case at least sometimes and that how far this is true remains a purely empirical question yet to be decided. In the meantime, I suggest that viewing animal social systems in purely functional terms as alliances may be a profitable antidote to a long-standing tendency within the behavioural sciences to view social systems as having a simple species-typical form.

Before going on to elaborate this, however, I must first begin with a brief summary of the biological context within which animals pursue their social strategies since this provides the essential framework within which we must interpret the functional aspects of coalitions.

THE BIOLOGICAL CONTEXT

I base my position on the straightforward Darwinian assumption that the evolutionary processes inevitably drive organisms to maximize their individual contributions to their species' gene pools. This assumption lies at the very heart of sociobiology and it is a simple consequence of the way in which the biological processes work. Whatever an animal is and does is part and parcel of its reproductive strategy, for every action necessarily has consequences for the extent to which it can contribute genes to future generations. This does not mean that behaviour is genetically determined, though it does mean that there must be some genetically inherited mechanisms that allow animals to determine both how closely they are matching an evolutionarily optimal solution and how to rectify the mismatch when they are too far off the mark. Animals need not know that the solution they aim for is an evolutionarily optimal one (in the sense that it maximizes their genetic fitness); their concerns are likely to be with more proximate goals that have been tuned over evolutionary time to act as cues for the appropriate evolutionary goals (see McFarland & Houston 1981).

Since social behaviour (and the social system that is built on that behaviour) is necessarily as much a component of an animal's reproductive strategy as any other aspect of its biology, we may assume that social behaviour is geared to maximizing its chances of reproducing successfully. However, in no species is reproduction in this sense a simple phenomenon. It depends on the organism successfully solving a number of problems of day-to-day biology that are concerned with survival, mating (or reproduction itself) and rearing (Goss-Custard, Dunbar & Aldrich-Blake 1972). Each of these in turn consists of a set of more specific problems: day-to-day survival, for example, consists of the two major subsets of problems concerned with (i) finding enough food of the right kind of nutritional quality to maintain bodily function and (ii) avoiding been caught by a predator.

In an ideal world, an animal would be able to maximize its contribution to its species' gene pool by solving each of these problems in the best possible way. In the real world, however, such ideal solutions are not often practicable — or, at least, they are practicable only at a cost that may well completely offset all the benefits of having solved that problem in the best possible way. Such costs are largely due to the fact that the whole suite of problems bearing on successful survival and reproduction has to be solved more or less simultaneously (or at best within a relatively short time frame).

The ideal solution to the problem of mating, for example, might be to mate with as many individuals of the appropriate sex as possible. But doing so ignores the fact that the animal has both to survive to mate another day and to ensure that the offspring it generates through mating mature into adults that can produce descendants for it, otherwise its contribution to future generations is likely to be very small. In many polygamous species, for example, males are forced to compete so fiercely for the opportunity to mate that they do not have much time to feed during the mating season and are obliged to draw on their fat reserves. In some species, males have to withdraw from the fray before the end of the mating season in order to feed (e.g. hartebeest: Gosling 1986); in other species, males are placed under such extreme stress that they are unable to survive the rigours of the following winter or dry season (e.g. feral goats: Dunbar, Buckland & Miller, in preparation). Moreover, the mere production of embryos is not enough, for unless those embryos mature into adults and breed for themselves, the whole effort put into mating will have been wasted. Biologically speaking, such wastage is a serious cost since the animal has expended time and effort producing offspring without contributing any of its genes to future generations.

The animal's difficulties in trying to find ideal solutions to these individual problems arise from the fact that not only is it difficult to do two things simultaneously with anything like the effectiveness that either requires, but also that the animal's own time and energy budgets are limited. There are only so many hours available each day, and each of the problems has its own demands on the animal's time. Energy too is limited, so that whatever excess an animal is able to set aside over and above its basic metabolic requirements has to be apportioned in such a way as to make the best use of what it has available. These constraints on an animal's freedom of movement mean that its solution to an individual problem will often be a compromise. In other words, it will try to achieve a set of solutions that optimizes its net contribution to the species' gene pool with respect to the biological constraints on its ability to reproduce effectively rather than trying to maximize the solutions to each problem on its own.

It is worth emphasizing here that there are two quite distinct steps in this argument. One is that organisms behave in such a way as to optimize a particular function; the other is that the function that an organism optimizes is genetic fitness (i.e. its contribution to the species' gene pool). The second of these can be replaced by any other function we like, and doing so would have no effect on the general claim encapsulated in the first statement. Thus, we might take exception to the idea that organisms live in a Darwinian world and suggest that they might instead be optimizing pleasure or economic wealth. Replacing one such criterion by another would not, however, alter the fact that organisms try to maximize something. In a Darwinian world, this is necessarily genetic fitness, though other criteria (e.g. numbers of offspring born, energy ingested per unit time, economic wealth, feelings of contentment) may well be staging posts on the way and may therefore be appropriate intermediate bases for analysis. Indeed, for the reasons discussed above, animals almost never try to maximize fitness directly: rather, they operate in terms of proximate goals like energy consumption which, over the course of evolution, have come to be correlated with genetic fitness (for further discussion, see Dunbar 1982a).

Sociality has two implications for animals in this context. The first is that it makes it possible to develop more effective solutions to key problems of survival (e.g. predation risk, food finding, defence against ecological competitors) by allowing them to share the costs among more individuals. This may — but need not — involve living together in the same group. Secondly, cooperative solutions invariably impose costs on the animals concerned, if only as a result of their having to reschedule activities in order to coordinate action with other individuals.

These costs are often exacerbated when cooperation requires the animals to live together. For one thing, they are more likely to get in each other's way when feeding; for another, they are more exposed to the 'predatory' behaviour of dominant group members who are better placed to steal food and other resources from them (see Harcourt 1987). In addition, group living may impose additional indirect costs on the animals. In order to provide each member with the foraging area required to provide all the food it needs, a group has to travel proportionately further each day than a single individual: in this context, there are almost no savings of scale to be gained from increases in group size. Not only may this expose animals to greater predation risks by forcing them to travel further from refuges, but it will also increase both the time that has to be devoted to travel and the time that has to be devoted to feeding in order to fuel that additional travel.

SOCIETIES AS NETWORKS OF RELATIONSHIPS

Hinde (1976) has pointed out that a social system is what we, as observers, abstract from the patterning and quality of the relationships that we observe among the animals in a group, relationships in turn being what we abstract from the patterning and quality of the interactions among these individuals. An important proviso, however, is that we should beware of restricting our view to the particular groups that we see at a given time and place. An animal's relationships with other conspecifics do not end abruptly at the boundary of the group it happens to be living in. Animals that belong to different groups often have antagonistic relationships with members of other groups in the neighbourhood, but this is far from universally true. There is accumulating evidence that animals know the identities of members of other groups and associate them with specific locations (e.g. vervets: Cheney & Seyfarth 1985). In other cases, certain groups have been found to exchange members more readily with each other than they do other (in some cases, spatially closer) groups (e.g. vervets: Henzi & Lucas 1980, Cheney & Seyfarth 1983; macaques: Colvin 1983, van Noordwijk & van Schaik 1985). These observations suggest that even in species that conventionally live in rather discrete stable groups, an animal's knowledge of and relationships with other individuals in the local population may be quite extensive.

One reason why we tend to see groups as being the essence of animal societies is that they are particularly obvious features of a population's organization. For the casual observer undertaking a census of a population, the spatio-temporal distribution of the animals is the one feature of their relationships that can be easily quantified. A recensus a year later might reveal no change in the distribution of group sizes, yet there may in the interval have been considerable movement of individuals between groups. A more detailed study of the patterns of interaction might have revealed that an individual animal has an extensive network of relationships that link it, directly or indirectly, with a large proportion of the animals in the population.

An unusually clear example of this is provided by gelada baboons (*Theropithecus gelada*). Most of the animals in a population live in discrete stable groups that consist of a single breeding male and a small number of reproductive females and their dependent offspring. A casual observer would conclude that these units wander widely and associate with each other more or less at random (see Crook 1966). More detailed analysis of the patterns of association between units, however, has revealed that they do not associate at random: some units associate together regularly, others rarely do so. Moreover, the one male reproductive units themselves are far from being homogenous groups: they exhibit marked structuring into cliques whose members associate more closely with each other (as indicated by the amount of time spent in physical proximity or time spent grooming) than they do with other members of the unit.

If we plot the relative frequencies with which individual gelada associate with each other, we find the kind of pattern illustrated in Fig. 1. Individuals are grouped into small coalitions of two or three individuals, with several of these groupings making up a reproductive unit. A number of such units (anything from two to thirty-one) are clustered into a higher level grouping called a 'band': these units share a common core ranging area and tend to associate together in loose herds significantly more often than they do with units of the other bands in the population (see Dunbar 1986). Within the set of bands that make up a population, a further super-cluster can be identified of bands that show a preference for associating in herds. Kawai *et al.* (1983) found that while units of some bands spent as much as 15−20% of their time together in the same herd, they might spend only 2−5% of their time associating with the units of other bands in the neighbourhood. Kawai *et al.* (1983) termed these super-clusters 'communities'.

Although Fig. 1 is structured in terms of clusters that form higher order clusters, we need to remember that these higher order clusters are a consequence of the grouping tendencies of the individual animals that make up the base of the pyramid. This is reflected in the way that individuals of different clusters interact. Regular associates groom together frequently, engage in vocal 'conversations' and form coalitions in mutual defence; less regular associates will barely acknowledge each other's presence, other than to peer carefully at the approaching animals until satisfied of their identity. A move by a regular associate will precipitate an immediate follow by its partner, whereas a similar departure by a more casual acquaintance is likely to be ignored (see Dunbar 1983a).

The particular structure of gelada society appears to arise from the successive fission of units over time. We know, for example, that individual reproductive units are genetically more homogenous than the bands to which they belong, and that the bands in turn are genetically more homogenous than the population as a whole (Shotake 1980). These genetic data conform well to the behavioural observations on rates of transfer and migration (Dunbar 1984). If a band represents the product of the successive fissioning of some ancestral unit and its daughter units

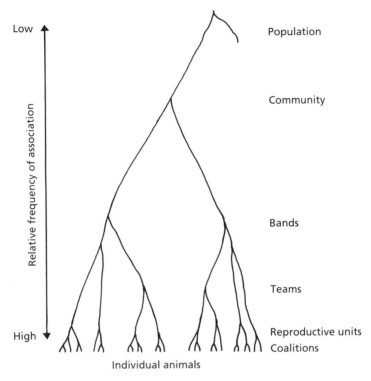

FIG. 1. Schematic representation of the pattern of relationships among individuals in a gelada baboon population. The figure illustrates the pattern that is produced by a simple cluster analysis of the frequencies with which individuals associate together in the same herd when foraging, based on more detailed analyses given by Kawai *et al.* (1983) and Dunbar (1986). (Reproduced with permission from Dunbar 1988.)

over time, then the hierarchical structure of a gelada community is something akin to a family tree. Moreover, it is a family tree whose ancestral origin is buried in the distant past.

This raises some interesting possibilities concerning the temporal dimension of relationships within gelada society. In our main study area in the Simen Mountains of northern Ethiopia, there are typically eight reproductive units in a band; with an average of three bands to a community, this means that there are about 24 units in the average community. It would take 4.5 'generations' to produce this number of units from a single ancestral unit by fission alone. Since reproductive units underwent fission only about once every 6.5 years in this population (Dunbar 1984), this represents a period of about 30 years. Thus, the network of relationships crystallized within a community takes us back well beyond the average lifespan of individual animals (12−14 years in this population; see Dunbar 1980). This is not to say that the animals have any conception of what these genetic relationships are: their knowledge is based solely on familiarity generated by frequent association in herds, with some of these individuals being more familiar than others. But that familiarity is a residual effect of the fact that their ancestors once belonged to the same band, and, at yet greater remove, to the same ancestral reproductive unit.

The gelada are by no means unique in this respect, though the hierarchical structuring of relationships in their case may be exceptional for its clarity. A similar structuring of relationships into a hierarchical series of clusters has been documented in the hamadryas baboon, *Papio hamadryas* (Kummer 1968, 1984). Although the gelada and hamadryas social systems appear to be superficially similar, it is only in respect of the structure and function of bands that they are in fact in any way the same (Dunbar 1983b). Even species like the vervet monkey (*Cercopithecus aethiops*), which ostensibly lives in a classic primate multimale group, has now been shown to exhibit comparable complexity in its social system. While the conventional view has been to assume that relationships among the members of these groups are more or less homogenous, more detailed study has demonstrated that these groups in fact consist of cliques of individuals that regularly combine to form coalitions against each other. In the vervets, these may consist of unrelated individuals or of individuals that are matrilineally related (Cheney 1983). In other species, such cliques are always matrilineal (e.g. many macaques: Koyama 1967; Sade 1972; Mori 1975). Moreover, the groups themselves are not such isolated entities as we have tended to suppose. Groups do exchange members from time to time and certain groups are more likely to do so than others (see above, p. 135). Figure 2 summarizes in schematic form the picture in the case of the vervets in so far as we currently understand it.

While it is true that most complex social systems so far documented have been described in primates, such complexity is by no means a uniquely primate characteristic. The intensive long-term study of elephants (*Loxodonta africana*) carried out in the Amboseli basin, Kenya, by Cynthia Moss has revealed a hierarchically organized clustering pattern similar to that described for primates (see Moss & Poole 1983). An individual female is a member of a family unit led by a matriarch. Each such unit contains several closely related adult females and their dependent offspring, each of which constitutes a mother–offspring group. Although a given family unit might be seen in association with almost every other family unit in the population at some time or another, the frequencies of association tend to cluster at definable points. Moss & Poole (1983) distinguished at least three other levels of grouping: two or three family units formed a bond group, with several bond groups making up a clan and several clans a sub-population, there being two sub-populations in the Amboseli basin. These grouping levels can be distinguished not only in terms of the animals' patterns of association, but, as with the gelada, in terms of the animals' behavioural responses towards each other. On meeting again after a period of separation, individuals that belong to the same family unit or bond group engage in intense greeting ceremonies in which they run together, back into each other, entwine trunks and flap their ears in a frenzy of excitement, all the while giving deep rumblings, trumpetings and screams. Members of different

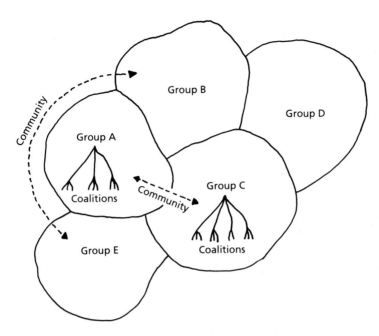

FIG. 2. Schematic representation of the relationships found in wild populations of vervet monkeys. A population consists of a number of groups of monkeys, each of which occupies a non-overlapping territory. Each group consists of twenty to thirty individuals who are organized into a number of cliques and coalitions which are reflected in the frequencies with which individuals groom each other and/or support each other in fights. Certain groups are more likely to exchange members with each other than they are with other groups in the population and these groups form a community. The groups in a community need not necessarily occupy adjacent territories and movement between them may have to occur across or around intervening groups' territories. (Based on analyses given by Cheney & Seyfarth 1983.)

bond groups, in contrast, approach each other quietly, often simply placing the tips of their trunks in each other's mouths before returning to their previous activities.

At the other end of the size spectrum, prairie dogs (*Cynomys ludovicianus*) have long been known to behave in rather similar ways. King (1955) found that a typical prairie dog town (some of which are several square kilometers in area) consists of a number of 'wards', each of which occupies its own sector within the town. King's main study town consisted of eight wards. Each ward in turn consisted of four to seven 'coteries': a coterie typically consisted of a single breeding male and two or three adult females, together with their dependent young, though some of the coteries contained additional males. King noted that members of the same coterie groomed each other frequently and engaged in intense 'kissing' ceremonies when they met. In contrast, members of different coteries are more antagonistic towards each other.

Structured social systems of this kind may also occur in birds, though they have not so far been widely reported. For present purposes, a single example must suffice, though I should emphasize that it is by no means the only example that could be given. Hegner, Emlen & Demong (1982) found that white-fronted bee-eater (*Merops bullockoides*) populations are divided into discrete colonies occupying isolated sand banks in a river bed. Each such colony is made up of a number of clans which own their own burrow systems within the colony's nest area. Each clan in turn consists of one or more breeding pairs, together with a small number of non-breeding individuals that act as 'helpers at the nest'. Clans actively defend their burrow system against non-members. They also have their own exclusive foraging areas, often situated at some considerable distance from the colony site. Thus, again, we find a hierarchically structured system in which groupings at one level are included within super-groups at a higher level (Fig. 3).

SOCIETIES AS ALLIANCE SYSTEMS

In considering such multi-level grouping patterns, we are, I think, inevitably reminded of the way in which many human social and political systems are organized. In many societies, a local unit of population (e.g. the village or parish) consists of several families living in close proximity and sharing a number of

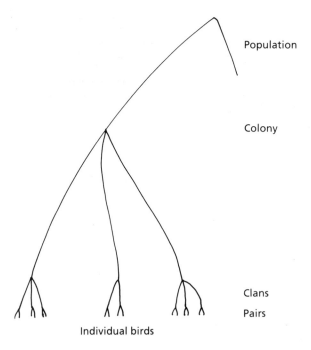

Fig. 3. Schematic representation of the relationships within a colony of white-fronted bee-eaters. A colony occupies a nesting site in a sandy bank. Each colony contains a number of clans, each of which defends a burrow within the colony nest site and a feeding territory away from the nest site; within each clan's burrow, individual breeding pairs have their own nest chambers. (Based on Hegner, Emlen & Demong 1982.)

common interests and problems. The village itself is part of a larger political unit (the district, county or clan) and these in turn are organized into yet larger units (the region, nation-state or tribe). In some cases, units are strictly hierarchical in a geographical sense: parishes are contained within counties which are contained within states. But in other cases they are not: members of a given clan may be dispersed throughout the territory of the tribe to which they belong. What marks out all these groupings as essentially similar is that they often seem to function as cooperative alliances for the solution of particular problems of common interest to the members of the group.

This is not to suggest that human and animal societies are organized in the same way or that they depend on the same cognitive or linguistic processes. Clearly, there are aspects to human societies that are unique, and that uniqueness often stems from the particular properties of human language. But these are differences in the mechanisms used to generate and cement social structures, not necessarily differences in the structures themselves or differences in the functions that those structures serve within the organism's life strategy.

Table 1 lists the various levels which have been identified in gelada social organization, together with the functions that they seem to subserve. The main problem faced by most primates (and particularly those feeding out on open plains as the gelada do) is predation risk (see van Schaik 1983; Dunbar 1988). To be able to exploit open grasslands which have few trees in which to escape predators at all effectively, the gelada form large herds. This was probably achieved through a general increase in the size of what had originally been a conventional *Papio* baboon troop (Dunbar 1986). Such large groups, however, force animals into close physical proximity. We know, for example, that inter-individual distances are severely compressed as herd size increases in the gelada and that herd sizes increase roughly in proportion to the predator-riskiness of the habitat sector (Dunbar 1986). This crowding increases tension and stress not only through competition for food and resting sites, but also as a result of repeated infringements of 'personal space' as animals move from one such site to another. Koyama, Fujii & Yonekawa (1981) describe similar effects in Japanese macaques (*Macaca fuscata*).

TABLE 1. Levels of grouping among gelada baboons and their probable functions. (From Kawai *et al.* (1983) and Dunbar (1986))

Grouping level	Function or context
Coalition	Minimizes harassment within unit, reduces reproductive suppression
Reproductive unit	breeding, rearing; minimizes stress within large herds
Team	Function unknown (?an incipient band)
Band	ecological unit and genetic grouping; facilitates formation of large herds as a defence against predators
Community	Uncertain: possibly facilitates ranging flexibility and provides males with an alternative source of reproductive units which they can take over

In the case of the gelada, the animals' response seems to have been for closely related females to form coalitions so as to buffer themselves against the worst effects without driving the other animals away altogether (since doing so would obviously make it difficult to benefit from the anti-predator advantages of large groups). The stability of these relationships over time has given rise to what we now see as one-male breeding units.

The gelada solution is thus a compromise arising out of the fact that their attempt to solve one problem (predation) creates an entirely new problem (stress) which in turn requires a further solution (the formation of stable groups by related females). But this in itself creates further problems. Once these groups grow in size as the females' daughters mature, there is a tendency for the same kinds of stresses due to crowding to be generated on a smaller scale. The females' response has been to form small-scale coalitions within the larger units. These coalitions are, again, based on matrilineal relationships among the females, so that a pattern emerges in which nuclear family groupings are formed within the extended families that constitute the one-male units.

The key point here is that although alliances and coalitions (I use the terms interchangeably) help individuals to solve particular problems, they are not without their costs. These either have to be set against the benefits to be gained from alliances (as an unavoidable cost) or some alternative sub-strategy has to be devised to minimize the worst effects. In certain cases, a flexible fission–fusion type of social system may allow animals to reduce these costs when there are no immediate benefits to be gained from a particular grouping strategy. Thus, the formation of tightly bonded groups of females which can come together in larger aggregrations allows the gelada to respond flexibly to the level of predation risk that they encounter as they move from one sector of their habitat to another. I shall say more about this in the following section; here, I simply want to draw attention to the fact that the different levels seem to be responses to specific problems that the animals encounter.

In terms of the ontogeny of gelada society, the band level of organization seems to provide a context in which individual reproductive units can form herds easily in response to predation risk. Bands do not exist in a physical sense (though they may well do so in the *minds* of the animals themselves), since they are simply the set of reproductive units that share a common ranging area and that therefore readily form herds with each other. In other words, a band is not a group that we can point to in a spatio-temporal sense (in the way that we can point to a reproductive one-male unit); rather, it emerges as a clustering of relationships in an analysis of the frequencies with which units are found together. The bands themselves seem to form super-groups termed 'communities' within which exchange of animals may be more likely to occur when animals move between bands (see Dunbar 1984). Such movement between bands may be facilitated by the fact that the constituent units are more familiar with each other; however, this kind of migration is a *consequence* and not a *cause* of the formation of community relationships. The existence of these relationships of greater toleration probably owes its origins to the animals' need to be able to search a much wider area than normal at times of periodic food shortage. However, so far no attempt has been made to identify the functional causes of community relationships, so that little can be made of this other than to note their presence and some of their consequences.

Disentangling the functional significance of such social systems is not easy and, so far, few attempts have been made to do so. We probably have a clear idea of what is involved for only one other primate, the hamadryas baboon. Yet even here, the significance of certain levels of society remains unclear. This might be because they do not have a definite function — they might simply be an 'emergent property' of the historical relationships among a set of animals. This appears to be the case, for example, for the 'teams' of two or three gelada reproductive units that have a particularly close relationship (see Fig. 1). So far as we can ascertain, a team has no particular function but rather represents the residual relationships between the products of the recent fission of a parent unit. In effect, a team is an incipient band, but has no special function of its own. (We do need to beware of asserting that a given phenomenon serves no function in an organism's life. Such claims are inevitably weak unless they are based on detailed tests that exclude specific functional explanations: it is all too easy to resort to a 'no function' explanation instead of carrying out a detailed analysis of the animals' behaviour. Although we need not suppose that every biological phenomenon has a function, in practice assuming that this is the case prompts more searching questions of the natural world than any other perspective does.)

By way of concluding this section, then, let me briefly outline one other example in which it has been possible to identify functional explanations for the evolution of a species' social system. As we have seen (p. 140, Fig. 3), the social system of the white-fronted bee-eater consists of a nested hierarchy of three types of groups: breeding pairs, clans and colonies. As in the case of the gelada, the bee-eater social system seems, in part at least, to be due to 'knock-on' effects created by an attempt to solve just one critical problem, namely a shortage of suitable sites (sand banks) where birds can excavate breeding burrows (Table 2). The scarcity of such sites leads to competition for control over them, and this seems to have led to the cooperative solution of colonial defence. But, as Emlen & Wrege (1986) show, the concentration of many individuals in one place creates new problems for the birds. Some females are given the opportunity to relieve themselves of part of the burden of reproduction by laying some (or all) of their eggs in other females' nests and leaving these individuals to cope with the costs of rearing the chicks to maturity. In addition, males are attracted by the concentration of receptive females: this causes severe harassment and stress, and a proportion of females may be fertilized by males other than their own mates as a result.

Emlen & Wrege (1986) argue that pairbonding is an attempt by the male to minimize this particular risk. The male's support then makes it possible for the female to reduce the risks of nest parasitism by other females because, with the male's support, she can now remain in her nest chamber throughout the critical

TABLE 2. Levels of social organization among white-fronted bee-eaters and their probable functional significance (based on Emlen (1982) and Emlen & Wrege (1986)

Grouping level	Function
Pair	Breeding unit: minimizes risk of harassment and nest-parasitism
Clan	Defence of nest chamber and feeding area; context facilitating 'helping-at-the-nest'
Colony	Defence of nesting sites

period when other females might lay their eggs there. Clans then seem to emerge as a means of cooperatively defending an actual nesting site (a burrow) within the colony. In addition, they make it possible for inexperienced or other unmated individuals to contribute genes to future generations by acting as 'helpers-at-the-nest' and helping related clan members to reproduce more effectively. Emlen (1982) was able to show that the frequency of helping was inversely related to habitat quality: helping was most common in those years when environmental conditions were too poor to allow some individuals to produce offspring successfully on their own.

DYNAMICS OF COMPLEX SOCIETIES

The complexity of multi-level social systems seems to arise from the fact that animals are trying to solve several different problems more or less simultaneously. If this is so, then we should expect to find a given level of grouping only where the relevant ecological or social problems are sufficiently prominent to have a significant impact on the animals' reproductive outputs. In other words, we should not expect to see all levels of a multi-level social system in every population of a given species. We could only expect every population to behave in the same way if either the underlying propensities are wholly genetically determined or if all ecological and demographic variables are constant across all habitats. Neither condition is likely to be true of large, slowly reproducing birds and mammals.

Gelada reproductive units, as we have already noted, form loose herds in order to be able to exploit rich grasslands on the flat, open plateau tops. The critical factor preventing lone units exploiting this resource is the risk of predation in an area that is devoid of refuges. Since habitats with abundant cliffs are more or less secure from cursorial predators, it is hardly surprising to find that a reproductive unit's willingness to travel alone is correlated with the slope of the habitat (Fig. 4). The band-level groupings that provide the basis for herd formation in gelada are thus less conspicuous in those habitats (and, indeed, in those sectors of a given habitat) where predation risk is negligible.

A second example is provided by the hamadryas baboon. Hamadryas one-male units also form bands. As in the case of the gelada, these bands seems to be a response to predation risk. Bands are a conspicuous feature of the hamadryas social system in Ethiopia where predators are moderately common (Kummer

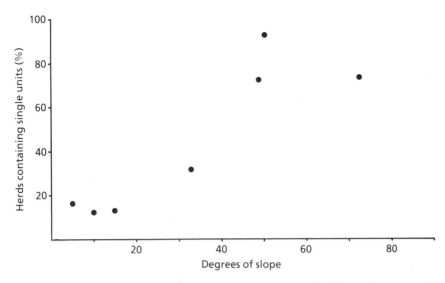

FIG. 4. The influence of predation risk on herd-formation among gelada baboons. The proportion of herds that consisted of a single reproductive unit (a measure of the units' willingness to travel alone) is plotted against the mean slope of the habitat from the horizontal for seven different habitat sectors in three study areas. The steeper the habitat, the less risk the animals face from cursorial predators and the more willing they are to travel alone. Removing food availability (measured as grass density) as a possible confounding variable increases the significance of the relationship. (From data given by Dunbar 1988; see also Dunbar 1986.)

1968; Sigg & Stolba 1981). In Saudi Arabia, however, where large predators have long since been exterminated, hamadryas units form bands much less readily; instead, the one male units tend to travel and sleep alone (Kummer *et al.* 1985).

Thus, in both cases, the occurrence of a particular form of social group depends on the extent to which the relevant environmental problem impinges on the daily lives of the animals. As the risk of predation recedes when gelada move onto the steep cliffs, the relative balance between the costs and benefits of grouping are tilted in favour of the costs. With no advantages to offset the stresses caused by the compacting of animals in large herds, the constituent reproductive units tend to drift apart and the herds break up.

In addition to such large-scale structural flexibility, these kinds of social systems may also be subject to important changes in their internal dynamics as demographic conditions change over time. I shall illustrate this with just two examples of the way in which an animal's options (and hence its preferences for particular social partners) can be radically altered by the demographic circumstances it happens to find itself in. In each case, the underlying rules that guide the animal's behaviour remain the same, but the optimum choice of social partner depends crucially on who happens to be available. The point I want to emphasize with these examples is that an animal's behaviour is not determined by some species-typical propensity to act in a certain way under all possible circumstances, but rather is the outcome of a certain amount of assessment of the options available and their relative costs and benefits. How an animal responds in this context will necessarily affect the form of the social system as we perceive it in the field.

Female gelada exhibit a hierarchical set of preferences for different types of potential allies. Attempts to model the females' decision problem suggest that the way in which a female ranks potential partners reflects the value of these classes of individuals in terms of the effect they are likely to have on her fitness (Dunbar 1984). Of the various potential allies available to a female, a mother or a daughter are by far the most profitable allies, followed by sisters. Do females behave in a way that conforms to these predictions? In fact, they do, showing a marked preference for close female relatives over less closely related individuals. However, not all females have such relatives available to them: the life-history characteristics of this population, for example, suggest that around 10% of the females can be expected to produce only a single daughter amid a string of sons, and in half these cases the mother will have died before the daughter is old enough to constitute an effective alliance partner. How should the daughter behave in such cases? The model suggests that females outside the circle of the immediate nuclear family will be of negligible value as allies, but that the unit's male may be a profitable option. Data from wild groups suggest that females who lack close relatives (mothers, sisters, daughters) do not often attempt to form coalitions with unrelated females; moreover, these are the only females who compete to monopolize the male as a partner (Dunbar 1984). In addition, a comparison of captive groups which have been together long enough to develop kinship structures with groups that have been convened from unrelated individuals confirms the importance of kinship as a determinant of the structure of relationships within the unit. In long-established groups, females form associations with their close female relatives, just as they do in the wild (Dunbar 1982b), but in convened groups of unrelated animals the females ignore each other and compete directly for access to the male (Kummer 1975). Both the structure of relationships within the group and the group's stability differ radically as a result.

A second example is provided by a study of mother–infant relationships in captive and free-ranging rhesus macaques (*Macaca mulatta*). Berman (1980) noted that mothers in the Madingley (Cambridge) colony became increasingly more tolerant of their infants straying from them with time, losing the restrictiveness which they had shown during the colony's early years and becoming more like the tolerant mothers found in the free-ranging Cayo Santiago population in the Caribbean. She pointed out that the key respect in which the two populations differed was the extent to which kinship lineages existed. The original Madingley colony consisted of unrelated females who were housed in groups. These groups were, however, allowed to grow naturally over the ensuing years, during which matrilines developed. Once extended families had evolved, females seemed to be aware that they could call on the support of their close relatives if their infants got into trouble and they were therefore more relaxed. Such differences in behavioural 'style' have been shown to have a marked effect on the nature of social relationships within groups (Simpson & Howe 1986; de Waal, this volume).

CONCLUSIONS

These various examples highlight two related points that we often overlook when discussing animal social systems. One is that the social systems of higher animals (and the primates in particular) are both very much more complex and more flexible than we have traditionally been inclined to suppose. Most species turn out on close study to exhibit such extreme flexibility that the whole notion of species-specific patterns of behaviour is called into question. This flexibility finds expression in multi-level societies in which individuals may have relationships with conspecifics at several different levels simultaneously. The second point that I have tried to highlight is that this complexity of social structure reflects the animals' need to solve many different ecological and social problems. What animals do in practice, however, depends on the way in which the relative costs and benefits of different patterns of behaviour fall out in a given context. And this will depend on the relative prominence of the various ecological and social problems that an animal has to contend with in order to reproduce successfully.

This flexibility of behaviour should discourage us from viewing animal societies as species-typical. One important reason for taking such a view is that it obviates any need to commit ourselves on the ontogenetic origins of behaviour. In most cases, we simply have no idea whether a given pattern of behaviour is genetically inherited in any simple sense or whether it is a purely phenotypic response to local environmental conditions. Since we do not know and since, from a functional (i.e. sociobiological) point of view, it makes no difference, there seems little advantage in becoming unnecessarily embroiled in the problems of the nature/nurture dispute. Ridding ourselves of this particular encumbrance seems to me an essential step towards a unified approach to the study of human and animal societies. Although there are important differences between animals and humans in terms of the cognitive processes on which their societies are founded, these differences are not generally relevant to the functional analysis of behaviour. Those differences are themselves, of course, an interesting topic for study, but we need to avoid the mistake of supposing that such differences have anything relevant to say about function.

ACKNOWLEDGEMENTS

This paper was prepared while I was in receipt of a University Research Fellowship from the University of Liverpool. I am indebted to the Department of Zoology in the University for providing me with support and facilities. I am especially grateful to Carel van Schaik and J.C. Coulson for their comments on the original draft of the manuscript.

REFERENCES

Berman, C.M. (1980). Mother–infant relationships among free-ranging rhesus monkeys on Cayo Santiago: with a comparison with captive pairs. *Animal Behaviour*, **28**, 860–73.

Cheney, D.L. (1983). Extrafamilial allowances among vervet monkeys. *Primate Social Relationships* (Ed. by R.A. Hind), pp. 278–86. Blackwell Scientific Publications, Oxford.

Cheney, D.L. & Seyfarth, R.M., (1983). Nonrandom dispersal in free-ranging vervet monkeys: social and genetic implications. *American Naturalist*, **122**, 392–412.

Cheney, D.L. & Seyfarth, R.M. (1985). Social and non-social knowledge in vervet monkeys. *Animal Intelligence* (Ed. by L. Weiskrantz), pp. 187–202. Clarendon Press, Oxford.

Colvin, J. (1983). Influences of the social situation on male emigration. *Primate Social Relationships* (Ed. by R.A. Hinde), pp. 160–71. Blackwell Scientific Publications, Oxford.

Crook, J.H. (1965). The adaptive significance of avian social organisation. *Social organisation of Animal Communities*. Symposia of the Zoological Society of London, 14, pp. 181–218. Academic Press, London.

Crook, J.H. (1966). Gelada baboon herd structure and movement: a comparative report. *Play, Exploration and Territory in Animals*. Symposia of the Zoological Society of London, 18, pp. 237–58. Academic Press, London.

Crook, J.H. & Gartlan, J.S. (1966). Evolution of primate societies. *Nature (London)*, **210**, 1200–3.

Dunbar, R.I.M. (1982a). Adaptation, fitness and the evolutionary tautology. *Current Problems in Sociobiology* (Ed. by King's College Sociobiology Group), pp. 9–18. Cambridge University Press, Cambridge.

Dunbar, R.I.M. (1982b). Structure of social relationships in a captive group of gelada baboons: a test of some hypotheses derived from studies of a wild population. *Primates*, **23**, 89–94.

Dunbar, R.I.M (1983a). Structure of gelada baboon reproductive units. IV. Integration at group level. *Zeitschrift für Tierpsychologie*, **63**, 265–82.

Dunbar, R.I.M. (1983b). Relationships and social structure in gelada and hamadryas baboons. *Primate Social Relationships* (Ed. by R.A. Hinde), pp. 299–307. Blackwell Scientific Publications, Oxford.

Dunbar, R.I.M. (1984). *Reproductive Decisions: An Economic Analysis of Gelada Baboon Social Strategies*. Princeton University Press, Princeton, New Jersey.

Dunbar, R.I.M. (1986). The social ecology of gelada baboons. *Ecological Aspects of Social Evolution* (Ed. by D. Rubenstein & R.W. Wrangham), pp. 332–51. Princeton University Press, Princeton, New Jersey.

Dunbar, R.I.M. (1988). *Primate Social Systems*. Croom Helm, Beckenham (Kent) and Cornell University Press, Ithaca.

Dunbar, R.I.M., Buckland, D.P. & Miller, D. (in preparation). Reproductive strategies of male feral goats.

Emlen, S.T. (1982). The evolution of helping. I. An ecological constraints model. *American Naturalist*, **119**, 40–53.

Emlen, S.T. & Wrege, P.H. (1986). Forced copulations and intra-specific parasitism: two costs of social living in the white-fronted bee-eater. *Ethology*, **71**, 2–29.

Gosling, M. (1986). The evolution of mating strategies in male antelopes. *Ecological Aspects of Social Evolution* (Ed. by D. Rubenstein & R.W. Wrangham), pp. 244–81. Princeton University Press, Princeton, New Jersey.

Goss-Custard, J.D., Dunbar, R.I.M. & Aldrich-Blake, P. (1972). Survival, mating and rearing strategies in the evolution of primate social structure. *Folia primatologica*, **17**, 1–19.

Gray, J.P. (1985). *Primate Sociobiology*. HARF Press, New Haven.

Harcourt, A.H. (1987). Dominance and fertility among female primates. *Journal of Zoology (London)*, **213**, 471–87.

Hegner, R.E., Emlen, S.T. & Demong, N.J. (1982). Spatial organisation of the white-fronted bee-eater. *Nature (London)*, **298**, 264–6.

Henzi, S.P. & Lucas, J.W. (1980). Observations on the inter-troop movement of adult vervet monkeys (*Cercopithecus aethiops*). *Folia primatologica*, **33**, 220–35.

Hinde, R.A. (1976). Interactions, relationships and social structure. *Man*, **11**, 1–17.

Kawai, M., Dunbar, R.I.M., Ohsawa, H. & Mori, U. (1983). Social organisation of gelada baboons: social units and definitions. *Primates*, **24**, 1–13.

King, J.A. (1955). Social behaviour, social organisation and population dynamics in a black-tailed prairiedog town in the Black Hills of South Dakota. *Contributions from the Laboratory of Vertebrate Biology of the University of Michigan, Ann Arbor*, **67**, 1–123.

Koyama, N. (1967). On dominance rank and kinship of wild Japanese monkey troop at Arashiyama. *Primates*, **8**, 189–216.

Koyama, T., Fujii, H., & Yonekawa, F. (1981). Comparative studies of gregariousness and social structure among seven feral *Macaca fuscata* groups. *Primate Behaviour and Sociobiology* (Ed. by B. Chiarelli & R. Corruccini), pp. 52–63. Springer-Verlag, Berlin.

Kummer, H. (1968). *Social Organisation of Hamadryas Baboons*. Karger, Basel.

Kummer, H. (1971). *Primate Societies*. Aldine-Atherton, Chicago, Illinois.

Kummer, H. (1975). Rules of dyad and group formation among captive gelada baboons (*Theropithecus gelada*). *Proceedings of the Fifth Congress of the International Primatological Society* (Ed. by S. Kondo, M. Kawai, A. Ehara & S. Kawamura), pp. 129–59. Japan Science Press, Tokyo.

Kummer, H. (1982). Social knowledge in free-ranging primates. *Animal Mind — Human Mind* (Ed. by D. Griffin), pp. 113–30. Springer-Verlag, Berlin.

Kummer, H. (1984). From laboratory to desert and back: a social system of hamadryas baboons. *Animal Behaviour*, **32**, 965–71.

Kummer, H., Banaja, A., Abo-Khatwa, A. & Ghandour, A. (1985). Differences in social behaviour between Ethiopian and Arabian hamadryas baboons. *Folia primatologica*, **45**, 1–8.

McFarland, D. & Houston, A.I. (1981). *Quantitative Ethology: A State Space Approach*. Pitman, London.

Mori, A. (1975). Signals found in the grooming interactions of wild Japanese monkeys of the Koshima troop. *Primates*, **16**, 107–40.

Moss, C. & Poole, J.H. (1983). Relationships and social structure of African elephants. *Primate Social Relationships* (Ed. by R.A. Hinde), pp. 314–25. Blackwell Scientific Publications, Oxford.

Nagel, U. (1979). On describing primate groups as systems: the concept of ecosocial behaviour. *Primate Ecology and Human Origins* (Ed. by I. Bernstein & E.O. Smith), pp. 313–40. Garland, New York.

Noordwijk, M. van & Schaik, C.P. van (1985). Male migration and rank acquisition in wild long-tailed macaques (*Macaca fascicularis*). *Animal Behaviour*, **33**, 849–61.

Schaik, C.P. van (1983). Why are diurnal primates living in groups? *Behaviour*, **87**, 120–44.

Sade, D.S. (1972). Sociometrics of *Macaca mulatta*. I. Linkages and cliques in grooming matrices. *Folia Primatologica*, **18**, 196–223.

Seyfarth, R.M. (1980). The distribution of grooming and related behaviours among adult female vervet monkeys. *Animal Behaviour*, **28**, 798–813.

Shotake, T. (1980). Genetic variability within and between herds of gelada baboons in central Ethiopian highlands. *Anthropologica Contemporanea*, **3**, 270.

Sigg, H. & Stolba, A. (1981). Home range and daily march in a hamadryas baboon troop. *Folia Primatologica*, **36**, 40–75.

Simpson, M.J.A. & Howe, S. (1986). Group and matriline differences in the behaviour of rhesus monkey infants. *Animal Behaviour*, **34**, 444–59.

Wilson, E.O. (1975). *Sociobiology: The New Synthesis*. Belknap/Harvard, Cambridge, Massachusetts.

Ecological correlates of pygmy chimpanzee social structure

F.J. WHITE

Department of Biological Anthropology and Anatomy, Duke University, Durham, North Carolina 27706, USA

SUMMARY

1 The fission−fusion nature of pygmy, or bonobo chimpanzee (*Pan paniscus*) social organization allows an examination of the influence of ecological and social factors on party size and composition.
2 Party size varies with the size of the food patch. The estimated amount of food removed proved a useful measure for smaller patch sizes. Party size increased with the size of the food patch and large food patches appeared to contain superabundant food.
3 Party composition changed with party size. The male proportion of the party increased with party size. Parties of pygmy chimpanzees were based on cores of females that regularly associated; males appeared to be attracted into larger parties in order to maintain proximity to females.
4 The major difference between the female-based social structure of *Pan paniscus* and the male-bonded *P. troglodytes* may be related to differences in patch size distributions.
5 The affiliative female homosexual behaviour of genito-genital rubbing observed in this species of chimpanzee was also correlated with food abundance.

INTRODUCTION

Living apes exhibit a heterogeneous collection of social systems. The smallest apes, gibbons, are almost all monogamous and territorial (Chivers 1980; Gittens & Raemaekers 1980; but see Haimoff 1986). Gibbons usually do not occur in large social groups. Adult orang-utans, like gibbons, do not form large social groups, but are essentially solitary (MacKinnon 1974), whereas gorillas and chimpanzees are both found in larger social groupings. Gorilla groups are relatively stable and consist of a dominant male, a number of other adult males, adult females and immatures (Harcourt *et al.* 1981). Both species of chimpanzees are found in large (up to eighty individuals), distinct communities. Community members associate in parties of flexible size and composition (Nishida 1979; Badrian & Badrian 1984; Goodall 1986). This type of social organization, generally referred to as fission−fusion, is relatively rare among primates. Although both *Pan paniscus*, the pygmy, or bonobo, chimpanzee, and the three subspecies of common chimpanzee, *P. troglodytes*, display a fission–fusion system, there are major differences in the type of fission–fusion social structure found in each species.

Ape species have drastically different body sizes and adaptations to different habitats. Each species displays a different social system. Differences in social systems have been related to major differences in feeding ecology (Wrangham 1979; Rodman 1984). The two species of chimpanzee show extensive overlap in body sizes, such that *P. paniscus* has the same mean and range of body weights as the well-studied *P. troglodytes schweinfurthii* (Jungers & Susman 1984). The differences in social system between *P. paniscus* and *P. troglodytes* may reflect adaptations to differences in the ecologies of these species.

The relationship between ecology and social systems in primates has been discussed and examined by many authors (for example, Crook 1970; Eisenberg *et al.* 1972; Clutton-Brock & Harvey 1977; Wrangham 1980; van Schaik 1983; Terborgh & Janson 1986). Many variables have been suggested as important selective factors that result in the development of different social structures in group-living primates. Differences in the costs and benefits to sociality in different populations could result in different primates developing different social systems. There are four major benefits and two major costs to sociality for primates. The benefits include; (i) reduced predation, (ii) increased access to limited resources such as food, water or resting sites, (iii) increased access to mates, and (iv) increased help with offspring. The costs include: (i) increased competition for limited resources, and (ii) increased competition for mates.

Of the four major factors that benefit individuals that live in groups, predation and access to helpers are probably not significant factors in chimpanzees, except in open country. Although large predatory carnivores have important influences on the behaviour of common chimpanzees in savannah habitats (Tutin *et al.* 1981), long-term observations of *Pan troglodytes schweinfurthii* in open forest at Gombe have shown that predation is not a significant mortality factor (Goodall 1983). Similarly, although allomothering and adoption do occur, they are not crucial for infant survival and development when adequate maternal care is available (Goodall 1986).

The remaining factors important in group living are the relative benefits and costs of feeding together, and access and competition for mates. Since differential parental investment between the sexes tends to result in competition among the members of the lower-investing sex for access to mates of the higher investing sex, in many mammals females are a limiting resource for males competing for reproductive success (Williams 1966; Trivers 1972). There is substantial evidence that females, in contrast, increase their reproductive success through effective foraging strategies (Sadlier 1969; Gaulin & Konner 1977). Therefore, it should be possible to explain the distribution of females by the distribution of food, and the distribution of males by the distribution of these females.

In a fission–fusion social system, such as the one displayed by pygmy, or bonobo chimpanzees, *Pan paniscus*, individuals are more able to vary whether or not they associate with others than are organisms with less flexible social structures. These variations in sociality presumably reflect how individual costs and benefits change with different ecological and social conditions. Because different factors affect the reproductive success of each sex, the relative cost-effectiveness of party membership should differ for males and females.

In this chapter I summarize data on the relationship between party size and food abundance in *P. paniscus* and compare the sociality of males and females across a range of party sizes. I will also compare these ecological correlates with data from a comparative study that examines why the two chimpanzee species have different social systems.

The pygmy, or bonobo, chimpanzee, *Pan paniscus*, is probably the least well known of all the Great Apes. Restricted in distribution to the primary and secondary forests of central Zaire, it is only recently that this species has become the focus of long-term field studies. There are two major *P. paniscus* study sites: at Wamba, where habituation of the study population has been facilitated by provisioning (for example, Kuroda 1979; Kano 1980) and at the Lomako Forest study site, where there has been no provisioning (for example, Badrian & Badrian 1984). In this chapter, I will present an overview of data examining the relationship between various ecological factors and the social system of *P. paniscus*. Data is only presented from the Lomako study site, as the effects that provisioning at Wamba may have on this relationship are presently unknown.

METHODS

The study site

The data presented here span 10 months from October 1984 to July 1985 and are from a 2-year field study of the behavioural ecology of *Pan paniscus* at the Lomako Forest Pygmy Chimpanzee Project study site in central Zaire. The study site covers a mosaic of forest types, with predominantly undisturbed polyspecific evergreen climax forest, as well as some areas of second growth. There are also some small areas of slope and swamp forest (White 1986). There is little apparent seasonality in rainfall, although rainfall may be less from June to August and from December to February.

Recording techniques

(a) When a party of *P. paniscus* was contacted, all individuals present were identified as far as possible. Changes in party composition, due to either fission or fusion, during observation were recorded when observed. One continuous sighting could, therefore, yield more than one measurement of party size or composition. (b) The location of the party relative to food patches was recorded. A *food patch*, as defined by White & Wrangham (1988) was any discrete location within which any food item could be found. Only one type of food, the pith of a *Haumania* vine, was not patchy in distribution but was ubiquitous throughout most forest types, although most abundant in secondary forest (White 1986; White & Wrangham 1988).

(c) The physical size of a food patch was estimated in several ways. (i) For trees (excluding strangler figs), the diameter at breast height (dbh) was recorded. (ii) The radius of the patch was recorded for trees (including strangler figs) and vines. (iii) Trees (including strangler figs) were divided, using several criteria, into large (greater than 50 cm dbh, taller than 5 m, radius larger than 10 m) and small (less than 50 cm dbh, radius less than 10 m).

(d) The amount of food taken from a patch by a party was estimated by recording the total amount of time that the party spent in a patch. However, not all individuals fed continually in a patch and therefore, the amount of food removed from a patch was not a simple function of the party size and the time spent at a patch. In order to calculate the amount of food taken, the number of individuals feeding at each 2-minute time point was estimated from observed feeding or feeding movements. This value was summed for the duration of the visit to the patch, and multiplied by two to give an estimate of the amount of food that a party removed from each patch, in chimpanzee-minutes.

(e) The recording of recognizable individuals present in parties produced a series of presence and absence data. Different party compositions also lasted for differing amounts of time. Therefore, the relative amount of time spent together in parties, as an association matrix, could be used as a measure of cohesion among different sets of individuals, as described by White (1986) and White & Burgman (in review). A second measure of cohesion was also measured and used to construct a second association matrix. This second association matrix was based on the amount of time that individuals spent in close feeding proximity (less than 5 m) during 2-minute sampling.

Other behaviours related to the cohesion between individuals, in particular genito-genital (GG) rubbing between females, as described by Thompson-Handler *et al.* (1984) were recorded when observed (White & Thompson-Handler, in review).

Treatment of data

The number of males and females observed in a party, as count data, were expected to follow a Poisson distribution and hence the means and variances were interrelated (Sokal & Rohlf 1981). Counts of party membership were, therefore, transformed using the equation $y' = $ square root of $(y + 0.5)$. The variances of the transformed data were compared, and in all cases were not significantly heterogeneous (Sokal & Rohlf 1981). Thus, the means could then be compared using analysis of variance (ANOVA).

Mantel tests were used to test specific hypotheses of the factors important in cohesion among individuals as reflected by the association matrices (White &

Burgman, in review). This is done by calculating the degree of correlation between the measured association matrix, and a hypothesis matrix. In the hypothesis matrix, pairs (for example male−male, or male−female) that are thought to have high levels of association are given large numbers (in this case, 1) and pairs thought to show low levels of association are given low numbers (in this case, 0). This degree of correlation is then compared to correlations calculated for many (500) randomized hypothesis matrices.

All possible pair combinations (i.e. male−male, female−female, and male−female) are examined separately. In this way, it is possible to test whether the observed correlation is greater than would be observed with a random distribution of numbers in the hypothesis matrix. The specific hypotheses, therefore, compared the relative cohesion among females, among males and between males and females. Patch size was measured as tree diameter (dbh) radius of patch, time spent by party in patch and food removed in chimpanzee minutes.

RESULTS

Party size and patch size

More than half of the food patches visited by *P. paniscus* were large trees greater than 50 cm dbh or 10 m radius (Table 1). Ground feeding on the ubiquitous *Haumania* vine accounted for about 15% of food patches visited. Before the relationship between party size and patch size — defined as the estimated amount of food removed — could be examined, it was necessary to see if the amount of food removed was independent of party size. It is possible that the number of 'chimpanzee-minutes' was a consequence of the number of animals present, and therefore, not an independent measure of the food available. If this is true, large parties should removed more food than small parties from patches of equivalent size. As has been shown elsewhere (White & Wrangham 1988), for small patches of less than 10 metres radius, the total number of chimpanzee minutes was a reliable measure of the amount of food available and patch size. In larger patches (greater than 10 metres radius), number of chimpanzee minutes spent was not an independent measure of patch size, but depended on the size of the party present in the patch.

The data indicate that party size increased with the size of the patch. The regressions (Table 2) show the relationships between the different measures of patch size and party size. Although there was no significant regression between party size and the dbh of the food patch (a measure only available for some trees), there was a significant regression between party size and the radius of the food patch and the amount of time spent in the food patch. The relationship between the total amount of food removed in chimpanzee minutes and party size was more complex (White & Wrangham 1988). In small food patches, there was a significant regression between party size and the total amount of food removed.

TABLE 1. Types of food patches visited (tree size is defined in the text) (from White 1986)

Food patch	% patch visits
Ground feeding (pith)	14·8
Vines	4·9
Small trees	15·6
Large trees	64·8

n = 132 food patches over 10 months

TABLE 2. Regressions of different measures of patch size with party size. R^2 values are given as percentages of total variance explained (from White 1986)

Parameter	R^2	R^2 (adjusted)	F	P
dbh	1·1	0·4	0·709	ns
radius	12·1	11·2	12·989	0·0005
time in food patch	18·9	17·8	17·940	0·0001

In larger patches, however, the amount of food removed depended on the party size present. This implies that food was not limited in large patches. Therefore, although party size in *P. paniscus* increased with the size of the food patch, large patches appeared to contain superabundant amounts of food.

Party size and party composition

Party membership often changed during sightings as well as from day to day (White, in press). The mean party size for all observations, including those of unknown communities, was 5·4 (n=164, S.D. 3·853, range 1 to 18). Observations were made of two separate communities and one splinter group. Of the two communities, only the Hedons were observed in a range of party sizes from small to large. This community was the most habituated and tolerant of observers. In contrast, the Rangers were relatively intolerant, unless in large parties. In consequence, all observations with reliable counts were of large parties. The splinter group, the Blobs, were always in relatively small parties. The mean party size for the Hedons was 7·15 (n=26, S.D. 4·505, range 2 to 17), for the Rangers 9·69 (n=26, S.D. 5·206, range 1 to 18), and 4·33 (n=87, S.D. 1·750, range 1 to 8) for the Blobs (White, in press).

Parties contained, on average, more females than males (Tables 3). There were twenty parties that contained at least two females but no males. There were no all-male parties. The mean number of males and females differed for parties from each community and the splinter group. The means of the counts were compared with an ANOVA (Sokal & Rohlf 1981). There were more females than males in parties of the Hedons (F = 27·38, P < 0·001) and the Blobs (F = 58·23, P < 0·001). There was no significant difference between the mean number of

TABLE 3. Mean party composition. Values for all observations include observations of unknown communities (from White 1986)

	Males	Females	Sample size
All observations	1·85	3·37	147
Hedons	1·38	4·46	26
Blobs	1·54	3·37	82
Rangers	3·69	4·54	26

males and the mean number of females in parties of Rangers ($F = 1·27$, N.S.). However, more of the known individuals were females. It is, therefore, possible that more females were observed in parties because there were more females in the population. A simple comparison of the observed sex ratio with an expected ratio calculated from the number of known individuals demonstrates some interesting differences (Table 4). In the Hedons and the Blobs, there were more females observed than the expected ratio. In the Rangers, there were more males observed than the expected ratio predicted. More detailed statistical analysis of the sex ratios in the Blobs has shown that the number of females in parties is greater than would be expected at random, even with the greater number of females known to be members of the Blob splinter group. Small sample sizes prohibit further testing of data for the Hedons and the Rangers (White 1986; White, in press).

The sex ratio, calculated as

$$\frac{\text{Number of males}}{\text{Number of males} + \text{number of females}},$$

was regressed against party size (White 1986; White, in press). There was a significant regression between this measure of sex ratio and party size in the Hedons ($F= 26.10$, $n=16$, $P<0·002$) and the Rangers ($F=11·15$, $n=23$, $P<0·003$). At larger party sizes, there were fewer females and more males, measured as a proportion of the total party size, in larger parties. This may explain the difference between the observed sex ratios of the Hedons and the Rangers, as well as the observed and expected ratios in this community. Observations of Rangers were all large parties, and large parties contain proportionally more males, thereby producing a sex ratio closer to equal numbers of males and females, and more males than expected, given the number of females among known animals.

Associations between individuals

As has been described in greater detail elsewhere (White 1986; White & Burgman, in review), it is possible to use techniques developed for numerical taxonomy to examine cohesion patterns between individuals. An association matrix based on similarity between each individual's presence record in parties was constructed. This index of similarity (equivalent to Jaccard's coefficient) was based on the

TABLE 4. Ratio of males to females (from White 1986)

	Known membership	Observed membership
Blobs	1 to 1	1 to 1·90
Rangers	1 to 1·57	1 to 1·23
Hedons	1 to 2·16	1 to 3·22

TABLE 5. Mantel test results for association matrix based on relative time spent together in parties (from White & Burgman, in review)

	Hedons	Blobs	Rangers
Female—female	0·714	0·964	0·078
Male—male	0·286	0·036	0·922
Male—female	0·300	0·028	0·984*

* denotes significance of $P<0.05$

amount of time that chimpanzee A and chimpanzee B were present together in a party, divided by the total time that either one was present.

It is possible to test whether the structure that one observes in data of this is due to particular factors (Schnell *et al.* 1985). The most significant factor in the structure of the association matrix, in the Hedons and the Blobs, was cohesion among females (Table 5). There was no cohesion among males, and there appeared to be a negative effect among males in the Blobs. Neither the Hedons nor the Blobs showed strong cohesions between males and females, to the point that there was almost a significant negative effect in the Blobs. However, in the Ranger data, the structure was more dependent on male—female and male—female cohesions, and not cohesions among females.

By displaying this same data in the form of a dendrogram (Figs 1a—c), it was possible to observe subgroupings of females that were regularly found together in the Hedons and the Blobs, whereas there was no similar grouping apparent in the Rangers. None of the dendrograms, including that of the Rangers, where a positive effect had appeared in the Mantel test, showed any subgrouping of only males. The Hedon and Ranger dendrograms both contained some subgroupings that consisted of males and females (White 1986; White & Burgman, in review).

A similar Mantel test was also carried out on another measure of cohesion among the same sets of individuals. This cohesion based on the amount of time spent feeding in close proximity (within 5 m). The results of this test differ to those of the test of party composition data (Table 6). In the Hedons and the Blobs, cohesion among females were still the most important factors in the data. There was a significant negative effect among males in the Blobs. There was no noticeable cohesion between males and females, to the extent that once more in the Blobs there may be a negative effect between the sexes. However, in the Rangers there was no longer any cohesion among males, but there was a significant cohesion between males and females. Therefore, although males were similar in their party membership, there were no all-male subgroupings in the dendrograms. This suggests that males do not form cohesive subgroups, even though there were important similarities in their membership in parties.

Males in the Rangers did not feed close together once they were within parties, but instead showed greater cohesion in this measure with females. These two results of party membership and feeding proximity taken together may indicate that, at the large party sizes observed in the Rangers, the same males were attracted to large parties, but they enter these parties in order to maintain proximity with females (White & Burgman, in review.)

DISCUSSION

The fission–fusion social organization of *Pan paniscus* is based around cores of females that regularly associate. As patch size increased, there was a concomitant increase in party size. However, as parties increased in size, they changed in composition. The proportion of males in the party increased; these males may have been attracted to the larger parties in order to maintain proximity to females. This social structure is radically different to the male-bonded social system of the common chimpanzee, *P. troglodytes* (Goodall 1986).

These observations support the theoretical proposals that female distribution should be dependent on the distribution of food. The observation that males are attracted more into larger parties also supports the proposal that male distribution is determined by female distribution in *P. paniscus*.

In a comparison of two populations of chimpanzees, it was found that *Pan paniscus* in the Lomako Forest and *P. troglodytes schweinfurthii* of Gombe showed major differences in the party size and patch sizes used (White & Wrangham, 1988). Party sizes were, on average, larger in the populations compared for *P. paniscus* than *P. troglodytes*. Larger party sizes within *P. paniscus* appeared to be related to the use of larger patches. *P. paniscus* used more large food patches, usually large trees, than *P. troglodytes schweinfurthii*. More of the patches used by *P. paniscus* contained superabundant food where feeding competition was either low or absent. Therefore, there appear to be ecological as well as social differences between the two chimpanzee species.

The small size of food patches of *P. troglodytes* has been proposed as an important factor in why females are unable to be more social (Wrangham 1975, 1979). Females are semi-solitary in order to minimize feeding competition and to maximize the benefits to offspring through improved nutrition. This does not appear to be the case in *Pan paniscus*, as females were highly cohesive. Therefore, reduced feeding competition has allowed female *P. paniscus* to be highly social. However, this does not explain what benefits these females gain from sociality.

One hypothesis proposed to explain this increased female sociality in *P. paniscus* is that it reflects an adaptation by females to feeding in large, predictable and abundant patches where food defence is a cost-effective strategy (White

(a)

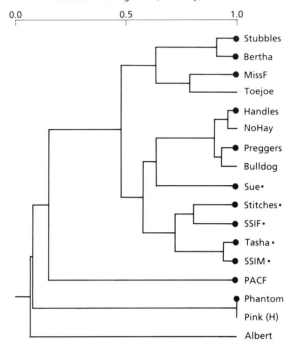

Hedons — duration of observation
Relative time together (similarity)

(b)

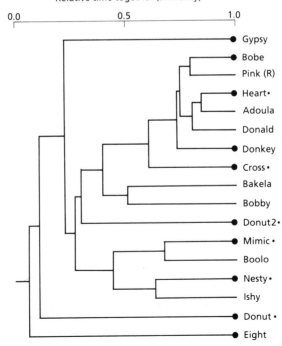

Rangers — duration of observation
Relative time together (similarity)

(c)

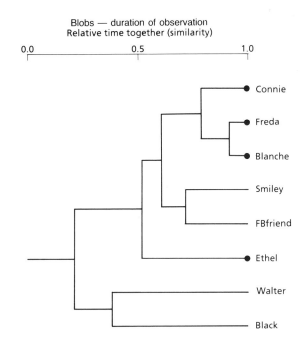

FIG. 1. Dendrograms of the relative amount of time that different individuals spent together. (a) Hedon community; note the subgroupings of male plus female, and females only. UPGMA r_{cs} = 0·959. (b) Ranger Community; note less higher order structure, with male plus female subgroupings UPGMA r_{cs} = 0·923. (c) Blob splinter group; tight subgroup of the females UPGMA r_{cs} = 0·976. Key: female ● (name); mother ● (name) •; male (name).

TABLE 6. Mantel test results for association matrix based on relative time spent in close feeding proximity (from White and Burgman in review)

	Hedons	Blobs	Rangers
Female−female	0·836	0·984*	0·546
Male−male	0·164	0·016*	0·454
Male−female	0·304	0·028	0·826

* denotes significance of $P<0·05$.

1986), whereas *P. troglodytes* feed in smaller food patches where feeding competition prohibits such cooperation (Wrangham 1979). This hypothesis is supported by the high degree of cohesion among females in presence of parties, maintenance of proximity, interactions, and observations of food calling by females (White 1986). Such highly differentiated female relationships and subgroupings, often occurred in the absence of males or infants (White 1986; White, in press) implying that female sociality was not dependent on males or on nursery parties. Defence of food patches has, however, yet to be demonstrated in the field.

Another hypothesis, proposed by Wrangham (1986), considers the utilization by *P. paniscus* of the ubiquitous *Haumania* vine. When feeding on this food type, feeding competition is presumably low, thus allowing females to be more social. The major factors selecting for sociality in this hypothesis are, however, social rather than ecological. The possible benefit to females of being more social is the avoidance of male harassment, by maintaining contact with protective males. Such a system would presumably not produce highly differentiated female relationships that are independent of male presence.

Other measures of cohesion between individuals were recorded during this study. The distribution of interactions also showed a high degree of cohesion among females but not among males (White 1986). The cohesive female homosexual behaviour of genito-genital (GG) rubbing has been shown to be highly correlated with patch size (White 1986; White & Thompson-Handler, in review). Most GG rubbings occurred at the start of a feeding bout, and the frequency of rubbing was not dependent on the party size present. There was, however, a significant correlation between the number of GG rubbings at the start of a feeding bout and the estimated amount of food removed during the subsequent feeding bout (Spearman rank correlation = 0·47, $P < 0·001$). Therefore, the cohesion among females, as reflected by the frequency of GG rubbing, was dependent on the amount of food that the animals were about to remove from a patch. Even females that travelled together before they reached a food patch would GG rub. This activity did not appear to function primarily as a greeting between individual females. Instead they appeared to enter a patch, assess the amount of food present, and GG rub accordingly. *P. paniscus* may, therefore, use this behaviour as an assurance at the start of a feeding bout that two females will at least tolerate, and possibly cooperate, to defend food patches.

ACKNOWLEDGEMENTS

I thank John Fleagle for continued advice, and I am grateful to Charlie Janson for assistance with analyses. Much of the results referred to here are from collaborations with Mark Burgman, Richard Wrangham and Nancy Thompson-Handler. Randy Susman, Nancy Thompson-Handler, Richard Malenky, Annette Lanjouw, Donald Gerhart, John Fleagle, and Charles Janson provided valuable help and stimulating discussion both in and out of the field. Computer assistance was provided by Scott Ferson and Kent Fiala. I thank William McGrew and Frans de Waal for helpful comments on the manuscript. This work was supported by a NSF doctoral dissertation improvement award and a grant from the Boise fund.

REFERENCES

Badrian, A.J. & Badrian, N.L. (1984). Group composition and social structure of *Pan paniscus* in the Lomako Forest. *The Pygmy Chimpanzee; Evolutionary Biology and Behavior* (Ed. by R.L. Susman), pp. 325–46. Plenum Press, New York.

Chivers, D.J. (1980). *Malayan Forest Primates: Ten Years' Study in Tropical Rainforest*. Plenum Press, New York.

Clutton-Brock, T.H. & Harvey, P.H. (1977). Primate ecology and social organization, *Journal of Zoology (London).*, 183, 1−39.

Coolidge, H.J. (1933). *Pan paniscus* pygmy chimpanzee from south of the Congo river. *American Journal of Physical Anthropology*, 18, 1−59.

Crook, J.H. (1970). The socio-ecology of primates. *Social Behaviour in Birds and Mammals* (Ed. by J.H. Crook). Academic Press, London.

Eisenberg, J.F., Muckenhirn, N.A. & Rudran, R. (1972). The relation between ecology and social structure in primates. *Science*, 176, 863−74.

Gaulin, S.J.C & Konner, M.J. (1977). On the natural diets of primates. *Nutrition and the Brain, Vol. 1* (Ed. by R.J. Wurtman & J.J. Wurtman). Raven Press, New York.

Gittens, S.P. & Raemaekers, J.J. (1980). Siamang, Lar and agile gibbons. *Malayan Forest Primates: Ten Years' Study in Tropical Rainforest* (Ed. by D.J. Chivers), pp. 63−105. Plenum Press, New York.

Goodall, J. (1983). Population dynamics during a 15-year period in one community of free-living chimpanzees in the Gombe National Park, Tanzania, *Zeitschrift für Tierpsychologie*, 61, 1−60.

Goodall, J. (1986). *The Chimpanzees of Gombe: Patterns of Behavior*. Harvard University Press, Cambridge, Massachusetts.

Haimoff, E.H. (1986). Preliminary observations on wild concolor gibbons (*Hylobates concolor concolor*) in Yunnan Province, People's Republic of China. *American Journal of Primatology*, 10, 405.

Harcourt, A.H., Fossay, D. & Sabater Pi, J. (1981). Demography of *Gorilla gorilla*. *Journal of Zoology (London)*, 195, 21−33.

Jungers, W.L. & Susman, R.L. (1984). Body size and skeletal allometry in African apes. *The Pygmy Chimpanzee; Evolutionary Biology and Behavior* (Ed. by R.L. Susman) pp. 131−71. Plenum Press, New York.

Kano, T. (1980). Social behavior of wild pygmy chimpanzees (*Pan paniscus*) of Wamba: a preliminary report. *Journal of Human Evolution*, 9, 243−60.

Kuroda, S. (1979). Pygmy chimpanzee association patterns in ranging. *Primates*, 24, 1−12.

MacKinnon, J. (1974). The behaviour and ecology of wild orangutans *Pongo pygmaeus*. *Animal Behaviour*, 22, 3−74.

Nishida, T. (1979). The social structure of chimpanzees of the Mahale Mountains. *The Great Apes* (Ed. by D.A. Hamburg & E. McCown), pp. 72−121. Benjamin Cummings, Palo Alto.

Rodman, P.S. (1984). Foraging and social systems of orangutans and chimpanzees. *Adaptation for Foraging in Non-human Primates* (Ed. by J.G.H. Cant & P.S. Rodman), pp. 134−60. (Columbis University Press, New York.

Sadlier, R.M.S. (1969). The role of nutrition in the reproduction of wild animals. *Journal of Reproduction and Fertility (suppl.)*, 6, 39−48.

Schaik, C.P. van (1983). Why are diurnal primates living in groups? *Behavior*, 87, 120−44.

Schnell, G.D., Watt, D.J. & Douglas, M.E. (1985). Statistical comparison of proximity matrices: applications in animal behavior. *Animal Behavior*, 33, 239−53.

Sokal, R.R. & Rohlf, F.J. (1981). *Biometry* (2nd edn.). Freeman & Co., New York.

Terborgh, J. & Janson, C.H. (1986). The socioecology of primate groups. *Annual Review of Ecology and Systematics*, 17, 111−35.

Thompson-Handler, N., Malenky, R.K. & Badrian, N. (1984). Sexual behavior of *Pan paniscus* under natural conditions in the Lomako Forest, Equateur, Zaire. *The Pygmy Chimpanzee; Evolutionary Biology and Behavior* (Ed. by R.L. Susman), pp. 347−68. Plenum Press, New York.

Trivers, R.L. (1972). Parental investment and sexual selection. *Sexual Selection and the Descent of Man, 1871−1971* (Ed. by B. Campbell), pp. 136−79. Aldine, Chicago.

Tutin, C.E.G., McGrew, W.C. & Baldwin, P.J. (1981). Responses of wild chimpanzees to potential predators. *Primate Behaviour and Sociobiology* (Ed. by R.S. Corruccini & A.B. Chiarelli). Springer, Heidelberg.

White, F.J. (1986). *Behavioral ecology of the pygmy chimpanzee*. Ph.D. thesis, State University of New York, Stony Brook.

White, F.J. (in press). Party composition and dynamics in *Pan paniscus*. *International Journal of Primatology*.

White, F.J. & Wrangham, R.W. (1988). Feeding competition and patch size in the chimpanzee species *Pan paniscus* and *Pan troglodytes*. *Behaviour*, **105(2)**, 148−64.

White, F.J. & Burgman, M.A. (in review). Pygmy chimpanzee social organization: multivariate analysis of associations.

White, F.J. & Thompson-Handler, N. (in review). Ecological correlates of homosexual behaviour in *Pan paniscus*.

Williams, G.C. (1966). Adaptation and Natural Selection. Princeton University Press, Princeton, New Jersey.

Wrangham, R.W. (1975). *The Behavioural ecology of chimpanzees in Gombe National* Park. Ph.D. thesis University of Cambridge, Cambridge.

Wrangham, R.W. (1979). On the evolution of ape social systems. *Social Science Information*, **18**, 335−68.

Wrangham, R.W. (1980). An ecological model of female-bonded primates. *Behavior*, **75**, 262−300.

Wrangham, R.W. (1986). Ecology and social relationships in two species of chimpanzee. *Ecological Aspects of Social Evolution in Birds and Mammals* (Ed. by D.I. Rubenstein & R.W. Wrangham), pp. 354−78. Princeton University Press, Princeton, New Jersey.

Ecological influences on systems of food production and social organization of South Turkana pastoralists

R. DYSON-HUDSON

Department of Anthropology, Cornell University,
Ithaca, New York 14853, USA

SUMMARY

1 This paper describes the food production system and social organization of the Ngisonyoka Turkana pastoralists of north-western Kenya, about 10 000 people living in an arid, topographically diverse ecosystem of about 7500 km^2, where rainfall is of low spatial and temporal predictability and plant productivity is primarily water controlled.

2 Each topographic region of Ngisonyoka territory has differences in plant communities, productivity, water sources, human competitors, predators and pests, and soil development. Most plant communities in the various topographic regions have well-developed herbaceous, shrub, and tree layers, each layer having a different pattern of plant growth and maturation after precipitation pulses.

3 The Ngisonyoka food production system copes with low and spatially unpredictable forage availability, and low fuel and capital availability, by labour-intensive herding livestock with relatively unconstrained movement within a territory with permeable boundaries. The people deal with the topographic and habitat diversity by herding five species of livestock, each of which exploits a different ecological niche.

4 The effects on animal production of seasonal and year-to-year variations in plant productivity are dampened through the use of multiple pathways of the food web, and by exchanging high protein for high calorie foodstuffs. The energy conversion of the Ngisonyoka food production system is comparable to that of other animal-based systems in arid regions, and has caused no discernible degradation of the ecosystem.

5 The social organization of Ngisonyoka pastoralists can best be understood as responses to the centrifugal forces of low plant productivity, heterogeneous topography, and limited access to water sources near some important grazing regions; and the centripetal forces of protection against human and animal predation, the need for cooperative well-digging, and the desire for sociality, in a topographically diverse territory with complex vegetation communities, where each region has different risks and resources, each vegetation layer within the community has a different phenology, and these vary with low predictability through time. It is characterized by flexibility and variability at all levels.

6 Ngisonyoka are members of three different types of residential groups: camps, and primary and secondary neighbourhoods. The main camp, which averages about thirteen family members, also includes the livestock to support these people. It moves frequently, following an unpredictable migratory path on the plains. When forage is abundant, all a herd owner's livestock are herded together in the main camp: when good forage and water for all species of livestock are not available in one area, species- and production-specific sub-herds are moved to satellite camps and pursue independent migratory orbits.

7 When forage is abundant, camps are clustered into primary neighbourhoods of some two to twenty-five camps, which may be grouped into secondary neighbourhoods of up to 1000 people. These provide social support and protection for the herd owner, but do not prevent him moving independently to where he believes the conditions to be best for his animals. When forage and water resources become scarce, the secondary neighbourhoods break up into individual primary neighbourhoods, or even independently moving camps.

8 Although proximate environmental variables clearly influence Ngisonyoka social organization, differences among individuals in well ownership, in family structure, in herding experience, and in capability, also influence behavioural responses. Individual behaviours are influenced by societal norms as well, and these also cannot be understood simply as adaptive responses to the features of the immediate environment.

INTRODUCTION

More than 25 years before the publication of John Crook's (1965) comparative study which focused on correlations between proximate environmental variables (particularly predation and the distribution and abundance of food resources) and variations in avian social organization, Julian Steward (1938) recognized that variations in social organization among aboriginal human groups in the Great Basin of the United States could best be interpreted as adaptive responses to differences in the distribution and abundance of their food resources. Although in later works, Steward specified that the appropriate level of analysis was the culture, his early work was compatible with the assumption that the individual, or individual family, was the unit of adaptive behaviour.

Later developments in cultural ecology and cultural evolution were dominated by essentialist descriptions of adaptation at the level of the society, or the culture, or the group (see, for example, Steward 1955; Bennett 1969; Carniero 1970; Rappaport 1984), rather than focusing on social organization as patterns generated by the aggregate of individual behaviours (Firth 1964). These 'ecological' and 'evolutionary' studies of human societies therefore had no relevance to the rapidly developing field of animal socioecology, which is rooted in Darwinian theory and requires that: 'When recognized, adaptation should be attributed to no higher level of organization than is demanded by the evidence' (Williams 1966, p. vii.)

In this paper, I shall describe the food production system and social organiz-ation of the Ngisonyoka section of Turkana pastoralists who live in north-western Kenya near the border of Ethiopia, the Sudan, and Uganda (Fig. 1). The study is part of the South Turkana Ecosystem Project, an ongoing investigation by natural and social scientists of the abiotic environment, the plants, the animals, the people, and the interactions between trophic levels, in the complex ecosystem which includes nomadic livestock herders. We use the concept of the ecosystem not to imply a homeostatic biological entity, but rather as an analytic concept to focus attention on the multiplicity of interaction within and between trophic levels: between different interacting species living together in a restricted space (Vayda & McCay 1975; Smith 1983; O'Neill et al. 1986).

This description of social organization of Ngisonyoka pastoralists is based on my 36 months of field work in Kenya, between August 1980 and August 1986, and on published research by other members of the team. In my research, I relied on direct observation, and on interviews carried out in the Turkana language, sometimes with the aid of an English-speaking interpreter.

The analysis of livestock movements is based on a 6-year study of Ngisonyoka pastoral families, carried out jointly with J.T. McCabe. Since each family moves frequently, at times divides its livestock holdings into several herds, each of which pursues a different migratory orbit, and may move its animals anywhere within an area of about 10 000 km^2 (see below), this study focused on a small sample of four families and all their livestock, recording where each herd of each family moved, when they moved, and why they moved. We could not make direct observations on the location of all the herds of these four families throughout the 6-year period of the study, because of the poor or non-existent roads, the harsh environment, the presence of bandits and hostile Pokot, and limitations of time and funding. We therefore used the indigenous knowledge system of Ngisonyoka herders to monitor herd movements (Dyson-Hudson 1981), supplementing direct observation with two forms of interviews. Bi-monthly surveys covering a wide variety of subjects (including camp locations) were carried out with each of four study families between July 1980 and November 1982 (McCabe 1985); and extended interviews exclusively about migratory moves, including questions about location of camps, grazing areas, water sources, and reasons for moving away from one site and to another, were carried out with each of the four herd owners and all their herd managers in October 1981, August–November 1982, November 1984, and August 1985. Maps which I had prepared locating about 1000 named land-marks throughout Ngisonyoka territory (including the major waterholes) enabled us to map camp sites, grazing areas, and water sources from this interview data. Ngisonyoka have a lunar calendar, so by using the months and lengths of stay specified by the herd owners and herd managers, and also by direct observations by us and by other project members, the times of migration events were recon-structed with reasonable accuracy (Dyson-Hudson & McCabe 1985, pp. 170–81.)

J.T. McCabe and I made independent, complementary studies of camp and neighbourhood composition. McCabe studied the variation among the four study

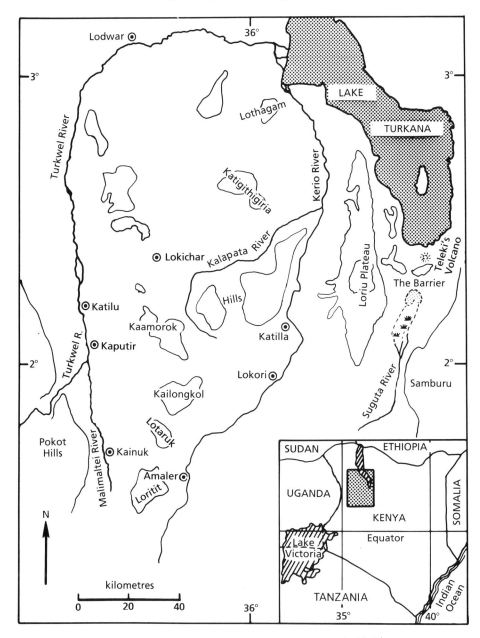

FIG. 1. Map of Ngisonyoka, Turkana District, Kenya (prepared by M.A. Little).

families in family and herd size and structure between July 1980 and November 1982. Between June and August, 1982, I made a survey of family and herd size and structure of a sample of approximately seventy-five Ngisonyoka families living in eight neighbourhoods. In subsequent years (June–August 1985, June–August

1986), I interviewed approximately twenty-five of these herd owners, to reconstruct changes in composition of camps and neighbourhoods to determine patterns of cooperation, and compare the relative stability of different kinds of social bonds (Dyson-Hudson 1985; McCabe 1985).

Inter-sectional relations, and processes of leadership and decision-making, were determined both by verbal reports during interviews, and by fortuitous observations made during the course of the above studies. Information about livestock, vegetation, and topography was obtained by direct observation, and from published work by other project members, as cited.

DESCRIPTION OF THE STUDY AREA AND GROUP

The Ngisonyoka, one of the nineteen sections of the Turkana tribe (Gulliver 1951) number about 10 000 people, and carry out most of their subsistence activities within an arid territory of approximately 7500 km^2 lying just north of the equator between 1° and 3° north longitude, at 36° east latitude (Fig. 1). Their food production system, which involves no capital investment and is not strongly tied into the national economy, is based on labour-intensive husbandry of five livestock species (cattle, camels, goats, sheep, and donkeys), and the use multiple resources (meat, blood, milk) from each species. Unlike African livestock herders who have been influenced by Islam, Turkana will eat the meat of an animal which dies of disease or starvation.

Although the present-day Turkana almost certainly are not descended from the earliest pastoralists in the region, people herding goats and sheep and cultivating grain have occupied the Lake Turkana basin since 3000 BC, and there were cattle in the area by the third millenium BC. Camels are a relatively recent acquisition, dating to the late eighteenth century (Lamphear 1986.)

Ngisonyoka depend almost entirely on food which comes directly or indirectly from these livestock. A 16-month nutritional study (August 1981 through November 1982), found that the four study families obtained 76% of their food energy directly from livestock through meat, milk, and blood; 16% through selling or bartering livestock for sugar, sorghum, and maize; and the remaining 8% from wild animals and plants (Galvin 1985). (Also, hides, horns, and other animal by products are used for utensils and shelters, while donkeys are important transporting water and the family effects.) However, Ngisonyoka herds do not seem to provide surplus food-energy for the people depending on them. The nutritional study showed that, although the diet of the four Ngisonyoka families was high in protein, it was low in calories: the food intake provided somewhat less than the estimated maintenance requirements of the population calculated from individual metabolic requirements, the age−sex structure of the population, and the observed activity budgets (Galvin 1985). The fact that infant mortality is lower among settled Ngisonyoka families in a mission-run irrigation scheme, who have a less variable and higher-calorie diet, suggests that a portion of the mortality in pastoral families is due to this low availability of food-energy (Brainard 1986.)

Livestock are necessary for a man to marry: although the numbers vary according both to the wealth of the groom and to the size of the bride's family, more than 100 animals may be given to the family of the bride in order to complete a marriage (Table 1). These substantial bridewealth payments are despite the fact that Ngisonyoka herds are relatively small. Although the mixed composition of livestock herds makes comparison difficult, two studies made in 1982 suggest that the number of livestock per person in Ngisonyoka pastoral families is low as compared with other East African pastoralists who also depend almost entirely on livestock, but who live in less arid regions (Table 2).

Since until very recently, the major alternative for pastoralists who failed was to replenish their herds through the high-risk strategy of raiding neighbouring groups, these data suggest that successful herd management has been a major factor in survival and reproduction among Ngisonyoka. To be successful, pastoralists have had to cope with the problems faced by all livestock keepers: providing the animals with water and nutritionally adequate forage, while protecting them from human and animal predators, and from other environmental hazards.

Livestock

Since each species of livestock has somewhat different forage and water requirements, and different energy costs of locomotion on level ground and up hill

TABLE 1. Number of livestock given in bridewealth by four Ngisonyoka herd owners (from Wienpahl 1984)

No. of herd owner	No. of wife	Camels	Cattle	Small stock	Donkeys
			Total given for bridewealth (number from own herd)		
	1	23 (21)	35 (31)	68 (all)	1 (all)
1	2	Just beginning to pay: five camels and three cattle so far			
	1	20 (all*)	40 (all*)	c.100 (all*)	0 —
2	2	30 (25)	40 (29)	c.200 (all)	0 —
	3	20 (14)	30 (all)	c.100 (all)	0 —
	4	Inherited from deceased elder brother			
3	1	30 (28)	45 (44)	many[†] (all)	11 (10)
	2	None yet			
4	1	25 (12)	15 (14)	70 (58)	0 —

* From his father's herd, who is still alive.
[†] 'You give goats without counting. I opened corral and gave them all the goats inside'.

TABLE 2(a). The numbers of livestock per person in Ngisonyoka herds (1982)

		Livestock/human ratio			
	No. of people	Goats and sheep	Cattle	Camels	Donkeys
Sample of 70 Awl*	879	6·74	1·00	0·88	ND
Total Ngisonyoka[†]	9560	8·91	1·03	1·03	0·55

TABLE 2(b). The ratios of cattle to people in East African groups with little or no agriculture (from Schneider 1979, p. 87)

Ethnic group	Ratio of cattle to people[‡]
Barabaig of central Tanzania	18·1
Samburu of Kenya	17·5:1
Maasai of Tansania	15:1
Rendille of northern Kenya	9·1
Uganda Pokot	8:1
Borana Galla of northern Kenya	6·5:1
Maasai of Kenya	6·1:1

* Based on information from a study of seventy Ngisonyoka awi made in June–August 1982 (Dyson-Hudson 1985).

[†] Based on a large-scale survey made by Ecosystems Analysis in 1982 (Ecosystems 1985).

[‡] These groups also herd other species of livestock, but their numbers are not recorded by Schneider.

(Table 3), the multiplicity of livestock species in Ngisonyoka herds adds to the complexity of herd management decisions. Cattle, donkeys, and sheep feed primarily on herbaceous plants; while camels browse both small and large woody species; and goats are true mixed feeders, eating both herbaceous and woody vegetation. In season, *Acacia tortilis* seed-pods provide important food for goats and sheep and occasionally the young camels (Coppock 1985).

Each of the five livestock species has different water requirements (Dyson-Hudson & McCabe 1985), with lactating females of any species needing 30% to 50% more water than do adult males and non-lactating females (NCR 1981). Adult small stock must drink every other day; and water must be brought from the wells on donkey back for the kids and lambs, who remain in the shade near the camp. Camels thrive on the water of mineral springs, and generally drink every fifth day during the dry season: when green forage is abundant they can go up to three weeks without drinking. Cattle thrive best on the 'cool' water of wells near the headwaters of the local drainage systems, and should be watered every day or at most every other day, although under exceptionally bad conditions they may be watered every third or even fourth day (McCabe 1985).

Since energy costs of locomotion vary with size (Schmidt Nielson 1984), the large livestock species have relatively low energy costs per unit weight in locomotion on level ground as compared with small stock; while the small livestock species

TABLE 3. Niche differentiation among the five species of Ngisonyoka livestock

Characteristics	Camels	Cattle	Goats	Sheep	Donkeys
Preferred forage[*]	Dwarf shrubs, shrubs, trees	Herbaceous plants	True mixed feeders	Herbaceous dwarf shrubs	Herbaceous dwarf shrubs
Adult mortality[†] 1980–1981 drought	37%	54%	37%	43%	No data
Mean weight (kg)[‡]	360(12)	183(15)	18(27)	20(24)	180(6)
Preferred water source	Springs	Wells	Both	Both	Both
Average distance to water[§]	6·8	6·4	3·7	3·7	—
Labour demands	Mature herder	Adult male	12-year-old	12-year-old	Not herded
Preferred watering interval	4 days	2 days	2 days	2 days	—
Maximum watering interval	10 days	3 days	3 days (5)	3 days (5)	No data
Energy cost: level ground ‖	Lowest	Low	High	High	Medium
Energy cost: uphill travel ¶	Highest	High	Lowest	Lowest	Medium

[*] Coppock (1985).
[†] McCabe (1985).
[‡] Coppock (1985). Number in parentheses is size of sample.
[§] Dyson-Hudson & McCabe (1985). Map distance between camp and well, not distance actually walk.
‖ This length of watering interval is rare, only in droughts when forage far from water.
¶ Schmidt Nielson (1984).

have relatively low energy costs per unit weight when climbing hills as compared with large livestock species.

Herding requirements of the different livestock species differ, and when he becomes an independent herd owner, a man tends to concentrate on the species he herded as a lad. Small stock are the easiest to manage, and usually are kept in mixed flocks of up to 300 goats and sheep managed by a boy, sometimes a girl, aged 8 to 18 years. Camels require a responsible adolescent herder: they can pace steadily for hours at a stretch, are not closely tied to particular water sources, and are not as gregarious as are cattle or small stock. A responsible young man must manage the cattle, which often are kept in large herds of 100–150 animals, use dry season pastures remote from the areas most suitable for small stock and camels, and (to get additional hours of foraging and additional moisture) sometimes graze at night and must be protected from large nocturnal predators. Donkeys usually are allowed to roam, and rounded up when needed.

Climatic conditions

Climatic conditions make it difficult for Ngisonyoka pastoralists to provide free water and nutritional forage for all livestock species on a daily basis throughout the year. Ngisonyoka territory is hot and dry with relatively little annual temperature variation: the mean ambient temperature is between 29° and 30°, with an annual range of 1° to 2°, and a diurnal range of 11° to 13° (Little & Johnson 1985). The average annual rainfall in South Turkana is about 270 mm. However, Ngisonyoka pastoralists are not able to predict with any degree of certainty how much rain will fall, when it will fall, nor where it will fall. At Lodwar, which is about 80 km north of the study area and therefore somewhat more arid, the mean annual rainfall over 56 years was 185·2 mm, with the highest recorded rainfall (498·3 mm) almost twenty-seven times the lowest recorded rainfall (18·6 mm). The coefficient of variation of annual rainfall at Lodwar is 0·58 (Table 4).

Seasonality in South Turkana depends primarily on an increased probability of convective storms during certain periods of the annual north–south migration of the Equatorial Trough. The *probability* of rainfall in South Turkana is greatest between late March and May, with a secondary peak in *probability* in November. However, the distribution of rainfall in time is of low predictability. In the 11 years for which there are data for Lokori near the eastern margin of Ngisonyoka territory, the wettest month of the year was January (1979), February (1978), March (1981), April (1970, 1983), May (1980), June (1974), July (1976), September (1975), and November (1977, 1982) (Fig. 2). The coefficient of variation of monthly rainfall at Lodwar over 56 years between 1923 and 1980 was 2·79 for September, the month with the most variation; and 0·96 for April, the month with least variation (Table 4).

There are statistical tendencies in the spatial distribution of rainfall. Although there are no rainfall stations in the highlands, vegetation indicates that more rain falls at higher elevations than on the plains. Also, rainfall data show a strong south–north gradient and a weak west–east gradient, with the south-western plains receiving the most rain (Ellis & Coppock 1985). However, since most rain falls in localized convective storms rather than broad frontal systems, there is a large stochastic element in the spatial distribution of rainfall in a given year. Heavy rainfall is associated with widespread storm systems; while the dry season showers which generate nutritious forage during critical periods are usually highly localized.

Coppock (1985) concluded that, although lack of nutrients, salinity, and/or soil pH may constrain plant production, South Turkana is primarily a water-controlled ecosystem 'in which precipitation pulses initiate biological processes of plant growth and reproduction that continue until the water ration is exhausted' (p. 20). Since forage availability is an important constraint on livestock survival and productivity, and lack of drinking water also can restrict livestock movements, the low predictability of rainfall is an important feature of the environment with which Ngisonyoka pastoralists must cope.

Topography

The Turkwel River to the north and west and the Kerio River to the south and east roughly demarcate Ngisonyoka territory, which exhibits great topographic diversity (Fig. 3; Dyson-Hudson & McCabe 1985). The *Turkwel Region* is the territory of the agricultural Ngikebotok section of Turkana. The north–south oriented *Central Mountains* rise to 2500 m elevation in the south, dwindling to small isolated hills in the north. This range is surrounded by four plains regions of

TABLE 4. Mean, standard deviation, and coefficient of variation of Lodwar rainfall over a 56–57 year period (1923–80)*†

	Monthly rainfall												Total annual rainfall
	Jan	Feb	Mar	Apr	May	Jun	Jul	Aug	Sept	Oct	Nov	Dec	
Number of years	57	57	58	58	58	57	57	57	57	57	57	57	56
Mean (mm)	8·88	8·04	20·78	48·23	23·82	8·191	17·83	8·99	3·414	8·62	16·83	10·71	185·24
Standard deviation	19·79	11·21	28·94	46·11	28·84	18·01	29·51	16·41	9·53	19·58	30·41	28·45	106·66
Coefficient of variation	2·23	1·39	1·39	0·96	1·21	2·20	1·66	1·86	2·79	2·27	1·81	2·66	0·58

* 1940 and 1941 rainfall data incomplete.
† Data from Dallyn (1981).

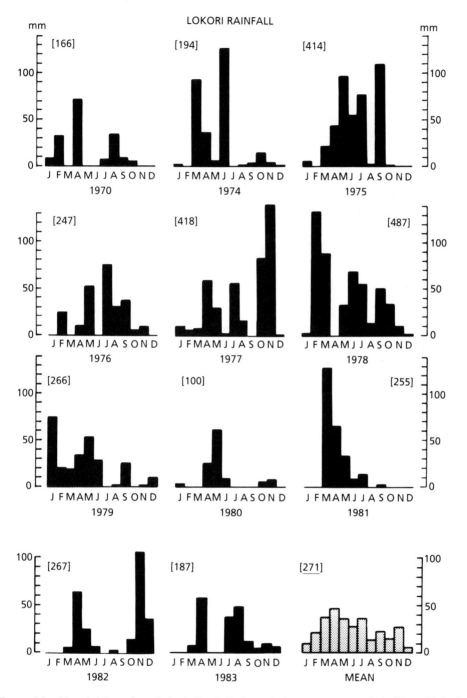

Fig. 2. Monthly rainfall totals at Lokori, South Turkana during an 11-year period. (From Little & Johnson 1985.)

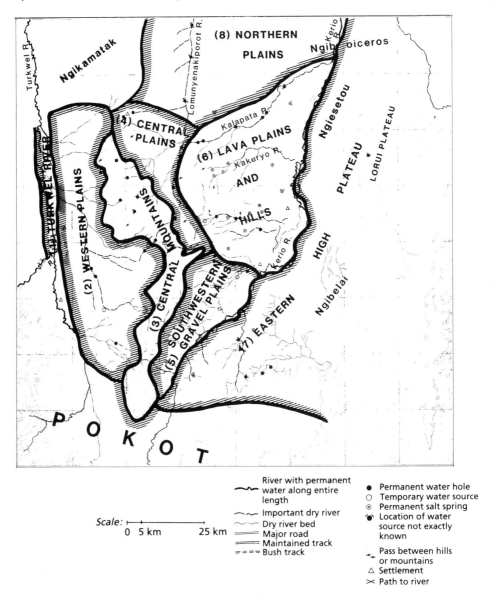

FIG. 3. Water resources and topographic regions of Ngisonyoka territory.

about 720–780 m elevation: the *Western Plains*, the *Central Plains*, the *Southeastern Gravel Plains and Hills*, and the *Northern Plains*. Between the Central Plains and the Kerio River to the east are extensive *Lava Plains* which are broken by large volcanic hills and plateaux. To the east and the south of the Kerio River, a series of highly dissected *High Plateaux* form an arc of about 100 km from Lake Turkana on the north to the Pokot border on the south. The Ngisonyoka share the southern portion of this highland region with the Ngibelai section of Turkana, and the Pokot, their traditional enemies, also sometimes use this region. The northern portion, the Loriu, is in the territory of the Ngiesetou section of Turkana, but is regularly used (with permission) by some Ngisonyoka pastoralists, particularly for grazing their cattle during the dry season.

Each of these topographic regions has differences in forage availability, water sources, human competitors, predators and pests, and soil development (Table 5).

Vegetation

In addition to low rainfall, soil development in Ngisonyoka territory is poor, so species diversity and net primary productivity of plants are low. In 1981−82, a 'good' year, the annual aboveground net primary production was estimated as 164 g/m^2 (Coughenour et al. 1985). However, each plant community exhibits great structural diversity, with grasses and forbs, dwarf shrubs, large shrubs, and trees interspersed. Furthermore variations in altitude, precipitation, topography, and soils lead to different plant communities in different topographic regions. Those parts of the Turkwel Region which are seasonally flooded are covered with dense perennial grasslands, and broad-leaved riverain forests. Despite their shallow soils, dense shrub-forest grows over much of the Central Mountains and the upper slopes of the highest mountains have productive stands of perennial grasses, with scattered bushes and trees. The High Plateaux have similar highland vegetation. The southern part of the Western Plains is dominated by dense bushland, which with regular burning is replaced by perennial grassland. The South-eastern Gravel Plains are somewhat less bushy with some areas of good ground cover of herbaceous vegetation. On the drier Central Plains, a narrow band of large Acacia trees grow along the sandy washes, with shrubs and stunted trees on the interfluves, and a ground cover primarily of shallow-rooted annual grasses and forbs. On the Northern Plains, which are even drier, annual grasslands predominate, with scattered large Acacia trees only along the larger water courses. Over much of the Lava Plains, the driest region of Ngisonyoka territory, the vegetation is widely dispersed, shrubby Acacia, with a scattering of annual grasses which provide sufficient forage for livestock only during years of good rainfall, although the hills and plateaux support a somewhat denser vegetation (Table 5) (see Ellis & Coppock 1985 for a more detailed description of the vegetation).

Each vegetation layer exhibits differences in phenology. For example, the relatively shallow-rooted grasses and forbs sprout after the rains begin, but mature and dry up within a few weeks after the rains end. In contrast, the leaves of large trees which tap the water stored in sandy drainage channels remain green during good years; though in drier years — particularly along the smaller water courses — they shed their leaves. In 1981−82, an estimated 67% of the above ground net primary production in the Ngisonyoka ecosystem was herbaccous plants, 13% shrub and tree foliage, 4% shrub and tree wood, 5% dwarf shrub, and 1% A. tortilis seed-pods (Coughenour et al. 1985).

TABLE 5. The characteristics of the eight topographic regions used by Ngisonyoka herders

	Turkwel River	Western Plains	Central Mountains	Central Plains	S.E. Gravel Plains	Lava Plains	Eastern High Plateaus	Northern Plains
Grazing	Available on floodplain in dry season only	Excellent in wet season and into dry season	Excellent dry season forage for cattle	Excellent after rains: poorer in dry season	Some after rains: sparse in dry season	Very sparse on plains: better in highlands	Good dry season grazing for cattle	Good grazing after good rainfall
Browsing	Available year-round	Good except in drought	Available year-round	Good except in drought years	Good except in drought years	Sparse except along Kerio	Available	Along larger sandy wash
Water	In Turkwel year-round: no springs	Very few permanent wells: no springs	No water in mountains: a few wells at base	Many wells on east margin: many springs in nearby Lava Plains	Much in Kerio: sparse near mountains: springs to south and to north	Many wells in rivers: also many mineral springs	Abundant and well distributed	Wells in largest rivers: many springs
Bandits	Since bandits are Turkana outlaws from groups to the east, north, and west, all regions are vulnerable							
Pokot	Vulnerability to attack by Pokot raiders depends on the distance from the Pokot border							
Other hazards	Tsetse: livestock trypanosomiasis	Buffalo common in bushy areas	Baboons hyaenas, large felids; ticks; rough ground hazard for camels	Hyaenas and jackals common	Baboons, hyaenas, large felids; ticks; cold off Kaikongkol Mountain	Very hot and dry; vegetation sparse	Baboons and other predators	No data

Water sources

Despite the aridity, water sources are relatively abundant and well distributed in Ngisonyoka territory (see Fig. 3), and water availability is more predictable on a year to year basis than is the distribution and abundance of the various kinds of forage. Both the Kerio and Turkwel Rivers rise in the humid Kenya highlands to the south: even when they do not flow, water can be readily found in pools or in wells dug in their sandy beds. Away from these two large drainage systems, permanent water occurs in mineral springs and wells. The large mineral springs, most of which occur in or adjacent to the Lava Plains, provide reliable water year-round, and do not appear to depend on local rainfall. In the other four plains regions, free water is primarily found in wells which are associated with sandy washes which have relatively small watersheds, rising in the Central Mountains or other highland regions. Where there is an impervious layer deep below the sandy river bed, water can be tapped throughout the year by digging the wells deeper and deeper. The depth to which these wells must be dug, and their recharge rate, depends on the amount of local rainfall and on the location of the wells in the drainage system. There are no wells in the Central Mountains; and in the washes which drain their footslopes, wells are owned by individual families, and people may have to dig up to 20 m in a dry year to reach water. In contrast, downstream on the barren Lava Plains near Lake Turkana, wells less than 2 m deep provide abundant water. There is, therefore, a discordance between availability of water and of forage which (because of higher rainfall) usually is more abundant at higher elevations.

Despite the relative abundance of water in Ngisonyoka territory, its distribution and abundance can restrict forage utilization. For example, camels do not thrive on the Western Plains because there are no mineral springs. And in dry years, only owners of wells near the lower slopes can pasture their cattle on the Central Mountains, and feed their milking goats on the abundant browse on footslopes. During a drought, some 'permanent' water sources dry up, and the energy demands of long treks to water exacerbate the nutritional stress of the livestock, particularly cattle.

Human competitors

There is a long history of raiding and warfare among East African pastoral groups (Fukui & Turton 1979). Hundreds or even thousands of livestock can change hands during a single raid between Turkana and their neighbours; and killing the enemy — women and children, as well as men — is also an important aim of raiding. The Pokot, a different ethnic group who live on the plains and in the highlands which surround Ngisonyoka territory on the south-east, south, and south-west (see Fig. 3), are traditional enemies of Ngisonyoka pastoralists. Al-though at times Pokot and their Ngisonyoka neighbours have had amicable relations, during the period of this study there was raiding back and forth across the border between these two groups. Since Pokot raid from bases in their own territory, the danger of attack is greatest in the south and west of Ngisonyoka territory, areas which have relatively good dry season grazing and abundant water (see Table 5).

During the study period (particularly before the Ngisonyoka Homeguard was organized in mid-1982), young Turkana bandits, primarily from regions to the north or west of Ngisonyoka territory, frequently stole Ngisonyoka livestock. Unlike Pokot raiders, bandits stole animals only for immediate consumption, taking a few of the fattest animals from a herd, and only killed people who seriously resisted their predation (see Table 5) (see also Dyson-Hudson & McCabe 1985; McCabe 1985).

Animal predators, pests and other hazards

Different animal predators, pests, and other hazards characterize each of the eight topographic regions (see Table 5). Hyaena, cheetah, and jackals are more common in the open Central Plains; while lions and baboons, which eat lambs and kids, are more common on the Central Mountains and their lower slopes; and buffalo, one of the most dangerous game animals, browse in the bushy country of the southern part of the Western Plains, and in the large canyons around the base of the Central Mountains. Ticks, which can debilitate small stock, are a major problem in the southern area of the Central Mountains and extend down to the bushy southern part of the Western Plain. Tsetse fly, the vector for livestock trypanosomiasis, occurs in the riverain forests along the Turkwel River.

Rough terrain, as found on the Central Mountains, is a hazard to camels which are clumsy on uneven ground and can fall and break a leg. Also the herd owners consider the cold which drains down from the high peaks of the mountain down to the South-eastern Gravel Plains to be bad for the camels.

Soil development

Low rainfall and poor soil development preclude crop cultivation in most of Ngisonyoka territory. Since extensive areas of alluvial soil which are suitable for naturally or artificially irrigated agriculture occur only along the mid and lower Kerio and Turkwel Rivers, on the margins of Ngisonyoka territory, most Ngisonyoka families do not combine nomadic animal husbandry with crop cultivation. The forage from the natural vegetation communities of these peripheral regions has not been heavily utilized by Ngisonyoka pastoralists, particularly the Turkwel because of tsetse fly. Since the areas of Ngisonyoka territory which are suitable for agriculture are not important grazing regions, pastoralists in Turkana have not suffered from the incursion of agricultural people to the degree that pastoralists in more humid regions of East Africa have.

Social factors

Several factors lead people to seek to live in the company of others. Environmental hazards of Pokot, bandits, and large animal predators, make it safer to live in groups. Also, collective effort is necessary to dig out wells, which fill with sand when the washes flow in spate after heavy rains. Although very rich men may sometimes choose to live alone, most Ngisonyoka value sociality and mistrust people who live apart.

ADAPTIVE STRATEGIES

There is no 'best area' nor 'best exploitation strategy' in Ngisonyoka territory. The low plant productivity, heterogeneous topography, and limited access to water sources near some important grazing regions lead to dispersion of livestock and their human caretakers. As noted, human and animal predation, the need for cooperative well digging, and the desire for sociality lead people to seek to live in groups. The food production system and the social organization of Ngisonyoka pastoralists can be interpreted as responses to these centrifugal and centripetal forces, in a topographically diverse territory with complex vegetation communities, where each region has different risks and resources, each vegetation layer within the community has a different phenology, and these vary with low predictability through time.

The food production system

The food production system of the Ngisonyoka can be viewed as based on adaptive strategies employed by people with virtually no access to capital and fuel energy, enabling them to cope with low plant productivity, and low spatial and temporal predictability, in a heterogeneous topography with spotty distribution of water sources. Unlike Western ranching and dairy operations which seek to maximize productivity and profit, the goals of the Ngisonyoka food production system are stability through biomass maintenance, and the use of multiple pathways of the food web (Coughenour *et al.* 1985).

Sparse plant productivity

Ngisonyoka deal with the problems caused by aridity, poor soil development, and low plant productivity by herding livestock which collect the plant productivity over a large area, concentrate it, and convert some of it into human food. In the Ngisonyoka ecosystem, a relatively high proportion of animal productivity is converted to human food by concentrating on milk rather than meat production, by using multiple resources from the livestock herds, and by exchanging high protein (meat) for high calorie foodstuffs (grain) through barter or sale.

Low capital and energy availability

Ngisonyoka livestock herders practise labour-intensive animal husbandry which requires no investment in capital equipment or fuel energy. The animals are guarded at all times while foraging, and corralled at night. Well water must be lifted into troughs on the river beds, a major labour demand during the dry season. Animals are milked morning and evening, with camels sometimes milked three or even four times a day. Bleeding, particularly of cattle, requires several people to capture and control the animals. (See Galvin 1985, pp. 119−25 for activity diaries and energy budgets of Ngisonyoka pastoralists.)

Since Ngisonyoka livestock husbandry is labour-intensive, successful herders need large families. The women's stated ideal family size is ten, while the completed family size in pastoral families is about five (Brainard 1986). This means that the Ngisonyoka pastoral population is expanding while their resource base is not increasing: clearly the Ngisonyoka ecosystem is not bounded, homeostatic, and self-regulating.

Variations in animal productivity

The use of multiple resources from livestock dampens the effects on animal production of seasonal and year-to-year variations in rainfall and in plant productivity. When available, milk is the major food of Ngisonyoka pastoralists. However, they also tap blood from the jugular vein of non-milking animals, particularly when there are few milking animals; and consume the meat either directly or by exchanging livestock for high calorie foods such as sorghum, maizemeal, and sugar. The first year of the study (1980−81) was the most severe drought that old men can remember: during that period maizemeal and meat of animals which died were the major food. Blood became important during the second study year (1981−82), when the livestock were in excellent condition but very few were giving milk since during the drought most of the females had aborted or their calves had died. During the third study year, milk provided 61% of the food energy of the four study families: a very high proportion of the adult female stock had calved/kidded/lambed, and milk was unusually abundant (Dyson-Hudson & McCabe 1985; Galvin 1985).

Although the use of multiple resources dampens variations in animal production, Ngisonyoka do not have effective mechanisms to enable their herds to track the unpredictable fluctuations in plant productivity of their ecosystem. After the herd crash during the drought of 1980−81 (when Ngisonyoka herds suffered an adult loss of 37% for camels, 58% for cattle, and 33% for small stock: McCabe 1985), a substantial portion of the plant productivity, particularly the ephemeral herbaceous vegetation, was not being utilized (Coughenour *et al.* 1985).

Habitat diversity and low predictability of forage in space and time

Ngisonyoka pastoralists deal with the topographic and habitat diversity in their

ecosystem by herding several species of livestock, each of which exploits a different ecological niche. Cattle and donkeys graze the shallow-rooted, ephemeral
herbaceous vegetation. Sheep and goats feed on grasses, forbs and dwarf shrubs:
of the two species of small stock, goats are true mixed feeders, while sheep
depend to a greater extent on herbaceous vegetation. Camels typically eat the
leaves of dwarf shrubs and large bushes, and of trees which can tap the ground
water (Coppock 1985).

   The pastoralists deal with the low spatial predictability of forage production by
relatively unconstrained movement within a territory with permeable boundaries.
They are not tied to particular places by material possessions such as permanent
houses, fields, and stored agricultural products, and are free to move their herds
and their families to those areas where environmental conditions best suit the
needs of each particular species of livestock, each family moving five to ten times
a year. When resources for all the species of livestock are not available at one
location, the herd owner may divide his animals into species and production
herds, separating the milking from the non-milking animals of each livestock
species, with each sub-herd seeking forage in a different area (Dyson-Hudson &
McCabe 1985).

Water sources and efficiency of locomotion

Ngisonyoka herding strategies take into account the differences among livestock
in water needs, and in energy requirements for locomotion. Small stock, which
are more efficient in vertical locomotion than are large livestock species, frequently
forage up and down the bush-covered, steep, small volcanic hills which dot the
plains regions of Ngisonyoka territory. However, they are less efficient in horizontal walking than are the large species, and also must drink every other day; so
when possible small stock are corralled not more than 4–6 km from the nearest
water source. The non-milking small stock, with lower water requirements, can be
corralled further from water — sometimes up to 10 km from the wells. They can
utilize forage on the footslopes of the Central Mountains which milking goats and
sheep cannot.

   In contrast camels, the livestock species which is most efficient in horizontal
locomotion but most inefficient in vertical locomotion, usually browse on the
plains and are taken to the highlands only in exceptionally bad years. The camel's
efficiency in horizontal locomotion, and ability to go for long periods without
water, enables herd owners to select water sources by quality rather than the
distance: for example, in February of 1979 one camel herd was taken 35 km each
way to drink from a mineral spring, rather than drinking water from the nearby
well (Dyson-Hudson & McCabe 1985).

   Although cattle occupy a niche which is not heavily exploited by the other
livestock species and therefore add significantly to the food resources of Ngisonyoka
pastoralists, they are not well suited to the Ngisonyoka ecosystem. The best dry
season pastures are in the Eastern High Plateau region, which is on the Pokot

border and therefore was shunned by most Ngisonyoka herd owners during much of the study period; and high on the Central Mountains, where there is no free water. Cattle grazing the high pastures of the Central Mountains must travel a vertical distance of 200 m or more during their walk of 8 km or more each way to water (McCabe 1985). Despite the relatively small size of Ngisonyoka cattle, in the dry years they are the first animal species to suffer from stress — caused both by the rapid deterioration in the quality of the shallow-rooted herbaceous vegetation on which they depend, and by the high energy costs of these relatively large animals travelling between water sources on the plains and pastures in the highlands.

Efficiency of energy conversion

The energy conversion of the Ngisonyoka pastoral system of about 25 megajoules of human food per hectare per year is low as compared with intensive agriculture, but is comparable to that of other animal-based systems in arid regions, such as Australian sheep ranching, or Sahelian nomadism (Coughenour *et al.* 1985). The Ngisonyoka livestock production system supports about 1·2 people/km^2 in their very arid environment; as compared, for example, the population density of 1·04 people/km^3 of the !Kung foragers and Bantu pastoralists reported by Lee (1969) in the somewhat more humid Dobe region of the Kalahari Desert. Project ecologists concluded that, although the Ngisonyoka Turkana 'have not greatly influenced the quantity of solar energy captured by the plant community nor that transferred to herbivores, they have directed solar energy through a food web so effectively as to permit the maintenance of a relatively high density and biomass of humans on marginal and variably productive landscapes, without inducing discernible degradation of the ecosystem' (Coughenour *et al.* 1985).

Social organization

The flexibility and variability at all levels of Ngisonyoka social organization makes it impossible to describe in the limited space available. So in discussing the territorial section, home area, camp, and the neighbourhood (those levels of organization which are most important in the day-to-day activities of Ngisonyoka pastoralists), I shall present people's ideals of organization and behaviour, as elicited in the course of our numerous interviews with Ngisonyoka pastoralists, and discuss how and under what circumstances actual behaviours varied from these ideals. These do not represent statistical norms or modes of behaviour. They are individual and/or societal norms of behaviour — cognitive models of desirable or appropriate behaviour, which are shared by, though not necessarily identical among, individuals in the group, and influence the behaviour of group members. The genesis of societal norms has a historical component, so they cannot be understood in simple adaptive terms. Among Ngisonyoka, they are enforced by social sanctions which range from disapproval to formal cursing by the elders, rather than by formal legal sanctions.

Territorial and quasi-territorial organization

There are three levels of territorial or quasi-territorial organization — the 'tribe', the section, and the home area.

The tribe. Today the primary significance of the highest level of territorial organization — Turkana — is as an administrative unit. However, for a century or more, most Turkana people have lived in a contiguous area, and shared a common language and common ethnic identity. Although it is possible to change tribal affiliation, this is very rare: there is only one case of change in tribal affiliation in my sample of about seventy-five Ngisonyoka herd owners.

The section. Because of the extreme dispersion and unpredictability of resources in South Turkana, individual grazing territories are not economically defendable (see Dyson-Hudson & Smith 1978). However, Turkana are divided into nineteen *sections* (Gulliver 1951), which allocate sub-populations of about 10 000 people to specific regions. Bounded sectional territories only occur in South Turkana, where the pattern of topographic diversity allows a wide variety of ecological zones to be included within a single bounded region. In North Turkana, where there are vast plains areas with large mountain massifs and rift valley walls near the margins, sections are not bounded territorial units, but rather each has rights to a region in the plains area, and to a region in the mountains (Gulliver 1951; Dyson-Hudson & McCabe 1985).

Although the ideal is that Turkana are all one people who do not fight among themselves, this varies among sections. Ngisonyoka have a long tradition of friendship and cooperation with Ngiesetou, their neighbours to the east; but tend to mistrust the Ngibelai to the south-east, some of whom, they believe, act as spies for Pokot raiders. There is enmity between North/West and South/East Turkana sections which dates back at least to the establishment of British military rule in the early twentieth century, when many South Turkana cooperated with the British, while in remote North Turkana most of the people either fought or fled northward into the mountains (Lamphear, unpublished manuscript).

The day-to-day subsistence activities of Ngisonyoka pastoralists are carried out primarily within the boundaries of their large sectional territory of about 7540 km^2, with free access to grazing limited to section members. This sectional territory is defended primarily through social boundary defence and reciprocity rather than perimeter defence and exclusion (see Cashdan 1983). The norm is that when the people of one South Turkana section want to graze their livestock in the territory of another section, animals must be sacrificed for the elders who then grant permission to graze. However, a study by N. Dyson-Hudson of Ngisonyoka violence documents that when the intruding herd is very large, or when grazing is scarce, permission may be refused and the intruders driven out by physical violence, or fear of witchcraft.

Although people can change sectional affiliation, this is rare. More often, herders establish affinal relationships or friendships (based on livestock exchanges) with people in a neighbouring section, and thereby gain access to a neighbouring sectional territory as well as their own.

The home area. The lowest level of territorial affiliation is a man's *home area* (singular: *ere*). This is the region to which a herd-owner prefers to return, although poor forage, fear of enemies, location in space, and/or the desire to stay together with other herders may prevent him from doing so. A man is most likely to have rights in a well in his home-area, and therefore to return there when water resources are limiting. The ideal is that a man's home area is the same as that of his father. However, about 10% of the sample of herd owners I am working with have changed their home-areas, moving from the Western Plains which they felt to be too vulnerable to Pokot attack, to an area in the Central Plains.

Residential groups

Ngisonyoka families as they migrate, primarily within their sectional territory, are members of three different types of residential groups: camps, and primary and secondary neighbourhoods.

The camp. The camp (singular: *awi*) is a unit of livestock management as well as a residential group, and includes shelters for the people and corrals for their animals. Ideally it is made up of a herd owner with his several wives and their children, together with all the family's livestock. However, the people of the *awi* must be able to provide all the labour necessary to care for the livestock; and the livestock of the *awi* must provide for all the needs of the people. Given the changes through time in family structure (Stenning 1958, and in herd structure (Dahl & Hjort 1976), the ages and numbers of a herd owner's family and livestock are unlikely to correspond to this ideal. Herd size is adjusted to labour resources by co-herding with a man whose family and livestock balance out the herd owner's needs, and by having marginal people move in and out of the *awi*, getting food when it is abundant, seeking food elsewhere when it is not. (Dyson-Hudson & McCabe 1985 give specific details of variations in organization among Ngisonyoka *awi*.)

Ideally, all the animals of a man's livestock are herded together in a single main camp (*awi ngunapolon*). Between about June and November of 1981, when forage was abundant all over Ngisonyoka territory, all the livestock of each of the study families came together in their main camps. However, when good forage and water for all species of livestock are not available in one area, some of the livestock will be moved to satellite camps (singular: *abor*). The cattle camp is the first to separate, moving to better pastures in the highland areas. The adult non-milking small stock are moved to take advantage of forage which is too far from water for the milking females to utilize. To prevent competition for browse with the milking camels, the non-milking camels also may be divided off into a separate *abor*, or corralled in the same *abor* with the non-milking goats. Most of the milch camels and/or milking small stock, and the donkeys for transporting water and goods, remain with the main *awi*, together with the herd owner, married women, infants, and young people needed to herd and help in the camp. The division of the *awi* into satellite camps is a direct response to forage conditions, but a man's ability to divide his herds is constrained by the availability of labour. Between July 1979 and February 1981, a period of intensifying drought, Angor (a herd owner with five grown brothers on whom he could rely) divided his livestock into six sub-herds, with separate satellite camps for the weak and the strong non-milking small stock, as well as for all the cattle and for the non-milking camels (Fig. 4a). In contrast Lori, who had only one (unreliable) younger brother in his *awi*, had one satellite camp for the non-milking camels, and depended on a distant agnate to herd his cattle (Fig. 4b). Angor was a successful herd manager, having parlayed a small foundation herd to a large livestock holding; whereas Lori's large foundation herd dwindled, and by 1983 was reduced to too few animals to support his family. However, Lori's failure to divide his livestock into satellite camps during the drought was probably evidence of his poor management abilities (in not establishing reliable herding partnerships which would have enabled him to do so), rather than the cause of his catastrophic livestock losses during the study period.

The main *awi* moves frequently, some eight to fifteen times a year. It follows an unpredictable migratory path on the plains, usually moving no more than 8–10 km, but making occasional long transitional moves of up to 50 km, when possible locating within 4–6 km of a major water source. Each satellite camp pursues a different migration path, usually moving more frequently than the main camp. Ideally the management unit remains the same, since each young man who manages a satellite camp 'must' regularly consult with and obey the herd owner. This, in fact, happens if the manager of the satellite camp is dependent on, respects, and/or fears, the herd owner. In that case, the satellite camps usually remain relatively close — within 5–15 km of the main *awi*. However, the herd manager sometimes takes advantage of distance to make independent decisions; pursuing an increasingly independent orbit enables a young man to make a gradual transition from dependent brother or son to independent herd owner. In such 'transitional' families, the satellite camps may be far — 50 km or more from the main *awi* — and information exchange is infrequent, even to the point where the herd owner has no idea where the satellite camp is (Dyson-Hudson & McCabe 1985).

The primary neighbourhood. For protection from human and animal predators, and to enhance social interactions, camps are clustered into primary neighbourhoods (singular *adakar*) of some two to twenty-five *awi* located within an area of about 0.5 km^2. Elders of the neighbourhood gather each day to talk, exchange information, ajudicate disputes, and make decisions — about ceremonies, marriages, and migratory moves.

The ideal is that after each move, the neighbourhood is reconstituted with the same membership. We were told that a man must ask permission of the elders of the neighbourhood if he wants to move away (but also that moving in with a new group of neighbours requires no permission). Our data showed that since migratory moves are not well coordinated, and there is ample opportunity for a herd

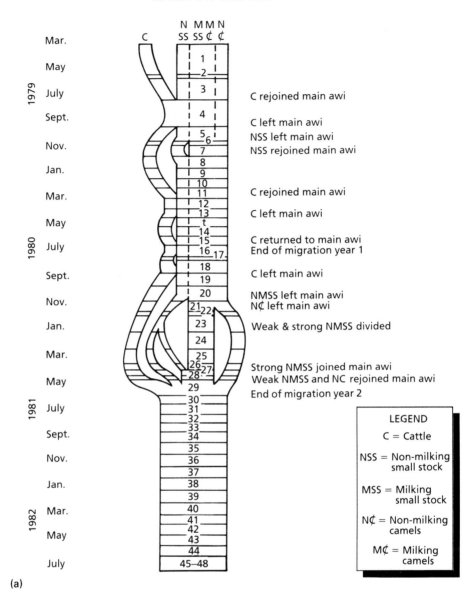

DIVISION AND COALESCING OF HERD OF ANGOR:
MARCH 1979–JULY 1982

(a)

owner to go his own way, and neighbourhoods are in fact transient, providing social support for the herd owner, but not restricting his decision to move where he believes the conditions to be best for his animals.

The secondary neighbourhoods. When forage is abundant, *adakar* are clustered into larger secondary neighbourhoods. For example, in November 1984, about seventy Ngisonyoka *awi*, more than 1000 people, were aggregated in an area of

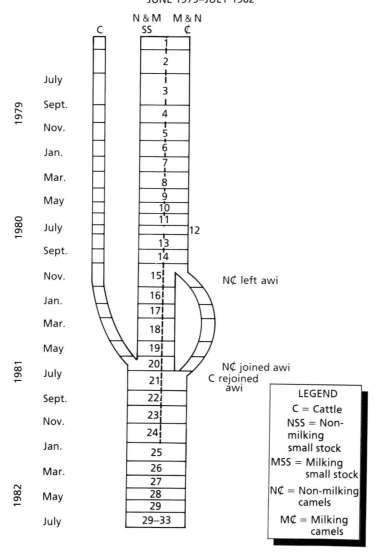

FIG. 4. The coalescing and dividing of the livestock herds of two Ngisonyoka herd owners during a period of drought (July 1979 through March 1981), followed by a period of abundant forage (July 1981 through July 1982). (a) A herd owner with five adult herd managers. (b) A herd owner with only one (unreliable) adult herding assistant.

250 km². The size of these loose associations is limited by the availability of forage and water. They exist when forage resources are sufficient to support relatively large numbers of livestock within a restricted area. When forage and water resources become scarce, as at the end of the 1979–1981 drought, the

secondary neighbourhood breaks up into individual primary neighbourhoods, or even independently moving camps (Dyson-Hudson & McCabe 1985).

We were told that moves of secondary neighbourhoods are directed by the major ritual specialist of the section (singular *emuron*), who is believed to have the power to predict events in the material world and to control them. However, we found that if the *emuron*'s directives conflicted with a herd owner's perceptions of what was best for his animals, they were ignored. For example in 1979, a number of Ngisonyoka herd owners took their livestock to the southern portion of the Eastern High Plateau region, despite the *emuron*'s admonition not to do so because of the danger of Pokot attacks (Dyson-Hudson & McCabe 1985).

Decisions about location in space. Among Ngisonyoka pastoralists, decisions about location in space depend on both environmental and social factors. The importance of sociality, which is a desired state and affords protection against predators, is substantiated by the fact that when forage became abundant in about June of 1981, rather than reducing the frequency of their nomadic moves, the herd owners chose to live in larger groups and move as frequently as or even more frequently than during the drought (see Fig. 4a, b) (Dyson-Hudson & McCabe 1985).

Individual variations and historical influences on behaviour. A detailed analysis of nomadic movements of individual Ngisonyoka families indicates that differences in well ownership, in family and herd size and structure, in perceptual and cognitive response and experience and so in herding skills, and in capacity to negotiate or cooperate with or confront would-be competitors, strongly influence the behaviour of individual herd owners (Dyson-Hudson & McCabe 1985). Furthermore the genesis of at least some societal (cultural) norms has a historical component, and so cannot be explained simply as adaptive responses to the immediate environment. For example, many Nilotic pastoralists, who probably share at least some common ancestors but now live in a wide range of environments, share a willingness to eat the meat of animals that died; while North African pastoralists who have been converted to Islam consider such meat to be 'unclean'.

COMPARISON WITH THE NEIGHBOURING KARIMOJONG

In the late 1950s, we made a study of the Karimojong of north-eastern Uganda (Dyson-Hudson 1966; Dyson-Hudson & Dyson-Hudson 1969, 1970), whose traditional social organization no longer exists because of the effects of Idi Amin's governance. The Karimojong and Turkana are representatives of a population which split, probably within the last few hundred years, and speak mutually understandable dialects of a common language (Lamphear 1986). Karimojong also used multiple resources from their livestock herds, which included cattle, small stock, and donkeys. They shared the cultural ideals of large families, of large bridewealth payments at marriage, and of movement as the major adaptive strategy of livestock management. However, their social organization was different from that of Ngisonyoka Turkana, in ways which reflected the difference in the distribution and abundance of their resources. Higher rainfall and more abundant agricultural land made it possible for all Karimojong families to supplement the food from their livestock herds by practising rain-fed agriculture. Their fields of sorghum, and their stored grain, were dense and abundant resources which were protected from permanent settlements, where most married women and old men lived year-round. Although the net primary productivity is higher than in Turkana, forage for Karimojong livestock was none the less dispersed and unpredictable in time and space. During the rains, when forage was abundant near the field areas, the animals were kept in corrals in the permanent settlements. During the dry season, the animals were kept in temporary stock camps, herded by the younger men and boys with the assistance of unmarried girls and young wives (Dyson-Hudson & Smith 1978).

DISCUSSION

Some aspects of the food production system and social organization of Ngisonyoka pastoralists fit Crook's model relating social attributes to ecological requirements of the individuals concerned. The use of many food energy pathways through the use of multiple resources from five species of livestock helps the people to survive in a region with low plant productivity and high variability in energy flow along individual pathways. Multiple livestock species occupying different ecological niches enable the people to utilize the great structural diversity within each vegetation community. Focusing on animal biomass maintenance rather than animal productivity leads to greater stability of food production for these people living in a highly variable environment. High mobility represents the peoples' attempt to utilize a topographically diverse region and cope with the low predictability of forage distribution. The residential flexibility enables pastoralists to aggregate and disperse according to resource availability and danger from human predators; and also for each individual herd owner to site his camps according to his own evaluation of risks and resources in the environment.

An understanding of proximate environmental variables, which include their pastoral production system, seems necessary for understanding the social organization of Ngisonyoka Turkana. However, since some societal norms appear to have been influenced by historical events which are difficult if not impossible to reconstruct, particularly in non-literate groups; and since social organization also reflects differences in individual behaviour which are based on differences among individuals in, for example, fertility, wealth and family structure, and herding skills; proximate environmental variables cannot provide a sufficient explanation for all aspects of Ngisonyoka social behaviour.

ACKNOWLEDGEMENTS

Support for my Turkana research to date was provided by National Science Foundation grants BNS 801 7800 and BNS 831 8533. The Norwegian Agency for International Development (NORAD), the Wenner-Gren Foundation for Anthropological Research, the Harry Frank Guggenheim Foundation; and the National Geographic Society. Geoffrey Chester, Dean, College of Arts and Sciences, Cornell University, provided the support which enabled me to complete this paper. Data collected by other members of the South Turkana Ecosystem Project is gratefully acknowledged. I wish to thank my Ngisonyoka friends, whose patience in answering my endless questions has made this work possible; and my Ngisonyoka assistants Nakapwan and Eewoi, whose cheerful goodwill has made the field work enjoyable.

REFERENCES

Bennett, J.W. (1969). *Northern Plainsmen: Adaptive Strategy and Agrarian Life.* Aldine, Chicago, Illinois.

Brainard, J. (1986). Differential mortality in Turkana agriculturalists and pastoralists. *American Journal of Physical Anthropology*, **70**, 525−36.

Carniero, R. (1970). A theory of the origin of the state. *Science*, **169**, 733−8.

Cashdan, E. (1983). Territoriality among human foragers: ecological models and an application to four bushman groups. *Current Anthropology*, **24**, 47−66.

Coppock, D.M. (1985). *Feeding ecology, nutrition, and energetics of livestock in a nomadic pastoral ecosystem.* Ph.D. thesis, Colorado State University, Fort Collins, Colorado.

Coughenour, M.B., Ellis, J.E., Swift, D.M., Coppock, D.L., Galvin, K., McCabe, J.T. & Hart, T.C. (1985). Energy extraction and the use of a nomadic pastoral ecosystem. *Science*, **230**, 619−25.

Crook, J.H. (1965). The adaptive significance of Avian social organization. *The Social Organization of Animal Communities* (Ed. by P.E. Ellis), pp. 181−218. Symposia of the Zoological Society of London, 14. Academic Press, London.

Dahl, G. & Hjort, A. (1976). *Having Herds: Pastoral Herd Growth and Household Economy.* Stockholm Studies in Social Anthropology. Department of Social Anthropology, University of Stockholm, Sweden.

Dallyn, J.P. (1981). *Irrigation in Arid Regions, Kenya: Irrigated Crop Production in Kenya.* Report to the Ministry of Agriculture, Republic of Kenya. AG: DP: KEN: 78·015.

Dyson-Hudson, N. (1966). *Karimojong Politics.* Oxford University Press, London.

Dyson-Hudson, R. (1981). Indigenous models of time and space as a key to ecological and anthropological monitoring. *The Future of Pastoral People* (Ed. by J.G. Galaty, D. Aronson, P.C. Salzman & A. Chouinard), pp. 353−8. International Development Research Centre, Ottawa.

Dyson-Hudson, R. (1985). Appendix III: South Turkana herd structure and livestock/human ratios. *South Turkana Nomadism: Coping with an Unpredictably Varying Environment* (By R. Dyson-Hudson & J.T. McCabe), pp. 331−65. HRAFLEX, New Haven, Connecticut.

Dyson-Hudson, R. & Dyson-Hudson, N. (1969). Subsistence herding in Uganda. *Scientific American*, **220** (2), 76−89.

Dyson-Hudson, R. & Dyson-Hudson, N. (1970). The food production system of a semi-nomadic society: the Karimojong of Uganda. *African Food Production Systems: Cases and Theory* (Ed. by P.F.M. McLoughlin), pp. 91−124. Johns Hopkins Press, Baltimore, Maryland.

Dyson-Hudson, R. & McCabe, J.T. (1985). *South Turkana Nomadism: Coping with an Unpredictably Varying Environment.* HRAFLEX press, New Haven, Connecticut.

Dyson-Hudson, R. & Smith, E.A. (1978). Human territoriality: an ecological reassessment. *American Anthropologist*, **80**, 21–41.

Ecosystems (1988). *Turkana District Resource Survey 1982–1984*. Report to Republic of Kenya, Ministry of Energy and Regional Development, Republic of Kenya.

Ellis, J. & Coppock, D.L. (1985). Appendix II: Vegetation patterns in Ngisonyoka Turkana. *South Turkana Nomadism: Coping with an Unpredictably Varying Environment* (By R. Dyson-Hudson & J.T. McCabe), pp. 315–30. HRAFLEX, New Haven, Connecticut.

Firth, R. (1964). *Essays on Social Organization and Values*. Athlone Press, London.

Fukui, K. & Turton, D. (1979). *Warfare among East African Herders*. Senri Ethnological Studies, no. 3, National Museum of Ethnology, Osaka, Japan.

Galvin, K. (1985). *Food procurement, diet, activities and nutrition of Ngisonyoka Turkana pastoralists in an ecological and social context*. Ph.D. thesis, State University of New York, Binghamton.

Gulliver, P.H. (1951). *A Preliminary Survey of the Turkana: A Report Compiled for the Government of Kenya*. Communication from the School of African Studies, New Series no. 26, University of Cape Town.

Lamphear, J. (1986). The persistence of hunting and gathering in a pastoral world. *SUGIA*, **7(2)**, 227–65.

Lamphear, J. (unpublished manuscript). *The Scattering Time: Turkana Responses to the Imposition of Colonialism*.

Lee, R.B. (1969). Kung Bushman subsistence: an input–output analysis. *Environment and Cultural Behavior* (Ed. by A.P. Vayda), pp. 47–79. Natural History Press, Garden City, New York.

Little, M.A. & Johnson, R.B. (1985). Appendix I: Weather conditions in South Turkana, Kenya. *South Turkana Nomadism: Coping with an Unpredictably Varying Environment* (By R. Dyson-Hudson & J.T. McCabe), pp. 298–314. HRAFLEX, New Haven, Connecticut.

McCabe, J.T. (1985). *Livestock management among the Turkana: a social and ecological analysis of herding in an East African pastoral population*. Ph.D. thesis, State University of New York, Binghamton.

NCR (1981). *Effects of environment on nutritive requirements of domestic animals*. National Academy Press, Washington, D.C.

O'Neil, R.V., DeAngelis, D.L., Waide, J.B. & Allen, T.F.H. (1986). *A Hierarchical Concept of Ecosystems*. Monographs in Population Biology, 23. Princeton University Press, Princeton, New Jersey.

Rappaport, R.A. (1984). *Pigs for the Ancestors: Ritual in the Ecology of a New Guinea People*. Yale University Press, New Haven, Connecticut.

Schmidt Nielson, K. (1984). *Scaling: Why is Animal Size so Important?* Cambridge University Press, Cambridge.

Schneider (1979). *Livestock and Equality in East Africa: the Economic Basis for Social Structure*. Indiana University Press, Bloomington.

Smith, E.A. (1983). Anthropology, evolutionary ecology, and the explanatory limitations of the ecosystem concept. *The Ecosystem Concept in Anthropology* (Ed. by E.F. Moran), pp. 51–86. AAAS Selected Symposium, 92. Westview Press, Boulder, Connecticut.

tenning, D. (1958). Household viability among the pastoral Fulani. *The Developmental Cycle in Domestic Groups* (Ed. by J. Goody). Cambridge University Press, Cambridge.

Steward, J. (1955). *Theory of Culture Change*. University of Illinois Press, Urbana, Illinois.

Steward, J. (1983). *Basin–Plateau Aboriginal Sociopolitical Groups*. Bureau of American Ethnology Bulletin, 120, Washington D.C.

Vayda, A.P. & McCay, B.J. (1975). New directions in ecology and ecological anthropology. *Annual Review of Anthropology*, **4**, 293–306.

Wienpahl (1984). *Livestock production and social organization among the Turkana*. Ph.D. in Anthropology, University of Arizona.

Williams, G.C. (1966). *Adaptation and Natural Selection*. Princeton University Press, Princeton, New Jersey.

The ecology of social relationships amongst female primates

C.P. VAN SCHAIK

Laboratory of Comparative Physiology,
University of Utrecht, P.O. Box 80086,
NL-3508 TB Utrecht, Netherlands

SUMMARY

1 The aim of this paper is to develop a framework in which to explain inter-specific variation in the patterning of female social relationships among diurnal primates (and hopefully some other mammals as well).

2 It is suggested that female social relationships reflect competition among females for food and safety. Competition for food can be of the scramble and of the contest type, and can occur both within and between groups.

3 For each meaningful combination of the three important components (between-group scramble is ignored), a set of predictions is derived for the social structure among the females in a group or society.

4 It is hypothesized that where high predation risk forces primates to live in cohesive groups, within-group competition predominates. This can be mainly by scramble or by both scramble and contest. In the former case, females develop individualistic and egalitarian ranking systems, in which female bonding varies, while in the latter they develop nepotistic and despotic ranking systems, accompanied by female residence.

5 The majority of folivores, and at least one gregarious insectivore, belong to the first type, whereas frugivores and omnivores belong to the second.

6 Among species of low vulnerability to predators, between-group competition predominates, and females should form nepotistic but egalitarian ranking systems, accompanied by female residence, provided that they can still live in sizeable groups. Primates on oceanic islands without predators, and some of the larger arboreal and the largest (semi-)terrestrial primates, are expected to show this pattern.

7 When high potential within-group competition precludes the formation of female groups, variable relationships are expected depending on the options open to males.

INTRODUCTION

There is very wide variation in the social organization of primate species. Ecological explanations have been sought for this. So far, these theories (see van Schaik & van Hooff 1983; Terborgh & Janson 1986; Wrangham 1987) have been successful mainly at the level of global features of social organization, such as the presence or absence of gregariousness, and, for gregarious species, group size and composition. This variation is even more pronounced at the level of social relationships, but there are no generally accepted theories accounting for the ecological determinants of social relationships among the members of a primate society (Silk 1987). Are relationships also determined by external conditions, or are they basically arbitrary variations related to chance events in the evolutionary history of the species?

Wrangham (1979) suggested that relationships among females should be linked more directly to ecological conditions than those among males or those between males and females. The two latter kinds depend primarily on the behaviour and spatial distribution of the females (cf. Emlen & Oring 1977), and are mainly related to mating competition and mate choice, respectively. Of course, we oversimplify if we look to the ecological conditions alone to account for the variability in female social relationships, because interactions between male and female strategies may result in different female relationships from those found in the absence of any male influence. This is most striking in species in which the male contribution to rearing offspring is considerable (for example, the callitrichids). However, if we limit ourselves to species with multi-female groups, female relationships form a very useful starting point for an enquiry into the role of ecological factors in social relationships, and this paper will concentrate on them.

The most influential attempt at an ecological theory for variation in female social relationships was made by Wrangham (1980). He distinguished two types of social structure, namely female-bonded (FB) and non-female-bonded structures. Females in FB societies are surrounded by genetic relatives, maintain strong grooming bonds, and are actively involved in agonistic encounters with other societies. Such societies were thought to have developed where females formed groups in order to cooperatively defend their food, occurring as defensible high-quality patches, against other groups and so needed relatives as reliable coalition partners. Thus, on this view, the evolution of group living in primates and female-bondedness are both ascribed to competition between groups (societies), and frugivores should live in FB-groups whereas folivores should be non-FB.

On the first score, the evolution of group living, Wrangham's theory turns out to have little empirical support (van Schaik 1983; Terborgh & Janson 1986; Dunbar 1988). First, it leaves unexplained why many non-FB primate species also live in groups (for example, hamadryas baboons, *Papio hamadryas*; red colobus, *Colobus badius*; gorilla, *Gorilla gorilla*). Second, in at least a number of species competition for food appears to increase monotonically with group size rather than being greatest in the smallest and perhaps the largest groups (van Schaik & van Hooff 1983). Third, a quantitative test in brown capuchins (*Cebus apella*) has shown that competition between groups is far less important than within groups (Janson 1985), something also suspected for various macaques and baboons (van Schaik, in preparation). Fourth, as a consequence of this, the predicted relationship between birth rate and group size is found in a few species only (van Schaik

1983): most species show declining birth rates as group size increases whereas an increase or a humped relationship would have been expected. The data on most species agree best with the idea that predation risk sets the lower limit to group size, whereas within-group competition, related to the sizes of food patches used (see, for example, Terborgh & Janson 1986), sets the upper limit. Female fitness is maximized at some intermediate group size.

Perhaps then, competition between groups was not the only, or even the major factor influencing female social relationships in most primate species. Indeed, many species also do not conform to the second part of the theory. Frugivores are expected to form FB groups, but not all do: chimpanzees (*Pan troglodytes*) and spider monkeys (*Ateles* spp.) live in fission−fusion societies in which males rather than females defend the area and adult females do not form strong bonds (see, for example, Goodall 1985, McFarland Symington 1987). Likewise, most of the highly frugivorous gibbons (*Hylobates* spp.) form mono-gamous pairs rather than multifemale groups; so do several Neotropical cebids and callitrichids. There are also various species without female emigration in which groups avoid each other rather than fight over area or food sources (for example, mangabeys, *Cercocebus albigena*: Waser 1976), or in which females take little or no part in between-group conflicts, which are often quite rare as well (for a review, see Cheney 1987; see also Terborgh & Janson 1986). In fact, female residence and natal dispersal by males is also found in the majority of solitary mammals (Waser & Jones 1983), in which between-group competition is notably absent. Finally, we would not expect steep ('despotic') linear hierarchies if conflicts between groups would be the major determinant of female social relationships (see below; cf. Vehrencamp 1983).

In its present form, therefore, Wrangham's (1980) theory can not provide an ecological explanation for female social relationships in primates. However, the approach is sound and I will merely propose two extensions. First, the notion of competition has to be diversified. Not only is there food competition between groups, but also, and often far more prominently, within groups; and in the latter case it can be of the scramble or contest type. Second, we also have to take into account that competition within groups may concern safety from predators.

To make clear why we should look to competition as the factor structuring female social relationships, consider the following argument. In a stable population, a female's fitness is approximated by the number of offspring she raises to reproductive maturity, in other words, her birth rate times the length of her reproductive career times the survival rate of her offspring. These variables are vitally affected by two factors. First, birth rate and offspring survival are strongly influenced by the quantity and quality of food a female can acquire (see, for example, Mori 1979; Wrangham 1979). Second, both her own survival, and hence the length of her reproductive career as well as that of her offspring, is strongly influenced by the level of safety she can maintain (see, for example, van Schaik 1983; Dunbar 1988). Therefore, a female's social relationships with other females in a group should serve these vital interests. It is suggested that among non-human primates, group living by females apparently evolved originally as a defence against predators (van Schaik 1983; Terborgh & Janson 1986; Dunbar 1988). If this is so and spatial clumping is enforced by extrinsic factors, we can expect females to form affiliative relationships when this raises their inclusive fitness by enhancing either their own competitive power or that of their relatives.

This paper is organized as follows. First, I define the main competitive regimes to which primate females are subjected and establish in which ecological conditions these are found. Next, I systematically explore the effects that each of these competitive regimes should have on female social relationships. Finally, I compare the predicted patterns with those found among primates, and look at more intermediate situations, which can realistically be expected in nature.

COMPETITIVE REGIMES

The three modes of feeding competition

Competition for food among free-moving animals can take on two forms: scramble and contest (Nicholson 1957). Scramble occurs when the net food intake of all individuals in a population is about equally affected by an increase in the populations's density. All animals share the same food supply, and none of them are able to obtain more than the others through overt behaviour, for example by evicting others from high-quality food sources or by staking out an exclusive territory. Whenever the distribution of the food and other factors allow it, competition will be by contest (also: interference), with territory owners or winners of interactions usurping a greater share of the critical resource than floaters or losers. In all but the most extreme conditions the competition experienced by an animal is a combination of both kinds.

If animals are living in groups, or are at least organized into distinct societies (i.e. unstable spatial coherence but stable membership), the situation becomes more complicated. Scramble and contest can now occur at two levels, both within the group or the society and between different groups or societies. In gregarious non-human primates competition therefore has four different faces: within-group scramble (WGS), within-group contest (WGC), between-group scramble (BGS) and between-group contest (BGC) (Janson & van Schaik 1988; see also Fig. 1). WGS occurs when animals foraging in a group have to share a limited supply of food with others and all of them suffer roughly equal reductions in foraging efficiency. It is expected when food is dispersed in patches that are either very small but cannot be monopolized or very large relative to group size, so at least among animals that graze or go after cryptic food in small patches (insects). WGC occurs when some group members are able to obtain a higher net food intake, hence when food occurs in well-defined patches, the access to which can be monopolized. Thus, it should at least be found among frugivores. WGC reflects

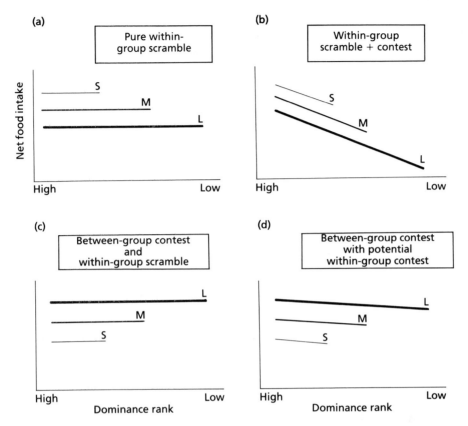

FIG. 1. Various combinations of the three relevant components of food competition among group-living mobile organisms to illustrate the main competitive regimes distinguished in Table 1, namely pure WGS (type A), WGC with a variable mix of WGS (type B), and strong BGC with WGS (type C) and strong BGC with potentially strong WGC (type D). S = small group; M = medium group; L = large group.

the dominance effect in competition and WGS reflects the group size effect with the dominance effect removed.

   Between-group contest occurs when the members of dominant groups obtain more food than the members of subordinate ones because the former aggressively displace the latter or are avoided by them, or because they defend a bigger or better territory. BGS occurs when groups overlap extensively in area and suffer a reduced foraging efficiency due to each other's removal of food, or when groups show mutual avoidance while occupying more or less exclusive ranges. Its quantitative importance depends on the potential for BGC, because, as between individuals, contest is likely to replace scramble whenever the conditions allow it. BGS basically represents the effect of population density on net food intake. Since I can see no way in which it may affect female social relationships, I shall ignore it in the rest of this paper.

Competitive regimes

The effects of competition on an individual female's net food intake will always be a combination of these three components and in some cases an additional component due to the advantages of social foraging. If we divide each component into just two categories, a week (W) or a strong (S) effect on net food intake, this gives us eight possible competitive regimes (Table 1). However, this number can be reduced for three reasons, the last two of which can be made clear only after we have examined the social consequences of the competitive regimes in detail. First, diurnal primates are unlikely to experience weak effects of all three components in nature (column 8, Table 1), since that would require a population density and a degree of spatial clumping that are low enough to preclude all food competition. Second, strong WGC and strong BGC have incompatible effects, so columns 6 and 7 can be eliminated. Third, the effects of strong WGC override those of WGS, so columns 2 and 3 are indistinguishable.

Hence, we can start our analysis with four competitive regimes (types A–D in Table 1), namely competition mainly by WGS (A), competition by WGC and a variable intensity of WGS (B), competition mainly by WGS and BGC (C), and competition mainly by BGC, usually with an unexpressed potential for strong WGC (D). Figure 1 illustrates these four regimes.

Ecological conditions and competitive regimes

Clearly, the next question to be answered is whether we can identify the characteristic ecological conditions for each of these competitive regimes. Strong competition within groups, be it through WGS or WGC, is generally predicated on group cohesiveness: the continuous close presence of conspecifics automatically exacerbates competition for food. Among primates, the occurrence of cohesive groups is clearly associated with high vulnerability to predators (see above). Thus, arboreal primates of small body size and (semi-)terrestrial primates of all but the largest body size form groups that are usually fairly cohesive.

Conversely, as predation risk decreases, animals tend to live in groups that are much less cohesive and often split up in parties of variable size and composition (see, for example, van Schaik & van Hooff 1983; Terborgh & Janson 1986). Party size in such fission–fusion groups varies directly with food supply and patch size,

TABLE 1. The eight possible (1–8) and four realized (A–D) competitive regimes amongst non-human primates (S = strong, W = weak effect on net food intake). See text for full explanation

Possible combination	1	2	3	4	5	6	7	8
Within-group scramble	S	W	S	S	W	W	S	W
Within-group contest	W	S	S	W	W	S	S	W
Between-group contest	W	W	W	S	S	S	S	W
Realized combination	A	B	B	C	D			

and thus inversely with the potential within-party competition (see, for example, McFarland Symington 1988). Thus, within-group competition is clearly limited in these species with a fission—fusion type of social organization.

Strong BGC should occur when it is both necessary due to a high population density relative to the environment's carrying capacity (K) and possible, i.e. when groups or societies are able to defend territories against their neighbours or to at least defend the access to valuable food sources.

This allows us to characterize the ecological conditions conducive to each competitive regime (see Table 2 for a summary). First, type A (predominance of WGS) is expected among those primates living in cohesive groups that rely on cryptic insects or foliage rather than fruit or whose food occurs in patches large enough to feed all group members. Second, type B (predominance of WGC) is expected among those primates living in cohesive groups whose food is distributed in clear-cut patches that are usually too small to accommodate all group members. Types C and D (strong BGC with and without strong WGS, respectively) are expected where primates live at a high density relative to K and have defensible resources or where the population density is lower but the resources are (at least seasonally) distributed in rare, but large patches. Unfortunately, although food does play a role in limiting primate populations (see, for example, Anon. 1981), little is known about the role of other factors. For instance, high rates of density-independent mortality generally lead to population levels well below K. They can be caused by periods in which only low-quality food is available or the weather is inclement, either or both of which may lead to very narrow energetic margins and occasionally mass-starvation (see, for example, Milton 1982). Also, in at least some species, predators may take more than just the 'doomed surplus', destined to die from starvation.

TABLE 2. Competitive regimes and ecological conditions in diurnal primates

Competitive regime	Type	Vulnerability to predators	Food distribution	Population density (N)	BGC potential
Strong within-group scramble only	A	High	Dispersed, or clumps>GS*	$N<K^{\dagger}$, or [$N\approx K$ and Low]	
Strong within-group contest only	B	High	clumped, clumps< GS	$N<K$, or [$N\approx K$ and Low]	
Strong within-group scramble and strong between-group contest	C	Intermediate/ high	Dispersed, or clumps> GS	$N\approx K$	High
				or	
			Scattered clumps> GS	$N<K$	High
Strong between-group contest only	D	Low	Clumped	$N\approx K$	High

* GS = group size.
† K = carrying capacity.

For type C groups we thus require either cohesive groups occurring at a high density, or cohesive groups with food distribution in rare, scattered, but large, patches. We simply do not know how frequently the first combination of factors occurs under natural conditions. The second condition might apply to species that specialize on large fruit trees, for example strangling figs, and forage for scattered food items in between.

Type D (strong BGC only) occurs when primates form groups of flexible party size that are nonetheless able to defend their range or high quality food sources against neighbouring groups. Such a condition may arise in two different kinds of species: (i) those of intermediate or large body size and a safe lifestyle, and (ii) those that are living in a predator-poor environment. The first condition is found among large primates (more than *c.* 30 kg) that are semi-terrestrial and forest-living, and among primates of perhaps more than *c.*8 kg living entirely in the canopy that are too large for the monkey-eating raptors. This size threshold could be lower for exclusively arboral primates living in Asia, where large raptors are absent. The second condition is found among primates of intermediate to large body size (more than *c.*3−6 kg; depending on degree of terrestriality) living on oceanic islands, where carnivores and large raptors are characteristically absent (see also Sondaar 1977).

We shall treat the two cases in turn; first, the situation when within-group competition is strong and, next, the one in which between-group competition prevails.

FEMALE RELATIONSHIPS WHERE WITHIN-GROUP COMPETITION IS STRONG

Theory

Social relationships

Let us now examine the social relationships expected under the first two competitive regimes, namely WGS alone and WGC plus WGS (types A and B). In a food contest situation, dominance rank (i.e. the ability to supplant other group members from resources) affects net food intake and so birth rate. Hence, dominance relationships will be consistent (i.e. aggression in a dyad is directed primarily from one partner to the other) and primarily transitive (i.e. the dominance hierarchy will tend to be linear or 'steep'). Since rank is so important, females who give agonistic support to maturing and adult relatives, in order to help them outrank other females, raise their inclusive fitness. Thus, the hierarchy will become nepotistic, because daughters will, as they mature, obtain ranks close to their mothers (Fig. 2). Ranks among adults will also tend to be stable because (i) relatives support each other, so that temporary variations in individual fighting power are not translated into rank changes, and (ii) the importance of rank to

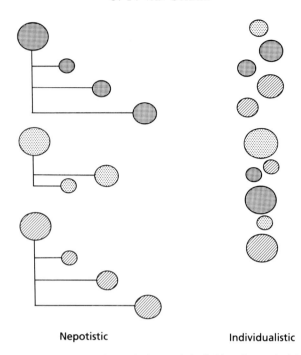

Nepotistic   Individualistic

FIG. 2. A schematic representation of a typical nepotistic (left) and a typical individualistic (right) female hierarchy. Size of the circles corresponds to female age; relatives have similar shading.

reproductive success (RS) will make animals very reluctant to give it up without escalated fights, (thus) making challenges to established higher-ranking group members risky (cf. Parker 1974).

In a food scramble situation a female's RS depends primarily on group size, and aggression over food is not often effective in increasing it. There is so little advantage to supporting female kin in attaining high rank that relatives do not support each other and agonistic coalitions are very rare. There may be short-lasting or context-dependent social advantages to dominance, for instance when it allows a female to protect young infants or to immigrate into a group. Dominance relationships will consequently be relatively unstable and inconsistent (i.e. within dyads there is a low asymmetry in the direction of aggression), and the dominance hierarchy will be weakly differentiated or 'shallow' and not necessarily linear. It is easier to recognize social strata than exact rank positions. These hierarchies will be individualistic, since in contrast to the nepotistic situation adult relatives need not occupy adjacent ranks (Fig. 2). The differences between the social effects of the scramble and contest situation are summarized in Table 3.

Let us now turn to the question of migration by females: how often and in which patterns should females decide to emigrate from their natal group? We assume that populations are relatively stable and at some level below K. In the food contest situation emigration by individual females from groups that have

TABLE 3. A summary of female social behaviour in primate groups experiencing either within-group scramble or within-group contest (+scramble) competition

	Female social relationships	
	Within-group scramble	Within-group contest (+ scramble)
Female reproductive success depends on:	Group size	(Group size +) dominance rank
Displacements over food	Rare	Common
'Spontaneous' aggression	Rare	Common
Agonistic support by relatives	Rare	Common
Dominance relationships	Inconsistent	Unidirectional
Stability of ranks	Low	High
Dominance hierarchy	Non-linear, egalitarian, individualistic	Linear, despotic, nepotistic
Groups mainly change by	Female migrations	Group splitting

reached a size well over the optimum should be rare. First, as shown by many experimental studies, a lone immigrant female invariably enters at the bottom of the hierarchy, because she has the double disadvantage of having to enter a new group and to fight alone against female coalitions. Second, without relatives she has no prospects of ever rising in rank, while there is some likelihood for this to occur in her lifetime if she remains with relatives (cf. Gouzoules, Gouzoules & Fedigan 1982). Hence, only low-ranking females without relatives are expected to emigrate. However, when the shortfall in RS relative to a small group becomes too great low-ranking females could emigrate together with relatives, either to set up their own group or to join existing groups. Because such female clusters threaten the rank positions of the incumbents if they migrate into an existing group, they are most likely to set up their own small group. Thus, if groups grow too large, they will often shrink by group splitting rather than by losing single female members.

In a food scramble situation, a female's RS depends on the size of the group she is in, regardless of whom she is with. This in itself is sufficient to lower the emigration threshold for females, since there is no social incentive to stay, unless other factors interfere (e.g. infanticide avoidance through female coalitions). Neither will there be pressure to emigrate together with relatives. Thus, whenever a female finds herself in a group well above or below the optimal size she should be willing to move, provided that migration costs (m) are not prohibitively high. The value of m depends on the risk of predation outside a group, and the ease of settling in existing groups or of finding others with whom to found a new group. We are assuming that there is space available, that females can reduce the risk of predation by teaming up with other females and can easily found new groups or enter existing ones, since group sizes are generally small.

Emigrants are most likely to be young females, for if a female moves into a group in which she can attain a higher birth rate, her benefit will be larger the earlier in her career she moves. If the expected increase in RS by migration exceeds m, females should migrate. However, it is important to note that WGS predicts conditional natal dispersal by females, not obligatory emigration. The proportion of migrants should depend on population growth, it being higher during phases of strong increase or decrease. Thus, within a given species, it should be possible to find some populations with a predominance of female residence and other with high rates of female dispersal.

Where females live in cohesive groups, in which they scramble, but also derive a net benefit from defending their range or food sources against other groups (type C), a slight change in social structure is expected. If females perform an essential role in between-group conflicts and have to perform cooperative attacks on other groups, they should prefer kin as allies (Wrangham 1980). Thus, females are expected to remain in their natal group in this case.

In conclusion, two interrelated factors determine the nature of female social structure, namely the type of food competition and the relatedness among the females. Table 4 shows the expected female social structure under the various regimes. This classification is only consistent in part with the FB versus non-FB dichotomy (Wrangham 1980). Type B clearly corresponds to the FB category.

TABLE 4. A summary of female social structure in cohesive multifemale primate groups

Type	Competition for food	Females resident (grooming bonds)	Hierarchy despotic/ egalitarian	Nepotistic/ individualistic	Examples (*tentative*)
A	Within-group* scramble	+ *or* −	Egalitarian	Individual	*Gorilla*(?), some spp. of *Saimiri, Colobus, Alouatta, Presbytis*
B	Within-group contest	+	Despotic	Nepotistic	Several spp. of *Macaca Cercopithecus, Papio, Cebus, Cacajao, Chiropotes, Lemur catta*, etc.
C	Within-group scramble + within-group contest	+	Egalitarian	Individualistic	Some *Presbytis* spp.?
D	Between-group contest	+	Egalitarian	Nepotistic	*Macaca nigra* group *Pan paniscus* (?)

* Competition for safety also of the scramble type, otherwise groups should turn into type B groups.

Dominance relationships are consistent, transitive and stable, giving rise to dominance hierarchies that are steep and linear ('despotic', see Vehrencamp 1983). In type A dominance relations are weakly consistent, often intransitive and relatively unstable over time, giving rise to dominance hierarchies that are weakly linear and shallow ('egalitarian'). Thus, type A corresponds to the non-FB category, but only if the females are the dispersing sex. If they remain resident, they would probably form grooming bonds, albeit weak ones. Type C groups, if they occur, would be intermediate between FB and non-FB, because females are expected to form grooming bonds in an egalitarian hierarchy.

An important prediction is that the combination of strong WGC and female natal dispersal should be extremely rare, only occurring if a factor other than food exerts an overriding selective pressure towards female dispersal.

Species experiencing different kinds of within-group food competition are also expected to differ with respect to patterns in aggression, allomothering, sex differences in juvenile mortality, and female life history. These predictions are still to be worked out in detail.

Scramble or contest for safety

Where there is a large variance in predation risk depending on the position in the group, we may expect contest for these positions, which will usually not involve overt agonistic behaviour. Females may increase their safety by being surrounded by as many others as possible, or by being close to certain individuals that are able to deter predators. Marginal predation (Hamilton 1971) is probably most important to terrestrial primates living in large groups on open plains. Thus, in savannah-living yellow baboons (*Papio cynocephalus*) high-ranking females tend to be in the centre (Collins 1984). It is probably less important among arboreal forest primates where groups are often small and where, due to the greater cover, predators can approach the group from more directions. However, among species living in larger and less compact groups, females may still contest for the safer positions, as in wedge-capped and brown capuchins (*Cebus olivaceus*: Robinson 1981; Janson 1985). Where groups fission at times of food scarcity, there should be contest for membership of the largest party, as in long-tailed macaques (*Macaca fascicularis*: van Noordwijk & van Schaik 1987). Females living in small groups in open areas, may contest for proximity to the much larger adult male, as in geladas (*Theropithecus gelada*: Dunbar 1980) and hamadryas baboons (Sigg 1980).

Among species in which within-group competition for food is mainly by scramble but competition for safety is by contest, we should expect a social structure closer to the contest than to the scramble type, although rank inheritance may become imperfect as some weak or aging relatives require more support or do not reap sufficient reproductive benefits from the support.

The primate evidence

Who scrambles for food?

Animals grazing on a continuous vegetation are obviously scrambling for food because a displaced animal does not encounter less food or food of a lower quality. Scramble is also expected where food is dispersed, clumps are small, and handling times are short, so that feeding sites cannot be monopolized, even provided that they can be recognized by the animals. Thus, no rank effect on the net intake of dispersed food is found among vervets (*Cercopithecus aethiops*: Whitten 1983), brown capuchins (Janson 1985), and yellow baboons (Altmann 1980; Post, Hausfater & McCusky 1980). Browsing arboreal folivores may also scramble when the spatial variation in the quality of feeding sites is low, and so displacements are ineffective. This may happen in the lean season when they feed on a poor food supply, and feeding efficiency can be raised by decreasing the costs rather than by increasing the quantity or quality of food. They also scramble when the amount they eat at any given patch is limited by detoxification of secondary compounds, rather than patch size. Unfortunately, no studies have yet examined variation in net food intake among female folivores. By contrast, competition is by contest when the spatial variation in quality of resources is high and clumps can be recognized and are large enough to defend. Then, dominants attain much higher intake rates on fruit (Whitten 1983; Janson 1985) or water (Wrangham 1981). As expected, aggression rates during foraging on dispersed food are far lower than during feeding on clumped food sources (see, for example, Janson 1985). Many authors have noted that food-related displacements are very rare among folivores.

Where feeding competition is by contest, high ranking females should have a higher birth rate than low ranking ones, provided the effect of age is controlled for. Although the relationship between dominance rank and reproductive success has recently received considerable attention in field studies, the complementary behavioural data are often lacking. However, among yellow baboons, food intake and energy budgets are not dependent on dominance rank and neither is birth rate, although age of first reproduction is lower for daughters of high ranking females (Altmann 1980). By contrast, among long-tailed macaques, low ranking females have a less favorable energy budget and also give birth at slightly lower rates than high ranking ones (van Noordwijk & van Schaik 1987). Likewise, when the degree of food clumping changed after provisioning by humans was terminated, the rank effect on birth rates among Japanese macaques (*Macaca fuscata*) decreased (Sugiyama & Ohsawa 1982). Among folivores, rank is often strongly correlated with age, and thus correlations with birth rate may be caused by either of the two. However, RS is often negatively correlated with rank because young, high ranking females lose many of their infants soon after birth in howler monkeys (*Alouatta palliata*: Jones 1980; Clarke & Glander 1984) and in langurs (*Presbytis entellus*: Hrdy 1977; Dolhinow, McKenna & vonder Haar Laws 1979).

Primate female hierarchies and dispersal

Because the bonding among females depends on their dispersal tendency we shall first review the effects that scrambling has on female emigration, and then the effects it has on female social relationships.

Natal emigration by single females has been recorded in many well-studied species, but female emigration is far more common among arboreal folivores than among species with other diets (Moore 1984; Pusey & Packer 1987). Female emigration is also common in Costa Rican squirrel monkeys (*Saimiri oerstedi*), which are strongly insectivorous (Boinski 1987). In a number of folivores all females usually emigrate from their natal group, an evolutionary development perhaps made possible by the lower migration threshold for females in scramble conditions. Although female emigration is rare among gregarious non-folivores, group splitting has often been reported in fast growing populations. These groups all tend to split along matrilines (see Moore 1984). Single female emigrants are either returning to their natal group from a group that broke away from it (references in Rasmussen 1981; unpublished observations) or are, as expected, females without relatives in the group (see, for example, Sugiyama & Ohsawa 1982). By contrast, among folivores, group splitting is rare compared with emigration by single females, but when it occurs is usually related to takeovers, i.e. social events (see, for example, Sugiyama 1967; Davies 1984). In fast growing populations of folivorous howler monkeys, high rates of female emigration and immigration are reported, but group splitting is not (Clark & Glander 1984; Crockett 1984).

Nepotistic ranking systems, well-defined female relationships, and a linear dominance hierarchy form a cluster of interrelated traits found in a great variety of diurnal, gregarious non-folivorous cercopithecines (Hinde 1983), at least one cebid (brown capuchins, Janson, personal communication) and even lemurs (*Lemur catta*: Taylor & Sussman 1985). The individualistic ranking system, with generally unstable ranks, poorly defined relationships and a frequent lack of linearity of the dominance hierarchy, is found among Costa Rican squirrel monkeys (S. Boinski, personal communication) and among many folivores of various taxonomic affinity which do not live in open areas: langurs, colobus, howlers, sifakas (*Propithecus* spp.), and gorillas (Jay 1965; Dunbar & Dunbar 1976; Oates 1977; Dolhinow, McKenna & vonder Haar Laws 1979; Harcourt 1979; Jones 1980; Davies 1984), regardless of whether they are close relatives or not. There exists appreciable variation in the strength of grooming bonds, which, at least in the cases where this is known, depends on genetic relatedness among the females. It is my impression that where females remain in their natal groups, grooming bonds are less strong in individualistic than in nepotistic societies, but quantitative comparisons are sorely needed.

Gregarious primates that scramble for food but may contest for safety were expected to be intermediate. Hamadryas females tend to disperse from their natal groups, contrary to expectation. In geladas high ranking females can monopolize access to the large harem male; the females have a peculiar ranking system in which relatives often occupy adjacent ranks but young females rank highest within families (Dunbar 1980). Although savannah baboons follow the rules of the nepotistic ranking system, Moore (1978) remarked that these rules seemed to be less rigid than among macaques.

FEMALE RELATIONSHIPS WHEN COMPETITION IS MAINLY BETWEEN GROUPS

Theory

Females experiencing low predation risk form less cohesive groups, if they form groups at all, and may tend to live at high densities. Hence, competition between groups or societies will turn into BGC whenever food sources are defensible. Since females are most directly affected by feeding competition (cf. Wrangham 1979), their social relationships should in this situation reflect the effects of BGC.

First, let us consider the situation that females can still form groups, even though these are often fissioned into parties. In the latter case, these parties should at least occasionally be able to coalesce and have contests with other groups or societies over access to prime food sources that are big enough to accommodate many or all group members. Female groups for BGC may be expected where large primates of low vulnerability to predators specialize on food that occurs in patches of variable size, but at least sometimes big enough to allow for sizeable female parties. They should also be expected on oceanic islands or other situations without large predators; these primates will generally be smaller and therefore be more likely to maintain groups on a fruit supply of a given mean patch size than larger species.

What social relationships should prevail among the females that cooperatively defend their range or access to prime food sources? We assume that females play an essential role in the defence and do not leave the task entirely to the males even though (some of) the latter are expected to take part as well. We further assume that the number of participating females is an important determinant of the outcome of the contest. We can note two things. First, in more or less cohesive groups strong BGC almost inevitably would imply strong WGC. This might prompt subordinates to leave the group and parasitize on the group's defence of the food, or to refrain from joining in the contests but still harvest some of the fruits of their labour. An alternative option that is even more damaging to the remaining group members would be to join another group and increase its contest power. Thus, the subordinates can force the dominants not to exert to the full their power to suppress the subordinates' food intake through WGC. This leads to female dominance hierarchies that are becoming fairly egalitarian rather than despotic (cf. Vehrencamp 1983; *pace* Wrangham 1980), despite the potential for great fitness differences within the group.

Second, females should prefer relatives as alliance partners (Wrangham 1980), because an alliance with kin is a greater contribution to inclusive fitness than one with non-kin (especially if this would imply that they are directed against kin) and because they are more stable in the long run (since defection against kin is more costly). Hence, female residence will be the most likely outcome, and the dominance hierarchy will be nepotistic (due to the small WGC that can remain) but egalitarian (type D in Table 4). Obviously, if groups exist whose food supply leads to WGS but yet allows for BGC through defence of feeding areas (territories), then social relationships should be of type C (see Table 4).

It is difficult to predict whether a gradual increase in BGC relative to WGC should lead to a gradual or to a sudden change towards an egalitarian system, but I suspect intermediate situations would show intermediate ranking systems.

When females no longer form cohesive groups, the range of outcomes is quite large. This is because the role of males in the female social structure can no longer be ignored. So far we dealt with cohesive groups of females, and the almost universal male strategy in response to female groups among primates is female defence polygyny (cf. Emlen & Oring 1977). When females are more dispersed, among primates the optimal male strategy is most likely to shift to resource defence polygyny, thus the establishment of territories by males, either alone or in alliance with other males.

Table 5 depicts the expected outcomes for the situation when the potential within-group competition leads to females being basically solitary for some or most of their time. They are still expected to defend their range against other females. Likewise, males are expected to become territorial among each other (cf. Wrangham 1979). Thus, various outcomes ensue depending on whether females and males, only males, or neither of the sexes can defend their range. It is impossible to give generalizations about the ecological conditions in which each of these outcomes is expected, although the size of food patches relative to female size, their density and the mobility of both males and females are clearly involved.

TABLE 5. A summary of female social structure when female diurnal primates forage alone or in ephemeral parties

Range defence feasible for:			
females	+	−	−
males	+	+	−
Social system	Monogamy/resource defence polygony	Fission−fusion society	Semi-solitary lifestyle
Female social structure	Resident, nepotistic despotic[*]	Non-resident, individualistic, despotic	Resident, nepotistic, despotic
Examples	Gibbons (*Hylobates* spp.)	Chimpanzee (*Pan troglodytes*)	Orang-utan (*Pongo pygmaeus*)

[*] Applicable only to polygynous situation

Yet, in all cases there will be some pressure towards female residence, either because mothers can assist their daughters in setting up a territory, or because they can share (part of) their range with daughters and tolerate them or even form alliances when food sources are available that can hold several individuals.

If both males and females can defend a territory, the result is either monogamy or resource defense polygyny, depending on how strong the advantage to the female of services by the male can be (see Table 5). If a single male can defend a range, while females cannot profitably do so and thus share it with other females, the result will be a nepotistic and despotic hierarchy among females who rarely meet. Alternatively, only a group of males is able to defend a territory containing a number of dispersed females. These males are likely to be relatives for the same reason as cooperatively defending female groups are. Groups of related males are formed most readily when maturing males remain in their natal area, and this provides strong pressure towards female natal emigration. Hence, dominance hierarchies among the females are expected to be individualistic, while still despotic. Finally, when neither males nor females can defend their range, discrete societies may disappear altogether. However, females may still benefit from staying near their natal range if mothers selectively tolerate their daughters or even form alliances with them if occasionally coalitions for access to food sources are possible, i.e. groups can be maintained for some time.

The primate evidence

Very few primates have been studied on oceanic islands. Some data are available on the macaques off the Sunda shelf. It is interesting to note that, where densities are known, they are very high (MacKinnon, MacKinnon & Chivers 1979; van Schaik & van Noordwijk 1985), consistent with the prediction. What little is known about their social behaviour supports the theory. Group size is clearly smaller than of comparable species in similar habitats on the Sunda shelf (MacKinnon *et al.* 1979; Whitten & Whitten 1982; van Schaik & van Noordwijk 1985), and groups are often fragmented into small parties, which in the case of the Simuleue macaques also have different composition than among their Sumatran counterparts. Sulawesi macaques have a social structure that is dramatically more egalitarian than that of other macaques (Thierry 1985). Although the aggression frequencies are not so low as among folivores, there is often symmetric aggression within dyads, resulting in inconsistent dominance relationships. In addition, the tension caused by agonistic interactions is defused by very frequent reconciliation behaviour. Aggression is often redirected towards non-group members. In contrast to all other macaques living in multi-male groups (cf. Herzog & Hohmann 1984), the males in these species have loud inter-group vocalizations, which suggests that BGC is important (Whitten & Whitten 1982; van Schaik & van Noordwijk 1985; Watanabe & Brotoisworo 1982). Physical attacks are most common during inter-group conflicts (Watanabe & Brotoisworo 1982).

The variability in social organization among large species with low vulnerability to predation is clearly greater than among the vulnerable species. For example, all possible outcomes expected by the above considerations are found among the apes (see Tables 4 and 5). Thus, female bonobos (*Pan paniscus*) form parties of varying sizes that may serve to cooperatively defend food sources and to which males attach themselves temporarily (White, this volume); gibbons are largely monogamous and territorial; among chimpanzees females are more or less solitary while cooperative groups of males defend a range; orang-utans (*Pongo pygmaeus*) are semi-solitary, and neither males nor females can maintain exclusive ranges. Gorillas, however, do not fit into this scheme, because unrelated females tend to form cohesive groups around a single adult male. More detailed discussion of ape social structure and their ecological determinants is provided by Wrangham (1979, 1986), van Schaik & van Hooff (1983), Rodman (1984) and Dunbar (1988).

DISCUSSION

Within-group competition

WGS is predicted to give rise to female social relationships that are very different from those found under WGC conditions. The main distinction is the one between an individualistic, egalitarian, and often non-FB ranking system and a nepotistic, despotic, FB one. An individualistic, despotic system is not expected to occur among primates. Such a system should obtain where animals in a contest situation have a short lifespan (no maternal support possible) or female dispersal is enforced by some other factor.

Arboreal folivores fit into the scramble pattern with regard to their egalitarian and individualistic social structure. Although the crucial test (measuring the effect of rank on female food intake) has not yet been performed, they also show very low rates of food-related aggression, no relationship between rank and RS, and their groups do not often split. The occurrence of very much the same pattern in the strongly insectivorous Costa Rican squirrel monkey suggests that it is not taxonomy or folivory *per se* that is responsible for this pattern.

Although the theory developed here is the only one proposed so far to account for this whole range of characters, several alternative hypotheses have been proposed to explain some of the differences in female relationships treated here. We shall consider them in turn. Hrdy (1977) refers to the WGS type dominance hierarchy as an altruistic one, which is found in harem systems with a high mean and a low variance in genetic relatedness among the females. Older females gain in inclusive fitness by giving up their ranks in favour of younger close relatives with higher expected future reproductive output whether these are daughters or not. Younger females thus outrank older, much heavier females. This argument assumes that it is the harem system rather than scramble competition that gave rise to the individualistic ranking system. If so, we can predict that this altruistic ranking system should also be found in other harem-living species regardless of diet but should not occur among folivores in multi-male groups, or in harems of immigrant females. Unfortunately, it is not clear whether other non-folivorous harem-living guenons (*Cercopithecus* spp.) have nepotistic ranking systems (Cords 1986) but harem-living geladas have some form of nepotistic ranking system (Dunbar 1980). The female hierarchies of langurs or other folivores (e.g. howlers) do not seem to differ where groups are multi-male (Jay 1965) or where females are immigrants (Jones 1980; cf. Sigg 1980). Thus, the comparative data do not support this hypothesis.

An alternative interpretation for the lack of female agonistic support to maturing daughters among folivores would be that they are incapable of it due to early senescence. Although folivore females may show earlier senescence, this possibility does not seem to be correct, because aging langur females who are low-ranking among the females are the most vigorous in defending the group against external threats such as other groups, predators and infanticidal males (Hrdy 1977), and not even middle-aged females are observed to support their (sub)adult daughters.

A simpler variation of this theme is that folivores refrain from aggression because the narrow energetic margins allowed by a folivorous existence necessitated the evolution of non-damaging conflict resolution (Jones 1980). Many authors have indeed ascribed the low levels of aggression among folivores to their diet. However, although high costs of aggression would raise the threshold of escalated fighting, they do not preclude the development of consistent dominance relation-ships within a dyad or of linear dominance hierarchies. Moreover, damaging fights are quite common among male folivores contesting access to the breeding group (see, for example, Sugiyama 1967; Crockett 1984). Thus, when a resource can be contested folivores are quite capable of escalated fights just like other mammals.

Competition between groups

One of the key points of this chapter is that BGC has not exerted a significant effect on female social relationships in most of the smaller gregarious primate species (*pace* Wrangham 1980). Yet, the primate literature is replete with accounts of between-group conflicts among several of these very species. How can this paradox be resolved? First, it is possible that BGC, although groups often meet, does not have any strong quantitative effects (cf. Janson 1985). Second, BGC may be an important component of food competition without having a strong effect on social relationships. Obviously, strong BGC does not radically change female relationships where within-group competition is by scramble, but it also need not always do this where within-group competition is by contest. For instance, as a result of female preferences for males that defend the food supply (cf. Smuts 1987), males may play the decisive role in interactions between groups in species with considerable sexual dimorphism in body size and canine length. At present, our knowledge of the role of males is mainly anecdotal. Finally, recent increases in population density may have caused an increased BGC, without, however, immediately changing the female social relationships since these largely represent evolved adaptations rather than direct behavioural choices. We would, for instance, expect not all females to be equally involved in between-group contest in such species, but rather that high-ranking females are most heavily involved since they derive the greatest benefits in a type B social structure. This is in fact found among the vervet monkeys of Amboseli (Cheney 1987), which is the only vervet population studied to date in which females actively participate in between-group conflicts. It is also inhabiting a shrinking habitat, and may thus live at a density that is higher than the carrying capacity, a situation likely to give rise to strict territoriality.

BGC may be the predominant mode in a group of species that are of small or intermediate body size, and thus clearly vulnerable to predators, and live in relatively small groups, yet seem to have relationships that were hypothesized to reflect a predominance of BGC. Known examples include white-fronted capuchins (*Cebus albifrons*: Janson 1986) and stump-tailed macaques (*Macaca arctoides*: de Waal, this volume). However, it is possible that here WGC is strongly reduced because they rely on large fruit trees to which access is communally defended. As Janson & van Schaik (1988) note, above a certain patch size contest seem to be reduced, and the intake of all individuals tends to be similar (cf. Janson 1985). Since at least white-fronted capuchins are known to specialize on large fig trees, this may account for the apparent predominance of BGC effects on their social relationships. However, nothing is known about stump-tailed macaques in the wild.

Testing the framework

Obviously, even the most ingenious theory will not make correct predictions for all primate species, and the present framework is bound to have its exceptions in cases where other factors override the social effects of competition for food or safety. It is not difficult to find cases that are seemingly at odds with the framework. We should, however, be aware that some of those may merely represent intermediate situations, which should of course predominate in nature rather than the extreme competitive regimes I considered here. For instance, the possible lack of strongly differentiated female relationships among guenons noted by Cords (1987), may perhaps be due to a relatively strong BGC component. Likewise, Srikosomatara & Robinson (1986) found an important effect of between-group competition among Venezuelan wedge-capped capuchins, although WGC is not negligible. In most cases, we simply do not know enough about the competitive regime of a species to decide, and the question is evidently how to test the framework. What we need are detailed field studies on a single species in which the competitive regime is characterized by precise estimates of the strength of the three components of food competition and on reproductive rates and survival or female fitness (see Janson & van Schaik 1988 for methods) and in which contest for safety is estimated. These data should then be compared with data on female social relationships: female dispersal or residence, consistency and stability of dominance relationships, linearity of the hierarchy, occurrence of grooming bonds, and coalitions and the female role in between-group conflicts.

Within species, there is usually a strong relationship between the density of a population and the occurrence of aggressive encounters between groups (Cheney 1987). Thus, there may be problems with testing the framework where recent ecological changes have led to increased population densities and thus to an increased importance of BGC. After all, for a proper assessment one needs to study the species at the density that prevailed during most of its evolutionary history.

Primate populations under study often increase, because the continuous presence of the researchers discourages hunting by natural predators, or because the latter are disappearing from nature reserves and parks due to human pressure and island-effects. Such recent increases in population densities may be responsible for some of the discrepancies between theory and observation. The high female emigration rates among the red howlers (*Alouatta seniculus*) observed by Crockett (1984) may also be an indication of high population growth, rather than a typical feature of the species. Likewise, the competition for group membership observed by Jones (1980) in a population of mantled howlers (*A. palliata*) inhabiting an isolated forest patch may reflect unusual crowding. Female residence, nepotistic hierarchies and group splitting would seem a much more plausible adaptive response to such conditions.

Changes in the distribution of food may also affect the results. The village langurs studied by Hrdy (1977) had more consistent and stable dominance relationships than those found among langurs in forested areas (see, for example, Jay 1965; Sugiyama 1967).

These reservations are clearly speculative. However, if they are substantiated, it means that testing the relationships proposed here will be increasingly difficult because the pristine habitats this requires are rapidly disappearing.

ACKNOWLEDGEMENTS

I thank Charles Janson, Willem Netto, Ronald Noë, Maria van Noordwijk, Han de Vries, and in particular Jan van Hooff for helpful discussion, Sandy Harcourt, and the editors for comments on the manuscript, Sue Boinski for allowing me to cite her unpublished data, and Greg Grether for linguistic advice. During the writing of this chapter the author held a senior fellowship of the Royal Netherlands Academy of Arts and Sciences.

REFERENCES

Altmann, J. (1980). *Baboon Mothers and Infants*. Harvard University Press, Cambridge, Massachusetts.
Anonymous (1981). *Techniques for the Study of Primate Population Ecology*. National Academy Press, Washington D.C.

Boinski, S. (1987). Mating patterns in squirrel monkeys (*Saimiri oerstedi*), implications for seasonal sexual dimorphism. *Behavioral Ecology and Sociobiology*, **21**, 13–21.

Cheney, D.L. (1987). Interactions and relationships between groups. *Primate Societies* (Ed. by B.B. Smuts, D.L. Cheney, R.M. Seyfarth, R.W. Wrangham & T.T. Struhsaker), pp. 267–81. Chicago University Press, Chicago, Illinois.

Clarke, M.R. & Glander, K.E. (1984). Female reproductive success in a group of free-ranging howler monkeys (*Alouatta palliata*) in Costa Rica. *Female Primates* (Ed. by M.F. Small), pp. 111–26, Alan Liss, New York.

Collins, D.A. (1984). Spatial pattern in a troop of yellow baboons (*Papio cynocephalus*) in Tanzania. *Animal Behaviour*, **32**, 536–53.

Cords, M. (1987). Forest guenons and patas monkeys: male–male competition in one-male groups. *Primate Societies* (Ed. by B.B. Smuts, D.L. Cheney, R.M. Seyfarth, R.W. Wrangham & T.T. Struhsaker), pp. 98–111, Chicago University Press, Chicago, Illinois.

Crockett, C.M. (1984). Emigration by female red howler monkeys and the case for female competition. *Female Primates* (Ed. by M.F. Small), pp. 159–73, Alan Liss, New York.

Davies, A.G. (1984). An ecological study of the red leaf monkey (*Presbytis rubicunda*) in the dipterocarp forest of northern Borneo. Ph.D. thesis, University of Cambridge.

Dolhinow, P., McKenna, J.J. & vonder Haar Laws, J. (1979). Rank and reproduction among female langur monkeys: Aging and improvement (They're not just getting older, they're getting better). *Aggressive Behavior*, **5**, 19–30.

Dunbar, R.I.M. (1980). Determinants and evolutionary consequences of dominance among female gelada baboons. *Behavioural Ecology and Sociobiology*, **7**, 253–65.

Dunbar, R.I.M. (1988). *Primate Social Systems*. Croom Helm, Beckenham.

Dunbar, R.I.M. & Dunbar E.P. (1976). Contrasts in social structure among black-and-white colobus groups. *Animal Behaviour*, **24**, 84–92.

Emlen, S.T. & Oring, I.W. (1977). Ecology, sexual selection, and the evolution of mating systems. *Science*, **197**, 215–23.

Goodall, J. (1985). *The Chimpanzees of Gombe, Patterns of Behavior*. Harvard University Press, Cambridge, Massachusetts.

Gouzoules, H., Gouzoules, S. & Fedigan, L. (1982). Behavioural dominance and reproductive success in female Japanese monkeys (*Macaca fuscata*). *Animal Behaviour*, **30**, 1138–50

Hamilton W.D. (1971). Geometry for the selfish herd. *Journal of Theoretical Biology*, **31**, 295–311.

Harcourt, A.H. (1979). Social relationships among adult female mountain gorillas. *Animal Behaviour*, **27**, 251–64

Herzog, M.O. & Hohmann, G.M. (1984). Male loud calls in *Macaca silenus* and *Presbytis johnii*; a comparison. *Folia primatologica*, **43**, 189–97

Hinde, R.A. (1983). *Primate Social Relationships, an Integrated Approach*. Blackwell Scientific Publications, Oxford.

Hrdy, S.B. (1977). The Langurs of Abu. Harvard University Press, Cambridge, Massachusetts.

Janson, C.H. (1985). Aggressive competition and individual food consumption in wild brown capuchin monkeys (*Cebus apella*). *Behavioral Ecology and Sociobiology*, **18**, 125–38

Janson, C.H. (1986). The mating system as a determinant of social evolution in capuchin monkeys (*Cebus*). *Primate Ecology and Conservation* (Ed. by J.R. Else & P. Lee), pp. 169–79. Cambridge University Press, Cambridge.

Janson, C.H. & Schaik, C.P. van (1988). Recognizing the many faces of primate food competition in primates: methods. *Behaviour*, **105**, 165–86.

Jay, P.C. (1965). The common langur of north India. *Primate Behavior* (Ed. by I. DeVore), pp. 197–249. Holt, Rinehart & Winston, New York.

Jones, C.B. (1980). The functions of status in the mantled howler monkey, *Alouatta palliata* Grav: Intraspecific competition for group membership in a folivorous neotropical primate. *Primates*, **21**, 389–405.

MacKinnon, J.R. MacKinnon, K.S. & Chivers, D.J. (1979). The use of forest space by a community of six species in peninsular Malavsia and of one in North Sulawesi. Paper presented at VIIth congress of International Primatological Society, Bangalore, India.

McFarland Symington, M. (1987). *Ecological and social correlates of party size in the black spider monkey (Ateles paniscus chamek).* Ph.D. thesis, Princeton University.

McFarland Symington, M. (1988). Food competition and foraging party size in the black spider monkey, *Ateles paniscus chamek. Behaviour*, **105**, 117–34.

Milton, K. (1982). Dietary quality and demographic regulation in a howler monkey population. *The Ecology of a Tropical Forest* (Ed. by E.G. Leigh, Jr, A.S. Rand, & D.M. Windsor), pp. 273–89. Smithsonian Institution Press, Washington D.C. .

Moore, J. (1978). Dominance relations among free-ranging female baboons in Gombe National Park, Tanzania, *Recent advances in primatology, Vol. 1* (Ed. by D.J. Chivers & J. Herbert), pp. 67–70. Academic Press, London.

Moore, J. (1984). Female transfer in primates. *International Journal of Primatology*, **5**, 537–89

Mori, A. (1979). Analysis of population changes by measurement of body weight in the Koshima troop of Japanese monkeys. *Primates*, **20**, 371–9.

Nicholson, A.J. (1967). Self-adjustment of populations to change. *Cold Spring Harbor Symposia in Quantitative Biology*, **22**, 153–73

Noordwijk, M.A. van & Schaik, C.P. van (1987). Competition among adult female long-tailed macaques. *Animal Behaviour*, **35**, 577–89.

Oates, J.F. (1977). The social life of the black-and-white colobus monkey, *Colobus guereza. Zeitschrift für Tierpsychologie*, **45**, 1–60.

Parker, G.A. (1974). Assessment strategy and the evolution of fighting behaviour. *Journal of Theoretical Biology*, **47**, 223–43.

Post, D.G. Hausfater, G. & McCusky, S.A. (1980). Feeding behavior of yellow baboons (*Papio cynocephalus*): relationship to age, gender and dominance rank. *Folia primatologica*, **34**, 170–95.

Pusey, A.F. & Packer, C. (1987). Dispersal and philopatry. *Primate Societies* (Ed. by B.B. Smuts, D.L. Cheney, R.M. Seyfarth, R.W. Wrangham & T.T. Struhsaker), pp. 250–66. Chicago University Press, Chicago, Illinois.

Rasmussen, D.R. (1981). Communities of baboon troops (*Papio cynocephalus*) in Mikumi National Park, Tanzania. A preliminary report. *Folia Primatologica*, **36**, 232–42.

Robinson, J.G. (1981). Spatial structure in foraging groups of wedge-capped capuchin monkeys *Cebus nigrivittatus. Animal Behaviour*, **29**, 1036–56.

Rodman, P.S. (1984). Foraging and social systems of orangutans and chimpauzees. *Adaptations for Foraging in Nonhuman Primates* (Ed. by P.S. Rodman & J.G.H. Cant), pp. 134–60. Columbia University Press, New York.

Schaik, C.P. van (1983). Why are diurnal primates living in groups? *Behaviour*, **87**, 120–44.

Schaik, C.P. van & Hooff, J.A.R.A.M. van (1983). On the ultimate causes of primate social systems. *Behaviour*, **85**, 91–117.

Schaik, C.P. van & Noordwijk, M.A. van (1985). Evolutionary effect of the absence of felids on the social organization of the macaques on the island of Simeulue (*Macaca fascicularis fusca*, Miller 1903). *Folia Primatologica*, **44**, 138–47.

Sigg, H. (1980). Differentiation of female positions in hamadryas one-male units. *Zeitschrift für Tierpsychologie*, **53**, 265–302.

Silk, J.B. (1987). Social behaviour in evolutionary perspective. *Primate Societies* (Ed. by B.B. Smuts, D.L. Cheney, R.M. Seyfarth, R.W. Wrangham & T.T. Struhsaker), pp. 318–29. Chicago University Press, Chicago, Illinois.

Smuts, B.B. (1987). Gender, aggression, and influence. *Primate Societies* (Ed. by B.B. Smuts, D.L. Cheney, R.M. Seyfarth, R.W. Wrangham & T.T. Struhsaker), pp. 400–12. Chicago University Press, Chicago, Illinois.

Sondaar, P.Y. (1977). Insularity and its effect on mammal evolution. *Major Patterns in Vertebrate Evolution* (Ed. by M.K. Hecht, P.O. Goody & B.M. Hecht), pp. 671–707. Plenum Press, New York.

Srikosamatara, S. & Robinson, J.G. (1986). Group size and use of space in wedge-capped capuchin monkeys. *Primate Report*, **14**, 67.

Sugiyama, Y. (1967). Social organization of hanuman langurs. *Social Communication among Primates* (Ed. by S.A. Altmann), pp. 221–36. Chicago University Press, Chicago, Illinois.

Sugiyama, Y. & Ohsawa, H. (1982). Population dynamics of Japanese monkeys with special reference to the effect of artificial feeding. *Folia Primatologica*, **39**, 238–63.

Taylor, L. & Sussman, R.W. (1985). A preliminary study of kinship and social organization in a semi-free-ranging group of *Lemur catta*. *International Journal of Primatology*, **6**, 601–14.

Terborgh, J. & Janson, C.H. (1986). The socioecology of primate group. *Annual Review of Ecology and Systematics*, **17**, 11–135.

Thierry, B. (1985). Patterns of agonistic interactions in three species of macaque (*Macaca mulatta, M. fascicularis, M. tonkeana*). *Aggressive Behavior*, **11**, 223–33.

Vehrencamp, S.L. (1983). A model for the evolution of despotic versus egalitarian societies. *Animal Behaviour*, **31**, 667–82.

Waser, P. (1976). *Cercocebus albigena*: site attachment, avoidance, and intergroup spacing. *American Naturalist*, *110*, 911–35.

Waser, P. & Jones, W.T. (1983). Natal philopatry among solitary mammals. *Quarterly Review of Biology*, **58**, 355–90.

Watanabe, K. & Brotoisworo, E. (1982). Field observations of Sulawesi macaques. Kyoto University Overseas Research Report of Studies on Asian Non-Human Primates, **2**, 3–9.

Whitten, A.J. & Whitten, J.E.J. (1982). Preliminary observations of the Mentawai macaque on Siberut island. Indonesia. *International Journal of Primatology*, **3**, 445–59.

Whitten, P.L. (1983). Diet and dominance among female vervet monkeys (*Cercopithecus aethiops*). *American Journal of Primatology*, **5**, 139–59.

Wrangham, R.W. (1979). On the evolution of ape social systems. *Social Science Information*, **18**, 335–68.

Wrangham, R.W. (1980). An ecological model of female-bonded primate groups. *Behaviour*, **75**, 262–300.

Wrangham, R.W. (1981). Drinking competition in vervet monkeys. *Animal Behaviour*, **29**, 904–10.

Wrangham, R.W. (1987). The evolution of social structure. *Primate Societies* (Ed. by B.B. Smuts, D.L. Cheney, R.M. Seyfarth, R.W. Wrangham & T.T. Struhsaker), pp. 282–96. Chicago University Press, Chicago, Illinois.

Section 3
Reproductive Strategy and Social Structure: The Influence of Life History, Demography and Social Relations

It was argued in the introduction to Section 2, and illustrated in the papers in that section, that the classic idea of the environment as a static habitat type grimly determining the shape of a society is an inappropriate one; resource structure has replaced environmental zone as a critical determining or influencing variable, and so made possible precise predictions about the nature of social relationships and organization.

The success (or indeed relevance) of socioecology to many biological and anthropological problems has not been limited, though, just by the way the natural environment has been defined. Another limitation, especially in terms of applicability of Darwinian theory and method to human sociality, has been the emphasis placed on the *natural* environment as a *direct* influence on social structure. This emphasis can be traced back to the beginnings of Darwinian thought.

The original conclusion drawn from Darwin's theory was that the fit survive. What these fit individuals (or species) had to survive were the rigours of their environments, taken to mean the actual physiographical constraints (temperature, altitude, rainfall, etc.), surviving ferocious predators and acquiring sufficient food to become even fitter. Fitness was survivorship, and this led ecologists to look for direct correlations between social behaviour and the availability of food (Lack 1968). Such broad correlations were indeed found, as such classic studies as that of Jarman (1974) showed. However, the increasing emphasis on individual fitness that came from the theoretical developments of the 1960s and 1970s led to a shift in interest. Survivorship is only a surrogate variable in modern Darwinian theory. The really important variable is maximization of reproductive success, and while those who live a long time have the opportunity to increase their reproductive success, it does not follow that this will automatically result in greater *relative* reproductive success. If there is a single major contribution that evolutionary biologists have made over the last 20 years it has been to replace, hopefully for ever, the simple idea that Darwinian fitness is necessarily associated with the common sense notion of physical fitness.

A focus on reproductive success rather than survivorship has had an important effect on perceptions of the environment and its role in determining social behaviour. In particular, the Darwinian environment is not just a list of physical or energetic attributes, but is *anything* in an animal's surroundings that influences, positively or negatively, the chances of reproductive success. This most certainly includes climatic and habitat-related phenomena, but it also includes many other factors. The individual is encased in layer upon layer of 'environments' from the microbial to the climatic, from the family to the population, from the small scale to the ecological community. Each of these may influence the adaptation and hence reproductive success of an individual. Furthermore, these environments may not impinge directly, but may be mediated through other elements — for example, the thermal environment may be significant only through the effect it has on the productivity of the resources available, and so act only indirectly on the fitness of an individual. Additional complications would also include the feedback nature of many organism–environment interactions, and of course, the fact that this multi-layered environment is never identical for any two members of the same species (see MacDonald & Carr, this volume).

In summary, from a socioecological point of view the environment cannot be treated simply as the habitat, nor as a general attribute of a species as a whole, but instead will vary from individual to individual, and, most importantly here, will include conspecifics and members of the social group as well. Indeed, these may be the most significant part of the selective environment of many species, as is indicated by the enormous power of runaway sexual selection in many species. In other words, it is the process of competition that is important, not necessarily what is competed for; socioecological approaches may in some cases give primacy to the social environment, rather than to the state of resources, as the principal factor determining social behaviour. A socioecological approach is not the same as an environmental one (Foley 1986).

The papers in this section examine the reproductive strategies and social organization of a variety of mammalian species in the light of this rather more subtle view of organism–environment relations. They focus on some of the factors influencing the behavioural, reproductive, and, in general, adaptive strategies available to individuals living in very specific contexts. In the first of these, Harcourt looks at the role of alliances in primate social organization. In many primate societies individuals build up cooperative ties with others through a variety of means, as a way of obtaining certain goals. These ties are exploited and utilized in specific contexts, so that the behaviour of an individual may be embedded in a complex web of shifting relationships. One of the main points Harcourt makes is that the scope for building and maintaining alliances, and indeed the motivation to do so, varies enormously both between and within species and populations. Each individual has its own environment, both social and in terms of actual *access* to resources, that determines the costs and benefits of maintaining an alliance. Alliances, therefore, grow out of the complexities of this web of social and ecological environments, not for species as a whole, but for individuals. Animal politics is micropolitics.

The notion that animals have a political life was elegantly documented by Frans de Waal in his studies of the Arnhem chimpanzee colony (de Waal 1982). De Waal explores the idea that other individuals represent the 'resources' of the social environment. In building up bonds animals are investing in the future. This, according to de Waal, will have an effect on immediate competitive interactions, for against the benefits of winning a competitive interaction must be put the costs to the long-term relationship. This view of the social environment as a set of resources to be both exploited *and* conserved provides important insights into why social groups involve not just dominance ranks and aggression, but also tolerance and very marked patterns of reconciliation.

Datta's chapter turns to the demographic context in which social strategies are carried out. She points out that the same strategy — that of acquiring dominance — pursued in different contexts may produce varying results (and presumably *vice versa*). As the specific demographic environment in which an animal may find itself will shift through time, as animals get old and die and new generations appear, the pay-offs of particular strategies would vary, and so too would the overall, observable social structure. One of the principal conclusions drawn by Datta is that differences in the social behaviour of separate species may be brought about by small changes in the demographic environment.

The fact that different individuals live in very different social environments is well illustrated by the marmosets. They live in social groups in which only the dominant female will breed; subordinate females will usually be completely infertile. Abbott shows the physiological mechanism responsible for this — not due to any direct influence of resource availability, but to the immediate social influence of a dominant female. The implication, from a Darwinian point of view, is that the most significant element in the total environment of a marmoset (and several other species), is the number and rank of other individuals.

What all these papers have in common is the subtle nature of total environmental pressures operating on individuals, populations and species, and thus shaping their evolution. This is, however, a relatively static view of evolution which regrettably is a circular and prolonged process involving a series of secondary adaptations to primary evolved characteristics. The long-term interaction between the characteristics of a species and the nature of its total environment is addressed by Harvey, Promislow & Read. The life history characteristics of species reflect the nature of both their social and ecological constraints; some species, Harvey *et al.* argue, have slowed down, while others have speeded up, and this is an adaptation to the type of mortality schedules suffered by a species and its effects on the costs of reproduction at different ages.

In the short term perspective of most of the papers in this volume, its life history characteristics represent the immediate environment to which an individual must adapt. But in a longer term the characteristics are themselves adaptations to environmental pressures, in particular to the ecological factors influencing mortality schedules. When the argument is extended to social species the individual adapts to the constraints of the environment in which it grows up — and that is principally a social environment. That social environment has evolved alongside and in response to life history parameters of the species which are themselves responses to more strictly ecological factors.

Although the papers in this section have addressed, for the most part, issues in non-human socioecology, there are important implications for anthropology. It could be argued that the re-orientation of the concept of the 'environment' in behavioural ecology may close the gap between humans and other animals. As discussed in the introduction to Section 2, one of the reasons for the failure of earlier ecological approaches to human social behaviour lay in the fact that social factors seemed far more significant in shaping the organization of a community. The papers in Section 3 show that the same may well be true of many animal species too; not that their environment and the nature of their resources is not important, but that their effect is mediated by proximate factors that relate more directly to social variables — and in particular, to reproductive behaviour. In other words, the natural environment may form the basis for a socioecological analysis, but the total environment of a social animal is far more complex.

REFERENCES

Foley, R. (1986). Anthropology and behavioural ecology. *Anthropology Today*, 2(3), 13–15.
Jarman, P.J. (1974). The social organization of antelope in relation to their ecology. *Behaviour*, 48, 215–67.
Lack, D. (1968). *Ecological Adaptations for Breeding in Birds*. Methuen, London.
Waal, F. de (1982). *Chimpanzee Politics*. Jonathan Cape, London.

Social influences on competitive ability: alliances and their consequences

A.H. HARCOURT

Department of Applied Biology, University of Cambridge,
Pembroke Street, Cambridge CB2 3DX, UK

SUMMARY

1 Primates and, less often, non-primates support kin in contests with group members, i.e. ally with them, so enabling them to gain access to resources that would otherwise be less easily obtainable. Both also use alliances for their own immediate benefit, as when larger groups of allies, who might or might not be kin, defeat smaller ones, and all gain access to the resource. The result is that competitive ability need no longer depend on intrinsic characteristics of individuals, but instead on the support received.

2 Individuals differ in their value as allies. For example, dominant animals give more powerful support than do subordinate ones. However, only primates appear to choose partners on the basis of their quality as potential allies. Thus in primate groups dominant animals are often preferred partners and even become a limiting resource over which individuals compete to establish supportive relationships.

3 Consequently, alliances among primates have led to the appearance of a whole set of competitive interactions involved in establishing, maintaining and preventing others establishing supportive relationships with other group members. Such social complexity has not been reported for non-primates, but dolphins and other toothed whales might prove an exception.

4 Alliances probably influence the stability of groups as, for example, it becomes advantageous to remain with allies and indeed intensify cooperation with them in the face of rival alliances. They might also influence the stability of dominance hierarchies, as dominant animals have their competitive ability continually reinforced by alliances with other group members.

5 Whether or not alliances are advantageous is constrained by habitat and demographic parameters, such as group size, and also possibly by the animals' information processing capacity, given the complexity of decisions in competitive interactions that involve alliances.

6 The analysis of alliance systems in humans needs to take advantage of the mammalian data, in particular of present understanding of the role of intra-group competition on the use of alliances as a competitive strategy. Alliances are not confined to humans and do not have a single evolutionary cause.

INTRODUCTION

'Extreme flexibility in rates of forming and dissolving alliances is a distinctive human trait, cultural in origin, that is universal and shows sexual asymmetry in its use' (Alexander & Noonan 1979, p. 438). 'Human sociality evolved ... during the long phase of our evolution when we lived as hunter/gatherers and as tribesmen ... in a context of small groups within which individuals were highly dependent on their neighbours ... for cooperation and aid ... in acquiring food resources, ... protection from conspecifics, ... and finding and keeping mates' (Chagnon 1982, p. 291). With alliances being understood as supportive relationships, these statements apply equally to non-human primates, and to some non-primates as well.

Animals cooperate in many different ways, protecting one another from predators, feeding each other's offspring and communally defending territories. They also establish supportive relationships with one another, giving and exchanging services in a manner very reminiscent of exchange of goods or services in human societies. The service that I will concentrate on is support in contests with other group members, or alliance formation. Alliances will be distinguished from long-term supportive relationships, which is what I understand anthropologists to mean by alliances. Supportive relationships in primates involve the exchange of several other services also, as in human partnerships, such as grooming (cleaning of the pelage), and care of the partner's offspring, the important point being that the services exchanged need not be identical.

My aims here are first to describe how animals use alliances for their relatives' and their own competitive advantage, and then to discuss how the competitive interactions of societies that use alliances and establish supportive relationships differ from those that do not, and hence how their social organization differs.

THE USE OF ALLIANCES AS A COMPETITIVE STRATEGY

Most analyses of competitive encounters among conspecifics concern contests between only two individuals, which is perhaps why anthropologists perceive humans as being unusual or even unique in their use of alliances. However, while non-primates rarely cooperate in fights, up to 30% of contests between any two individuals in a primate group involve the participation of other group members (Harcourt, in press). The evolution of cooperation and the nature of its effects on social organization depend on the nature of its benefits (Emlen 1984), on whether they are gained immediately or only after a delay (Trivers 1971; Wrangham 1982) and on the identity of the beneficiary (Brown 1983). I therefore start with a discussion of how animals use alliances for their relatives' and their own benefit.

The most common use of support in contests for a relative's benefit is to protect immature offspring from larger or more dominant group members, a behaviour that occurs in most social mammals and birds. In addition, however, support is also given in the context of competition for resources in both non-primate (Fig. 1) (Scott 1980) and primate species (Fig. 2) (Harcourt & Stewart 1987) to enable the recipient of support to gain access that might otherwise be denied it. Since cost is almost inevitably involved in supporting another individual in a contest, animals support close kin more frequently than they do distant kin and risk more when they do so (Fig. 3) (Harcourt, in press). The benefits of such support are not necessarily confined to immature animals. For example, adult vervet monkey *Cercopithecus aethiops* females continue to be supported by their mothers, apparently to such an extent that those still with mothers in the group have higher reproductive success than those without (Fairbanks & McGuire 1986).

When the only beneficiary of support is the recipient, animals support only relatives. However, the supporter, as well as the recipient, can benefit from an alliance, as when two animals together defeat a third whereas neither alone could, or support given now is reciprocated later. Direct benefit to the supporter means that consanguinity between supporter and recipient is not a necessary condition for the alliance to be advantageous as a competitive strategy. However, even where an alliance is used by the supporter to obtain direct benefit, cooperation with kin will usually be more advantageous than cooperation with non-kin. By helping kin, the supporter will benefit indirectly through shared genes as well as directly; cheating by the recipient will be less of a cost; and the alliance might be more reliable and efficient because relatives are more familiar and more likely to agree to cooperate. Nevertheless, relatives are not always available, with the result that, for example, 38% of thirteen breeding partnerships of male lions *Panthera leo* contained non-relatives (Packer & Pusey 1982).

Lions provide a classic example of mutual advantage from alliance formation: they acquire prides of females more easily and retain them for longer, the more partners they have (Fig. 4) (Bygott, Bertram & Hanby 1979). Birds too appear to

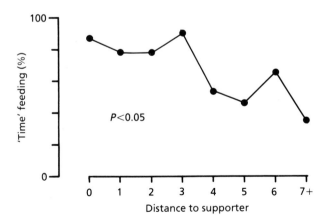

FIG. 1. Immature swans spend more time feeding when near supporters (parents) than when far from them (distance in 'swan lengths'), 17–236 records per distance category, except 7+ where $n = 3$; $r_s = -0.76$; $P<0.05$. (Data from Scott 1980, Fig. 2.)

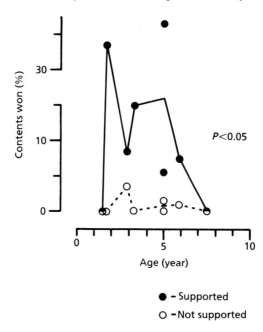

FIG. 2. Immature gorillas gain access to food in contests with dominant opponents more often when the contests are intervened in (supported) than when they are not. Median individual's number of contests = 12 with intervention, 106 without; *P*<0.05, Sign test. (Data from Harcourt & Stewart 1987, Fig. 5.)

cooperate in contests in order to obtain direct benefit for themselves. In contests among acorn woodpeckers *Melanerpes formicivorus*, winning groups were twice the size of losing ones, and apparently unrelated acorn woodpeckers switched groups depending on which was winning at the time (Hannon *et al.* 1985); and pigeons *Columba livia* protect their mate's access to food (Lefebvre & Henderson 1986).

Reciprocal cooperation involves a delay in reception of benefit. For various reasons, direct evidence for reciprocation in primate groups is poor (Silk 1982: Harcourt 1987a). Nevertheless, the sum of evidence is compelling. Adult male baboons *Papio cynocephalus* who have independently immigrated into the same group and are probably unrelated will ally with each other against other males in competition over oestrous females (Packer 1977; Smuts 1985). In six of six cases in a wild baboon group where one male successfully persuaded another to support it in its contest over an oestrous female, the soliciting male, not the one who supported it, mated with the female acquired as a result of the alliance (Packer 1977). In other words, only the supported male benefited at the time of the alliance. The explanation for the apparent altruism of the supporter is that male baboon alliances are probably reciprocal; frequent supporters are supported frequently. Macaque (*Macaca mulatta*) males help one another in contests over

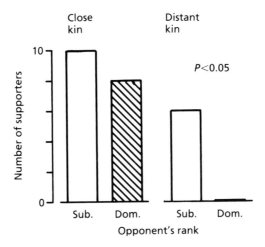

FIG. 3. Gorillas more often support close kin ($r \geq 0.05$) than distant kin ($r < 0.05$) and more often support them against dominant opponents, i.e. risk themselves; $P < 0.05$, Sign test. (From Harcourt, in press.)

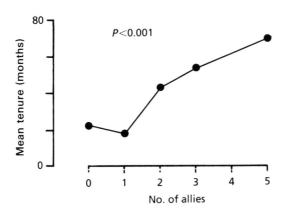

FIG. 4. Male lions hold prides for longer when with several allies. Two to ten records per number of allies; $r_s = 0.71$, $P < 0.001$. (Data from Bygott, Bertram & Hanby 1979, Fig. 1a.)

females, but here the males are brothers or half-brothers (Meikle & Vessey 1981). Hunte & Horrocks (1987) also found that among adult and immature free-ranging vervet monkeys, frequent supporters were supported frequently. More specifically, they showed that individual immatures supported most often those immatures who supported them the most; (there were no records of adult females supporting one another).

Similar special relationships, 'friendships', between male and female baboons are well known (Smuts 1985). While 'friendship' doubled the male's chances of mating with the female, 'friends' were by no means assured of paternity: only four of eight females subsequently had offspring that were probably fathered by their 'friends' (Smuts 1985). Nevertheless, males new to a group often tried to establish 'friendships' with females whom they had not mated as a means of gaining entry to a group (Smuts 1985). Adult gelada baboon females *Theropithecus gelada* might experience even greater delay in receiving a direct benefit from giving support: old females with matrilineal relatives in their group apparently have higher competitive ability than those without, whereas supportive relatives make no difference to younger adult females (Dunbar 1984).

The baboon 'friendships' illustrate well the idea of reciprocal exchange of different services. 'Friends' spent much time near one another and grooming, and the male received mating rights in exchange for protecting the female and her infant, Smuts suggested. Seyfarth & Cheney (1984) provided more direct evidence that one helpful act, in this case grooming, caused another, a readiness to provide support in a fight. Unrelated vervet monkeys paid attention for longer to a group member's distress call if that group member had groomed them in the previous 1.5 hours than if it had not.

In summary, primates and some non-primates use cooperation to improve their own and their relatives' immediate and long-term competitive ability. When both partners mutually benefit, alliances are formed with non-kin, sometimes even when considerable delay in receiving the benefit is involved. Briefly, alliances allow competitive ability to become independent of personal characteristics, such as health and size, that would otherwise confer competitive advantage. As described so far, a species that uses alliances as a means of competition might not have a social organization much different from one that does not. However, alliances can in fact enormously increase the complexity of interactions among group members.

ALLIANCES AND SOCIAL COMPLEXITY

Dominance hierarchies

Alliances confer a temporary improvement in competitive ability on the participants, in other words they temporarily raise their dominance rank. Since dominance rank correlates with reproductive success under certain conditions (Harcourt 1987b), it should pay animals to use alliances to raise the dominance rank of their kin, particularly their offspring (Chapais & Schulman 1980). All individuals would so benefit, but not all animals are equally efficient supporters (see West-Eberhard (1975) for general discussion). In the case of alliances, dominant animals can provide more frequent and successful support than can subordinate ones (Fig. 5; Cheney 1977; and also Datta 1983, 1986). Therefore, offspring of dominant mothers have their competitive ability raised more than do the offspring of subordinate mothers. The result is that offspring assume a dominance rank adjacent to their mother, hence the phenomenon of what has been

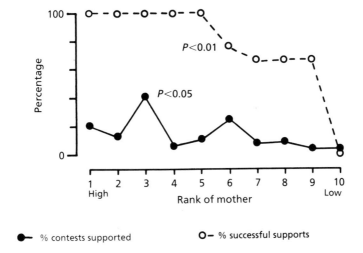

FIG. 5. High ranking baboon mothers support their offspring in a greater proportion of cases when they are threatened (solid circles) than do low ranking mothers and are more successful in stopping the threat (open circles). $r_s = 0.73$ and 0.93; (data from Cheney 1977, Table 2.)

called 'inheritance' of rank in several Old World monkey species (macaque: Datta 1983; baboon: Cheney 1977; vervet; Horrocks & Hunte 1983). (Further details of the process, for example the necessity for support of offspring when they are threatening, and when they are dominant to their opponents, can be found in the above references, and also in Datta 1986 and Harcourt & Stewart 1987). The appearance of an inheritance-of-rank system as a result of support in contests is reported in some non-primate species also, notably so far the spotted hyaena *Crocuta crocuta* (Frank 1986) and the Bewick swan *Cygnus columbianus* (Scott 1980).

Cooperation in the form of alliances influences not only the nature of dominance hierarchies, converting size-related hierarchies to inheritance-of-rank systems (see above), but it also influences their stability. Where an extensive network of supportive relationships exist, as in cercopithecine groups, the advantages of cooperation with high-ranking animals could reinforce the hierarchy as group members support dominant animals for the resultant reciprocated benefits. On the other hand, alliances can diminish the stability of competitive relationships, when they enable two otherwise subordinate animals to outcompete a dominant one (Goodall 1987, Chapter 15; de Waal 1978).

Preferences for allies of high quality

If support is given to improve only the recipient's competitive ability, and the supporter benefits only indirectly by consanguinity with the recipient, the most effective support is often that given to young, close relatives of high reproductive value (Hamilton 1964; Chapais & Schulman 1980; Charlesworth & Charnov 1981). However, when alliances, or long-term supportive relationships, are mutually beneficial, young relatives are no longer necessarily the best partners. One of my younger sisters also works at Cambridge University. However, she is below me in the University hierarchy, with the result that if I want to advance up the hierarchy, I need to form a supportive relationship with a Cambridge professor, not with my sister. Primates use the same principle.

Up to now, differences between primates and non-primates have been mostly quantitative: primates ally more frequently. By contrast, active choice of partners on the basis of their abilities appears to be a qualitative difference separating primates from non-primates. Primates deliberately try to establish supportive relationships with, for example, powerful potential allies, but no non-primate has yet been reported to do this.

In several species of primate, dominant individuals receive more grooming than do subordinate ones (Seyfarth 1977). More than that, some individuals choose non-kin as partners in preference to kin, if the non-kin are dominant. Vervet monkeys are a classic 'inheritance-of-rank' species, with females achieving a rank adjacent to that of their relatives. Nevertheless, in three wild groups, adult females in the bottom half of the linear dominance hierarchy allied with females in the top half as often as with each other (Fig. 6) (Seyfarth 1980). Put simply, they allied with powerful (dominant) non-kin as well as with their own subordinate kin. Since an alliance involves aggression to one of the contestants, as well as support of the other, animals have to be careful about the contests in which they take part. Thus unrelated adult female macaques support the more dominant of two contestants against the other, but only as long as the opponent is subordinate to themselves; they tend to ignore contests between dominant animals (Fig. 7)

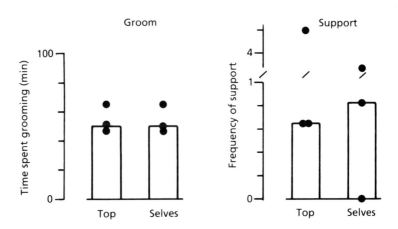

FIG. 6. Vervet females in the bottom half of the dominance hierarchy ally with females in the top half (Top), to whom they are probably unrelated, as frequently as with one another (Selves). Points are each of the three groups' mean individual's frequency of giving support in contests; histograms show the middle point. *n* = 7–8 females per group. (Data from Seyfarth 1980, Fig. 4.)

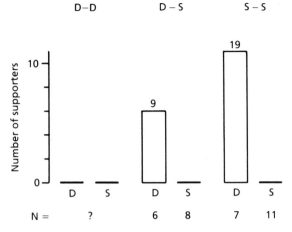

FIG. 7. Adult female rhesus macaques consistently support the more dominant (D) of two unrelated contestants, when either one (D-S) or both (S-S) are subordinate to themselves, but ignore contests when both are dominant (D-D). n = number of contestants; values above bars = number of incidents of support. (Data from Chapais 1983, Table 10.6.)

(Chapais 1983). It seems that the females use a divide-and-rule tactic against females subordinate to themselves (see also p. 235), but stay out of fights among dominant animals because of the high cost of retaliation (Chapais 1983).

Because dominant animals are more efficient and effective supporters than are subordinate ones (see above, Fig. 5), the most dominant group member should be the preferred ally, other things being equal (Cheney 1977; Seyfarth 1977). Thus Cheney (1977) found that 81% of incidents of support from immatures to adult females in a baboon group were directed to the three top ranking of the ten females. In the establishment of long-term supportive relationships, as opposed to temporary alliances, individuals not only support the most dominant females, they also groom them. Twelve of seventeen females in three wild vervet groups who had two or more females dominant to themselves in the group, groomed a female two or more ranks above them more than they groomed the adjacently ranked female above them (Fig. 8; Seyfarth 1980). A choice of partners on the basis of their dominance status is not confined to Old World monkey females: in a wild chimpanzee *Pan troglodytes* population, dominant males received more grooming than did subordinate ones (Simpson 1973). That such services might be given to elicit future support from the best allies is indicated by the fact that when actively soliciting support, as opposed to giving it, baboons apparently direct their overtures to the more dominant among the dominant females (Fig. 9) (Walters 1980).

To receive reciprocation, it is not necessary to save the drowning man (Trivers 1971); saving his drowning child will probably do instead; and if the saving is a calculated act, saving the boss's child could be the best ploy. In the same way that dominant females are preferred alliance partners, so also are their offspring. Six

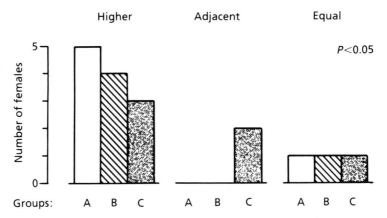

FIG. 8. Adult females in each of three wild vervet groups groomed a female at least two ranks above them in the hierarchy (Higher) more often than they groomed the adjacently ranked female (Adjacent). Histograms are number of females per group that groomed a 'higher' ranking female more often than they groomed the dominant female 'adjacent' to them in the hierarchy, the number that groomed the 'adjacent' female more often, and the number that groomed the 'adjacent' and a 'higher' female equally often (Equal). $P<0.05$, Sign test. (Data from Seyfarth 1980, Fig. 4.)

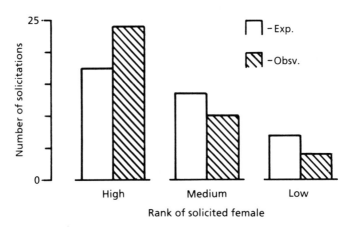

FIG. 9. When soliciting support from adult females dominant to their opponent, adolescent baboons solicit it most from the most dominant females. $n = 38$ solicitations. (Data from Walters 1980, p. 80.)

of seven infants of dominant macaque mothers were helped by animals other than their mother, but none of six infants of middle or low-ranking mothers (Berman 1980). And in Cheney's (1977) study of wild baboons, eight of nine immatures supported the offspring of mothers higher ranking than their own mother, whereas only three supported offspring of lower-ranking mothers (Fig. 10). Moreover, all support to recipients who were younger than the supporter was to offspring of higher-ranking mothers (Fig. 10). Since the younger immatures were subordinate to their supporter, and thus of no use to it as an ally, therefore their mothers, or

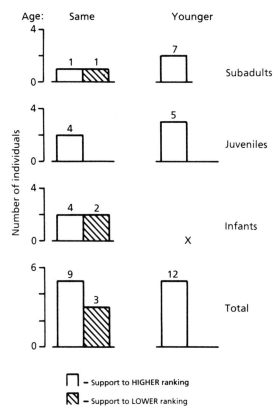

FIG. 10. Immature baboons support the immature offspring of females higher ranking than their mother (open histogram) more often than they do the offspring of lower-ranking females (hatched histogram), even when the recipients are in the age class younger than their own (younger) and hence subordinate to them. Numbers above histograms are number of incidents of support. (Data from Cheney 1977, Figs. 1b, 2b.)

other relatives, were the target of the alliance, Cheney suggested. (See also the case of 'Curry' in de Waal, this volume.) Evidence that primates are capable of such second-order social calculations is provided by Cheney & Seyfarth's (1986) study of redirected aggression. Adult vervet monkeys did not retaliate only against those that had threatened them, nor only against the relatives of those that had threatened them; they threatened the relatives of those who had threatened their own relatives. The aim of such indirect aggression is presumably to inhibit future aggression, in the same way that the aim of indirect support is to promote future support (Cheney & Seyfarth 1986).

Although power is an obviously useful quality in an ally, there are others. A reliable ally might be more useful than a powerful but untrustworthy one, and relatives are normally more reliable than are non-relatives.

Of eight free-ranging immature male rhesus monkeys who had a dominant non-kin and a subordinate close kin ranking adjacent to them, seven chose the subordinate kin (Colvin 1983). Individuals of similar age and similar rank could well have similar needs; furthermore, costs of establishing supportive relationships with them could be less than those incurred trying to establish such a relationship with a very dominant animal (see next section). Consequently, Colvin (1983) and de Waal & Luttrell (1986) argue, individuals often choose as partners group members ranked close to them in preference to dominant group members.

In general then, primates appear to chose partners on the basis of their qualities as allies. Scores of papers in the non-primatological literature treat the subject of dominance hierarchies and competitive ability. In few is there any indication that alliances influence competitive ability; none suggest that quality of ally is a significant factor.

Competition for social resources

Where alliances are used as a tactic to influence ability in competition over resources, a second level of competition is introduced into the social group, because the means to the end, the ally, itself becomes an object of competition. One alliance thus affects others as group members compete to form alliances and prevent others establishing rival alliances.

Given that primates support group members in contests in exchange for future services, one way to compete for reciprocated support is to provide more or better services than do rivals (Brown 1982). Attempts to ally with dominant animals rather than subordinate ones and to form supportive relationships with them can be interpreted in this way (Figs 6–10). The result is a situation in which primates tend not to be altruistic to their least serious competitors, those subordinate to them, as might be expected (Alexander 1974). Instead, they support and otherwise cooperate with their most serious and most frequent competitors, i.e. those dominant or adjacently ranked to them.

The other way to prevent rival alliances is to inhibit competitors from forming long-term supportive relationships. Primates use this tactic. In three-quarters of eighty-nine cases of prevention of grooming between others in a free-ranging rhesus macaque group, a female of intermediate rank prevented one subordinate to her from grooming one dominant to her. Consanguinity did not alter the pattern of intervention: subordinate relatives and non-relatives alike were prevented from grooming dominant ones (Chapais 1983). The most dominant individuals are potentially the most powerful allies, and in three wild vervet groups, the females over whom there was the most competition for grooming were the most dominant (Fig. 11) (Seyfarth 1980).

While a rival alliance that included a dominant female might be the most dangerous, alliances among subordinate animals can be effective, especially if the target of the alliance itself lacks allies (Cheney 1983; Goodall 1987, Chapter 15; de Waal 1978). It might therefore be necessary for individuals to prevent friendly relationships among subordinate group members. Grooming has been interpreted as a means of establishing supportive relationships, and both Silk (1982), studying

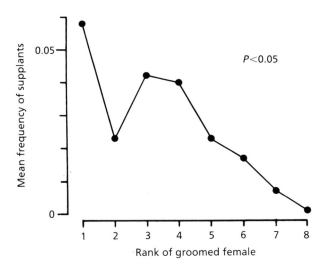

FIG. 11. Adult female vervet monkeys prevent others grooming high ranking females more often than they prevent them grooming low ranking females. Points are the mean of three groups' rate per minute of grooming at which the grooming was interrupted. $r_s = 0.90$, $P<0.05$. (Data from Seyfarth 1980, Fig. 1.)

bonnet macaques *M. radiata*, and de Waal & Luttrell (1986), with rhesus macaques, found that in over 80% of cases when one animal prevented grooming among two others, the two were both subordinate to the intervener (see also p. 231).

   To conclude, alliances force primates to become consummate social tacticians, as they compete for social resources in addition to environmental ones. With alliances as part of the primate social repertoire, both aggressive and friendly behaviours between two individuals can have ramifications throughout the group. An attacker can suffer retaliatory attacks from its opponent, from its opponent's relatives, and from unrelated group members if the opponent is high-ranking, or the offspring of a high-ranking animal. Even friendly interactions can incite aggression from others, as group members compete to maintain supportive relationships in the face of rival alliances. Present information indicates that in non-primate groups, neither alliances nor their consequences are anywhere near as complex.

DEMOGRAPHIC CONSEQUENCES OF COOPERATION

If cooperation has widespread social consequences, these in turn affect demographic parameters, since the structure of a population is based in large part on the nature of the social interactions among its members. The bias towards establishment of supportive relationships with dominant group members could affect, for instance, family stability, as members of low-ranking families attempt to form alliances with high-ranking families instead of with each other (Cheney 1983). In one population of wild vervet monkeys, high-ranking adult females allied among themselves over five times more frequently than did low-ranking females, groomed one another for twice as long, and were in proximity to one another for over five times as long (Seyfarth 1980). However, a captive group of rhesus macaque females showed no relation between family rank and time spent near family members (de Waal & Luttrell 1986). Cooperation could influence the stability of groups also, because of the advantages of remaining with cooperative partners (Wrangham 1980, 1982), and indeed of intensifying cooperation with them to prevent them leaving to join competing groups (Brown 1982). Of course, effects on stability of either relationships or especially groups are probably not the consequence of the behaviour on which selection has acted; they are an epiphenomenon. In anthropological terms, therefore, alliances and the establishment of supportive relationships result in emergent properties in the society (see Hinde 1976).

Alexander (1974) hypothesized that humans were an exception to the general rule that either anti-predation or feeding strategies were responsible for grouping. Instead, intergroup aggression and the advantages of cooperative defence against conspecifics both maintained groups and led to the evolution of complex sociality. Wrangham (1980) argued the same for female primates. However, the arguments lack generality. Why is cooperation apparently more advantageous to human males than to other male mammals, among which cooperative defence of females seems to be rare? Stacey (1982) has made a start on the analysis of the conditions under which cooperation among males might be advantageous. He uses arguments very similar to Vehrencamp's (1983) and Emlen's (1984) concerning differences among the males in competitive ability and opportunities for subordinate males to breed elsewhere, but he almost entirely omitted from consideration the effect of other males on any one male's chances of fertilization. Why also do so many female mammals that are not primates remain in the area or group of their birth (Greenwood 1980) when cooperation in defence of food sources against other groups appears to be so rare among them? Given that even among solitary mammals, females tend to be philopatric, it seems more likely that cooperation is a consequence, not a cause, of grouping (cf. Pusey & Packer 1987; Dunbar, this volume; White, this volume; van Schaik, this volume).

Lastly on the subject of demographic consequences of cooperation, offspring sometimes help parents feed subsequent broods, and Emlen, Emlen & Levin (1986), have argued that it would pay parents to produce more of the sex that helps, or in Fisherian terms, more of the sex in which less is invested. In primates, biased sex ratios in relation to dominance rank have been found. However, it is difficult to relate these to the extent of cooperation. In the first place, little has been written about sex differences in the distribution of cooperation between parents and offspring; secondly, effects of mother's rank on son's reproductive success are poorly documented; and thirdly, the sex bias goes in both directions in relation to rank (Clutton-Brock & Iason 1986).

CONSTRAINTS ON COOPERATION

Most analyses of the evolution of cooperation assume that animals formed groups before they cooperated, rather than that they grouped in order to cooperate (discussions in Alexander 1974; Brown 1982, 1983; Vehrencamp 1983; Emlen 1984; Dunbar, this volume; van Schaik, this volume). Whatever the causes of grouping, it is clear that alliances as a form of cooperation will not evolve unless defence competition (interference competition) occurs. Thus alliances will not be seen unless resources are worth defending and are defensible (Wrangham 1980; Davies & Houston 1984). Resources therefore need to be in relatively small, rich patches. In addition, the resource needs to be divisible (Wrangham 1980, 1982), for if it is not, the supporter (the altruist), gains nothing by its act. To some extent such an analysis seems to explain the distribution of cooperation across taxa. Herbivores generally do not have defensible food resources and alliances are very rare among ungulates; some carnivores have defensible resources and alliances can occur frequently enough for dominance rank to be inherited (Frank 1986); and omnivorous and frugivorous primates, who have defensible resources, might ally more frequently as a means of gaining access to resources than do primarily folivorous primates (Wrangham 1980; Stewart & Harcourt 1987; White, this volume; van Schaik, this volume).

While the environment and the way in which a species exploits it influence the pay-offs of using cooperation as a competitive strategy, they are not the only constraints. Demography plays a part also, as can be illustrated by an analysis of cooperation in gorilla (*Gorilla gorilla*) groups (Harcourt & Stewart 1986). In Harcourt & Stewart's study, the environment could certainly have had an influence: the gorillas' main food item, foliage, was widespread, abundant and generally of low quality, meaning that it was not defensible nor worth defending. As a consequence the benefits of using support in contests to raise permanently the competitive ability of kin might not have outweighed the costs.

In addition, however, Harcourt and Stewart argued that demographic constraints existed. Gorilla groups were extremely small; therefore immatures were disparate in age; the costs of raising immatures to a rank discordant with their size were thus high; consequently, immature gorillas did not achieve a rank adjacent to their mother's. The other result of small group size, combined with the large size of the adult male and the social constraint of his absolute dominance over all females, was that the adult male could effectively inhibit the unequal distribution of alliances that appears to be the basis of baboon, macaque and vervet social relationships. Thus female gorillas very rarely supported one another in contests and showed no indication of a preference for dominant partners. The relative importance of environment and demography remains to be elucidated in this case, but demographic constraints have been suggested to influence cooperation in lions (Packer 1986, though see Caro, this volume), some bird species (Ligon 1983), and other primates (Hausfater *et al.* 1987; Datta, this volume).

Primates are notable both for the frequency and complexity of their cooperative interactions and also for the cohesiveness and stability of their groups (Wrangham 1983). Given the broad range of environments in which primates occur, all of which overlap extensively with non-primates, it is not obvious why cooperation should be more competitively advantageous to primates than to non-primates. However, competitive interactions that involve cooperation are hugely more complex than simple dyadic (two-animal) contests. Therefore the capacity to process information might be a limiting factor in the use of cooperation as a competitive strategy (Trivers 1971; Axelrod & Hamilton 1981). Primates have brains that are larger, relative to body size, than those of any other taxon except the toothed whales (*Odontoceti*) (Jerison 1983; Ridgway 1986); simians have larger brains than prosimians, and possibly a more complex system of social interaction (Jolly 1966); and frugivores have larger brains than folivores (Clutton-Brock & Harvey 1980), and might form more frequent and complex alliances (Wrangham 1980; van Schaik, this volume). Whether primates made use of a large brain selected for other reasons, such as processing environmental information (Clutton-Brock & Harvey 1980; Milton 1988) or whether the benefits of cooperation selected for information processing capacity and hence brain size (Harcourt 1988) has yet to be determined.

Alliances are far more rarely reported among non-primates than among primates, at least partly I suspect because non-primatologists have not looked for them (Harcourt 1988). The foregoing analysis of the constraints operating on the evolution of cooperation as a competitive strategy allows prediction of where in non-primates support in contests as a form of cooperation might be found. The species will live in stable social groups, especially large ones, and sometimes feed on rich, localized resources. Coatimundies *Nasua narica* (Russell 1983) might be one example. Toothed whales which, in addition, are so obviously intelligent and have such a complex system of communication are another, yet so far only protective support has been reported among odontocetes (Conner & Norris 1982; Tyack 1986).

ALLIANCES AND HUMAN SOCIAL ORGANIZATION

The term 'alliance' among social anthropologists involves in its meaning an exchange of services over some period of time. In other words it is what I have called a 'long-term supportive relationship'. The work reviewed here demonstrates that any emphasis on alliances, in the anthropologists' sense, as a distinguishing feature of humans is mistaken: non-human primates commonly form highly complex supportive relationships with one another. If there is a distinguishing line, it is between primates and non-primates. I suspect too that Alexander's (1974) emphasis on inter-group conflict being the selection force in the divergence between humans and primates is also mistaken, as might be Wrangham's (1980) stress on its importance in current patterns of dispersion in primate populations (see discussion on p. 236). Little evidence exists to show that inter-group competition is more likely than intra-group competition to lead either to grouping, or to complex sociality and forms of cooperation. For instance, very few of the alliances discussed in this paper resulted from inter-group conflict.

The emphasis on inter-group conflict appears to have led Alexander & Noonan (1979) to the explicit suggestion that alliances are sex-biased in humans, presumably meaning that males use them in their social interactions more than do females. Early studies of wild primate groups were based largely on descriptions of male interactions and dominance hierarchies. The reason was that male behaviour was so much more overt than female behaviour. It turned out that understanding of female relationships was just as crucial to appreciation of primate social systems. It will be surprising if more studies by female anthropologists do not reveal just as complex and frequent a use of alliances among women as is currently perceived to occur among men. Certainly cooperation of various forms occurs among women in both traditional (Irons 1979) and industrialized societies (Essock-Vitale & McGuire 1985). Indeed the human pair bond is valuably seen as a mutually beneficial relationship (Smuts 1985), rather than as a means of enforcing male care or mate fidelity (Alexander & Noonan 1979). Within groups of children too, cooperation appears to be used as a competitive strategy by both sexes in ways remarkably similar to what is reported among non-human primates. For example, when children supported others in their threats to a third, the frequency with which they did so was proportional to the dominance status of the supported antagonist (Strayer & Noel 1986). In conclusion, while the use of cooperation as a competitive strategy is crucial in human evolution, it is not confined to humans, nor to one section of humanity, and almost certainly does not have a single evolutionary cause.

ACKNOWLEDGEMENTS

I thank the National Geographic Society and the Harry Frank Guggenheim Foundation for funding my field-work; and the Governments of Zaire and Rwanda for permitting it. The paper benefited considerably from criticism by Drs Kelly Stewart, Robin Dunbar and William McGrew.

REFERENCES

Alexander, R.D. (1974). The evolution of social behavior. *Annual Review of Ecology and Systematics*, **5**, 325–83.

Alexander, R.D. & Noonan, K.M. (1979). Concealment of ovulation, parental care, and human social evolution. *Evolutionary Biology and Human Social Behavior: an Anthropological Perspective* (Ed. by N.A. Chagnon & W. Irons), pp. 436–53. Duxbury, North Scituate, Massachusetts.

Axelrod, R. & Hamilton, W.D. (1981). The evolution of cooperation. *Science*, **211**, 1390–6.

Berman, C.M. (1980). Early agonistic experience and rank acquisition among free-ranging infant rhesus monkeys. *International Journal of Primatology*, **1**, 153–70.

Brown, J.L. (1982). Optimal group size in territorial animals. *Journal of Theoretical Biology*, **95**, 793–810.

Brown, J.L. (1983). Cooperation — a biologist's dilemma. *Advances in the Study of Behavior*, **13**, 1–37.

Bygott, J.D., Bertram, B.C.R. & Hanby, J.P. (1979). Male lions in large coalitions gain reproductive advantages. *Nature,* **282,** 839–41.

Chagnon, N.A. (1982). Sociodemographic attributes of nepotism in tribal populations: man the rule-breaker. *Current Problems in Sociobiology* (Ed. by King's College Sociobiology Group), pp. 291–318. Cambridge University Press, Cambridge.

Chapais, B. (1983). Dominance, relatedness and the structure of female relationships in rhesus monkeys. *Primate Social Relationships* (Ed. by R.A. Hinde), pp. 208–19. Blackwell Scientific Publications, Oxford.

Chapais, B. & Schulman, S.R. (1980). An evolutionary model of female dominance relations in primates. *Journal of Theoretical Biology,* **82,** 47–89.

Charlesworth, B. & Charnov, E.L. (1981). Kin selection in age-structured populations. *Journal of Theoretical Biology.* **88,** 103–19.

Cheney, D.L. (1977). The acquisition of rank and the development of reciprocal alliances among free-ranging immature baboons. *Behavioral Ecology and Sociobiology,* **2,** 303–18.

Cheney, D.L. (1983). Extrafamilial alliances among vervet monkeys. *Primate Social Relationships* (Ed. by R.A. Hinde), pp. 278–86. Blackwell Scientific Publications, Oxford.

Cheney, D.L. & Seyfarth, R.M. (1986). The recognition of social alliances by vervet monkeys. *Animal Behaviour,* **34,** 1722–31.

Clutton-Brock, T.H. & Harvey, P.H. (1980). Primates, brains and ecology, *Journal of Zoology, London,* **190,** 309–23.

Clutton-Brock, T.H. & Iason, G.R. (1986). Sex ratio variation in mammals. *Quarterly Review of Biology,* **61,** 339–74.

Colvin, J. (1983). Familiarity, rank and the structure of rhesus male peer networks. *Primate Social Relationships* (Ed. by R.A. Hinde), pp. 190–200. Blackwell Scientific Publications, Oxford.

Conner, R.C. & Norris, K.S. (1982). Are dolphins reciprocal altruists? *American Naturalist,* **119,** 358–74.

Datta, S.B. (1983). Relative power and the acquisition of rank. *Primate Social Relationships* (Ed. by R.A. Hinde), pp. 93–103. Blackwell Scientific Publications, Oxford.

Datta, S.B. (1986). The role of alliances in the acquisition of rank. *Primate Ontogeny, Cognition and Social Behaviour* (Ed. by J.G. Else & P.C. Lee), pp. 219–26. Cambridge University Press, Cambridge.

Davies, N.B. & Houston, A.I. (1984). Territory economics. *Behavioural Ecology,* 2nd edn. (Ed. by J.R. Krebs & N.B. Davies), pp. 148–69. Blackwell Scientific Publications, Oxford.

Dunbar, R.I.M. (1984). *Reproductive Decisions. An Economic Analysis of Gelada Baboon Social Strategies.* Princeton University Press, Princeton, New Jersey.

Emlen, S.T. (1984). Cooperative breeding in birds and mammals. *Behavioural Ecology* (Ed. by J.R. Krebs & N.B. Davies), pp. 305–39. Blackwell Scientific Publications, Oxford.

Emlen, S.T., Emlen, J.M. & Levin, S.A. (1986). Sex–ratio selection in species with helpers-at-the-nest. *American Naturalist,* **127,** 1–8.

Essock-Vitale, S.M. & McGuire, M.T. (1985). Women's lives viewed from an evolutionary perspective. II. Patterns of helping. *Ethology and Sociobiology,* **6,** 155–73.

Fairbanks, L.A. & McGuire, M.T. (1986). Age, reproductive value, and dominance-related behaviour in vervet monkey females: cross-generational influences on social relationships and reproduction. *Animal Behaviour,* **34,** 1710–21.

Frank, L.G. (1986). Social organisation of the spotted hyaena (*Crocuta crocuta*). II. Dominance and reproduction. *Animal Behaviour,* **34,** 1510–27.

Greenwood, P.J. (1980). Mating systems, philopatry and dispersal in birds and mammals. *Animal Behaviour,* **28,** 1140–62.

Goodall, J. (1987). *The Chimpanzees of Gombe.* Belknap Press/Harvard, Cambridge, Massachusetts.

Hamilton, W.D. (1964). The genetical evolution of social behaviour. *Journal of Theoretical Biology,* **7,** 1–52.

Hannon, S.J., Mumme, R.L., Koenig, W.D. & Pitelka, F.A. (1985). Replacement of breeders and within-group conflict in the cooperatively breeding acorn woodpecker. *Behavioral Ecology and Sociobiology,* **17,** 303–12.

Harcourt, A.H. (1987a). Cooperation as a competitive strategy in primates and birds. *Animal Societies:*

Theories and Facts (Ed. by Y. Ito, J.L. Brown & J. Kikkawa), pp. 141–57. Japan Scientific Societies Press, Tokyo.

Harcourt, A.H. (1987b). Dominance and fertility in female primates. *Journal of Zoology*, **213**, 471–87.

Harcourt, A.H. (1988). Alliances and social intelligence in primates. *Social Expertise and the Evolution of Intellect* (Ed. by R. Byrne & A. Whiten), pp. 132–52. Oxford University Press, Oxford.

Harcourt, A.H. & Stewart, K.J. (1986). High dominance rank in primate groups requires help from others. *The Individual and Society* (Ed. by L. Passera & J-P. Lachaud), pp. 93–100. Université Paul Sabatier, Toulouse.

Harcourt, A.H. & Stewart, K.J. (1987). The influence of help in contests on dominance rank in primates: hints from gorillas. *Animal Behaviour*, **35**, 182–90.

Hausfater, G., Cairns, S.J. & Levin, R.N. (1987). Variation and stability in the rank relations of nonhuman primate females: analysis by computer simulation. *American Journal of Primatology*, **12**, 55–70.

Hinde, R.A. (1976). Interactions, relationships, and social structure. *Man (NS)*, **11**, 1–17.

Horrocks, J. & Hunte, W.A. (1983). Maternal rank and offspring rank in vervet monkeys: an appraisal of the mechanisms of rank acquisition. *Animal Behaviour*, **31**, 772–82.

Hunte, W & Horrocks, J.A. (1987). Kin and non-kin interventions in the aggressive disputes of vervet monkeys. *Behavioral Ecology and Sociobiology*, **20**, 257–63.

Irons, W. (1979). Investment and primary social dyads. *Evolutionary Biology and Human Social Behavior: An Anthropological Perspective* (Ed. by N.A. Chagnon & W. Irons), pp. 181–213. Duxbury, North Scituate, Massachusetts.

Jerison, H.J. (1983). The evolution of the mammalian brain as an information processing system. *Advances in the Study of Mammalian Behavior* (Ed. by J.F. Eisenberg & D.G. Kleiman), pp. 113–46. The American society of Mammalogists, Pittsburgh, Pennsylvania.

Jolly, A. (1966). Lemur social behavior and primate intelligence. *Science*, **153**, 501–6.

Lefebvre, L. & Henderson, D. (1986). Resource defense and priority of access to food by the mate in pigeons. *Canadian Journal of Zoology*, **64**, 1889–2.

Ligon, J.D. (1983). Cooperation and reciprocity in avian social systems. *American Naturalist*, **121**, 366–84.

Meikle, D.B. & Vessey, S.H. (1981). Nepotism among rhesus monkey brothers. *Nature*, **294**, 160–1.

Milton, K. (1988). Foraging behaviour and the evolution of primate cognition. *Machiavellian Intelligence: Social Expertise and the Evolution of Intellect in Monkeys, Apes and Humans* (Ed. by R. Byrne & A. Whiten), pp. 285–305. Oxford University Press, Oxford.

Packer, C. (1977). Reciprocal altruism in olive baboons. *Nature*, **265**, 441–3.

Packer, C. (1986). The ecology of sociality in felids. *Ecological Aspects of Social Evolution* (Ed. by D.I. Rubenstein & R.W. Wrangham), pp. 429–51. Princeton University Press, Princeton, New Jersey.

Packer, C. & Pussey, A. (1982). Cooperation and competition within coalitions of male lions: kin selection or game theory? *Nature*, **296**, 740–2.

Pusey, A.E. & Packer, C. (1987). Dispersal and philopatry. *Primate Societies*. (Ed. by B.B. Smuts, D.L. Cheney, R.M. Seyfarth, R.W. Wrangham & T.T. Struhsaker), pp. 250–66. University of Chicago Press, Chicago, Illinois.

Ridgway, S.H. (1986). Dolphin brain size. *Research on Dolphins* (Ed. by Bryden, M.M. & Harrison, R.), pp. 59–70. Clarendon Press, Oxford.

Russell, J.K. (1983). Altruism in coati bands; nepotism or reciprocity. *Social Behavior of Female Vertebrates* (Ed. by S.K. Wasser), pp. 263–90. Academic Press, New York.

Scott, D.K. (1980). Functional aspects of prolonged parental care in Bewick's swans. *Animal Behaviour*, **28**, 938–52.

Seyfarth, R.M. (1977). A model of social grooming among adult female monkeys. *Journal of Theoretical Biology*, **65**, 671–98.

Seyfarth, R.M. (1980). The distribution of grooming and related behaviours among adult female vervet monkeys. *Animal Behaviour*, **28**, 798–813.

Seyfarth, R.M. & Cheney, D.L. (1984). Grooming, alliances and reciprocal altruism in vervet monkeys *Nature*, **308**, 541–3.

Silk, J.B. (1982). Altruism among female Macaca radiata: explanations and analysis of patterns of grooming and coalition formation. *Behaviour,* **79,** 162–88.

Simpson, M.J.A. (1973). The social grooming of male chimpanzees. In: *Comparative Ecology and Behaviour of Primates* (Ed. by R.P. Michael & J.H. Crook), pp. 411–505. Academic Press, London.

Smuts, B.B. (1985). Sex and Friendship in Baboons. Aldine, New York.

Stacey, P.B. (1982). Female promiscuity and male reproductive success in social birds and mammals. *American Naturalist,* **120,** 51–64.

Stewart, K.J. & Harcourt, A.H. (1987). Gorillas: variation in female relationships. *Primate Societies* (Ed. by B. Smuts, D.L. Cheney, R.M. Seyfarth, R.W. Wrangham & T.T. Struhsaker), pp. 155–64. University of Chicago Press, Chicago, Illinois.

Strayer, F.F. & Noel, J.M. (1986). The prosocial and antisocial functions of preschool aggression: an ethological study of triadic conflict among young children. *Altruism and Aggression.* (Ed. by C. Zahn-Waxler, E.M. Cummings & R. Iannotti), pp. 107–31. Cambridge University Press, Cambridge.

Trivers, R.L. (1971). The evolution of reciprocal altruism. *Quarterly Review of Biology,* **46,** 35–57.

Tyack, P. (1986). Population biology, social behavior and communication in whales and dolphins. *Trends in Ecology and Evolution,* **1,** 144–50.

Vehrencamp, S.L. (1983). A model for the evolution of despotic versus egalitarian societies. *Animal Behaviour,* **31,** 667–82.

Waal, F.B.M. de (1978). Exploitative and familiarity-dependent support strategies in a colony of semi-free living chimpanzees. *Behaviour,* **66,** 268–310.

Waal, F.B.M. de & Luttrell, L.M. (1986). The similarity principle underlying social bonding among female rhesus monkeys. *Folia Primatologica,* **46,** 215–34.

Walters, J. (1980). Interventions and the development of dominance relationships in female baboons. *Folia Primatologica,* **34,** 61–89.

West Eberhard, M.J. (1975). The evolution of social behavior by kin selection. *Quarterly Review of Biology,* **50,** 1–33.

Wrangham, R.W. (1980). An ecological model of female-bonded primate groups. *Behaviour,* **75,** 262–300.

Wrangham, R.W. (1982). Mutualism, kinship and social evolution. *Current Problems in Sociobiology* (Ed. by King's College Sociobiology Group), pp. 269–89. Cambridge University Press, Cambridge.

Wrangham, R.W. (1983). Social relationships in comparative perspective. *Primate Social Relationships* (Ed. by R.A. Hinde), pp. 325–34. Blackwell Scientific Publications, Oxford.

Dominance 'style' and primate social organization

F.B.M. DE WAAL

Wisconsin Regional Primate Research Center,
University of Wisconsin, 1223 Capitol Court,
Madison, Wisconsin 53715–1299, USA

SUMMARY

1 Social primates must use non-dispersive methods of resolving conflicts inherent in group living. In this study the emphasis is on the costs and regulation rather than the outcome of social conflict.

2 Long-lived animals such as primates are not likely to sacrifice friendships readily by hostile behaviour; recognition of such partnerships complicates models of animal competition. The present paper provides examples of constraints on competition which lend support to the development of more complex models.

3 In experiments with rhesus monkeys the social hierarchy remained intact but the order of access to resources was not explained by it. Thus in addition to dominance other facets such as tolerance and motivation determine the outcome of competition. Consequently, the establishment of tolerant relations with dominants is an alternative route for a subordinate to gain access to resources; one striking example is discussed.

4 The typical method of conflict resolution varies with the species, with gender within species, and sometimes with the level in the hierarchy. For example, a clique of unrelated 'upper class' females in a group of rhesus monkeys showed a high tendency of reconciliation and social tolerance among themselves, and a low tendency toward other females. Such differences can be summarized as contrasting dominance 'styles', which may range from despotic to lenient.

5 In male chimpanzees the existence of a well-recognized hierarchy reduces damaging fights, and provides a format for conflict resolution which allows the males to preserve bonds in spite of fierce competition. Female chimpanzees, which live more dispersed in the natural habitat, do not show a clear dominance hierarchy nor the same potential for coping with serious competition.

6 Reciprocal beneficial interactions were observed in three captive primate species — suggesting that 'one good turn deserves another'. However, while reciprocity in harmful behaviour was not observed in two species of macaque, such retaliation did occur in chimpanzees. This could be related to the relatively loose hierarchy of chimpanzees. This leads to the paradox that mechanisms of reconciliation, which underlie the flexibility of the chimpanzee hierarchy, also allow a revenge system.

INTRODUCTION

Many animal species have a strong need for social companionship, as reflected in their bored and depressed appearance if held in isolation. At first sight, this gregarious tendency may seem sufficient to explain the formation of cohesive groups. Yet, such an explanation immediately raises two questions. Firstly, what is the evolutionary origin of the attraction to conspecifics? Why is it strong enough in some animals to result in group formation, while it is virtually absent in others, or only expressed under particular conditions? The answer to this question requires an evaluation of the environmental pressures under which the species evolved, a topic addressed by a number of authors in this volume. Secondly, and this is the problem I wish to discuss, we need to know how animals deal with the competition inherent in group life. Social attraction is not enough to hold a community together. Since food and other resources are limited, intra-group competition is inevitable. If dispersal is the simplest way of resolving conflict, non-dispersive forms of conflict settlement are the key to cohesive group life.

Obvious as this may sound, it is not the perspective from which competitive relationships have been studied in the past. From Kawai's (1958) classical 'sweet potato tests' to Hausfater's (1975) analysis of sexual competition, to mention just two studies, the focus in primatology, and other fields, has been on the outcome rather than the costs of social conflict. As a result, there exists a massive literature concerning the pay-offs of social dominance, particularly, of course, in terms of reproductive success (reviewed for primates by Fedigan 1983 and Shively 1985). The narrow focus on this theoretical issue has led to dangerous simplifications. Instead of clear-cut winners and losers, we actually observe endless ambivalencies, compromises, and a great deal of tolerance among social primates. 'Limited wars' are usually explained on the basis of the risk of incurred damage (see, for example, Maynard Smith & Price 1973), yet this is only the direct physical cost of aggression; we also need to consider the burden aggression places on long-term social relationships. Competitive advantages are often sacrificed for the sake of peace.

I would not go as far as Crowe (1984, p. 75) in speaking of a 'capitalistic bias which has held Western biology together', but do believe that there has been a fundamental one-sidedness in concept formation. At the theoretical level animal life is still all too often treated as the gladiators' show with which Huxley (1888) compared it. As a result, we lack a vocabulary to describe and analyse the psychological subtleties regulating competition, and know remarkably little about the principles of peaceful coexistence.

Primates are constantly faced with the dilemma, familiar among people as well, that one sometimes cannot win a fight without losing a friend. In other words, when two individuals compete over a particular resource they have to take into account not only the risk of injury, and the value of the resource itself, but also the value of their relationship. Since social bonds are best regarded as long-term investments (Kummer 1979), we can expect primates to avoid jeopardizing these investments by hostile behaviour. Most existing models of animal competition ignore this important dilemma, making them inapplicable to primates, and indeed to intelligent, long-lived mammals in general. The assumption of partner value represents a major complication for game-theoretical approaches to animal conflict, well beyond the complications that were introduced by the assumption of individual recognition (van Rhijn & Vodegel 1980). In view of our incomplete knowledge, it would seem premature to specify alternative models, although stimulating first attempts have been made by Vehrencamp (1983) and Hand (1986). The present paper supports these attempts by providing examples of constraints on competition as a consequence of primate social organization.

Table 1 summarizes the terminological differentiation minimally needed, in my opinion, to come to grips with the complex proximate mechanisms that enable animals to maintain social relationships in a competitive world. The background and empirical justification of several of the concepts, such as reconciliation and formalized dominance, have been treated in detail elsewhere (see Table 1 for references).

THE INTERPLAY OF DOMINANCE AND TOLERANCE

The disastrous consequences of out-and-out competition were convincingly demonstrated more than half a century ago by Zuckerman's (1932) observations of the London Zoo colony of hamadryas baboons (*Papio hamadryas*). One hundred and thirty baboons were released into a rockwork enclosure of 30 by 20 m. The animals were presumably strangers to each other, and, to make matters worse, the sex ratio was approximately three males to one female. In the 6.5 years following the foundation of the colony, over 70% of the population died of stress and injuries. The massacre resulted from fights between males over females, and the dragging around of females by male 'overlords', sometimes preventing the females from feeding for days on end. Proportionally female mortality was higher than male mortality; the decimated colony included only 14% of the original females compared to 38% of the males, resulting in a sex ratio of seven males to one female. After a fight in which one of the last surviving females lost her life — an event described as particularly repellent 'from the anthropocentric point of view' — zoo officials finally interfered by removing the remaining females (Zuckerman 1932, p. 219).

Science had to wait for the studies by Kummer and co-workers on the same species to understand what had gone wrong. Both in the wild and in well-established captive colonies, male hamadryas baboons recognize each other's bonds with females. If a female is released in a cage with two males, they will fight over her. But if the female is first put together with only one male, while the other can watch them from an adjacent pen, the outcome is different. The female needs to be seen with the one male only briefly for the other to respect the pair-bond upon introduction into their cage. Even big, totally dominant males are inhibited from fighting (Kummer, Götz & Angst 1974). Sigg & Falett (1985) demonstrated similar inhibitions in hamadryas baboons during competition over food.

TABLE I. Basic terminology suggested to describe and analyse the proximate mechanisms of competitive relationships among primates, divided into three aspects: social dominance, competitiveness and mechanisms of tension regulation

Dominance
  Formal dominance   Dominance expressed in ritualized communication signals and greeting rituals of which the direction does not vary with social context (de Waal 1986a)

  Agonistic dominance   Dominance expressed in the outcome of agonistic encounters

  Competitive ability   Ability to claim a resource by means of force or the threat of force

Competitiveness
  Competitive tendency   Readiness to use one's competitive abilities

  Respect for possession   Inhibition to contest a resource already possessed by another (Kummer, Götz & Angst 1974; Sigg & Falett 1985)

  Social tolerance   Low competitive tendency, especially by dominants towards subordinates

Tension regulation
  Reconciliation   Friendly reunion of former adversaries not long after their conflict (de Waal, in press)

  Reassurance   Acts serving to reduce the anxiety of another individual

  Appeasement   Acts serving to reduce or forestall aggression by another individual

Dominance style
  Dominance style refers to the nature of the entire relationship between dominant and subordinate in terms of the above concepts, that is, the way two individuals deal with each other in the face of competition and potential aggression. Styles may range from 'despotic' and 'strict' to 'mellow' and 'indulgent' for the dominant, and from 'terrified' and 'obedient' to 'trusting' and 'relaxed' for the subordinate

The development of such 'respectful' relationships was prevented in the London Zoo colony, due to crowding, lack of familiarity, and an abnormal sex ratio. The resulting society had no winners, only losers, because even survivors of the intense competition over sexual partners lost everything they had fought for. This sort of society simply does not serve *any* of its members. We must conclude, therefore, that the evolution of checks and balances on escalated fighting is in the interest even of individuals with superior fighting abilities.

Selfish competition and group life are antithetical; the two do coexist, but in an uneasy manner. To understand this precarious equilibrium we need to change our focus from the traditional 'who gets what?' questions to the consequences that conflicts have for long-term relationships, and the way animals cope with these consequences. This change in perspective can be illustrated by comparing two simple test paradigms to study competitive relationships. The traditional method of investigation has been to present severely deprived animals with a monopolizable resource (e.g. one drinking nipple) and record the priority order among them. This procedure forces a group of animals into a hierarchical pattern by reducing variations in motivation and opportunities for friendly interaction. In this paradigm social tolerance is regarded as a confounding factor that needs to be eliminated. The procedure has been applied to both non-primates (see Syme 1974 for a review) and primates (see, for example, Varley & Symmes 1966; Boelkins 1967; Richards 1974; Weisbard & Goy 1976; O'Keeffe, Lifshitz & Linn 1983).

Whereas this test is intended to measure the competitive abilities of individuals, we designed a different test to measure their competitive tendencies. A large captive group of rhesus monkeys (*Macaca mulatta*) was deprived of water for only 3 hours, after which they were presented with a water-filled basin large enough for simultaneous drinking by up to four adults or eight juveniles. Most monkeys visited the basin many times per test, drinking together with some individuals and excluding others in ever-changing combinations (Fig. 1). This 'chaotic' pattern of resource competition may be more characteristic of rhesus monkeys than the orderly pattern of coming and going that can be produced under more severe test conditions. It is important to note, though, that the clear-cut formal hierarchy among the monkeys remained the same. That is, submissive baring of the teeth occurred without a single exception in the same dyadic directions as observed in other situations. This conspicuous facial expression, also known as the 'fear grin' (Fig. 2), is 100% unidirectional among rhesus monkeys,

FIG. 1. Tolerant relationships are easily observable in a mildly competitive situation with a shareable resource. Three rhesus monkeys drinking together from one basin, while a juvenile shows impatience by pulling a tail. (Photo courtesy of F.B.M. de Waal.)

Fig. 2. A rhesus monkey bares her teeth in response to a dominant who tries to push her out of a huddle group. This submissive signal is completely unidirectional within each dyad, making it the most reliable expressions of status. The term 'formal dominance' in this paper refers to such ritualized expressions of status. (Photo courtesy of F.B.M. de Waal.)

which implies that its direction is independent of variations in social context (de Waal & Luttrell 1985). The outcomes of agonistic and approach–retreat interactions, too, were unchanged during the drinking tests; 98% of the over one thousand displacements among adult monkeys were as expected from the formal hierarchy.

The drinking order itself, in contrast, showed a great deal of flexibility. During their encounters around the basin the monkeys had globally two options: to compete or to share. Almost half of the interactions were of a non-competitive nature; the monkeys drank simultaneously, or a dominant drank while a subordinate waited within arm's reach. Yet, such tolerance was not uniformly distributed over the hierarchy. The hierarchy could be subdivided into two classes; the 'upper class' as a whole excluded the 'lower class', but within each class there was a fair amount of social tolerance, both among matrilineally related and unrelated individuals. Much of the flexibility in drinking priority could be explained on the basis of this unequal distribution of tolerance. Formal dominance predicted over 90% of dyadic drinking priorities between members of different classes, but less than 60% between members of the same class, close to the 50% expected by chance (de Waal 1986c).

This experiment demonstrates that, even though the dominance hierarchy remains fully intact, the order of access to a resource is not necessarily explained by it. High-ranking monkeys, while having no trouble claiming the resource, make only limited and selective use of this capacity. The conclusion is that, in addition to dominance, social tolerance and variations in motivation determine the outcome of competition. It should be stressed that our test conditions can hardly be dismissed as unusual, as they probably come closer to the conditions under which dominance relationships must have evolved than the conditions of traditional dominance tests.

The influence of social tolerance opens an interesting alternative route for individuals to gain access to resources. Good relationships with particular dominants may yield pay-offs which subordinates cannot obtain by means of aggressive strategies. Thus, Weisbard & Goy (1976) demonstrated that female stumptail macaques (*Macaca arctoides*) may rise in the drinking order of their group following parturition. Since this rise takes place in the absence of any corresponding changes in the directionality of agonistic behaviour, it is probably due to a change in social tolerance towards females carrying a newborn infant.

In our rhesus group we encountered a striking example of a tolerance-based advantage during competition over water. A female, named Curry, belonging to one of the lowest ranking matrilines, had formed a close bond with a female one year older, named Bizzy, who belonged to one of the highest ranking matrilines. This bond, formed when both females were very young, is still recognizable now that they are adult. That is, Bizzy and Curry spend a lot of time together, groom each other, and are attracted to each other's offspring. Such bonds between age-mates are typical; similarity in age is one of the main factors underlying association patterns among adult rhesus females (de Waal & Luttrell 1986). Not typical, however, is the effect this relationship has had on Curry's behaviour during drinking tests. We have data on her third through her sixth year of life (during which period we videotaped forty-six drinking tests), which I will compare with data on other females in the same age range. To understand this analysis, one should know that female macaques 'inherit' the rank position of their mother. Daughters of dominant females normally achieve high status; daughters of subordinate females normally end up low in the hierarchy. Walters & Seyfarth (1987) review the literature on matrilineal hierarchies, and discuss their social (as opposed to genetic) basis.

On the mean, Curry has been 14·8 rank positions ahead of her mother in the drinking order (i.e. the order in which individuals obtain their first drink from the basin), and the smallest yearly average rank difference between her and her

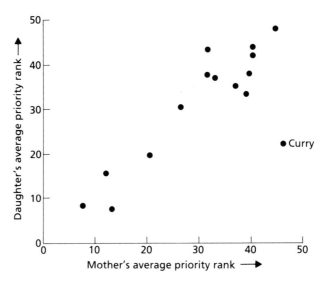

FIG. 3. Mothers and daughters tend to occupy similar ranks in the drinking order among rhesus monkeys. This graph illustrates the correlation between the mean priority ranks of 4-year-old daughters and their mothers ($R_s = 0.68$, $P < 0.01$). One daughter, Curry, is exceptional in that she ranks far ahead of her low-ranking mother.

mother has been eleven positions. For the other twenty-one young females the mean difference with their mother's rank has been only -1.6 (the negative sign indicating that on average they drank after their mother), with individual means ranging between -13.3 and 6.8. Figure 3 illustrates Curry's exceptional position during her fourth year of life. She is way out of line with the matrilineal rank system in which daughters and mothers occupy similar ranks.

This effect has come about without any change in Curry's formal or agonistic rank position. Like all rhesus females, she directs aggression to and receives submission from females ranking below her mother, but she does not dominate in this respect any of the females ranking above her mother. The explanation for Curry's high rank in the drinking order appears to be that she developed the habit of drinking with her friend Bizzy at a young age, after which other females of Bizzy's matriline became increasingly tolerant of her. Presently, even some unrelated upper class females seem to have grown used to this 'interloper', as reflected in their failure to chase her away when they meet her near the water basin.

Unfortunately, due to the functional perspective that predominates in ethology these days, social dominance has become almost synonymous with priority of access to resources (see, for example, Popp & DeVore 1979). In primates, this connection is so problematic that Fedigan (1982, p. 97) concluded, 'Much as IQ is often defined as 'that which IQ tests measure', food test dominance could be defined as that which food tests measure'. Bernstein (1981, p. 421) warned: 'Functional definitions often run into problems when many different activities produce the same consequences'. Although rhesus monkeys are not particularly known for their reluctance to use force to obtain resources (Fig. 4), Curry's example dramatizes the importance of a distinction, even in this species, between process and result, i.e. between the behaviour regulating the outcome of competition and the outcome itself. Rather than inferring from her success in the drinking tests that Curry has a high dominance status, it seems more appropriate to assign her a high position in terms of received tolerance as there is no evidence that she can forcibly claim priority over females ranking above her mother.

RECONCILIATION BEHAVIOUR

One obvious reason for an individual to refrain from competition is fear of retaliation; the risk of an injurious fight has to be weighed against the value of the resource. Occasionally, such considerations lead to rather amusing situations

FIG. 4. Whereas dominants may show tolerance towards certain individuals, towards others their competitive style can be totally different. A juvenile rhesus monkey passively accepts an inspection of her cheekpouches by a dominant female, who will remove any items of interest to her. (Photo courtesy of F.B.M. de Waal.)

among animals: two foraging adult male baboons, both armed with dangerous canines, may walk straight past a small food item that each male would no doubt have picked up had he been alone (Sigg & Falett 1985). Yet, even individuals proven capable of fully dominating others may show restraint. They probably do so because frequent competition, whatever the short-term gains, is costly in the long run. For the student of social organization it is crucial to determine in which relationships competition is avoided or mitigated, because it is these relationships that hold the social fabric together.

Reconciliation after aggression is an alternate way of reducing the costs of competition. De Waal (in press) reviews evidence from both observational and experimental studies that primates remember with whom they fought, and that they seek to restore friendly contact with these individuals afterwards. The attraction between former adversaries is selective, i.e. the probability of post-conflict contact between them increases both absolutely and relative to contact with other individuals. It results in special forms of interaction, rarely seen outside this context. Chimpanzees, for example, make up by means of embracing and mouth-to-mouth kissing (de Waal & van Roosmalen 1979). We may regard reconciliation as a backup mechanism to social tolerance. That is, if for some reason a tolerant attitude cannot be maintained, and aggression does occur, its damaging effects can be 'undone' by reassurance and appeasement behaviour. Since, in this view, social tolerance and reconciliation both reflect the value individuals attach to their relationships, a correlation between the two measures is to be expected.

To illustrate this, let me return to the class division in our rhesus group. In addition to the relatively high drinking tolerance found within each class, we discovered that unrelated upper class females (i) associate two to three times more with each other than with unrelated females of the lower class (de Waal & Luttrell 1986), and (ii) show a remarkably high tendency to reconcile after fights among themselves, almost as high as the conciliatory tendency among kin (de Waal 1986b). In other words, patterns of affiliation, social tolerance and conflict resolution seem to from a single complex, which, in this case, can be characterized by saying that the clique of top-ranking females varies its dominance style according to the class membership of subordinates. These females treat subordinates of the lower class in a much less lenient and less conciliatory manner than subordinates of their own class.

It is easy to see how such a difference in dominance style would lead to a centralized social organization, first described for free-ranging Japanese macaques (*Macaca fuscata*), in which the hierarchical top acts as a cohesive nucleus, while lower-ranking monkeys occupy more peripheral positions (see, for example, Imanishi 1960; Yamada 1966). Inasmuch as this organization is based on differentiation across the hierarchy in the tendency to peacefully settle disputes, it is also not surprising that it manifests itself most clearly at moments of competition, such as during the food provisioning done by Japanese fieldworkers. Questions as to the generality of these phenomena remain, however. For example, is the central/peripheral or upper/lower class division also recognizable during competition over a dispersed food source? Is the social stratification of our rhesus group maintained over time? Is it obvious in other rhesus groups tested under similar conditions? We are presently investigating these questions to arrive at a generalizable model linking mechanisms of tension regulation to macaque social organization.

Another angle from which to study this link is by interspecies comparisons. Thierry's (1984, 1985) work on three macaque species suggests a relation between the frequency of reassurance behaviour, the intensity of aggression, and the symmetry of contests. One of the species, the tonkeana macaque (*Macaca tonkeana*), seems particularly pacific and relaxed in its relationships. Similarly, current research by our team on stumptail macaques shows that these monkeys, in comparison to rhesus macaques, (i) reconcile a greater proportion of their conflicts (de Waal & Ren, 1988), (ii) have a richer repertoire of reassurance gestures, (iii) groom more frequently, and (iv) show greater tolerance during drinking tests. Thus, in spite of the general resemblance between the social organizations of macaque species, certain members of the genus have a considerably 'softer' dominance style than others, and the norm value for the homeostasis between dispersive and cohesive processes may be set differently in each species.

Mechanisms of tension regulation shape social organization. On the basis of the dominance style and reconciliation behaviour of captive stumptail monkeys we may speculate that this species lives in close-knit groups in its natural habitat. We do not actually know how these monkeys live, but it would not make sense for a loosely organized or dispersedly living species to possess the observed powerful mechanisms of physical reassurance and conflict resolution.

CHIMPANZEE SOCIAL ORGANIZATION

Elsewhere (de Waal 1986a), I have argued that primates show status awareness, that is, they evaluate their dominance relationships with others and possess special signals to communicate about them. Submissive displays serve to reassure the dominant that his or her position is not in danger, which, in turn, allows the dominant to treat the subordinate in a non-aggressive manner. Such mechanisms of mutual reassurance between dominants and subordinates may almost 'cancel' some of the competitive assymmetries as reflected in, for example, a flexible drinking order among macaques whereas their formal hierarchy remains rigidly the same (see previous sections). In this view, dominance is *not* an abstraction by the human observer (cf. Altmann 1981), in other words dominance relationships are evaluated and recognized by the animals themselves. Primates clearly seek to establish dominance over others, and formalize the resulting asymmetrical relationship by means of ritualized communication, e.g. teeth-baring among rhesus monkeys (see Fig. 2); bowing and pant-grunting among chimpanzees (Fig. 5).

It is doubtful whether competitors can be unified in the absence of such a 'vertical' organization. Thus, group cohesiveness can probably be achieved only

FIG. 5. Status differences among male chimpanzees are communicated in ritual encounters. A dominant male (left) makes himself look as large as possible, walking bipedally and raising his hair, whereas the subordinate makes himself smaller and utters a series of panting grunts. The two males in this photograph are in reality approximately the same size. (Photo courtesy of F.B.M. de Waal.)

when competition is either reduced because of abundant resources or coped with in a formalized hierarchy. Both possibilities can be observed in captive chimpanzees (*Pan troglodytes*), females showing the first, males the second basis of intra-sexual association. Table 2 compares behavioural data on a chimpanzee colony living on a large island, at the Arnhem Zoo (Netherlands), with data on the wild chimpanzees of Gombe National Park and the Mahale Mountains, both in Tanzania. The females in the Arnhem colony bond among themselves and occasionally combine forces in aggressive coalitions. They do so to such an extent that they form a political power to be reckoned with by the males (de Waal 1982). In this respect, these females act very differently from their wild counterparts, which live largely solitary lives, dispersed over the forest. Yet, two other gender differences are unaffected by the living conditions. Under both conditions reconciliations are significantly less common among females than among males, and the female hierarchy is inconsistent and vague. For example, after 6 years of study of the Arnhem colony — with an estimated 6000 hours of observation — there remain a dozen female–female pairs in which status rituals have *never* been observed.

If formalized dominance relationships facilitate peaceful conflict settlement, as argued above, the dispersal of female chimpanzees in the wild may be explained

TABLE 2. Social relationships of adult males and females in the large Arnhem Zoo colony of chimpanzees compared with wild populations studied in Gombe National Park and the Mahale Mountains, Tanzania

	Arnhem Zoo colony	Natural habitat
Hierarchy	In both environments dominance is highly formalized among adult males, but infrequently expressed among females (Bygott 1974; Goodall 1986; Nishida, in press; de Waal 1986a)	
Coalitions	In both environments males follow flexible, opportunistic coalition strategies aimed at high status (Riss & Goodall 1977; de Waal 1982, 1984; Nishida 1983)	
	Females unite forces in defence against aggression and to influence status struggles among males (de Waal 1982, 1984)	Female coalitions are virtually unknown
Reconciliations	In both environments male–male aggression is more often followed by reassurance behaviour than is female–female aggression (de Waal 1986a; Goodall 1986)	
Social bonding	Strong association among males in both environments	
	Female bonding is as strong as among males; less association between the sexes (de Waal 1986a)	Females lead largely solitary lives with their dependent offspring (Nishida 1979; Goodall 1986)

by the virtual absence of a hierarchy among them. Food competition in the forest, with its irregular supply, drives females apart, so to speak, because they are not equipped to cope with it. One might challenge this explanation by arguing that the potential to establish a hierarchy is present in female chimpanzees but has as yet not been observed because in captive settings there is food abundance, whereas in the natural habitat competition can easily be avoided. This argument is not convincing for two reasons. Firstly, scarcity of resources is by no means a prerequisite for the establishment of a hierarchy. The primatological literature contains countless examples of captive monkey groups with clear-cut (female) hierarchies in spite of food abundance. Secondly, even if competition is induced on a regular basis female chimpanzees fail to establish formalized dominance relationships. At both the Gombe and Mahale field sites investigators provisioned chimpanzees with attractive food, but no clear-cut hierarchy emerged (see, for example, Bygott 1974).

The conclusion is that female chimpanzees lost or did not evolve mechanisms of social dominance comparable to those of their male conspecifics. Asymmetries in the outcome of aggressive encounters may be observed among females; what is lacking are the patterns of conflict resolution and reconciliation associated with the ritualization of dominance relationships. In other words, female chimpanzees have an agonistic hierarchy, but not a formalized one. One may speculate that formalization is less pronounced because a solitary life style suits their foraging needs while they benefit little from each other's presence (Wrangham 1986). Only when the costs of competition are reduced, as in a zoo environment with its food abundance, do female chimpanzees show social bonding. Male chimpanzees, in contrast, form bonds regardless of the degree of competition. Both in captivity and in the wild they overcome strong sexual and status rivalries. Males in the Arnhem colony show the same degree of association and grooming among themselves as females in spite of a rate of aggression that is twenty times higher than that among females (de Waal 1986a).

We also found that the probability of damaging fights among males increased by a factor of almost five if a formal dominance relationship (as expressed in greeting rituals with bowing movements and pant-grunting; see Fig. 5) was lacking, and that re-establishment of formal dominance resulted in a sharp increase in grooming and reconciliation behaviour among males (de Waal 1982; 1986a). Thus, the male hierarchy regulates conflict in such a manner that bonds are preserved, the net effect being that their hierarchy *unifies* the competitors. Nishida (1979, p. 93) has expressed a similar view when stating: 'It is likely that complex sequences of threat–submission–reassurance may strengthen the male bond among chimpanzees'.

It has been suggested that chimpanzee males have evolved these unifying social mechanisms as a response to intercommunity aggression. According to Wrangham (1986), writing about great apes, 'vulnerability to conspecifics is the principal source of social bonds, and defensive groups are therefore formed when foraging constraints permit'. Male chimpanzees of the same community form a network of coalitions. Against external enemies, they operate with great solidarity (see Goodall 1986 for a review), but amongst themselves they are divided into changing partnerships (Riss & Goodall 1977; de Waal 1982, 1984; Nishida 1983). This adds important reasons for effective conflict management. Coalition partners have to 'agree' on the division of the pay-offs of their coalition, and even rivals cannot afford to hold grudges as they may in the future need one another against a third party.

In summary, female chimpanzees show a potential for social bonding, but not for coping with serious competition, whereas male chimpanzees do possess the necessary coping mechanisms, which they need for both internal and external 'political' reasons. Let us consider the latter reasons in some detail, taking as an example Nikkie, an adult male in the Arnhem colony. Nikkie achieved alpha status with the support of an older male, Yeroen (Fig. 6); together they successfully challenged a strong third male. Nikkie dominated his coalition partner both formally and agonistically, but avoided conflicts with him. This restraint allowed Yeroen to have more matings with oestrus females than normal for a second-ranking male. Furthermore, if a conflict between the two coalition partners

FIG. 6. Nikkie demonstrating his bond with Yeroen by mounting him from behind during a tense confrontation with their common rival (both males are screaming). Nikkie completely depended on the older male to maintain his alpha position. When Yeroen broke the coalition, after 3 years of support, Nikkie immediately lost his rank. (Photo courtesy of F.B.M. de Waal.)

did break out, it was usually the dominant, Nikkie, who quickly tried to repair the bond. He did so in response to the intimidation displays and charges of their common rival. In other words, Nikkie's dependence on support greatly reduced his competitiveness towards his partner. The need for this restraint became evident years later when Nikkie's tolerance towards Yeroen reached a low point. In this period their coalition collapsed, followed by a total destabilization of the male hierarchy with, eventually, fatal consequences (de Waal 1986d).

The more sophisticated a social system, the greater the number of constraints on competition. Yeroen's competitive success, in terms of mating behaviour, during the period of Nikkie's reign was not based on his competitive ability in the strict sense, as defined in Table 1. It was based on social manipulation. Yeroen simply 'dropped' Nikkie when the pay-offs of their relationship were no longer to his advantage, indicating that he had exploited the young male's need for a partnership. At this level of complexity the terminology in Table 1 becomes insufficient, and new concepts have to be introduced, such as 'power' (the ability to influence group processes, which ability does not necessarily correspond with formal rank) and 'transactions' (the linkage between continuation of a cooperative relationship and the benefits derived from it).

POSITIVE AND NEGATIVE RECIPROCITY

The discrepancies between dominance and resource acquisition can be so striking in chimpanzees that one almost gets the impression that formalized status has become a façade, useful to avoid tensions and to canalize aggression, but not necessarily the first and foremost determinant of resource allocation. Subordinates frequently obtain food by begging from dominants (see, for example, Goodall 1963; Nishida 1970), and, at least in the Arnhem colony, particular subordinates may even assertively take over preferred sitting places or social partners from individuals who clearly dominate them, both formally and agonistically (Table 3). Except in the ritual greeting ceremonies that signify formal status, the great plasticity of the chimpanzee hierarchy, noted already by Maslow (1940), is visible in all social domains. For example, over 95% of the aggression in captive rhesus macaques is as predicted from the formal hierarchy (i.e. directed by dominants to subordinates) compared to less than 80% in the Arnhem chimpanzees (Noë, de Waal & van Hooff 1980; de Waal & Luttrell 1985).

   Such plasticity has tremendous consequences, which are of interest in connection with the evolution of human social systems. It has been argued that a strict hierarchy forms an obstacle to reciprocal altruism, that is, that it hampers an equal exchange of benefits between dominants and subordinates (Trivers 1971). With respect to the exchange of harmful behaviour the influence of dominance style may be even more important, as reflected in our finding that chimpanzees show *revenge*, whereas macaques do not.

   We determined the degree of reciprocity in the distribution of agonistic interventions among adults of both sexes in captive groups of three primate species: rhesus and stumptail macaques, and chimpanzees (de Waal & Luttrell, 1988). An agonistic intervention was defined as one individual interfering with an ongoing conflict between two others, by supporting one of the two. Such interventions benefit one party while harming another, so every *pro* choice is also a *contra* choice. Since symmetrical relationship characteristics, such as kinship and time spent in association, may account for reciprocity of behaviour, the data were statistically controlled for these characteristics. Table 4 provides the partial correlations remaining after this data adjustment. The correlations are between given and received intervention behaviour in dyads among adult individuals. Probabilities have been tested by means of a random matrix permutation procedure (Dow, Cheverud & Friedlaender 1987).

TABLE 3. Characteristics of competitive interactions between the three top ranking males and the three top ranking females of the Arnhem chimpanzee colony. Formal dominance is measured by submissive greeting rituals involving pant-grunting; agonistic dominance reflects the outcome of agonistic encounters, and spatial/social priority is measured by the outcome of non-agonistic competition over places to sit (e.g. shelter) and social partners (excluding the females' own offspring). (Data from Noë, de Waal & van Hooff 1980)

Relationship aspect	Number of interactions	% dominance by Male	Female
Formal dominance	142	100	0
Agonistic dominance	92	80	20
Spatial/social priority	47	19	81

TABLE 4. Dyadic reciprocity in three primate species in captivity. The table provides partial Pearson correlations of agonistic interventions by individual A to individual B, with such interventions by B to A. Data have been statistically adjusted for symmetrical relationship characteristics, i.e. matrilineal kinship, proximity relations, and same-sex combination. (From de Waal & Luttrell 1988)

Intervention type	Rhesus	Stumptail	Chimpanzee
Pro choices	0·281**	0·182*	0·545**
Contra choices	−0·186**	−0·285**	0·315*

* $P<0.05$
** $P<0·01$

The data demonstrate significant reciprocity of beneficial interventions (pro choices) in all three species. Since effects of symmetrical traits are excluded we may assume that this reciprocity is based on cognitive processes, i.e. that these primates are able to keep mental records of the support received from others and to adjust their own supportive behaviour according to the rule of 'one good turn deserves another'. The distribution of harmful interventions (contra choices), on the other hand, is anti-reciprocal among macaques, i.e. if individual A often intervenes against B, B rarely intervenes against A. Only in chimpanzees do contra-choices show a reciprocal distribution, i.e. if individual A often intervenes against B, B does the same to A. Chimpanzees appear to follow the rule: 'an eye for an eye, a tooth for a tooth'.

Further analysis indicates that the anti-reciprocity of contra-interventions in macaques is due to an intimidating effect of high rank — the macaques are reluctant to intervene against dominants. Among the chimpanzees interventions against the formal hierarchy are quite common; middle ranking individuals actually direct more interventions against individuals positioned above them than against subordinates. Complete social reciprocity, which takes both beneficial and harmful acts into account, apparently requires a relatively loose hierarchy, that is, a hierarchy in which subordinates are not totally inhibited from protesting against or attacking dominants.

The possibility of revenge introduces powerful sanctions to a social system; a way for subordinates to influence rules of conduct within their community. If we follow Boehm's (1979, p. 21) definition of a moral system as 'a basic system of conditioning which limits and shapes human hedonic and retaliatory impulses so that a reasonably predictable and cooperative life may result', chimpanzees appear to come very close to such an elaborate reward-and-punishment system.

Again, it is possible that the Arnhem chimpanzees demonstrate a potential unknown from this species in the natural habitat. Since wild females, for example, rarely interfere with conflicts initiated by adult males, they may not 'square accounts' with males to the extent observed in our captive colony. On the other hand, we can hardly assume a species to possess remarkable behavioural potentials that are never used in the environment in which they evolved. Our knowledge of chimpanzee communication and social life is still very fragmentary, and we can only hope that these apes survive in the wild long enough for us to fully understand the natural functions of their mental capacities and social sophistication.

CONCLUSION

The competitive tendency of primates is strongly modified by their dependence on community life in general, and their need for support and friendship of certain individuals in particular. Among animals with such long memories, peaceful coexistence requires compromises, restraints, and reconciliations after fights. A change in research focus from the outcome to the social costs of competition is therefore proposed. The principal mechanism to cope with competition is formalized dominance as expressed in ritualized status communication combined with mechanisms of reassurance and appeasement. The formal hierarchy serves a cohesive function, and does not necessarily predict competitive outcomes and the direction of aggression. This discrepancy between different aspects of dominance is of particular importance in chimpanzees. It affects social organization at every level, including reciprocity relationships.

The unifying function of hierarchies requires further study, one perspective being a differentiation among dominance 'styles'. The ways in which dominants settle conflicts with subordinates ranges from despotic to tolerant, and these differences seem to correlate with differences in group cohesiveness. If cohesiveness is a response to environmental pressures, such as predation and inter-group competition, dominance style probably is a function of the environment as well. Thus, we may assume that female chimpanzees have little need to overcome competition in the natural environment, as reflected in the virtual absence of a formalized hierarchy among them in both captivity and the wild. Male chimpanzees and stumptail macaques, on the other hand, must heavily depend on each other's presence, as they exhibit very powerful behavioural mechanisms of conflict resolution.

ACKNOWLEDGEMENTS

I thank Mary Schatz and Jackie Kinney for typing the manuscript, and Linda Endlich for drawing the figure. The manuscript has benefited from the comments of one anonymous referee. Writing of this chapter was supported by grant No. RR00167 of the National Institutes of Health to the Wisconsin Regional Primate Research Center. This is publication no. 26–029 of the Center.

REFERENCES

Altmann, S. (1981). Dominance relationships: the Cheshire cat's grin? *Behavioral and Brain Sciences*, **4**, 430–1.

Bernstein, I. (1981). Dominance: the baby and the bathwater. *Behavioral and Brain Sciences*, **4**, 419–57.

Boehm, C. (1979). Some problems with altruism in the search for moral universals. *Behavioral Science*, **24**, 15–24.

Boelkins, R. (1967). Determination of dominance hierarchies in monkeys. *Psychonomic Science*, **7**, 317–18.

Bygott, D. (1974). *Agonistic behaviour and dominance in wild chimpanzees*. Ph.D. thesis, University of Cambridge.

Crowe, B. (1984). Ideological constraints on evolutionary theory. *Social Cohesion* (Ed. by P. Barchas & S. Mendoza), pp. 65–84. Greenwood Press, Westport.

Dow, M., Cheverud, J. & Friedlaender, J. (1987). Partial correlation of distance matrices in studies of population structure. *American Journal of Physical Anthropology*, **72**, 343–52.

Fedigan, L. (1982). *Primate Paradigms*. Eden Press, Montreal.

Fedigan, L. (1983). Dominance and reproductive success in primates. *Yearbook of Physical Anthropology*, **26**, 91–129.

Goodall, J. (1963). My life among wild chimpanzees. *National Geographic*, **124**, 272–308.

Goodall, J. (1986). *The Chimpanzees of Gombe: Patterns of Behavior*. Belknap (Harvard University Press), Cambridge, Massachusetts..

Hand, J. (1986). Resolution of social conflicts: dominance, egalitarianism, spheres of dominance, and game theory. *Quarterly Review of Biology*, **61**, 201–20.

Hausfater, G. (1975). *Dominance and Reproduction in Baboons* (*Papio cynocephalus*). Karger, Basel.

Huxley, T.H. (1888). Struggle for existence and its bearing upon man. *Nineteenth Century*, February 1888.

Imanishi, K. (1960). Social organization of subhuman primates in their natural habitat. *Current Anthropology*, **1**, 390–405.

Kawai, M. (1958). On the system of social ranks in a natural troop of Japanese monkeys. *Primates*, **1**, 111–48. English translation in *Japanese Monkeys* (Ed. by B.K. Imanishi & S. Altmann). Emory University, Atlanta.

Kummer, H. (1979). On the value of social relationships to nonhuman primates: a heuristic scheme. *Social Science Information*, **17**, 687–705.

Kummer, H., Götz, W. & Angst, W. (1974). Triadic differentiation: an inhibitory process protecting pair bonds in baboons. *Behaviour*, **49**, 62–87.

Maslow, A. (1940). Dominance-quality and social behaviour in infra-human primates. *Journal of Social Psychology*, **11**, 313–24.

Maynard Smith, J. & Price, G. (1973). The logic of animal conflict. *Nature*, **246**, 15–18.

Nishida, T. (1970). Social behaviour and relationships among wild chimpanzees of the Mahali Mountains. *Primates*, **11**, 47–87.

Nishida, T. (1979). The social structure of chimpanzees of the Mahale Mountains. *The Great Apes* (Ed. by D. Hamburg & E. McCown), pp. 73–121. Benjamin Cummings, Menlo Park.

Nishida, T. (1983). Alpha status and agonistic alliance in wild chimpanzees. *Primates*, **24**, 318–36.

Nishida, T. (in press). Social conflicts among adult female chimpanzees. *Understanding Chimpanzees* (Ed. by P. Heltne). The Chicago Academy of Sciences, Chicago, Illinois.

Noë, R., de Waal, F. & Hooff, J. van (1980). Types of dominance in a chimpanzee colony. *Folia primatologica*, **34**, 90–110.

O'Keeffe, R., Lifshitz, K. & Linn, G. (1983). Relationships among dominance, interanimal spatial proximity and affiliative behaviour in stumptail macaques (*Macaca arctoides*). *Applied Animal Ethology*, **9**, 331–9.

Popp, J. & DeVore, I. (1979). Aggressive competition and social dominance theory: synopsis. *The Great Apes* (Ed. by D. Hamburg & E. McCown), pp. 317–38. Benjamin Cummings, Menlo Park.

Rhijn, J. van & Vodegel, R. (1980). Being honest about one's intentions: an evolutionary stable strategy for animal conflicts. *Journal of Theoretical Biology*, **85**, 623–41.

Richards, S. (1974). The concept of dominance and methods of assessment. *Animal Behaviour*, **22**, 914–30.

Riss, D. & Goodall, J. (1977). The recent rise to the alpha-rank in a population of free-living chimpanzees. *Folia primatologica*, **27**, 134–51.

Shively, C. (1985). The evolution of dominance hierarchies in nonhuman primate society. *Power, Dominance and Nonverbal Behaviour* (Ed. by S. Ellyson & I. Dovidio), pp. 67–87. Springer-Verlag, Berlin.

Sigg, H. & Falett, J. (1985). Experiments on respect of possession in hamadryas baboons (*Papio hamadryas*). *Animal Behaviour*, **33**, 978–84.

Syme, G. (1974). Competitive orders as measures of social dominance. *Animal Behaviour*, **22**, 931–40.

Thierry, B. (1984). Clasping behaviour in *Macaca tonkeana*. *Behaviour*, **89**, 1–28.

Thierry, B. (1985). A comparative study of aggression and response to aggression in three species of macaque. *Primate Ontogeny, Cognition and Social Behaviour* (Ed. by I. Else & P. Lee). Cambridge University Press, Cambridge.

Trivers, R. (1971). The evolution of reciprocal altruism. *Quarterly Review of Biology*, **46**, 35–57.

Varley, M. & Symmes, D. (1966). The hierarchy of dominance in a group of macaques. *Behaviour*, **27**, 54–75.

Vehrencamp, S. (1983). A model for the evolution of despotic versus egalitarian societies. *Animal Behaviour*, **31**, 667–82.

Waal, F. de (1982). *Chimpanzee Politics*. Jonathan Cape, London.

Waal, F. de (1984). Sex-differences in the formation of coalitions among chimpanzees. *Ethology and Sociobiology*, **5**, 239–55.

Waal, F. de (1986a). Integration of dominance and social bonding in primates. *Quarterly Review of Biology*, **61**, 459–79.

Waal, F. de (1986b). Conflict resolution in monkeys and apes. *Primates — the Road to Self-Sustaining Populations* (Ed. by K. Benirschke), pp. 341–50. Springer-Verlag, New York.

Waal, F. de (1986c). Class structure in a rhesus monkey group; the interplay between dominance and tolerance. *Animal Behaviour*, **34**, 1033–40.

Waal, F. de (1986d). The brutal elimination of a rival among captive male chimpanzees. *Ethology and Sociobiology*, **7**, 237–51.

Waal, F. de (in press). Reconciliation among primates: a review of empirical evidence and theoretical issues. *Primate Social Conflict* (Ed. by W. Mason & S. Mendoza). Alan Liss, New York.

Waal, F. de & Luttrell, L. (1985). The formal hierarchy of rhesus monkeys: an investigation of the bared-teeth display. *American Journal of Primatology*, **9**, 73–85.

Waal, F. de & Luttrell, L (1986). The similarity principle underlying social bonding among female rhesus monkeys. *Folia primatologica*, **46**, 215–34.

Waal, F. de & Luttrell, L. (1988). Mechanisms of social reciprocity in three primate species: symmetrical relationship characteristics or cognition? *Ethology and Sociobiology* **9**, 101–18.

Waal, F. de & Ren, R. (1988). Comparison of the reconciliation behavior of stumptail and rhesus macaques. *Ethology* **78**, 129–42.

Waal, F. de & Roosmalen, A. van (1979). Reconciliation and consolation among chimpanzees. *Behavioral Ecology and Sociobiology*, **5**, 55–66.

Walters, J. & Seyfarth, R. (1987). Conflict and cooperation. *Primate Societies* (Ed. by B. Smuts, D. Cheney, R. Seyfarth, R. Wrangham & T. Struhsaker), pp. 306–17. University of Chicago Press, Chicago, Illinois.

Weisbard, C. & Goy, R. (1976). Effect of parturition and group composition on competitive drinking order in stumptail macaques (*Macaca arctoides*). *Folia primatologica*, **25**, 95–121.

Wrangham, R. (1986). Ecology and social relations in two species of chimpanzees. *Ecological Aspects of Social Evolution: Birds and Mammals* (Ed. by D. Rubenstein & R. Wrangham), pp. 352–78. Princeton University Press, Princeton, New Jersey.

Yamada, M. (1966). Five natural troops of Japanese monkeys in Shodoshima Island. *Primates*, **7**, 315–61.

Zuckerman, S. (1932). *The Social Life of Monkeys and Apes*. Kegan Paul, London.

Demographic influences on dominance structure among female primates

S.B. DATTA

MRC Unit on the Development and Integration of Behaviour,
Cambridge University,
Madingley, Cambridge CB3 8AA, UK

SUMMARY

1 Proximate constraints on the processes underlying the acquisition and maintenance of rank have received little attention as causes of variation in the pattern of dominance among females in female-bonded Old World monkeys.

2 Simple simulations, based on realistic rules derived from studies of the propensities and mechanisms underlying rank relationships among rhesus macaque females, are used to illustrate the effect of demographic constraints on female dominance patterns.

3 The rules are applied to two different populations, in which females differ in life-history variables, and hence in the age and size structure of the populations, and in the availability of kin.

4 The results suggest that demographic factors alone can have a considerable impact on female dominance patterns, and can generate differences comparable to observed intra-species differences between provisioned and wild macaques, and to inter-species differences between provisioned macaques, wild savannah baboons, and wild gelada baboons.

INTRODUCTION

Many attempts to understand inter-specific variation in primate social organization are attempts to understand the ultimate adaptive significance of differences in social structure and organization. As such they often carry the implicit assumption that observed differences reflect species-specific differences in behaviour and propensities, and have a genetic basis. While this assumption may often be correct, it can also be misguided, since phenotypic differences, whether between individuals, social groups, populations or species, may not have a genetic basis, but may reflect differences in circumstances, or constraints on underlying processes.

This is increasingly acknowledged in the case of intra-species variation in non-human primates. For instance, ecological conditions, and even random factors, can act as proximate causes of variation in group size and composition (Altmann & Altmann 1979; Dunbar 1979). Group composition, through effects on partner availability, can in turn affect the occurence of social behaviours such as play (White 1977), which can also be affected by the time and energy constraints

imposed by ecological conditions (Lee 1983). Such variation in group composition
and social experience may have consequences for individual development and
social organization, although data on such consequences are lacking, and too little
is generally known about social and developmental processes to predict the effects
of given socioecological conditions.

The potential flexibility of social behaviour and social organization raises an
important issue for comparative studies, since understanding of which selection
pressures have shaped particular aspects of primate behaviour and social organization, and of how these differ across species, is ultimately involved. In many
cases, intra-species variability has probably been underestimated, and 'species-
specific' social characteristics too narrowly defined, because a species has been
studied in only a narrow range of socioecological conditions. There is the related
possibility that some apparent inter-species differences in behaviour and social
organization reflect differences in the socioecological conditions in which species
have been observed, rather than differences in underlying behavioural strategies
or propensities.

In this chapter, I suggest one way in which the patterning of dominance
relationships among females, often treated as a species characteristic, might be
affected by socioecological conditions. Ecological factors, by affecting demographic processes, can affect group composition, which can constrain the ability of
females to acquire and maintain rank. Such indirect, proximate effects of ecological conditions may account for some differences in observed patterns of female
dominance relationships, and could affect the success a female can achieve in
intra-group competition with other females.

DOMINANCE RELATIONS AMONG FEMALES

Most primates live in matrilineal social groups. Female kin tend to associate for
life, while males migrate between groups. While females probably benefit in
various ways from living together (see, for example, Wrangham 1980), they are
also likely to suffer costs, due to intra-group competition. Such competition is
often manifested in the form of agonistic dominance hierarchies among females.

Does high rank benefit females? Although a positive, significant, correlation
between high female rank and reproductive success has not been found in all
studies, where a significant correlation has been demonstrated, it has been found
to be positive rather than negative (Silk 1987). Where the distribution of important
resources is such that they can be monopolized, high-ranking females may have
the edge, particularly if resources are relatively scarce (Wrangham 1981; Gouzoules,
Gouzoules & Fedigan 1982; Whitten 1983). Female reproductive success may also
be affected more directly, by female–female reproductive competition, and here,
too, high-ranking females may have the advantage. For instance, among wild
savannah baboons (Wasser 1983) and wild gelada baboons (Dunbar & Dunbar
1977), harassment from higher-ranking females appears to reduce fertility or
cause pregnancy failure in subordinate females. The offspring of low-ranking

females may also suffer increased mortality due to aggression from high-ranking females (Silk *et al.* 1981a).

High rank could, therefore, be a valuable asset to a female in intra-group competition. This raises the issue of what determines female rank. Traditionally, the answer has been couched in terms of empirical rules which predict the position a female is likely to achieve in the adult female dominance hierarchy of her group. For instance, among provisioned, free-ranging groups of rhesus macaques on Cayo Santiago (Sade 1967; Missakian 1972), and of Japanese macaques in Japan (Kawamura 1965; Koyama 1967), the adult rank of any female born into the group is highly predictable from three empirical rules: (i) Mothers dominate daughters for life; (ii) Younger sisters dominate older sisters as adults, and (iii) Females inherit their mother's rank relative to others in the group. So impressive is the predictive power of these rules for these well-studied populations that the resulting pattern of dominance relationships is widely regarded as characteristic of rhesus and Japanese macaques. In other female-bonded, multi-male species, 'violations' of the rules, particularly of the mother−daughter and sister−sister rules, appear common, for example in olive baboons (Moore 1978), yellow baboons (Hausfater *et al.* 1982), bonnet macaques (Silk, Samuels & Rodman 1981b), crab-eating macaques (Angst 1975), and Barbary macaques (Paul & Kuester 1987). In female-bonded species with a harem structure, such as gelada baboons (Dunbar 1980) and Hanuman langurs (Hrdy & Hrdy 1976), females apparently have an 'age-graded' dominance hierarchy (contrasted, by Hrdy & Hrdy 1976, with the 'nepotistic' hierarchy of macaques and baboons), in which females in their reproductive prime dominate both younger and older females.

These differences have often been regarded as reflecting 'species' differences, perhaps in underlying female propensities or strategies, and some attempts made to seek functional explanations for them (see, for example, Hrdy & Hrdy 1976; Moore 1978; see also Discussion, below). However, such explanations cannot easily account for the existence of intra-species variation, often on a par with inter-species variation. For example, violations of the nepotistic rules, similar to those seen in savannah baboons and in other macaques, occur in captive rhesus macaques (Chikazawa *et al.* 1979), and in wild, unprovisioned Japanese macaques (Furuichi 1984; D. Sprague, personal communication), the archetypal nepotistic species.

One possibility is that differences in circumstances, for instance in group composition, have a proximate effect on female dominance patterns. Although this idea is not new (see, for example Chikazawa *et al.* 1979; Chapais & Schulman 1980; Hausfater, Cairns & Levin 1987), exactly which demographic features affect female dominance relationships, and how, is still largely unexplored. Recently, Hausfater, Cairns & Levin (1987) modelled the relationship between demography and dominance structure in female primates, making the important demonstration that group composition, in conjunction with given rules about the acquisition of rank, can have a significant impact on aspects of hierarchical structure. However, the model ignores the possibility that social processes, such as alliances, can have a critical influence on female dominance. It is also unclear in the model what determines the likelihood that a female will outrank or remain subordinate to another. This likelihood is varied arbitrarily, independently of group composition. However, if the likelihood of rank changes is affected by factors such as the availability of coalition partners, it may be highly dependent on group composition.

A more process-oriented approach is taken below. Using simple simulations, based on realistic rules derived largely from work on the mechanisms and propensities underlying dominance relationships among rhesus macaque females, I consider how the pattern of dominance relationships among females might alter with changes in group composition.

UNDERLYING RULES AND PROPENSITIES

In this section some underlying features of dominance relationships among rhesus and Japanese macaques, which form the basis of the rules used in the simulations below, are described briefly.

Preference for high rank

Females appear to have a strong urge to achieve and maintain high rank. Subordinate females seize opportunities to improve their rank, while females whose rank is challenged do not give it up willingly, but often fight back vigorously (see, for example, Varley & Symmes 1966; Marsden 1968; Datta 1983a, b; Chapais 1985; de Waal 1987).

Determinants of dominance–subordinance

Older, larger females tend to dominate younger, smaller females. However, younger females can dominate older, larger females, provided that they have powerful allies against the latter, and that the size difference between opponents is not too great (see, for example, Koyama 1967; Datta 1983a, 1988).

Females appear to assess (cf. Parker 1974) relative power or fighting ability, power being determined by a combination of physical factors, such as body-size, which affect individual fighting ability, and social factors such as alliances. Dominants tend to be challenged and outranked in circumstances where their power is expected to be reduced relative to the challenger, for example when: (i) an individual has an ally feared by (dominant to) the dominant; (ii) the dominant has lost a powerful ally; (iii) the dominant has suffered a decline in individual fighting ability, for example due to old age or illness.

Who allies are, and when they are powerful

A number of tendencies come into play in 'decisions' about who to support against whom (Datta 1983c). In general, females support closer relatives, who are also closer associates, against less close relatives (Kaplan 1977; Massey 1977). If relatives are equally related to the interferer, there is a tendency to support the younger, smaller one in disputes, perhaps because it is more vulnerable (see, for example, Datta 1988). There is also a tendency to support high-ranking animals against low-ranking animals, and, to improve or maintain her own rank, a female may form an alliance with a high-ranking non-relative, or more distant relative, against a close relative (Datta 1983c; Chapais 1985).

Only allies dominant to (hence feared by) an individual's opponent are likely to be powerful or 'effective' in helping that individual improve or maintain rank (Datta 1986). Such allies are also the ones most likely to act, since subordinate allies risk attack by the opponent (Kaplan 1978; Datta 1983b), and tend to avoid taking the risk (Datta 1988).

THE SIMULATIONS

Two populations

Figure 1 illustrates the descendants of a female M in two populations which differ in female life-history variables. In Population A (Fig. 1a) a median female reaches sexual maturity at 4 years, and produces an infant who survives to sexual maturity every 12 months. Assuming that the birth sex ratio is 50:50, and that males and females are produced in alternate years, the interval between daughters is 2 years. In Population B (Fig. 1b) a median female reaches sexual maturity at 6 years, and gives birth every 24 months to an infant who survives to sexual maturity. Hence the interval between daughters is 4 years. M has a lifespan of 16 years in both populations, but this is varied below.

Since proximate effects of environment on life-history variables are well established (see, for example, Dunbar 1987; Cords & Rowell 1987), these populations could represent the same species in different environments. For instance, Population A could represent rhesus macaques in an extremely favourable environment, such as that of Cayo Santiago. Here there are no predators, little disease, no extremes of climate, and the monkeys are well-fed due to abundant provisioning. Most females on Cayo Santiago achieve sexual maturity at 3·5 years, give birth to their first infant at 4 years, and produce an infant every 12 months (see, for example, the matriline in Datta 1988; Rawlins & Kessler 1986). Population B could represent rhesus macaques in a much harsher environment such as that of the Himalayas. Here, females reach sexual maturity at 5·5 years, giving birth to their first infant at 6 years; and birth intervals range from 24 to 36 months (Melnick & Pearl 1987).

Matrilines in Population B are also strikingly similar to those reported for wild populations of some other Cercopithecines, such as yellow baboons at Amboseli (Altmann & Altmann 1979; Hausfater, Altmann & Altmann 1982), gelada baboons in Ethiopia (Dunbar 1984) and hamadryas baboons in Ethiopia (Sigg et al. 1982).

(a)

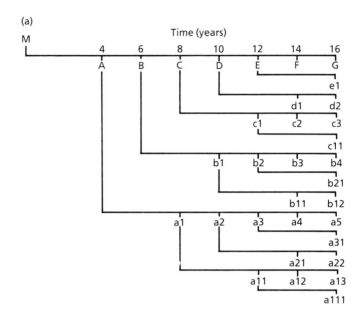

(b)

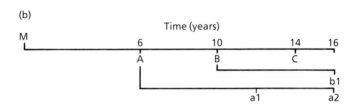

FIG. 1. The female descendants of a female, M, in Population A (A), and in Population B (B), when M dies at 16 years.

In all these populations median females give birth to their first infant at about 6 years, and there is an interval between surviving offspring of about 2 years.

Mother-daughter dominance relationships

Background information

A daughter who outranks her mother may be able to prevent herself being outranked by her younger sisters, since the mother is a very important ally of the younger sister against the older (Datta 1988). If conditions are favourable, daughters do outrank mothers. Favourable conditions arise in two ways. First, if the daughter manages to acquire an ally who is dominant to the mother. This is probably very rare in naturalistic conditions, such as those considered here, where there appears to be virtually no interference in mother–daughter disputes (Datta 1981).

Of the very few interventions seen in well-established groups, those by other daughters dominant to a given daughter favour the mother (Datta 1981). This may contribute to the inhibition of more subordinate daughters against challenging their mother (cf. also Horrocks & Hunte 1983), and may explain why challenges and outrankings tend to come from the highest ranking daughter. If the highest ranking daughter does challenge the mother, the few data available suggest that relatives dominant to the mother and daughter (such as the mother's mother and dominant sisters) tend to favour the mother (Datta 1981).

The second possibility is that the daughter manages to exceed the mother in individual fighting ability. This is unlikely to happen while the daughter is still very young, but is more likely if the daughter is adult, and the mother is old or ill. In naturalistic groups it is indeed in these circumstances that mothers are out-ranked by daughters (Missakian 1972; Moore 1978; Hausfater, Altmann & Altmann 1982). Even a very old and feeble mother may, however, be able to buffer herself against decline in personal fighting ability if she has powerful allies against her daughter.

Finally, it should be mentioned that the outranking of an old mother by one daughter often is not followed by outrankings by other adult daughters (see, for example, Missakian 1972). It is possible that the outranking daughter remains in alliance with the mother against her subordinate sisters, so inhibiting their rise.

Simulation rules and conditions

Taking cognizance of the above, the following rules were applied to M and her daughters in Populations A and B:

(a) A daughter has no allies dominant to her mother.

(b) A sexually mature daughter exceeds an 'old' mother of 13 years in individual fighting ability, and can outrank her on her own.

(c) The highest ranking sexually mature daughter challenges the mother.

(d) Relatives dominant to the mother and daughter (such as the mother's sisters) help the mother against her rebellious daughter.

(e) The outranking daughter remains an ally of the mother against other daughters, who are thereby prevented from outranking the mother.

These rules were applied under a number of conditions (C1–C4) which, in effect, vary the number of dominant sisters that M has when, at the age of 13 years, she is challenged by her highest-ranking daughter (5-year-old C in Population A, Fig. 2; 7-year-old A in Population B, Fig. 3). The number of such sisters can be calculated using information about the conditions, together with the simulation rules for sister–sister dominance relationships (see below).

C1: M's mother dies at 16 years; M is the eldest daughter, like A in Figs 2 and 3a.

C2: M's mother dies at 13 years; M is the eldest daughter, like A in Figs 2 and 3b.

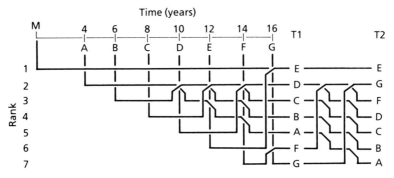

FIG. 2. Intra-family ranks among M and her daughters in Population A. Younger sisters reverse dominance with older sisters at sexual maturity (4 years), as long as another family member is dominant to those sisters. T1: just after M's death; T2: just after the youngest daughter reaches sexual maturity.

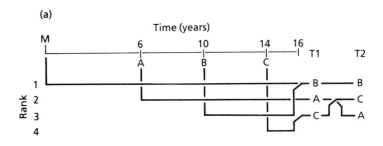

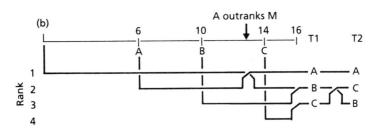

FIG. 3. Intra-family ranks among M and her daughters in Population B, when: (A), M remains dominant to all her daughters; and (B), M is outranked at 13 years by her highest-ranked daughter, A. Younger sisters reverse dominance with older sisters at sexual maturity (6 years), as long as another family member is dominant to those sisters. T1: just after M's death; T2: just after the youngest daughter reaches sexual maturity.

$C3$: M's mother dies at 16 years; M is the middle daughter, like D in Fig. 2, and B in Fig. 3a.
$C4$: M's mother dies at 16 years; M is the youngest daughter, like G in Fig. 2, and C in Fig. 3a. M's sisters die at 16 years.

Results

As Fig. 4 shows, in Population A, M has at least one dominant sister in three out of four conditions (C1–C3). This is in contrast to Population B, where M has a dominant sister in only one out of four conditions (C1). If M has any dominant sisters, she has powerful support against a rebellious daughter. Thus, in three out of four conditions, mothers in Population A successfully resist a rebellious daughter, while in only one out of four conditions do they do so in Population B.

Several factors not considered here could widen this difference. If mothers in a harsh environment are under nutritional stress (Altmann 1980), then adult female mortality may be higher, and female lifespan shorter, in Population B. Thus C2 might be relatively more common in Population B than in Population A, and even in C1, M's contingent of dominant sisters is more susceptible to erosion in Population B than in Population A. Second, if higher-ranking relatives other than sisters (e.g. aunts and nieces, Datta 1981) occasionally interfere on behalf of the mother, they are more likely to be available in Population A than in Population B (see Fig. 1), and this may increase the chances of a Population A mother winning against her daughter in C4. Finally, the age difference between a mother and her highest ranking daughter is, paradoxically, likely to be greater in Population A than in Population B, due to the greater likelihood of sister–sister reversals (see below). The size difference between a mother and her highest ranked daughter in Population A is thus also likely to be greater, and may inhibit the daughter from dyadic challenge, unless the mother is very old.

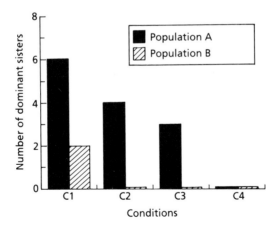

FIG. 4. The number of dominant sisters a mother, M, has, who could help her against a rebellious daughter. This is shown for Populations A and B, under a number of conditions C1–C4 (see text), which affect the number of dominant sisters. If M has any dominant sisters, she defeats the rebellious daughter, and retains her rank.

Sister−sister dominance relations

Background information

A female begins smaller than, and subordinate to, her older sister. Depending on
the age difference between them, the size advantage of the older sister may
persist for several years. As long as the older sister exceeds or equals the younger
in intrinsic fighting ability, and is willing to defend her position, it may be
injurious for a younger sister to challenge her on purely dyadic terms. Even fully
grown females refrain from challenging older dominant sisters on such terms
(Datta, 1988). Even when older sisters decline in intrinsic fighting ability, younger
sisters may be at a disadvantage, because by then older sisters may have daughters
to buffer their positions against such decline. Without help from others, therefore,
a younger sister is unlikely to be able to outrank an older sister.

Younger sisters do normally have allies against older sisters from birth. Close
relatives such as the mother and other sisters are the main allies, but only if they
dominate the older sister (Datta 1988). If subordinate, they avoid intervening,
possibly because of the risk of retaliation by the older sister. If the mother dies,
any sisters dominant to the older sister become the younger sister's principal allies
(Datta 1988). Help may be given because the younger sister, especially when very
young, is vulnerable in disputes with older sisters. Younger sisters appear to
exploit this circumstance to escalate disputes against older sisters.

Even with help, younger sisters appear unable to dominate older sisters till
they reach 60−70% of their body size. The precise ages of the sisters at reversal
therefore depend on the relative age of sisters: a younger sister a year old can
outrank an old sister 2 years old, but not one several years older (Datta 1988).
Provided they have suitable help, however, females can dominate all older sisters,
irrespective of relative age, by sexual maturity (Koyama 1970; Missakian 1972;
Sade 1972). As long as the younger sister is distinctly smaller than the older, she
may need help to maintain dominance. Thus removal or death of an important
ally, such as the mother, can lead to re-reversal (unpublished data).

Simulation rules and conditions

Taking account of the above, the following rules were applied to sisters (M's
daughters) in Populations A and B:
(a) If a female has a mother or sister dominant to an older sister, she has
powerful support against the latter. A mother outranked by the older sister (see
above) ceases to be a powerful ally of the younger sister against the older sister.
(b) If a female has powerful support against an older sister, she outranks the
latter at sexual maturity (4 years in Population A, 6 years in Population B).
(c) For simplicity, assume that a sexually mature younger sister who has outranked
her older sister is able to maintain this rank dyadically (but see below).

Using these rules, the proportion of sister—sister pairs in which the younger sister dominated the older was determined for Populations A and B at two times, T_1, just after the mother's death, and T_2, just after the youngest of the sisters reached sexual maturity. This was done for three conditions, which had the effect of varying the number of close relatives (mother and other sisters) dominant to, and hence powerful allies against, an older sister:

S_1: As C_1 above. Thus M remains dominant to all her daughters till she dies, at 16 years, in both Populations A and B (Figs 2 and 3a respectively).

S_2: As C_2 above. Thus M, who dies at 16 years, is outranked at 13 years in Population B (Fig. 3b), but not in Population A (Fig. 2).

S_3: M remains dominant to all her daughters in Populations A and B, dying at 13 years.

Results

Fig. 5 shows the proportion of pairs in which the younger sister dominates the older, at T_1 and T_2, for Populations A and B, for the three conditions. (The proportion at T_2 is greater than at T_1 because, after M dies, younger sisters continue to be helped against older sisters by other sisters dominant to the latter as suggested in Datta 1988.) Clearly, the proportion of sister pairs with reversal is likely to be lower in Population B than in Population A under all conditions.

Several factors not considered here are likely to increase differences between the two populations. First, if adult females have a shorter lifespan in Population B (see above), S_2 and S_3 will be more common in Population B than in Population A. Second, if younger sisters need help to maintain their rank until they are fully grown (which may not be for a few years after sexual maturity), re-reversals may be more common in Population B, where younger sisters will more often be left without an ally against an older sister. Third, the proportion of pairs in which the younger sister dominates the older in Population A at any time is probably underestimated here, since age differences in many pairs are small. This would allow younger sisters to outrank some older sisters well before maturity, and the greater availability of allies would allow these ranks to be maintained (Datta 1988).

Consequences for inter-matriline dominance relations

What determines whether one matriline will outrank another, or remain dominant to another, and how does demography affect the likelihood that matriline ranks will remain stable, and that daughters will inherit maternal rank? This is not considered in detail here, because there are few precise data on the determinants of intermatriline rank. However, the following would be expected.

First, by analogy with individuals, the direction of dominance between two matrilines is expected to depend partly on their relative intrinsic fighting ability, which will depend on the number and kinds of individuals in the matriline. As

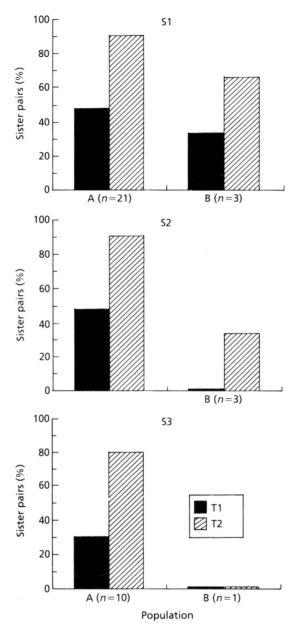

FIG. 5. Percentage of pairs in which the younger sister dominates the older, in Populations A and B, at two times, T1: just after the mother's death; and T2: just after the youngest sister reaches sexual maturity. The percentage is shown under three different conditions, S1–S3 (see text), which affect the number of allies the younger sister has against the older.

long as a higher-ranked matriline exceeds or equals a lower-ranked matriline in fighting ability, the relationship is expected to be stable. However, if the asymmetry clearly shifts in favour of the lower-ranked matriline, for example due to the death of females in the higher-ranked matriline (see, for example, Gouzoules 1980), or the simultaneous maturation of a cluster of females in the lower-ranked matriline (Samuels & Henrickson 1983) the higher ranked matriline might be challenged and overthrown.

Demographic processes could affect the stability of inter-matriline ranks through their effect on matriline size. In a small matriline, each individual represents a greater proportion of the power, or potential power, of the matriline than in a large matriline. Any alterations in group membership will therefore have a disproportionately large effect in small matrilines, so shifts in relative power, and rank reversals between matrilines, might be more common in Population B. However, if matrilines buffer their ranks through reciprocal alliances with other matrilines (see, for example, Walters 1980; Chapais 1983; Cheney 1983), the difference between the two populations might be reduced. For instance, even where matrilines are small, as among yellow baboons at Amboseli, rank relations can be highly stable for long periods (Samuels, Silk & Altmann 1987). Dyadic relationships, even between large matrilines, can be unstable: among rhesus monkeys on Cayo Santiago (Population A), a rare rank reversal (K. Cushing personal communication) between matrilines (the 'genealogies' of Sade *et al.* 1976) occurred in Group J, which was unusual in having only two.

Second, effects on the inheritance of maternal rank. Although the mother–daughter and sister–sister rules are violated in a number of populations (see above), the maternal inheritance rule is typically more stable in the same populations (see, for example, Silk, Samuels & Rodman 1981b; Hausfater, Altmann & Altmann 1982). This may be partly due to the fact that a greater number and variety of allies participate in disputes between families than in disputes within families, giving a wider alliance base. Only very close relatives are generally involved in intra-family fights (Datta 1988), but because of reciprocal alliances between families, other relatives such as aunts, nieces and cousins (Berman 1980, Datta 1983c), and even unrelated females (see, for example, Chapais 1983) are often involved in inter-family fights. Thus inter-family dominance relationships may be better buffered against demographic variation than intra-family dominance relationships, and even an individual without close relatives may be able to inherit maternal (or family) rank. For instance, on Cayo Santiago (Population A) a young orphan without siblings inherited his family rank with help from his high-ranking uncle and aunts against lower-ranking families (personal observation).

Violations of the maternal inheritance rule are expected to be more common in Population B than in Population A. First, if matriline ranks are comparatively weakly buffered in Population B (see above), then high-ranking females may often be only weakly dominant to low-ranking females, and this may hamper the transmission of maternal rank. In a possible example of this among captive rhesus monkeys, the daughters of a weakly dominant alpha female outranked females in the lower-ranked family younger than, or the same age as, themselves, but, unusually, failed to outrank the low-ranking adult female by sexual maturity. The alpha female seemed unable to intimidate the low-ranking female when the latter attacked the alpha female's daughters; she also behaved as if wary of escalation, and eventually stopped supporting her older daughters against the low-ranking female (unpublished data). Second, larger age and size differences among females in Population B might make it more difficult for the daughters of high-ranked females to outrank older low-ranked females. Finally, orphans may be less likely to have a relative or family friend to help them achieve family rank in Population B than in Population A.

DISCUSSION

Group composition and female dominance relationships

The pattern of dominance relationships among provisioned free-ranging populations of rhesus or Japanese macaque females, predicted by three famous empirical rules (see Introduction), has been considered characteristic of the species. However, using realistic rules about underlying mechanisms and propensities, the above suggests that demographic processes, through their effect on group composition, could significantly affect the likelihood of violation of these rules, and hence the pattern of dominance. The possibility that substantial variation in the pattern of dominance could occur without any change in underlying mechanisms, or in female propensities and strategies, points to the dangers of inferring these from observed patterns, or 'surface structure' (Hinde 1976). Such patterns do not represent underlying strategies or propensities directly, but are the outcome of interactions between the latter, and the constraints and opportunities presented by a given environment.

The significance of group composition in the present case lies in its effect on the power asymmetries among females, which affect the potential costs to a female of escalating a dispute (cf. Parker 1974), and hence decisions about whether to fight for dominance, or to accept subordinance. These asymmetries depend both on differences in the individual fighting ability of females, dependent on age and size, and on differences in the availability of suitable allies, who are generally close relatives dominant to the opponent. Life-history variables such as age at sexual maturity, inter-birth intervals, mortality rates, and lifespan, affect both these components of power, with the result that asymmetries tend to favour certain females in some conditions and others in others.

For instance, in conditions where females achieve sexual maturity earlier, inter-birth intervals are relatively short, and mortality rates are low (Population A), mothers are likely to be able to remain dominant over their daughters, and younger sisters are likely to be able to outrank older, larger sisters. However, in conditions where females are older at sexual maturity, inter-birth intervals are longer, and mortality rates are higher (Population B), the balance shifts in favour of daughters, who can now often outrank mothers, and in favour of older sisters, who can now often retain dominance over younger sisters.

The model also suggests that the acquisition of maternal or family rank is less likely to be affected by demographic variation than the acquisition of intra-family rank, partly because the number and kinds of allies involved in inter-family disputes is relatively large, making this alliance-base less sensitive to demographic events. Nevertheless, the potential for inter-matriline dominance reversals, and for independence from maternal rank, is expected to be greater in Population B than in Population A. Thus rank differences between matrilines may be more likely to even-out over time in Population B than in Population A, where rigidly nepotistic dominance hierarchies, more stable over time, are expected.

What significance, if any, might these differences have for females? If there is a positive association between female rank and reproductive success, then a female's success relative to other females in her group, insofar as it is linked to the rank she and her descendants achieve and are able to maintain relative to other females, could be affected by demographic conditions. For example, females may have more chance of overthrowing higher-ranked matrilines in Population B than in Population A, while younger daughters have a better chance of achieving and maintaining high intra-family rank in Population A than in Population B.

Usefulness and limitations of model

In theory, demographic variation, in conjunction with the simulation rules used here, could account for a considerable amount of variation in female dominance patterns observed among female-bonded Old World monkeys. Within groups, micro-demographic differences (such as whether the mother is alive or dead) may explain why the rules of inheritance are violated in some dyads but not in others (e.g. cases mentioned in Angst 1975; Silk, Samuels & Rodman 1981b; Datta 1988). There are few data on inter-population differences in female dominance patterns, but recent data from a wild population of Japanese macaques (similar to Population B) suggest that the dominance pattern among females differs from that of provisioned populations (Population A) of the same species in at least some of the expected ways (Furuichi 1984; David Sprague, personal communication).

Demographic factors could also be important sources of some apparent inter-species differences in dominance patterns. For instance, savannah baboons appear to differ from rhesus and Japanese macaques in showing a higher incidence of daughters outranking mothers, and a lower incidence of younger sisters outranking older sisters (Moore 1978; Hausfater, Altmann & Altmann 1982). However, in existing comparisons of macaques and baboons, group demography has been a confounding variable, since the traditional macaque model is based on Population A groups, while the traditional baboon model is based on Population B groups. In

Population B, rhesus-like female propensities and strategies are expected to lead to an increased incidence of daughters outranking mothers, and a decreased incidence of younger sisters outranking older sisters, exactly what is seen in savannah baboons. This interpretation of the difference is strengthened by evidence that the rank-related propensities and strategies of female savannah baboons, as well as the mechanisms underlying rank relationships, are very similar to those of rhesus females (see, for example, Walters 1980; Johnson 1987; Samuels, Silk & Altmann 1987). Where sufficiently detailed information is available, there is excellent correspondence between rankings predicted using the simulation rules above, and actual rankings. For instance, in Hausfater, Altmann & Altmann (1982), a female, AL, was dead by 1977, leaving three daughters, ALd1, ALd2 and ALd3, who were at most 7, 4 and 2 years old (respectively) at their mother's death. It would be predicted that, at these ages, their ranks would be ALd1>ALd2>ALd3 (though the authors do not provide this information). It would also be predicted that ALd1 would help ALd3 to outrank ALd2, but that she would not be outranked by either of her younger sisters. Indeed, by 1981, the sisters ranked in the order ALd1>ALd3>ALd2; their ages were 11, 6 and 8 years, respectively.

At first sight, the pattern of dominance relationships among females in harem-living female-bonded Old World monkeys, such as gelada baboons and hanuman langurs, appears very different from those among macaques and savannah baboons. In both these species, females in their reproductive prime apparently outrank both older and younger females; langurs: Hrdy & Hrdy (1976); geladas: Dunbar (1980). In principle, such a hierarchy could arise from operation of the simulation rules used in this chapter, especially in conjunction with Population B, which could apply to gelada baboons (Dunbar 1984), and, possibly in an extreme form (since infanticide by males is expected to increase age differences between mothers and daughters, and between sisters), to langurs. In Fig. 3b, for instance, if A outranks her mother M when M is 13 years old, the rank order in the family when M is 15 years old will be A>M>B>C, and the corresponding ages will be 9 years>15 years>5 years>1 year. This accords with Dunbar's (1980) idea that prime gelada females achieve higher ranks because of greater aggressiveness (or intrinsic power), and would be expected in conditions where all females prefer high rank, but, because of demographic constraints, are unable to buffer themselves effectively against loss of intrinsic power. However, it differs from Hrdy & Hrdy's (1976) adaptive explanation for the difference between macaque and langur female dominance patterns. They suggested that in uni-male groups of langurs, average relatedness among females in a group might be higher than in multi-male groups of macaques, and the inclusive fitness of females might be best served by altruistically deferring to female relatives of higher reproductive value, rather than selfishly maintaining high rank. While present data are inadequate to decide between these explanations, there are some difficulties with Hrdy and Hrdy's hypothesis. First, the ranking by reproductive value idea may be untenable (see Horrocks & Hunte 1983 for a recent critique). Second, langur females may be less closely related than supposed. While females in the same age cohort may share a father, females further apart in age (such as sisters), who are likely to differ most in reproductive value, are also least likely to share a father, if there are frequent male takeovers, long inter-birth intervals, and infanticide by incoming males.

There are some obvious limitations to the present model. For example, it assumes that females have a strong propensity to achieve and maintain high intra-group rank, and that this remains uninfluenced by changes in ecological conditions. However, if the quality and distribution of important resources, such as food, are such that dominants either cannot monopolize them, or do not profit by doing so, high rank may have relatively little value, and females may be less inclined to incur the potential costs of acquiring and maintaining high rank, or of helping their relatives and friends to do so. This could result in competitive interactions including coalitions, being rare, and the direction of dominance among females being difficult to discern, inconsistent, or in order of individual fighting ability (this last resembling the pattern seen in hierarchies where prime females hold the highest ranks, as above). Thus the effects of demographic factors could be confounded with those of other factors, such as the strength of the preference for high rank, which could be inherited as well as affected by immediate ecological circumstances.

To conclude, demographic constraints on the processes by which females acquire and maintain rank are potentially important sources of variation in the patterning of female dominance relationships. However, except in rare cases, too little is known about female propensities, or about the processes determining rank, to differentiate between conflicting hypotheses about the causes of patterns of dominance. Data from more species are needed on the processes underlying rank acquisition and maintenance, on the correlates of intra-specific variation in dominance relationships, and on the effects of demographic variation on female dominance relationships.

REFERENCES

Altmann, J. (1980). *Baboon Mothers and Infants*. Harvard University Press, Cambridge, Massachusetts.

Altmann, S.A. & Altmann, J. (1979). Demographic constraints on behaviour and social organisation. *Primate Ecology and Human Origins* (Ed. by I.S. Bernstein & E.O. Smith), pp. 47–64. Garland Press, New York.

Angst, W. (1975). Basic data and concepts in the social organisation of *Macaca fascicularis*. *Primate Behaviour, Vol. 4* (Ed. by L.A. Rosenblum), pp. 325–88. Academic Press, New York.

Berman, C.M. (1980). Early agonistic experience and rank acquisition among free-ranging infant rhesus monkeys. *International Journal of Primatology*, 1, 153–70.

Chapais, B. (1983). Dominance, relatedness, and the structure of female relationships in rhesus monkeys. *Primate Social Relationships: an Integrated Approach* (Ed. by R.A. Hinde), pp. 209–19. Blackwell Scientific Publications, Oxford.

Chapais, B. (1985). An experimental analysis of a mother–daughter rank reversal in Japanese macaques (*Macaca fuscata*). *Primates*, 26, 407–23.

Chapais, B. & Schulman, S. (1980). An evolutionary model of female dominance relationships in primates. *Journal of Theoretical Biology*, 82, 47–89.

Cheney, D.L. (1983). Extra-familial alliances among vervet monkeys. *Primate Social Relationships: an*

Integrated Approach (Ed. by R.A. Hinde), pp. 278–86. Blackwell Scientific Publications, Oxford.

Chikazawa, D., Gordon, T., Bean, C. & Bernstein, I.S. (1979). Mother–daughter dominance reversals in rhesus monkeys (*Macaca mulatta*). *Primates*, **20**, 301–5.

Cords, M. & Rowell, T.E. (1987). Birth intervals of *Cercopithecus* monkeys of the Kakamega Forest, Kenya. *Primates*, **28**, 267–70.

Datta, S.B. (1981). *Dynamics of dominance among free-ranging rhesus females.* Ph.D. thesis, University of Cambridge.

Datta, S.B. (1983a). Relative power and the acquisition of rank. *Primate Social Relationships: an Integrated Approach* (Ed. by R.A. Hinde), pp. 93–102. Blackwell Scientific Publications, Oxford.

Datta, S.B. (1983b). Relative power and the maintenance of rank. *Primate Social Relationships: an Integrated Approach* (Ed. by R.A. Hinde), pp. 103–11. Blackwell Scientific Publications, Oxford.

Datta, S.B. (1983c). Patterns of agonistic interference. *Primate Social Relationships: an Integrated Approach* (Ed. by R.A. Hinde), pp. 289–98. Blackwell Scientific Publications, Oxford.

Datta, S.B. (1986). The role of alliances in the acquisition of rank. *Primate Ontogeny, Cognition and Social Behaviour* (Ed. by J.G. Else & P.C. Lee), pp. 219–27. Cambridge University Press, Cambridge.

Datta, S.B. (1988). The acquisition of rank among free-ranging rhesus monkey siblings. *Animal Behaviour*, **36**, 754–72.

Dunbar, R.I.M. (1979). Population demography, social organisation, and mating strategies. *Primate Ecology and Human Origins* (Ed. by I.S. Bernstein & E.O. Smith), pp. 65–88. Garland Press, New York.

Dunbar, R.I.M. (1980). Determinants and evolutionary consequences of dominance among female gelada baboons. *Behavioural Ecology and Sociobiology*, **7**, 253–65.

Dunbar, R.I.M. (1984). *Reproductive Decisions.* Princeton University Press, Princeton, New Jersey.

Dunbar, R.I.M. (1987). Demography and reproduction. *Primate Societies* (Ed. by B.B. Smuts, D.L. Cheney, R.M. Seyfarth, R.W. Wrangham & T.T. Struhsaker), pp. 240–9. The University of Chicago Press, Chicago, Illinois.

Dunbar, R.I.M. & Dunbar, E.P. (1977). Dominance and reproductive success among female gelada baboons. *Nature*, **266**, 351–2.

Furuichi, T. (1984). Symmetrical patterns in non-agonistic interactions found in unprovisioned Japanese macaques. *Journal of Ethology*, **2**, 109–19.

Gouzoules, H. (1980). A description of genealogical rank changes in a troop of Japanese monkeys (*Macaca fuscata*). *Primates*, **21**, 262–7.

Gouzoules, H., Gouzoules, S. & Fedigan, L. (1982). Behavioural dominance and reproductive success in female Japanese monkeys (*Macaca fuscata*). *Animal Behaviour*, **30**, 1138–51.

Hausfater, G., Altmann, J. & Altmann, S.A. (1982). Long-term consistency of dominance relations among female baboons (*Papio cynocephalus*). *Science*, **217**, 752–5.

Hausfater, G., Cairns, S.J. & Levin, R.N. (1987). Variability and stability in the rank relations of non-human primate females: Analysis by computer simulation. *American Journal of Primatology*, **12**, 55–70.

Hinde, R.A. 1976. Interactions, relationships and social structure. *Man*, **11**, 1–17.

Horrocks, J.A. & Hunte, W. (1983). Rank relations in vervet sisters: a critique of the role of reproductive value. *American Naturalist*, **122**, 417–21.

Hrdy, S.B. & Hrdy, D.B. (1976). Hierarchical relations among female hanuman langurs (Primates: Colobinae, *Presbytis entellus*). *Science*, **193**, 913–15.

Johnson, J.A. (1987). Dominance rank in juvenile olive baboons, *Papio anubis*: the influence of gender, size, maternal rank and orphaning. *Animal Behaviour*, **35**, 1694–1708.

Kaplan, J.R. (1977). Patterns of fight interference in free-ranging rhesus monkeys. *American Journal of Physical Anthropology*, **49**, 241–50.

Kaplan, J.R. (1978). Fight interference and altruism in free-ranging rhesus monkeys. *American Journal of Physical Anthropology*, **47**, 241–9.

Kawamura, S.J. (1965). Matriarchal social ranks in the Minoo-B troop: a study of the rank system of Japanese monkeys. *Japanese Monkeys* (Ed. by K. Imanishi & S.A. Altmann), pp. 105–12. Emory University Press, Atlanta, Georgia.

Koyama, N. (1967). On dominance rank and kinship of a wild Japanese monkey troop in Arashiyama. *Primates*, **8**, 189–216.

Koyama, N. (1970). Changes in dominance rank and division of a wild Japanese monkey troop in Arashiyama. *Primates*, 11, 335–90.

Lee, P.C. (1983). Ecological influences on relationships and social structures. *Primate Social Relationships: an Integrated Approach* (Ed. by R.A. Hinde), pp. 225–30. Blackwell Scientific Publications, Oxford.

Marsden, H.M. (1968). Agonistic behaviour of young rhesus monkeys after changes induced in the social rank of their mothers. *Animal Behaviour*, 16, 38–44.

Massey, A. (1977). Agonistic aids and kinship in a group of pigtail macaques. *Behavioural Ecology and Sociobiology*, 2, 31–40.

Melnick, D.J. & Pearl, M.C. (1987). Cercopithecines in multimale groups: Genetic diversity and population structure. *Primate Societies* (Ed. by B.B. Smuts, D.L. Cheney, R.M. Seyfarth, R.W. Wrangham & T.T. Struhsaker), pp. 121–34. The University of Chicago Press, Chicago, Illinois.

Missakian, E.A. (1972). Genealogical and cross-genealogical dominance relations in a group of free-ranging rhesus monkeys (*Macaca mulatta*) on Cayo Santiago. *Behaviour*, 45, 224–40.

Moore, J. (1978). Dominance relations among free-ranging female baboons in Gombe National Park, Tanzania. *Recent Advances in Primatology, Vol. 1* (Ed. by D.J. Chivers & J. Herber), pp. 67–70. Academic Press, London.

Parker, G.A. (1974). Assessment strategy and the evolution of fighting behaviour. *Journal of Theoretical Biology*, 47, 223–43.

Paul, A. & Kuester, J. (1987). Dominance, kinship and reproductive value in female Barbary macaques (*Macaca sylvanus*) at Affenberg Salem. *Behavioural Ecology and Sociobiology*, 21, 323–31.

Rawlins, R.G. & Kessler, M.J. (1986). *The Cayo Santiago macaques: History, Behaviour and Biology*. State University of New York Press, Albany.

Sade, D.S. (1967). Determinants of dominance in a group of free-ranging rhesus monkeys. *Social Communication Among Primates* (Ed. by S.A. Altmann), pp. 99–114. University of Chicago Press, Chicago, Illinois.

Sade, D.S. (1972). A longitudinal study of the social behaviour of rhesus monkeys. *The Functional and Evolutionary Biology of Primates* (Ed. by R.H. Tuttle), pp. 378–98. Aldine, Chicago, Illinois.

Sade, D.S., Cushing, K., Cushing, P., Dunaif, J., Figueroa, A., Kaplan, J.R., Lauer, C., Rhodes, D. & Schneider, J. (1976). Population dynamics in relation to social structure on Cayo Santiago. *Yearbook of Physical Anthropology*, 20, 253–62.

Samuels, A. & Henrickson, R.V. (1983). Outbreak of severe aggression in captive *Macaca mulatta*. *American Journal of Primatology*, 5, 277–81.

Samuels, A., Silk, J.B. & Altmann, J. (1987). Continuity and change in dominance relations among female baboons. *Animal Behaviour*, 35, 785–93.

Sigg, H., Stolba, A., Abegglen, J.J. & Dasser, V. (1982). Life history of hamadryas baboons: physical development, infant mortality, reproductive parameters and family relationships. *Primates*, 23, 473–97.

Silk, J.B. (1987). Social behaviour in evolutionary perspective. *Primate Societies* (Ed. by B.B. Smuts, D.L. Cheney, R.M. Seyfarth, R.W. Wrangham & T.T. Struhsaker), pp. 318–29. University of Chicago Press, Chicago, Illinois.

Silk, J.B., Clark-Wheatley, C.B., Rodman, P.S. & Samuels, A. (1981a). Differential reproductive success and facultative adjustment of sex ratios among captive female bonnet macaques. *Animal Behaviour*, 29, 1106–20.

Silk, J.B., Samuels, A. & Rodman, P.S. (1981b). Hierarchical organisation of female *Macaca radiata*. *Primates*, 22, 84–95.

Varley, M. & Symmes, D. (1966). The hierarchy of dominance in a group of macaques. *Behaviour*, 27, 54–75.

Waal, F.B.M de (1987). Dynamics of social relationships. *Primate Societies* (Ed. by B.B. Smuts, D.L. Cheney, R.M. Seyfarth, R.W. Wrangham & T.T. Struhsaker), pp. 421–30. University of Chicago Press, Chicago, Illinois.

Walters, J.R. (1980). Interventions and the development of dominance relationships in female baboons. *Folia Primatologica*, 34, 61–89.

Wasser, S.K. (1983). Reproductive competition and cooperation among female yellow baboons. *Social behaviour of female vertebrates* (Ed. by S.K. Wasser), pp. 350–90. Academic Press, New York.

White, L.E. (1977). The nature of social play and its role in the development of the rhesus monkey. Ph.D. thesis, University of Cambridge.

Whitten, P.L. (1983). Diet and dominance among female vervet monkeys (*Cercopithecus aethiops*). *American Journal of Primatology*, **5**, 139–59.

Wrangham, R.W. (1980). An ecological model of female-bonded primate groups. *Behaviour*, **75**, 262–300.

Wrangham, R.W. (1981). Drinking competition in vervet monkeys. *Animal Behaviour*, **29**, 904–10.

Social suppression of reproduction in primates

D.H. ABBOTT

MRC/AFRC Comparative Physiology Research Group, Institute of Zoology,
Regent's Park, London NW1 4RY, UK

SUMMARY

1 Social contraception can be a major consequence of sociality in primates. It is more prevalent among females, where by various behavioural and physiological means, social dominants raise more offspring than their social subordinates.

2 Social suppression of reproduction is taken to its extreme in marmoset and tamarin monkeys where only the dominant (Rank 1) female breeds in each group. In marmoset monkeys, only dominant females receive ejaculatory mounts from males and ovulation is suppressed in all subordinate females.

3 This social suppression of fertility is rapidly removed when either the dominant female dies or subordinate females are housed away from their dominant female. Infertility is rapidly reinstated when subordinate social status is reimposed on isolated females.

4 Suppression of ovulation is due to inadequate gonadotrophic stimulation from the anterior pituitary gland which in turn is due to inhibited secretion of gonadotrophin releasing hormone (GnRH) from the hypothalamus in the brain.

5 High blood levels of cortisol or prolactin or low body weight are not responsible for the ovarian failure. However, increased negative feedback sensitivity to oestradiol and increased endogenous opioid inhibition of GnRH may explain the suppressed gonadotrophin and GnRH secretion.

6 The subordinate female marmoset monkey may not only serve as a good example of the physiological impairments imposed during social suppression of reproduction in a female primate, it may also help in our understanding of stress-related infertility in women.

INTRODUCTION

In primates, reproduction is suppressed by a myriad of causes. In seasonally breeding species, fertile matings and births are either restricted to certain times of the year or they noticeable peak at such times (Michael & Zumpe 1976; Van Horn 1980). Outside the annual breeding and birthing seasons, the occurrence of fertile matings and births is absent or reduced.

Insufficient availability of food also reduces the reproductive output of female primates by extending the contraceptive effects of lactation. This then increases the female's inter-birth intervals and reduces the number of offspring born to females over a given time (Lee 1987).

Lactation, itself, is a well-known reproductive suppressant in many primate species and in women, and its ability to inhibit ovulation has been elegantly demonstrated by McNeilly, Glasier & Howie (1985). Low social status may transcend all of the above limitations on primate reproduction and impose a further and separate contraceptive effect.

Suppression of reproduction in subordinate males

In laboratory studies of socially-living male primates, the frequency of copulations with females is significantly reduced in subordinate males because of aggressive interactions with dominant males, e.g. the talapoin monkey, *Miopithecus talapoin* (Keverne, Meller & Martinez-Arias 1978; Keverne *et al.* 1984), the squirrel monkey, *Saimiri sciureus* (Coe & Levine 1981) and the marmoset monkey, *Callithrix jacchus* (Abbott & Hearn 1978). This suppression of sexual behaviour is well ingrained into the behaviour of subordinate male talapoin monkeys because even if all other males are removed from their social group, subordinates do not perform ejaculatory mounts with females for several weeks (Keverne 1983).

The male lesser mouse lemur (*Microcebus murinus*) takes this suppression of subordinate male reproduction one stage further. Not only is the sexual behaviour of subordinate males inhibited, but dominant males also employ a volatile urinary pheromone to inhibit the rise of blood testosterone levels in subordinate males during the breeding season (Schilling & Perret 1987). The inhibition of testosterone secretion probably reflects the inhibition of gonadotrophin releasing hormone (GnRH) from the hypothalamus in the brain of subordinate males, and the inhibition of GnRH is apparently due to elevated circulating levels of prolactin, a hormone known to suppress the reproductive system in primates (Schilling & Perret 1987). This pheromonally-induced physiological suppression is induced in isolated subordinate males by pumping in the scent of dominant males' urine into their cages. Furthermore, males only demonstrated this inhibition of testosterone secretion to the scent of dominant male's urine and not to that of subordinates (Schilling, Perret & Predine 1984).

In these examples from studies of captive primate social groups, the likelihood of subordinate males fathering offspring is reduced by behavioural or pheromonal inhibition from dominant males.

In field studies of male primates, a male's reproductive success is usually assessed from his mating success with females because the mating systems of highly social primates commonly preclude a determination of paternity. However, field studies, which correlated a positive relationship between male social rank and male mating success (e.g. baboons, *Papio cynocephalus*: Hausfater 1975; *P. anubis*: Packer 1979; Sapolsky 1982; Macaques, *Macaca mulatta*: MacMillan 1982; *M. fuscata*: Takahata 1982) have been matched by field studies which clearly demonstrated no such relationship (e.g. baboons, *P. anubis*: Harding, 1980; Scott 1984; Berkovitch 1986; *P. ursinus*: Saayman, 1971; chimpanzee, *Pan troglodytes*: Tutin 1979). Dewsbury (1982), MacMillan (1982) and Berkovitch (1986) have all suggested a possible explanation for this major contradiction in field study findings. Many of the studies which found positive relationships between male social rank and male mating success included immature males along with adult males in their analyses. Those studies which found no relationship only included adult males in their analyses. At least in male baboons, Berkovitch (1986) does not hesitate to eliminate immature males from any analysis of the effects of social rank on male mating success because immature males contributed so little to the number of ejaculations with females around the likely days of conception: the adult males alone showed no rank-related mating success. MacMillan (1982), in a study of rhesus macaques, similarly excluded immature males and found no correlation between male social rank and mating success.

However, simple associations between male social rank (based on approach-avoidance and supplant interactions among males) and ejaculation frequency with females during their most probable days of conception may prove difficult to attain because of the complex social and sexual systems of primate societies. Male primates use a large array of tactics to gain access to sexually receptive females (Packer 1979; Tutin 1979, Strum 1982). In baboons, high male rank may aid access to females close to ovulation (Berkovitch 1986), but male baboons also experience frequent changes in social status (Hausfater 1975; Berkovitch 1986). Perhaps as our understanding of the subtleties and complexities of male social rank and mating tactics improves, a better evaluation of male social status and mating success in free-living primates will be possible. DeVore (1965) still summarizes this problem well: 'A male's dominance rank, as revealed by his access to incentives such as oestrous females, depends not so much on his status as an individual as upon his participation in a mutually supportive group of males — the central hierarchy ... the present evidence suggests that the same males which are most active in protecting the group and policing it internally are also the most effective breeders, despite the fact that they may be subordinate as individuals to other males in the troop.'

Suppression of reproduction in subordinate females

The suppression of reproduction in subordinate female primates is more striking because, unlike subordinate males, subordinate females can experience complete infertility (e.g. Abbott 1986; Abbott et al. 1986). The impaired reproductive ability of subordinate females is manifest in many ways in captive and free-living primates. Subordinates have been shown to have deficient physiological responses (Keverne, Eberhart & Meller 1982), delayed puberty (Epple & Katz 1980), inhibition of ovulation (Abbott et al. 1981) delayed first conception (Meikle, Tilford & Vessey 1984; Wilson, Walker & Gordon 1983), reduced numbers of offspring born (Harcourt 1987) and increased infant mortality (Wilson, Walker & Gordon 1983). However, this emphasis on the association between low female social status and reproductive failure has its critics who question (i) whether social dominance induced reproductive suppression occurs in wild populations of primates and (ii) whether any of the above-reproductive failures have detrimental effects on the overall reproductive success of subordinate females during their entire lifespan (Gouzoules, Gouzoules & Fedigan 1982; Fedigan 1983; Gray 1985). Certainly, this discontent has justifiably highlighted the quantitative inadequacies of some studies (see, for example, Drickamer 1974) and the overemphasis subsequently given to such findings by other authors. On the other hand, where quantifiable evidence exists, low social status has been shown to lead to reduced reproductive output (Harcourt 1987). Critical examination of the link between reproductive suppression and low social status is important because this link has been repeatedly demonstrated in laboratory populations of primates (e.g. marmoset: Abbott 1984; talapoin: Keverne *et al.* 1984), but has proved difficult to substantiate to the same degree in wild populations (Harcourt 1987), in an analogous fashion to the problem in subordinate male primates.

The Callitrichid monkeys provide a good opportunity to carry out a critical examination of the link between subordinate social status in females and the suppression of reproduction. In both wild groups (Garber, Moya & Malaga 1984; Stevenson & Rylands in press; Terborgh & Goldizen 1985) and captive groups (Abbott 1984; Epple & Katz 1984; French, Abbott & Snowdon 1984) of Callitrichid monkeys, only one female produces offspring. Only in captive groups has the breeding female been identified as the socially dominant female from quantitative behavioural parameters (Abbott & Hearn 1978). This failure by all but one female to breed is in contrast to many other primates, where social suppression of reproduction is a matter of degree and increases in intensity with decreasing social status: it is not complete (see, for example, Keverne *et al.* 1984) and not all subordinate females are rendered infertile (Harcourt 1987). Consequently, the extreme nature of social contraception in subordinate female marmoset monkeys (*Callithrix jacchus*) makes this primate an excellent candidate in which to examine social suppression of reproduction.

Social suppression of reproduction in subordinate female marmoset monkeys

To cover in detail social suppression of reproduction in male and female primates is too extensive an undertaking for this paper. Because of the more clear-cut associations between low social status and reproductive failure in female primates and the good example provided by subordinate female marmoset monkeys, this paper will concentrate specifically on the social suppression of reproduction in subordinate female marmoset monkeys. It will describe work on the behavioural and endocrinological failures which contribute to the suppression of reproduction in captive subordinate female marmoset monkeys. It will not deal with (i) the ecological constraints which may have an important bearing on the development of social suppression of reproduction or (ii) the competitive ability required to attain and maintain social dominance status, as both topics are addressed generally by a number of authors in this volume.

METHODS

Subordinate females in social groups

The monkeys in this 2-year study were all captive-born and unrelated and were housed in the Institute of Zoology under the conditions described by Hearn (1983). A total of sixteen social groups were used, but only a maximum of eight groups were maintained at any one time. The turnover in the number of social groups was due to colony management practices (five groups) and to the instability of some groups beyond 6–10 months from group formation (three groups). The social groups were put together from offspring in family groups or members of separate male–female pairs when the animals were between 15 and 26 months old (post-pubertal–adult: Abbott & Hearn 1978).

At the formation of each group, usually four males and four females were placed into either (i) an observation room (2·9 x 2·2 x 1·7 m) fitted with tree branches and a nest box, or (ii) an exercise cage (2 x 1 x 2 m) connected by flexible ducting to a home cage of 100 x 50 x 75 cm (Hearn et al. 1975). Each group remained in their initial housing for 2–4 weeks before they were permanently housed in a home cage with access to an exercise cage for 1 day per week.

The development of this social group system was necessary in order to systematically study the social suppression of reproduction in subordinate female marmoset monkeys. In this way, the social group composition and size was standardized and subordinate females were anovulatory. In family groups of marmosets, which is the usual way of keeping these monkeys in captivity (Hearn et al. 1975), (i) the ages of daughters are highly variable, (ii) pubertal and post-pubertal daughters are commonly ejected from their families by their mother or their sisters, and (iii) not all daughters are anovulatory (Abbott 1984).

Eight of the sixteen social groups were observed initially during the first 3 days following group formation and then later between 2–5 and 6–9 months after group formation (Abbott 1986). To aid the identification of individuals, each monkey's large white ear tufts were distinctly coloured (Abbott & Hearn 1978). There was no behavioural response to this procedure. The social dominance hierarchy of each group was formed within 2–3 days of group formation. Dominant (Rank 1) or subordinate (Ranks 2 and below) social status of each monkey was determined from the quantitative analysis of behavioural recordings of aggressive and submissive interactions between group members (Abbott & Hearn 1978; Abbott 1984). Dominants directed aggression at other animals and received little aggressive and mostly submissive behaviour in return. When the converse was observed, animals were designated as subordinates. In marmoset monkeys, each sex forms a discrete hierarchy (Abbott 1979). A marmoset's rank within its own sex or intra-sexual hierarchy is a vital determinant of its reproductive performance (Abbott & Hearn 1978; Abbott 1978, 1979, 1984). A marmoset's rank in its group overall, the inter-sexual hierarchy, is not so important.

By 2 weeks after group formation, one or two subordinate females and one or two subordinate males were usually removed from each social group because of excessive aggression from an animal of higher rank and the same sex (as previously found by Abbott & Hearn 1978). Thus, as far as subordinate females are concerned, each group contained at least one, but not more than three anovulatory subordinate female(s) beyond this initial 2-week period. In the wild, groups of marmoset and tamarin monkeys usually contain between three and fifteen animals (Stevenson & Rylands, in press). Thus, the total of four to seven animals in our captive groups was well within the normal range.

The repeated observational checks made, 2–9 months after group formation, on the eight groups initially observed consistently confirmed the original assignment of dominant or subordinate social status. During both 3-month blocks of these later observations, the males and females of the eight social groups were housed as single sex groups. They were only re-united for each 100 minute observation period which took place one to two times per week up to a total of 20 hours per group per 3-month block (Abbott 1986). The sexual behaviour of the males and females in the eight groups under observation was only recorded in the observations made between 2 and 9 months after the groups were formed. The behavioural definitions used have been reported in detail elsewhere (Abbott 1984, 1986; Kendrick & Dixson 1984).

Ovarian function and the collection of blood samples

The important hormones involved in controlling ovarian function in female primates are illustrated in Fig. 1. In the female marmoset monkeys used in this study, ovarian function was monitored by the measurement of circulating progesterone concentrations in blood samples collected twice weekly. Blood samples were withdrawn from the femoral vein in unanaesthetized females between 0900 and 1130 h and details of the procedure and subsequent processing of the plasma are described by Abbott et al. (1981). Females were routinely bled throughout the time they spent in social groups. The twice weekly frequency of blood sampling was sufficient to cover the approximately 19–20-day period of plasma progesterone elevation during the luteal phase of the 28-day ovarian cycle (Harlow et al. 1983).

The luteal phase of each ovarian cycle was defined as lasting from the elevation of plasma progesterone concentrations above 10 ng/ml until concentrations decreased again below that level (Harlow et al. 1983). The day of ovulation was designated as the day before plasma progesterone exceeded 10 ng/ml. Harlow et al. (1983) have shown that this 10 ng/ml value represents the earliest measurable increase in progesterone which is consistently associated with a luteal phase. However, as a twice weekly blood sampling schedule was employed in this study, the estimation of the day of ovulation is only accurate to plus or minus a day.

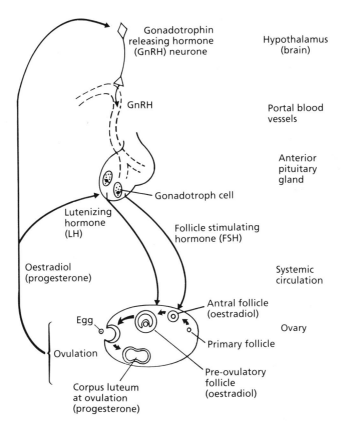

FIG. 1. Diagrammatic representation of the important hormones involved in regulating ovulation in female primates. GnRH is secreted from the terminals of neurones in the hypothalamus into portal blood vessels draining into the anterior pituitary gland. In the anterior pituitary gland, GnRH stimulates the synthesis and release of LH and FSH from gonadotroph cells into the blood stream. LH and predominantly FSH stimulate follicular development in the ovaries. Oestradiol, and to a lesser extent progesterone, secreted from the developing follicles, exert a negative feedback inhibitory control on further secretion of GnRH, LH and FSH. This negative feedback control loop is overridden by the large amounts of oestradiol secreted from the pre-ovulatory follicles which initiate a massive discharge of LH, and to a lesser extent FSH (and possible GnRH) which triggers ovulation. This stimulatory effect of oestradiol is known as positive feedback. The follicular tissue which remains at the ovulation site is transformed into a corpus luteum which secretes large quantities of progesterone, important for implantation and maintaining early pregnancy. If no pregnancy occurs, the corpus luteum degenerates at the end of each cycle.

Ultrasound scanning of ovarian function and pregnancy

Follicular development, the presence of corpora lutea and the detection of early pregnancy in female marmosets were monitored every 1−5 days using a Semens Sonoline SL ultrasound scanner. A 5 MHz probe was used set at a power output rating of 25%. The above reproductive parameters were visualized by this non-invasive method in unanaesthetized monkeys. The four subordinate females receiving GnRH treatment were monitored throughout the treatment period and five untreated subordinates were monitored for 1–2 months as controls.

Ovarian cycle length and termination of pregnancy

Ovulation and ovarian cycle length were controlled artificially in dominant females in fourteen of the sixteen groups. To accomplish this, dominant females were given an injection of a prostaglandin F2α analogue, Cloprostenol (Summers, Wennick & Hodges 1985) 14–20 days after ovulation. The prostaglandin treatment resulted in luteolysis and the termination of the luteal phase of the cycle or early pregnancy (Summers Wennick & Hodges 1985). No prostaglandin treatment was given to the dominant females in the remaining two groups and no other methods for controlling ovarian cycle length or terminating pregnancy were used.

Experimental manipulation of social status

In order to assess the effects of changing social status on ovarian function, four subordinate females, which had been anovulatory for at least 8 months, were removed from their separate social groups, and were housed singly for 36–77 days. After this time, all four females were returned to new social groups where they resumed subordinate social status. Blood samples were collected as before throughout this period, and in addition were collected every 1–2 days during the first 10 days of single housing.

Administration of GnRH

In order to assess the pituitary LH responses to exogenous GnRH stimulation (there is no FSH assay for the marmoset) and whether or not this gonadotrophin activation could induce ovulation, four subordinate females, remaining in their social groups, were infused with 1 µg GnRH either every 75 min ($n=1$) or 105 min ($n=3$) for between 6 and 93 days. GnRH only maintains its stimulatory effect on pituitary gonadotrophin secretion when it is administered in a pulsatile manner (Belchetz *et al.* 1978). GnRH was infused using a miniaturized syringe pump (I.S. Sutherland, G.R. Chambers & D.H. Abbott, unpublished data) housed inside a small backpack worn by each female (Ruiz de Elvira & Abbott 1986). The subcutaneous infusion site was linked to the pump by a small polyvinyl cannula. Induction of LH secretion and ovulation in female subordinates remaining in their groups would directly implicate suppressed hypothalamic secretion of GnRH as the cause of the reproductive failure.

Hormone assays

Progesterone was measured directly in unextracted plasma by radioimmunoassay (Summers, Wennick & Hodges 1985). Bioactive LH was measured using a modified mouse Leydig cell bioassay (Harlow *et al.* 1984; Hodges *et al.* 1987). Both assays had previously been validated for use in the marmoset monkey.

Statistical analysis

Heterogeneity of variance observed in the plasma concentrations of LH reported in Table 1 was reduced by log transformation of the data. A one-way analysis of variance for repeated measures was then carried out using the least mean squares method (Statistical Analysis System, SAS; Helwig & Council 1979). Comparisons of individual means were made *post hoc* using Duncan's Multiple Range Test (SAS).

RESULTS

Social suppression of sexual behaviour

Only dominant females ($n=8$) solicited sexual behaviour from males. Not surprisingly, therefore, dominant females received between 93–100% of all male mounts and 100% of all ejaculations. Apart from two subordinate females in one of the eight groups, which received a total of 7% of all male mounts in that group, dominant and subordinate males paid little sexual attention to subordinate females. The lack of subordinate female sexual behaviour could therefore easily explain the lack of offspring produced by these females. However, the social suppression of reproduction in these females has a deeper underlying cause than behavioural inadequacy.

Social suppression of ovulation

Only dominant females underwent normal ovulatory cycles, as illustrated in Fig. 2 by a typical example from one of the 16 groups. Subordinate females either did

TABLE 1. Plasma luteinizing hormone concentrations (mean ± sem) in dominant (rank 1) and subordinate (ranks 2 and 3) female marmoset monkeys in four social groups. The values cover a 1–5 month blood sampling period, carried out at least 2 months after group formation. The four groups chosen were typical of those containing more than one subordinate female

	Plasma LH (miu/ml)		
Social group	Rank 1	Rank 2	Rank 3
3	$34\cdot1 \pm 10\cdot2(22)^a$	$2\cdot9 \pm 0\cdot6^*(16)$	$2\cdot2 \pm 0\cdot2^*(26)$
6	$19\cdot6 \pm 3\cdot8(38)$	$2\cdot0 \pm 0\cdot0^*(20)$	$3\cdot6 \pm 0\cdot4^*(31)$
8	$29\cdot6 \pm 7\cdot6(9)$	$6\cdot4 \pm 1\cdot3^*(19)$	$2\cdot9 \pm 0\cdot2^*(18)$
14	$46\cdot5 \pm 14\cdot7(40)$	$4\cdot1 \pm 0\cdot8^*(18)$	$4\cdot3 \pm 1\cdot1^*(21)$

a = number of values.
* $P<0\cdot05$ vs. rank 1 (Duncan's multiple range test).

not ovulate (twenty-four out of twenty-eight: two typical examples are shown in Fig. 2) or underwent inadequate ovulatory cycles (one cycle: three females; two cycles: one female), where plasma progesterone concentrations remained above 10 ng/ml for only 1–10 days, in contrast to the 19–20 days of the normal cycle.

However, such social suppression of ovulation only remained in force while subordinate females remained with their dominant females. The following six examples demonstrate the effect of either the death of the dominant female in the group ($n=2$) or the removal of subordinate females from their groups into single housing ($n=4$) on the ovarian activity of subordinate females. The effects of losing the dominant female in a group are shown in Fig. 3. In the first example, when the dominant female died following a post-partum haemorrhage (Fig. 3a), the only subordinate female in the group ovulated 9 days later and conception occurred in this cycle. In the second example, the dominant female died shortly after group formation from a 'wasting syndrome' (Fig. 3b). Following this, the highest ranking of the two subordinates ovulated and conceived after 2–12 days, as indicated by her continually elevated plasma progesterone concentrations following the death of the dominant female (the immediate post-ovulation progester-

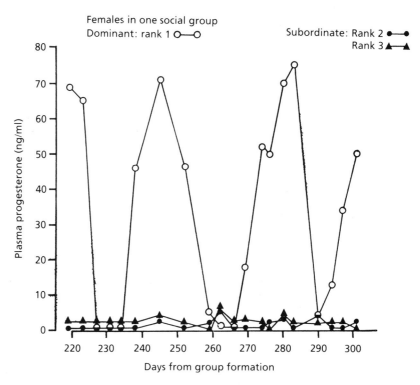

FIG. 2. A typical example of plasma progesterone concentrations in a dominant female and her two subordinates in one of the sixteen social groups. The period of time shown starts at the end of a luteal phase for the dominant female and continues through her next three ovarian cycles, finishing in the mid-luteal phase of the third cycle.

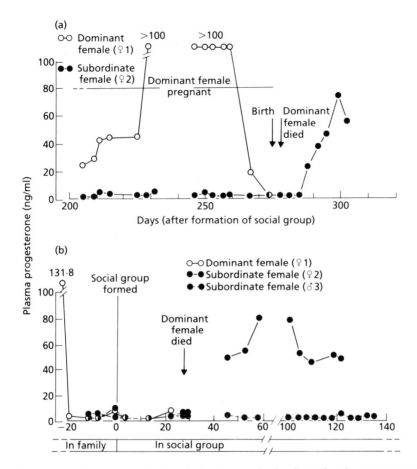

FIG. 3. Plasma progesterone concentrations in dominant and subordinate female marmoset monkeys in (a) a social group where the dominant female died following a post-partum haemorrhage and in (b) a social group where the dominant female died from a 'wasting syndrome' shortly after the group was set up.

one rise was missed because of a temporary interuption of blood sampling). The remaining subordinate did not ovulate. In both cases, the rapid conceptions went to term and normal healthy infants were born and raised. No similar situations have been attempted experimentally by removing healthy dominant females from their groups.

On the four remaining occasions, when subordinate females were removed from their social groups and housed singly, three out of the four females ovulated within 18 days of removal. Figure 4 provides an example from one of those three females. The fourth female ovulated after 46 days. On return to subordinate social status, all four females immediately stopped ovulating, as illustrated in Fig. 4. The female shown in Fig. 4 (224W), also illustrates the premature collapse of

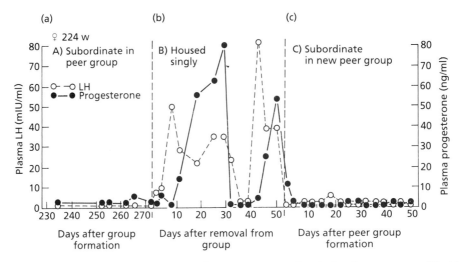

Fig. 4. Plasma progesterone and luteinizing hormone concentrations in female marmoset 224W while (a) subordinate in a social group (established 9 months previously), (b) housed singly (for 53 days), and (c) during and after becoming subordinate in a new social group. (Reproduced with permission from Abbott 1988.)

the luteal phase demonstrated by three of the four females on return to subordinate social status.

Thus, these examples of removing the inhibiting influence of a dominant female from subordinate females clearly implicated dominant females in mediating the social suppression of ovulation observed in subordinates. What was the endocrinological cause of the socially suppressed ovulation?

Social suppression of pituitary LH secretion

One obvious answer to the above question was that reduced gonadotrophic stimulation of the ovaries was responsible for the ovulatory failures in subordinate females. With this in mind, plasma concentrations of bioactive LH were determined in dominant and subordinate females in four of the sixteen groups, to give an indication of whether or not gonadotrophin secretion from the anterior pituitary gland of subordinate females was reduced. As Table 1 indicates, plasma LH concentrations in all female subordinates were significantly lower than those in dominant females ($F(2,9)=34.48$; $P<0.0001$). Such reduced pituitary LH output might well explain the suppressed ovulation in female subordinates. This explanation is further supported by the rapid rise in plasma LH concentrations in the socially isolated subordinate females prior to their first ovulations, as typically illustrated by female 224W in Fig. 4. Plasma LH values frequently remained high in these isolated females, but they fell precipitously following the return of the isolated females to subordinate status in a new group. This socially-induced fall in pituitary LH secretion was probably responsible for the premature termination of the luteal phase of the cycle experienced by three of the four females and is illustrated in Fig. 4.

Social suppression of hypothalamic GnRH secretion

The most logical explanation for inadequate pituitary LH secretion is suppressed GnRH secretion from the hypothalamus. This was investigated by administering GnRH to four subordinate females, remaining in their social groups, by means of a miniaturized syringe pump housed in a backpack jacket. The GnRH treatment lasted for between 6 and 93 days and overcame the social suppression of pituitary LH secretion in all four subordinates. The best example of the magnitude of the LH rise is illustrated by female 315W in Fig. 5.

Furthermore, the GnRH treatment also stimulated ovarian follicular growth in all four treated subordinates, as observed in ultrasound scan images of their ovaries. A typical example of the increased size of follicles observed in the ovaries of subordinate females receiving GnRH treatment is shown in Fig. 6a. It is clear that follicular growth up to the size of pre-ovulatory follicles (2.5 mm and over: O.L. Wilson, unpublished results) was occurring in GnRH-treated subordinates, which is quite different to the smaller follicle sizes normally observed in untreated subordinates (Fig. 6b illustrates the follicular profile of a typical subordinate female). However, only one ovulation and short-term pregnancy was induced in a subordinate female receiving GnRH treatment (315W illustrated in Fig. 5), as confirmed by prolonged elevation of plasma progesterone concentrations and ultrasound scan images of early pregnancy (Abbott 1987). It is hoped that a more reliable induction of ovulation and pregnancy in subordinate female marmosets will occur when the GnRH administration is optimized. Nonetheless, this evidence strongly suggests that social suppression of GnRH secretion is responsible for the reproductive failures of subordinate female marmosets.

DISCUSSION

In the extreme form of social suppression of reproduction which occurs in subordinate female marmoset monkeys, both behavioural and endocrinological mechanisms operate to impose infertility. The behavioural mechanisms result in sexual inadequacy. The endocrinological mechanisms result in the suppression of ovulation. This paper has clearly shown that the suppression of ovulation is due to suppressed pituitary LH secretion and that the suppression of LH secretion is due to suppressed hypothalamic GnRH secretion. The infertility state experienced by subordinate female marmosets is rapidly reversed and re-imposed because the infertility is entirely dependent on subordinate social status.

What is the physiological cause of this shut-down of the reproductive system of subordinate female marmosets? It has proved easier to discount various factors which might be the cause than it has been to identify which factors are actually involved. For example, high plasma concentrations of cortisol or prolactin, or low

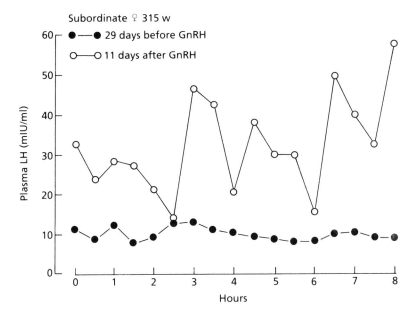

FIG. 5. Plasma luteinizing hormone concentrations in subordinate female marmoset 315W, 29 days before gonadotrophin releasing hormone (GnRH) treatment and 11 days after the start of GnRH treatment.

body weight have all been eliminated as potential reproductive suppressants (Dubey & Plant 1985; Bowman, Dilley & Keverne 1978; Yen 1986) because dominant and subordinate females were found to be identical in all three respects (Abbott *et al.* 1981; Abbott 1986).

However, recent work has succeeded in demonstrating that there may be two physiological causes of reproductive suppression in subordinate female marmosets. One manifests as an increased negative feedback effect of oestradiol, secreted from the ovaries, on the secretion of LH from the pituitary (see Fig. 1). It seems that even the very small amounts of oestradiol secreted from the anovulatory ovaries of subordinate female marmosets have a large inhibitory influence on subordinates' LH secretion (Abbott 1988). This was suggested by the fact that (i) the LH response of subordinate females to ovariectomy was delayed and reduced and that (ii) low-dose oestradiol implants succeeded in suppressing plasma LH concentrations in ovariectomized subordinates but not in ovariectomized dominants.

The other physiological inhibitor seems to be independent of ovarian oestradiol negative feedback. This was suggested by the fact that (i) plasma LH concentrations in long-term ovariectomized subordinate females are significantly lower than in long-term ovariectomized dominant females (Abbott 1988) (ii) this LH deficit in ovariectomized subordinates can be removed by housing them singly away from their dominant female and can be re-imposed by placing these singly

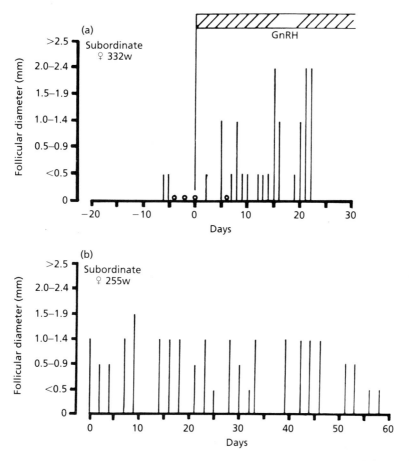

Fig. 6. Maximum follicular diameter measured from the ultrasound scan images of the ovaries of (a) subordinate female 332W before and during the first 3 weeks of gonadotrophin releasing hormone treatment and (b) subordinate female 255W receiving no treatment. o = no follicles discernable in either ovary.

housed females back with their dominant female in their original groups (Abbott 1988), and that (iii) ovariectomized subordinate females demonstrated an LH rise following the administration of the opiate antagonist, naloxone, whereas ovariectomized dominants and intact subordinates did not (Abbott et al. 1987). There may therefore be a brain opioid-mediated suppression of GnRH and LH secretion in subordinate females separate from the increased oestradiol negative feedback suppression mechanism.

However, this remains to be fully confirmed. Certainly in women and in subordinate male talapoin monkeys, endogenous opioid peptides have been implicated in the suppression of GnRH and LH secretion (Martensz et al. 1986; Yen 1986). Our continuing research work is focusing on the precise physiological causes of reproductive suppression in subordinate female marmoset monkeys.

The external cue for the complete suppression of reproduction in subordinate female marmosets may be pheromones present in the urine, the ano-genital scent glands, or both, of dominant females. Male lesser mouse lemurs (*Microcebus murinus*) are reproductively suppressed merely by the scent from the urine of dominant males which they have never met (Schilling & Perret 1987; Schilling, Perret & Predine 1984). Rylands (1985) has suggested that the scent marking behaviour of breeding female marmosets over the gouged feeding holes on their group's feeding trees may serve to transmit a contraceptive-like pheromone. Epple & Katz (1984) have shown that scent transfer from a dominant female saddle-back tamarin (*Saguinus fuscicollis*) and her group to a pair-housed subordinate female (originally from that group) and her male mate, disrupted ovarian function in the separately housed female. However, while these results nicely link pheromones with reproductive suppression in subordinate female Callitrichid monkeys, the definitive experiments have still to be carried out.

Turning to other primate species, suppression or delay of ovulation in subordinates following behavioural harassment from dominants has been blamed for the increased time taken to conceive in subordinate female talapoin monkeys (Abbott *et al.* 1986) and subordinate female gelada baboons (*Theropithecus gelada*: Dunbar 1980) and for the increased menstrual cycle length in subordinate female olive baboons (*Papio anubis*: Rowell 1970). Impaired ovarian function may underlie other associations between low social status and reduced reproductive function, especially in instances where females' first conception is delayed (e.g. rhesus macaques: Meikle *et al.* 1984) or where reduced numbers of offspring are born (e.g. several species: Harcourt 1987).

The extreme type of socially-induced female infertility found in marmoset monkeys is also found in other mammals, such as the silver-backed jackal (*Canis mesomelas*: Moehlman 1983), dwarf mongoose (*Helogale parvula*: Rood 1980), prairie vole (*Microtus ochrogaster*: Carter, Getz & Cohen-Parsons 1986), and naked mole-rat (*Heterocephalus glaber*: Jarvis 1981; Brett 1986). Certainly in the case of the jackal, the development of this suppression of female reproduction may be due to two factors: (i) daughters benefiting from delaying their departure from their natal group because of the difficult problems of establishing a separate breeding group and territory from that of their natal group, and (ii) breeding females requiring help from non-reproductive animals to rear their young in order to maximize the chances of the young surviving (this topic is dealt with in detail by MacDonald in this volume). Whether these are the common causes for the development of extreme reproductive suppression in subordinate female mammals remains to be seen.

The socially-induced GnRH-related infertility in subordinate female marmoset monkeys may also help in our understanding of the causes of GnRH-related infertility in women which are linked with psychological stress (Yen 1986). Stress-related infertility in women is becoming increasingly common (Dunbar 1985). The physiological disturbances underlying the infertility require further investigation. In the marmoset monkey, the reliable suppression of ovulation due to the stress of social subordination may aid studies into the neuroendocrine mechanisms which translate the 'stress state' into suppressed GnRH secretion and infertility.

ACKNOWLEDGEMENTS

The author would like to thank his colleagues, including J.K. Hodges, L.M. George, O.L. Wilson, M.C. Ruiz de Elvira, S.F. Ferrari and B.R. Ferreira for their collaboration in some of the work reported here; I.S. Sutherland and G.R. Chambers (National Institute of Medical Research, Mill Hill, London) for designing and building the miniaturized syringe pump; M.J. Llovett and the animal technical staff for their care and maintenance of the animals; T.J. Dennett and M.J. Walton for preparation of the illustrations and T. Grose for typing the manuscript. The manuscript benefited from the criticism of G.R. Smith, P.M. Summers, G.H. Manley and an anonymous referee. This work was supported by grants from the Association for the Study of Animal Behaviour, the Nuffield Foundation and an MRC/AFRC Programme Grant to Professor J.P. Hearn.

REFERENCES

Abbott, D.H. (1978). The physical, hormonal and behavioural development of the common marmoset, *Callithrix jacchus jacchus*. *Biology and Behaviour of Marmosets* (Ed. by H. Rothe, H.J. Wolters & J.P. Hearn), pp. 99–106. Eigenverlag Hartmut Rothe, Göttingen.

Abbott, D.H. (1979). *The sexual development of the common marmoset monkey, Callithrix jacchus jacchus*. Ph.D. thesis, University of Edinburgh.

Abbott, D.H. (1984). Behavioural and physiological suppression of fertility in subordinate marmoset monkeys. *American Journal of Primatology*, **6**, 169–86.

Abbott, D.H. (1986). Social suppression of reproduction in subordinate marmoset monkeys (*Callithrix jacchus jacchus*). *A Primatologia No Brasil*, Vol. 2, (Ed. by M.T. De Mello), pp. 1–16. Sociedade Brasileira de Primatologia, Brasilia.

Abbott, D.H. (1987). Behaviourally mediated suppression of reproduction in female primates. *Journal of Zoology, London*, **213**, 455–70.

Abbott, D.H. (1988). Natural suppression of fertility. *Reproduction and Disease in Captive and Wild Animals* (Ed. by J.P. Hearn & G.R. Smith), Symposia of the Zoological Society of London, 60, pp. 7–28. Oxford University Press, Oxford.

Abbott D.H. & Hearn, J.P. (1978). Physical, hormonal and behavioural aspects of sexual development in the marmoset monkey, *Callithrix jacchus*. *Journal of Reproduction and Fertility*, **53**, 155–66.

Abbott, D.H., Keverne, E.B., Moore, G.F. & Yodyinguad, U. (1986). Social suppression of reproduction in subordinate talapoin monkeys, *Miopithecus talapoin*. *Primate Ontogeny, Cognition and Social Behaviour* (Ed. by J.G. Else & P.C. Lee), pp. 329–41. Proceedings from the 10th Congress of the International Primatological Society, 3. Cambridge University Press, Cambridge.

Abbott, D.H., McNeilly, A.S., Lunn, S.F., Hulme, M.J. & Burden, F.J. (1981). Inhibition of ovarian function in subordinate female marmoset monkeys (*Callithrix jacchus jacchus*). *Journal of Reproduction and Fertility*, **63**, 335–45.

Abbott, D.H., O'Byrne, K.T., George, L.M., Lunn, S.F. & Dixson, A.F. (1987). *Naloxone-Induced LH Secretion and the Suppression of Fertility in Subordinate Female Marmoset Monkeys*. Proceedings of the International Conference on Brain Opioid Systems in Reproduction, Cambridge, 1987. Abstr. C34.

Belchetz, P.E., Plant, T.M., Nakai, Y., Keogh, E.J. & Knobil, E. (1978). Hypophysial responses to continuous and intermittent delivery of hypothalamic gonadotrophin-releasing hormone. *Science*, **202**, 631–3.

Berkovitch, F.B. (1986). Male rank and reproductive activity in savanna baboons. *International Journal of Primatology*, **7**, 533−50.

Bowman, L.A., Dilley, S.R. & Keverne, E.B. (1978). Suppression of oestrogen-induced LH surges by social subordination in talapoin monkeys. *Nature (London)*, **275**, 56−8.

Brett, R.A. (1986). *The ecology and behaviour of the naked mole-rat (Heterocephalus glaber Ruppell) (Rodentia: Bathyergidae)*. Ph.D. thesis, University of London.

Carter, C.S., Getz, L.L. & Cohen-Parsons, M. (1986). Relationships between social organisation and behavioural endocrinology in a monogamous mammal. *Advances in the Study of Behaviour*, **16**, 109−45.

Coe, C.L. & Levine, S. (1981). Psychoneuroendocrine relationships underlying reproductive behaviour in the squirrel monkey. *International Journal of Mental Health*, **10**, 22−42.

Devore, I. (1965). Male dominance and mating behaviour in baboons. *Sex and Behaviour* (Ed. by F.A. Beach), pp. 266−89. Wiley, New York.

Dewsbury, D.A. (1982). Dominance rank, copulatory behaviour and differential reproduction. *Quarterly Review of Biology*, **57**, 135−9.

Drickamer, L.C. (1974). A ten-year summary of reproductive data for free-ranging *Macacca mulatta*. *Folia Primatol*, **21**, 61−80.

Dubey, A.K. & Plant, T.M. (1985). A suppression of gonadotrophin secretion by cortisol in castrated male rhesus monkeys *(Macaca mulatta)* mediated by the interuption of hypothalamic gonadotrophin releasing hormone release. *Biology of Reproduction*, **33**, 423−31.

Dunbar, R.I.M. (1980). Determinants and evolutionary consequences of dominance among female Gelada baboons. *Behavioural Ecology and Sociobiology*, **7**, 253−65.

Dunbar, R.I.M. (1985). Stress is a good contraceptive. *New Scientist*, **1439**, 16−18.

Epple, G. & Katz, Y. (1980). Social influences on first reproductive success and related behaviours in the saddle-back tamarin (*Saguinus fuscicollis*, Callitrichidae). *International Journal of Primatology*, **1**, 171−83.

Epple, G. & Katz, Y. (1984). Social influences on oestrogen excretion and ovarian cyclicity in saddle back tamarins *(Saguinus fuscicollis)*. *American Journal of Primatology*, **6**, 215−27.

Fedigan, L.M. (1983). Dominance and reproducive success in primates. *Yearbook of Physical Antropology*, **26**, 91−129.

French, J.A., Abbott, D.H. & Snowdon, C.S. (1984). The effect of social environment on oestrogen excretion, scent marking and sociosexual behaviour in tamarins *(Saguinus oedipus)*. *American Journal of Primatology*, **6**, 155−67.

Garber, P.A., Moya, L. & Malaga, C. (1984). A preliminary field study of the moustached tamarin monkey *(Saguinus mystax)* in North Eastern Peru: questions concerned with the evolution of a communal breeding system. *Folia Primatologica*, **42**, 17−32.

Gouzoules, H., Gouzoules, S. & Fedigan, L. (1982). Behavioural dominance and reproductive success in female Japanese monkeys. *Animal Behaviour*, **30**, 1138−50.

Gray, J.P. (1985). *Primate Sociobiology*. HRAF Press, New Haven, Connecticut.

Harcourt, A.H. (1987). Dominance and fertility among female primates. *Journal of Zoology, London*, **213**, 471−87.

Harding, R.S.O. (1980). Agonism, ranking and the social behaviour of adult male baboons. *American Journal of Physical Anthropology*, **53**, 203−16.

Harlow, C.R., Gems, S., Hodges, J.K., & Hearn, J.P. (1983). The relationship between plasma progesterone and the timing of ovulation and early embryonic development in the marmoset monkey *(Callithrix jacchus)*. *Journal of Zoology, London*, **201**, 273−82.

Harlow, C.R.; Hearn, J.P. & Hodges, J.K. (1984). Ovulation in the marmoset monkey: endocrinology, prediction and detection. *Journal of Endocrinology*, **103**, 17−24.

Hausfater, G. (1975). Dominance and reproduction in baboons *(Papio cynocephalus)*. *Contributions to Primatology*, **7**, 1−150.

Hearn, J.P. (1983). The common marmoset *(Callithrix jacchus)*. *Reproduction in New World Primates* (Ed. by J.P. Hearn), pp. 181−215. MTP Press Ltd., Lancaster, England.

Hearn, J.P., Lunn, S.F., Burden, F.J. & Pilcher, M.M. (1975). Management of marmosets for biomedical research. *Laboratory Animals*, **9**, 125−34.

Helwig, J.T. & Council, K.A. (1979). *SAS User's Guide, 1979 edn.* SAS Institute Inc., Cary, North Carolina.

Hodges, J.K., Cottingham, P.G., Summers, P.M. & Yingnan, L. (1987). Controlled ovulation in the marmoset monkey *(Callithrix jacchus)* with human chorionic gonadotrophin following prostaglandin-induced luteal regression. *Fertility and Sterility*, **48**, 299–305.

Jarvis, J.U.M. (1981). Eusociality in a mammal: Co-operative breeding in naked mole-rat colonies. *Science*, **212**, 571–3.

Kendrick, K.M. & Dixson, A.F. (1984). A quantitative description of copulatory and associated behaviour of captive marmosets *(Callithrix jacchus)*. *International Journal of Primatology*, **5**, 199–212.

Keverne, E.B. (1983). Endocrine determinants and constraints on sexual behaviour in monkeys. *Mate Choice* (Ed. by P.P.G. Bateson), pp. 407–20. Cambridge University Press, Cambridge.

Keverne, E.B., Eberhart, J.A., & Meller, R.E. (1982). Dominance and subordination: concepts or physiological states? *Advanced Views in Primate Biology*. (Ed. by O. Chiarelli, pp. 81–94. Springer-Verlag, Berlin.

Keverne, E.B., Eberhart, J.A., Yodyinguad, U. & Abbott, D.H. (1984). Social influences on sex differences in the behaviour and endocrine state of talapoin monkeys. *Progress in Brain Research*, **61**, 331–47.

Keverne, E.B., Meller, R.E. & Martinez-Arias A.M. (1978). Dominance, aggression and sexual behaviour in social groups of talapoin monkeys. *Recent Advances in Primatology, Vol. 1* (Ed. by D.J. Chivers & J. Herbert), pp. 533–47. Academic Press, London.

Lee, P.C. (1987). Nutrition, fertility and maternal investment in primates. *Journal of Zoology, London*, **213**, 409–422.

Martensz, N.D., Vellucci, S.U., Herbert, J. & Keverne, E.B. (1986). B-endorphin in the cerebrospinal fluid of male talapoin monkeys related to dominance status in social groups. *Neuroscience*, **18**, 651–8.

McMillan, C.A. (1982). Male age and mating success among rhesus macaques. *International Journal of Primatology*, **3**, 312.

McNeilly, A.S., Glasier, A. & Howie, P.W. (1985). Endocrine control of lactational infertility. I. *Maternal Nutrition and Lactational Infertility* (Ed. by J. Dobbing), pp. 1–16. Nestle Nutrition, Vevey/Raven Press, New York.

Meikle, D.B., Tilford, B.L. & Vessey, S.H. (1984). Dominance rank, secondary sex ratio and reproduction of offspring in polygynous primates. *American Naturalist*, **124**, 173–88.

Michael, R.P. & Zumpe, D. (1976). Environmental and endocrine factors influencing annual changes in sexual potency in primates. *Psychoneuroendocrinology*, **1**, 303–13.

Moehlman, P.D. (1983). Socioecology of silver-backed and golden jackals, *Canis mesmomelas* and *C. auveus. Recent Advances in the Study of Mammalian Behaviour* (Ed. by J.F. Eisenberg & D.G. Kleiman), pp. 423–53. Special Publications of the American Society of Mammalogists.

Packer, C. (1979). Male dominance and reproductive activity in *Papio anubis. Animal Behaviour*, **27**, 37–45.

Rood, J.P. (1980). Mating relations and breeding suppression in the dwarf mongoose. *Animal Behaviour*, **28**, 143–50.

Rothe, H. (1975). Some aspects of sexuality and reproduction in groups of captive marmosets *(Callithrix jacchus)*. *Zeitschrift für Tierpsychologie*, **37**, 255–73.

Rowell, T.E. (1970). Baboon menstrual cycles affected by social environment. *Journal of Reproduction and Fertility*, **21**, 133–41.

Ruiz de Elvira, M.C. & Abbott, D.H. (1986). A backpack system for long-term osmotic minipump infusions into unrestrained marmoset monkeys. *Laboratory Animals*, **20**, 329–34.

Rylands, A.B. (1985). Tree-gouging and scent-marking by marmosets. *Animal Behaviour*, **33**, 1365–7.

Saayman, G.S. (1971). Behaviour of the adult males in a troop of free-ranging chacma baboons *(Papio ursinus)*. *Folia primatologia*, **15**, 36–57.

Sapolsky, R.M. (1982). The endocrine stress-response and social status in the wild baboon. *Hormones and Behaviour*, **16**, 279–92.

Schilling, A. & Perret, M. (1987). Chemical signals and reproductive capacity in a male prosimian primate *(Microcebus murinus)*. *Chemical Senses*, **12**, 143–58.

Schilling, A., Perret, M. & Predine, J. (1984). Sexual inhibition in a Prosimian primate: a pheromone-like effect. *Journal of Endocrinology*, **102**, 143–51.

Scott, L.M. (1984). Reproductive behaviour of adolescent female baboons *(Papio anubis)* in Kenya. *Female Primates: Studies by Women Primatologists* (Ed. by M.F. Small), pp. 77–100. Alan R. Liss, New York.

Stevenson, M.F. & Rylands, A.B. (in press). The marmoset monkeys, genus *Callithrix. Ecology and Behaviour of Neotropical Primates, Vol. 2.* (Ed. by R. Mittermeier & A.F. Coimbra-Filho), Academia Brasiliera de Sciencies, Rio de Janeiro.

Strum, S.C. (1982). Agonistic dominance in male baboons: An alternative view. *International Journal of Primatology,* **3,** 175–202.

Summers, P.M., Wennick, C.J. & Hodges, J.K. (1985). Cloprostenol-induced luteolysis in the marmoset monkey *(Callithrix jacchus). Journal of Reproduction and Fertility,* **73,** 133–8.

Takahata, Y. (1982). The socio-sexual behaviour of Japanese monkeys. *Zeitschrift für Tierpsychologie,* **59,** 89–108.

Terborgh, J. & Goldizen, A.W. (1985). On the mating system of the co-operatively breeding saddle-backed tamarin *(Saguinus fuscicollis). Behavioural Ecology and Sociobiology,* **16,** 293–9.

Tutin, C.E.G. (1979). Mating patterns and reproductive strategies in a community of wild chimpanzees *(Pan troglodytes schweinfurthii). Behavioural Ecology and Sociobiology,* **6,** 29–38.

Van Horn, R.N. (1980). Seasonal reproduction patterns in primates. *Progress in Reproductive Biology, Vol. 5.,* pp. 181–221.

Wilson, M.E., Walker, M.L. & Gordon, T.P. (1983). Consequences of first pregnancy in Rhesus monkeys. *American Journal of Physical Anthropology,* **61,** 103–10.

Yen, S.S.C. (1986). Chronic anovulation due to CNS – hypothalamic – pituitary dysfunction. *Reproductive Endocrinology. Physiology, Pathophysiology and Clinical Management,* (2nd edn.) (Ed. by S.S.C. Yen & R.B. Jaffe), pp. 500–45. W.B. Saunders, Philadelphia.

Causes and correlates of life history differences among mammals

P.H. HARVEY, D.E.L. PROMISLOW AND A.F. READ

Department of Zoology, University of Oxford, South Parks Road,
Oxford OX1 3PS, UK

SUMMARY

1 Life history variation among mammals is correlated with species differences in brain and body size and, largely for historical reasons, these factors have taken explanatory precedence in the literature.

2 Cross taxonomic differences in the timing of life-history events are correlated with each other independently of size in such a way that some mammals have speeded up lives in comparison with others. These differences relate to mortality schedules: for their sizes, those species with high rates of mortality are the ones that live fast.

3 Living fast has the evolutionary advantage of resulting in higher reproductive output. Ecologists must focus on the reasons why mammals should ever have evolved slowed down lives. We suggest that delayed onset of first reproduction among species with low non-reproductive mortality rates might result from the costs of reproduction decreasing with increased age.

INTRODUCTION

This chapter is born of frustration. Out there in the natural world live mammals displaying an astonishing diversity of life histories, but here in academia we have little understanding of the reasons for that diversity. There have been different approaches to the study of life history variation, but each has its own set of problems. An enormous amount of effort has gone into modelling. But Henry Horn's 1978 claim remains true: 'Unfortunately, most of the theoretical discussions of tactics of life-history are overloaded with turgid mathematical formalism', and worse 'many of the important papers are unintelligible even to the authors of other important papers'. Fortunately, there are elegant exceptions which have focused our attention on the likely importance of trade-offs in life history evolution (see Charlesworth 1980). For example, producing many young one year may reduce the chances of an adult's survival to the next year. Differences in the costs of reproduction presumably lead to different optimal life history solutions and, hence, to species differences in life history patterns. Experimental manipulations are necessary to test such ideas (Partridge & Harvey 1985; Reznick 1985). Brood sizes have been manipulated, particularly among birds, resulting in some parents being persuaded to raise more and others less offspring. It might seem simpler to ask whether individuals which produce more young in a year have lower survival rates, but the results of such comparisons are confounded by differences between good and bad individuals; healthy animals get the best of both worlds by producing more young and having higher survival rates. Experimental studies can control for individual differences in what animal breeders might call 'condition'. Nevertheless, inter-specific comparative studies have their place, and our intention in this chapter is to demonstrate that they have a much more important place than is usually realized. Our focus will be across species and higher level taxa, rather than across individuals within a species.

The timing of life history events differs among mammals so that some species live at faster rates than others (Stearns 1983): species with short gestation lengths have earlier ages at first reproduction, shorter inter-birth intervals, and shorter maximum recorded lifespans. Such patterns in mammalian life history variation are interpreted by many as responses to r- versus K- selection (see, for example, Western 1979; Eisenberg 1981; Martin 1981; Western & Ssemakula 1982; Ross 1987). r-selected species are normally under density-independent regulation, while K-selected species are usually under density-dependent population control (MacArthur & Wilson 1967). It is important to avoid misinterpreting the $r-K$ axis as one of environmental unpredictability (see Turelli & Petry 1980; Boyce 1984), as does one recent paper on primate life history patterns (Ross 1987). Even when the theory is interpreted correctly, the fast−slow continuum of mammalian life history patterns need not reflect an $r-K$ axis (Stearns 1977). For example, when density dependent factors operate equally on all age classes in a population, the effects are indistinguishable from r-selection: 'if we were to compare two populations which differed only in that one was held in check by this sort of density dependence, and the other was density independent, no differences in the outcome of selection would be expected' (Charlesworth 1980, p. 255). As Charlesworth goes on to emphasize, models that incorporate demographic factors in the evolution of life history variation need to specify the age dependence of those factors; $r-K$ selection does not specify such age dependence.

Life history variation has been shown to be highly correlated with body size, so that large species have slower lives: they take longer to gestate, wean, mature and die. It has become common practice to argue that life history variation can be best understood by interpreting it in relation to size (Western 1979; Western & Ssemakula 1982; Peters 1983; Calder 1984). Population biologists understand some of the selective forces resulting in body-size differences among species (Clutton-Brock & Harvey 1983), and many differences in life history may merely be a consequence of size differences. After all, smaller-bodied species can grow to adult size more rapidly. Although we outline that argument in this chapter, we shall suggest that size is by no means the complete answer. Different components of life histories covary independently of size (see also Stearns 1983; Harvey & Clutton-Brock 1985, Harvey & Read 1988). For example, we shall show that those mammals with long gestation lengths for their body size also have late ages at weaning for their body size. Why should species differ in the rate at which they

live *after size effects have been removed?* Part of the answer lies in mortality schedules experienced by natural populations (Harvey & Zammuto 1985; Sutherland,
Grafen & Harvey 1986). Here, for the first time, we shall show that those species
with high mortality rates for their body sizes have short gestation lengths, as well
as weaning and maturing earlier for their sizes.

Two further questions must be answered before we can begin to understand
the selective forces resulting in life history differences. First, why do species differ
in their mortality schedules? Second, why should some species live slow lives for
their body sizes? There will, presumably, always be selection to increase reproductive output.

LIFE HISTORY VARIATION

The life history of an individual can be defined in terms of gestation length, age at
weaning, age at maturity, inter-birth interval, litter size (number of young per
litter), and lifespan. Lifespan varies among individuals of a species and mortality
curves are necessary to describe its distribution, while the other parameters vary
relatively little. This does not mean that intra-specific variation in life histories is
unimportant, merely that it is small compared with inter-specific differences.

Patterns of life history variation differ among orders of mammals, but one
fundamental distinction must be made from the start. Marsupials have very short
gestation lengths, generally between 1 and 3 weeks (Lee & Cockburn 1985). This
short gestation period is associated with the birth of extremely small young, never
weighing more than 1 g. For example, although opossums (*Didelphis virginiana*)
grow to be about 4 kg, after a 13-day gestation period, they weigh just 0.13 g at
birth (twenty could fit into a teaspoon: Nowak & Paradiso 1983). In comparison,
the European hare (*Lepus europaeus*) also weighs about 4 kg as an adult, but
after a 43-day gestation period it produces young weighing about 123 g. Why do
adults of the same body size produce young differing in size by a multiple of about
one thousand? The difference is associated with placentation. The helpless
marsupial young suckles usually (but not invariably) in a pouch, while the gestating
placental mammals derive considerable amounts of nutrition through the placenta
before being born to suckle. The distinction between marsupial and placental
mammals provides a valuable source of comparative information, and has elicited
considerable attention elsewhere (Russell 1982; Lee & Cockburn 1985; Hayssen,
Lacy & Parker 1985). This chapter focuses on the eutherian (placental) mammals,
among which there is ample variation.

Our first distinction was between placental mammals and the marsupials, and
our second is among species of eutherian mammals. The young differ in the
extent to which they are developed at birth. The extremes are precocial and
altricial. Precocial mammals are born with their eyes open, they have fur, and
they can walk or swim. Eisenberg (1981) recognizes various grades along a
precocial–altricial axis, depending on whether the neonate can walk, stand,

crawl, cling, or is merely 'embryonic', and on whether eyes are open and fur is present. Species from the same order tend to be have very similar grades according to Eisenberg's classification. Differences among grades are associated with the degree of development of the brain at birth (Bennett & Harvey 1985), with litter size (Pagel & Harvey 1988a), and with gestation length in relation to adult body size (Martin 1984; Martin & McLarnon 1985, Pagel & Harvey 1988a). We must turn to adult body size as an explanatory variable before we can compare other life history variables in a meaningful way.

LIFE HISTORY VARIATION AND SIZE

Life history variables are correlated with each other across eutherian mammals. Gestation lengths, ages at weaning and maturity, inter-birth intervals, and life-spans are positively correlated with each other, but negatively correlated with litter size. Why should life history measures covary?

Perhaps the most widespread belief is that, in some sense, body and brain size differences dictate life history variation. For example, small mammals cannot give birth to large young, and maximum gestation length may thus be related to body size. On the other hand, large mammals need large brains to coordinate their activities and, since neuronal division may be a pacemaker for development and no neuronal division occurs after birth (Martin 1983), minimum gestation length may also be related to brain size (Sacher & Staffeldt 1974). Larger-bodied species take longer to grow to full size, and age at weaning or maturity may therefore be limited by the time it takes to reach adult size. It has also been argued that the brain is a homeostatic organ and that, since larger-bodied species have larger brains, they can also live for longer (Sacher 1959). These ideas have been discussed at length and many of them criticized in detail elsewhere (Harvey & Read 1988). In the following paragraphs, we summarize the major issues.

How do life history differences relate to size differences among species? Most relationships are fairly well described by power functions (Calder 1984), so that life history variables (L) increase with body size (S) according to

$$L = aS^b. \tag{1}$$

a and b are usually estimated by calculating a line of best fit on logarithmically transformed variables

$$\log(L) = \log(a) + b \cdot \log(S). \tag{2}$$

The three most common best fit procedures are model 1 regression, major axis and reduced major axis. There is considerable debate in the literature about the most appropriate procedure to use (see, for example, Harvey & Mace 1982; Seim & Saether 1983; Pagel & Harvey 1988b). Indeed, each of the procedures makes unreasonable assumptions about the likely distribution of error variance (Rayner 1985) which can lead to the acceptance of statistical artifacts as real patterns (Pagel & Harvey 1988b). Nevertheless, when a wide array of species or higher level taxa are employed, the correlation between logarithmically transformed life history variables and the logarithms of body weights can be very high (often above 0·97) and under such circumstances all three statistical methods give similar estimates of a and b. Across higher level taxa, about 80% of the variance in life history variables can be accounted for by body size differences (Harvey & Read 1988) Life history timings such as gestation length, and ages at weaning, maturity and maximum recorded lifespan scale on body size with an allometric exponent (b) of between 0·1 and 0·4 (see, for example, Kihlström 1972; Western 1979; Eisenberg 1981; Western & Ssemakula 1982; Harvey & Clutton-Brock 1985; Gittleman 1986). The reason for the allometric exponents of life history timings lying in such a narrow range has not been explained in terms of any optimality model. Although we do not know why b should take these values (indeed one of the major failures of such allometric analyses has been our inability to explain allometric exponents), Lindstedt (1985) has been moved to claim that these exponents converge on 0·25 and that there is a Periodengeber (similar to the Zeitgeber) which controls the pace of life according to this scaling principle with body size (see also Lindstedt & Swain 1988). The fact that b differs among life history timing parameters and among taxa for any particular parameter (Harvey & Read 1988) leads us to be wary of such speculations which describe biological variables as constants.

Small body size is correlated with earlier ages at maturity, shorter inter-birth intervals and larger litters (Harvey & Read 1988) and, therefore, with high potential reproductive rates. Why, then, are mammals not as small as possible? Clutton-Brock & Harvey (1983) suggested that large size might be favoured for inter-specific and intra-specific competition, anti-predator defence, hunting efficiency, the ability to survive on a diet of lower quality food (e.g. leaves), and the ability to produce larger neonates. If the latter were true, we might expect adult females to be larger in body size than males, although this is rarely the case among mammals (Ralls 1976).

Sacher (1959) and Sacher & Staffeldt (1974) have argued that the brain is an important variable in life history evolution. Their argument is based on two main findings: (a) Neonatal brain size is correlated with gestation length, independently of neonatal body size, and (b) adult brain size is highly correlated with lifespan, independently of adult body size. More recent correlational evidence from primates, carnivores and rodents makes the picture far less clear. Harvey & Read (1988) summarize the evidence and conclude that 'some life history variables in some taxa may be more closely linked to brain size than to body size'.

The situation is further complicated by the suggestion that brain size may merely stand as a surrogate variable for an even tighter correlation. Two pieces of evidence bear on this argument. First, Economos (1980) shows that the size of the adrenal gland is more highly correlated with lifespan than is brain size. Unlike the body but like the brain, adrenals vary little is size among individuals. Species estimates of body size are associated with large variance (some animals are obese while others are emaciated), and the brain may therefore represent a more accurate estimate of some other aspect of size (even the adrenals!) than does the body. The second correlate concerns differences in age at maturity among primates: once the effect of age at maturity is removed by partial correlation, the relationship between lifespan and brain size disappears (Harvey & Clutton-Brock 1985). Age at maturity may be independently related to brain size and to lifespan, thus leading to a correlation between brain size and lifespan. Confounding variables are an important problem for the comparative biologist (Clutton-Brock & Harvey 1979, Pagel & Harvey 1988c).

LIFE HISTORY COVARIATION

Life history variables are correlated with each other and, as we have pointed out, this is generally assumed to be because they have evolved in response to changes in body and brain size. But we also mentioned one relationship that occurs independently of size, that between lifespan and age at maturity among primates. Harvey & Clutton-Brock's (1985) survey of life history variation among primates revealed other relationships between life history variables that were significant even when the effects of size were removed by partial correlation. Here we focus on one of the most interesting of those relationships and ask how it generalizes across eutherian mammals, but before we do that we need to take a brief diversion into the problem of taxonomic levels of analysis.

Closely related species should not be treated as providing independent points for statistical analysis (for a review see Pagel & Harvey 1988c). Briefly, closely related species share much more in common than their recent ancestry: they tend to be similar in morphology, diet, behaviour, and physiology. By treating them as independent points for analysis, the problem of confounding variables discussed in the last section is made much more serious (Clutton-Brock & Harvey 1979; Ridley 1983). Before applying the comparative method, we need to identify independent evolutionary events so that we can be sure that we are dealing with evolutionary convergence (Ridley 1983). A pragmatic solution to the problem is to perform comparative analyses at the lowest taxonomic level at which most variance in the data is located (Harvey & Mace 1982). For example, among primates this was at the subfamily level, leading Harvey & Clutton-Brock (1985) to perform their analyses across subfamily averages.

We might expect to find that those primate subfamilies which have long gestation periods for their body size would produce relatively precocial (well developed) young which are weaned earlier for their body size. That is, increased pre-natal maternal investment would be associated with decreased post-natal investment (for such a prediction, see Eisenberg 1981). In fact, the partial correlation between gestation length and age at weaning with body size removed is positive. This means that those primate subfamilies which have long gestation lengths for their body size, wean their young relatively late. A similar finding has been reported by Lack (1968) for birds: those species with long incubation periods for their body sizes have relatively late ages at fledging. Is this pattern general across mammals?

After searching the literature for information on gestation length, age at weaning, and body size across mammal species, we compiled data for females belonging to 770 species from ninety families and nineteen orders. Because the physiology of reproduction of eutherian or placental mammals differs from that of the marsupials and monotremes, we have excluded the latter from our analysis. The highest level at which we could perform a comparative analysis was across orders. So that speciose genera and genera-rich families did not weight our analysis, generic points were first calculated as the average of logarithmically transformed species values. Family points were then estimated as the mean of constituent generic points, and order values as the mean of constituent family points. This gave us seventeen order values for gestation length, age at weaning, and body size. Relative gestation lengths and ages of weaning were then calculated for each order as the deviations from best fit lines relating gestation length and age at weaning to body weight. Orders with relatively long gestation lengths also had relatively late ages at weaning (Fig.13.1; $r=0.6$, $P<0.01$).

The above analysis demonstrates that those orders with slowed down pre-natal development also have prolonged postnatal development; we can detect no obvious trade-off between the two components of development. What are the selective forces responsible for differences in rates of development? Comparisons between the different orders and other comparisons within orders provide no obvious ecological or behavioural correlates of relative gestation length or age at weaning (Harvey & Read 1988). Why has there been such a notable lack of success? One clue to understanding the pattern reported in Fig. 13.1 comes from Harvey & Zammuto's (1985) comparison of age at first reproduction with life expectation at birth across mammal species. They found that species with early ages at first reproduction for their body size were those with short life expectations for their body size. Could it be the patterns of mortality in natural populations are related to other components of the speed of life among mammals? We can only answer that question if we consider mortality curves and attempt to relate species differences in patterns of mortality to other life history variables.

LIFE HISTORY VARIATION AND MORTALITY SCHEDULES

If population size is to remain constant and immigration from a population is balanced by emigration, then the birth rate (number of births per annum) within the population must equal death rate (number of deaths per annum). As a consequence, some components of life history variation across species must inevitably be correlated with each other (see Sutherland, Grafen & Harvey 1986). If the population birth rate increases, then so must the death rate. If an increased death rate is not balanced by an increased birth rate, then a population will go extinct. Life expectation at birth differs among natural populations of mammals

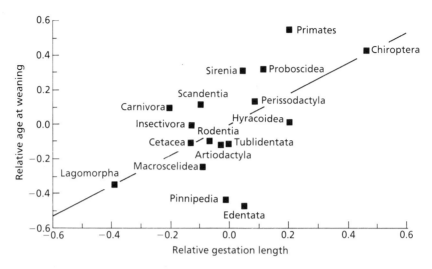

FIG. 1. The relationship between age at weaning and gestation length *after the effects of body size have been removed*, across seventeen orders of mammals. The seventeen order points are derived from data on 770 species (see text).

that have approximately stable age distributions (Millar & Zammuto 1981). What components of life history differ between such populations? Two obvious candidates are age at maturity and litter size, since these are likely to be important components of r_{max} (Ross 1987). Populations with higher mortality rates could maintain themselves by having earlier ages at maturity or larger litters.

Smaller-bodied species have higher rates of mortality but, after body size effects are removed, differences in life expectation are balanced by differences in age at maturity but not by differences in litter size (Harvey & Zammuto 1985; Sutherland, Grafen & Harvey 1986). But life expectations, whether they are at birth or at maturity, are *ad hoc* summary statistics of a mortality schedule. Can we define more comprehensive summary statistics that capture the essence of mortality schedules? In this section we provide one such statistic, and examine the relationships between that statistic and life history variation. Are high mortality rates associated with life history variables other than age at maturity?

We have examined twenty-four mortality curves from natural populations of mammals (from females of a wide variety of species, mostly listed in Harvey & Zammuto 1985), and searched for suitable summary statistics. Since there are so few species in our sample but they are from a wide range of taxa, our analyses are performed at the species level. The relationship between the logarithm of the number of individuals from a cohort still alive and age of the individuals can be described reasonably well by best fit third-order polynomial curves.

$$\log N_t = \log N_0 + a(t) + b(t)^2 + c(t)^3, \qquad (3)$$

where N is population size, t is age. This relationship, although explaining an average 98% of the variance in N_t, is somewhat unsatisfactory because it involves the estimation and comparison of three constants (a, b and c).

However, when age is measured in years, we have found that the relationship between log (population size) and log ($t+1$) is approximately linear for the first 5 years or more of life for almost all the species in the sample, (see, for example, Fig. 2). Therefore, it is possible to compare the mortality rates of large and small species for the first 5 years of life by a single variable, d, defined by

$$\log N_t = \log N_0 + d \cdot \log(t+),\qquad (4)$$

which is equivalent to

$$N_t = N_0 \cdot (t + 1)^d.\qquad (5)$$

d is negative since the numbers in a cohort decrease with time, and a large negative value represents a high mortality rate, meaning that d is positively related to survival rates. d is positively correlated with body size ($r=0\cdot679$, $n=36$, $P<0\cdot01$), age at maturity ($r=0\cdot789$, $P<0\cdot01$), gestation length ($r=0\cdot766$, $P<0\cdot01$), and negatively to litter size ($r_s=0\cdot733$, $P<0\cdot01$; Spearman correlation calculated here since litter size cannot be transformed to approximate a normal distribution). When body size effects are removed by partial correlation on logarithmically scaled variables, d is related to gestation length ($r=0\cdot507$, $P<0\cdot05$), age at weaning ($r=0\cdot489$, $P<0\cdot05$), age at maturity ($r=0\cdot548$, $P<0\cdot01$), and litter size ($r_s=-0\cdot423$, $P<0\cdot05$; r_s calculated as the Spearman rank correlation between the residuals of litter size regressed on body weight and d regressed on body weight). *Species with low survival rates for their body size have short gestation lengths, larger litters, and earlier ages at weaning and first reproduction for their body sizes.* Age at weaning

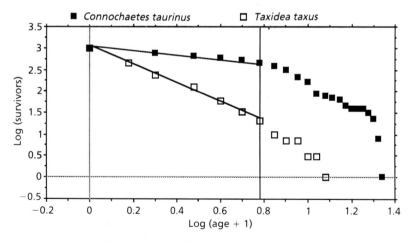

Fig. 2. The numbers of individuals left at different ages (years) from a cohort (of 1000) for the gnu (*Connochaetes taurinus*) and the American badger (*Taxidea taxus*). Both axes are scaled logarithmically. For the first 5 years of life (log 6 = 0.778), the relationship between log (survivors) and log (age+1) is approximately linear.

is a component of age at maturity and might be expected to show a similar correlation. The time between age at weaning and age at maturity is also highly correlated with d after body-size effects have been removed ($r=0.560$, $P<0.01$).

We pointed out above that our description of mortality included only the first 5 years of life. Most small mammals rarely live even that long. Does this mean that patterns as well as rates of mortality differ for larger-bodied species? We have come to an important question: do animals live long enough to become senescent in natural populations?

If we describe individuals which have reached sexual maturity as adult, then we could define senescence as an increase in adult mortality rate with age. However, once cohorts are reduced to few individuals, empirical estimates of mortality rates begin to vary wildly. Indeed, the chance of mortality will be 1·0 the year that the last animal in a cohort dies. Accordingly, we have looked for senescence as an increase in mortality rate (measured as the proportion of animals surviving between successive years) from the age at maturity up until the first year when less than ten individuals from the original cohort are left. In five of the twenty-four species there was a significant increase in mortality with age at the $P=0.05$ level (an example is given in Fig. 3), and in a further three at the $P=0.10$ level of significance. With the exception of *Tamias striatus*, only the heavier species in our sample exhibited senescence (Promislow, in preparation).

The use of the statistic d in the above analysis can be criticized because it only covers the first 5 years of a cohort's existence, and those 5 years may or may not encompass the whole of a cohorts' existence, depending on body size. An alternative analysis on the same data which calculated a, b and c in formula (3) above demonstrated that each of the variables is related to body size (Promislow & Harvey, in preparation). Furthermore, the same analysis demonstrated that, when expected mortality curves (based on species body sizes to estimate a, b and c) are compared with actual curves, (i) size-corrected juvenile and adult mortality rates are positively correlated and (ii) species with relatively high mortality rates also have shorter gestation lengths, earlier ages at weaning, and earlier ages at maturity.

DISCUSSION

The thrust of this chapter has been broad. We have attempted to produce analyses which incorporate data from as diverse an array of placental mammals as possible. The patterns revealed must be interpreted at this level, which leads us to sweep aside details of how closely related species differ in life histories.

We have reported two major findings. First, different mammal orders can be placed along a slow—fast life history continuum independently of body size: those species with relatively fast pre-natal development also have speeded up post-natal lives (see also Stearns 1983). The speeded up post-natal development may not be a straightforward consequence of increased parental investment since relative periods of adolescence (time between weaning and reproductive maturity with body size effects removed) are also shorter (Read & Harvey, in preparation). Our

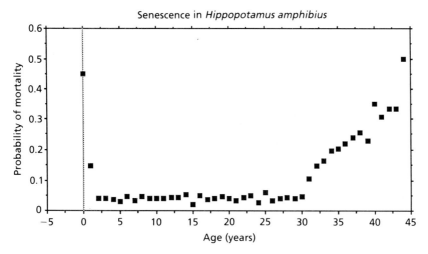

FIG. 3. Senescence in the hippopotamus. The probability of survival between successive years is plotted against age. Young and old individuals have high probabilities of mortality.

second finding is that these differences in life history patterns are related to patterns of mortality in natural populations. Those species which live fast for their sizes also have high mortality rates for their sizes. This latter analysis points to the types of ecological factors that might be associated with interspecific differences in life history patterns after body size effects have been removed. One recent study provides an interesting correlation which may be interpreted along these lines.

Ross (1987) examined patterns of life history variation among primate species. Her main conclusions are that the intrinsic rate of natural increase (r_{max}) of a species is related independently to two factors: body size and habitat. Large species have low r_{max}, while species living in tropical rain forest have low r_{max} for their body size compared with species from secondary forests and savannah (Fig. 4). However r_{max} is calculated from three components: age at first reproduction, litters per year, and reproductive lifespan. When we analysed the data presented by Ross, we found that species differences in both age at first reproduction and birth rates correlated with habitat after size effects had been removed ($t_{55}=2.782$ and 2.471, $P=0.007$ and 0.017, respectively). However, there was no correlation between environment and species' differences in relative lifespan ($P=0.472$). It is unlikely that the results arise from a taxonomic artifact since the same association between relative r_{max} and habitat occupied is found (albeit nonsignificantly) among species from five of six variable subfamilies represented in the sample.

Ross interprets her result in terms of $r-K$ selection, arguing that tropical rain forests provide a more stable environment than savannah or secondary forests. The findings reported earlier in this chapter suggest an alternative explanation: mortality rates may be lower in tropical rain forests than in savannah or secondary

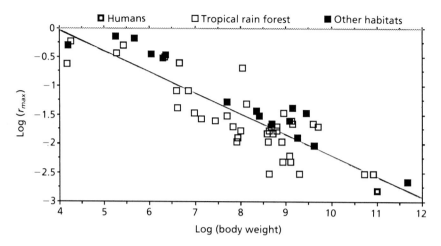

FIG. 4. Relationship between log (r_{max}) and log (body weight) for primate species categorized by habitat. After Ross (1987).

forests. The distinction, then, is between environmental predictability and environmental harshness. It is necessary to distinguish between these alternatives but, whatever the cause, the patterns reveal an *ecological correlate of life history variation which is independent of body size*. Indeed, in her sample, Ross could detect no size differences between species occupying the two habitat categories.

We return to the missing piece of the jigsaw: why do some mammals slow down their lives when, apparently, their body size allows more rapid reproduction? It seems to us that the problem should be taken in parts. For example, we might ask why mammals with low mortality rates for their size have later ages at first reproduction? One answer might be that the costs of reproduction per unit time decrease with increased age, so that fecundity increases with age. Imagine the extreme case of a mammal species that can produce and wean two offspring the year after its birth (i.e. at age 1) or four offspring a year later, but that the costs of reproduction are so high that the animal dies after the young have weaned. If non-reproductive mortality between age 1 and 2 is nil then an individual will be selected to delay reproducting until age 2. However, if non-reproductive mortality is so high that very few animals survive from age 1 to age 2, an individual will be selected to breed at age 1. There is evidence from a number of species that, at around the time of reproductive maturity, the costs of reproduction decrease with increased age and that fecundity increases with age (Promislow & Harvey, in preparation).

ACKNOWLEDGEMENTS

We thank Mark Pagel for discussion. D.E.L.P. was supported by a Rhodes Scholarship, A.F.R. by a Commonwealth Scholarship.

REFERENCES

Bennett, P.M. & Harvey, P.H. (1985). Brain size, development and metabolism in birds and mammals. *Journal of Zoology (London)*, A207, 491−509.

Boyce, M.S. (1984). Restitution of *r*- and *K*- selection as a model of density-dependent natural selection. *Annual Review of Ecology and Systematics*, 15, 427−48.

Calder, W.A. (1984). *Size, Function and Life History.* Harvard University Press, Cambridge, Massachusetts.

Charlesworth, B. (1980). *Evolution in Age-Structured Populations.* Cambridge University Press, Cambridge.

Clutton-Brock, T.H. & Harvey, P.H. (1979). Comparison and adaptation. *Proceedings of the Royal Society of London, Series B*, 205, 547−65.

Clutton-Brock, T.H. & Harvey, P.H. (1983). The functional significance of body size among mammals. *Advances in the Study of Mammalian Behavior* (Ed. by J.F. Eisenberg & D.G. Kleiman), pp. 632−33. Special Publications of American Society of Mammalogists, 7.

Clutton-Brock, T.H. & Harvey, P.H. (1984). Comparative approaches to investigating adaptation. *Behavioural Ecology: an Evolutionary Approach* (Ed. by J.R. Krebs & N.B. Davies), pp. 7−29, 2nd edn. Blackwell Scientific Publications, Oxford.

Economos, A.C. (1980). Brain-lifespan conjecture: a re-evaluation of the evidence. *Gerontology*, 26, 82−9.

Eisenberg, J.F. (1981). *The Mammalian Radiations.* Chicago University Press, Chicago, Illinois.

Gittleman, J.L. (1986). Carnivore life history patterns: allometric, phylogenetic and ecological associations. *American Naturalist*, 127, 744−71.

Harvey, P.H. & Clutton-Brock. T.H. (1985). Life history variation in primates. *Evolution*, 39, 559−81.

Harvey, P.H. & Mace, G.M. (1982). Comparisons between taxa and adaptive trends: problems of methodology. *Current Problems in Sociobiology* (Ed. by King's College Sociobiology Group), pp. 343−61. Cambridge University Press, Cambridge.

Harvey, P.H. & Read, A.F. (1988). How and why do mammalian life histories vary? *Evolution of Life Histories: Pattern and Process from Mammals* (Ed. by M.S. Boyce), pp. 213−32. Yale University Press, New Haven, Connecticut.

Harvey, P.H. & Zammuto, R.M. (1985). Patterns of mortality and age at first reproduction in natural populations of mammals. *Nature*, 315, 319−20.

Hayssen, V., Lacy, R.C. & Parker, P.J. (1985). Metatherian reproduction: transitional or transcending? *American Naturalist*, 126, 617−32.

Horn, H.S. (1978). Optimal tactics of reproduction and life-history. *Behavioural Ecology: an Evolutionary Approach* (Ed. by J.R. Krebs & N.B. Davies), pp. 411−29. Blackwell Scientific Publications, Oxford.

Kihlström, J.E. (1972). Period of gestation and body weight in some placental mammals. *Comparative Biochemistry & Physiology*, 43, 673−79.

Lack, D. (1968). *Ecological Adaptations for Breeding in Birds.* Methuen, London.

Lee, A.K. & Cockburn, A. (1985). *Evolutionary Ecology of Marsupials.* Cambridge University Press, Cambridge.

Lindstedt, S.L. (1985). Birds. *Non-Mammalian Models for Research in Aging* (Ed. by F.A. Lints), pp. 1−21. Karger, Basel.

Lindstedt, S.L. & Swain, S.D. (1988). Body size as a constraint of design and function. *Evolution of Life Histories: Pattern and Process from Mammals.* (Ed. by M.S. Boyce), pp. 97−105. Yale University Press, New Haven, Connecticut.

MacArthur, R.H. & Wilson, E.O. (1967). *The Theory of Island Biogeography.* Princeton University Press, Princeton, New Jersey.

Martin, R.D. (1981). Field studies of primate behaviour. *Symposia of the Zoological Society of London*, 46, 287−336.

Martin, R.D. (1983). Human brain evolution in an ecological context. *52nd James Arthur Lecture on the Evolution of the Human Brain.* American Museum of Natural History, New York.

Martin, R.D. (1984). *Scaling Effects and Adaptive Strategies in Mammalian Lactation.* Symposia of the Zoological Society of London, **51**, 81–117.

Martin, R.D. & McLarnon, A.M. (1985). Gestation period, neonatal size and maternal investment in placental mammals. *Nature*, **313**, 220–3.

Millar, J.S. & Zammuto, R.M. (1983). Life histories of mammals: an analysis of life tables. *Ecology*, **64**, 631–5.

Nowak, R.M. & Paradiso, J.L. (1983). *Walker's Mammals of the World*, 4th edn. Johns Hopkins University Press, Baltimore.

Pagel, M.D. & Harvey, P.H. (1988a). How mammals produce large brained offspring. *Evolution* **42(5)**, 948–57.

Pagel, M.D. & Harvey, P.H. (1988b). The taxon level problem in mammalian brain size evolution: facts and artifacts. *American Naturalist* **132(3)**, 344–59.

Pagel, M.D. & Harvey, P.H. (1988c). The comparative method. *Quarterly Review of Biology.*

Partridge, L. & Harvey, P.H. (1985). The costs of reproduction. *Nature*, **316**, 20.

Peters, R.H. (1983). *The Ecological Implications of Body Size.* Cambridge University Press, Cambridge.

Promislow, D.E.L. & Harvey, P.H. (in preparation). Living fast and dying young: a comparative analysis of life history variation among mammals.

Ralls, K. (1976). Mammals in which females are larger than males. *Quarterly Review of Biology*, **51**, 245–76.

Rayner, J.M.V. (1985). Linear relations in biomechanics: the statistics of scaling functions. *Journal of Zoology*, **206**, 415–39.

Reznick, D. (1985). Costs of reproduction: an evaluation of the empirical evidence. *Oikos*, **44**, 257–67.

Ridley, M. (1983). *The Explanation of Organic Diversity: the Comparative Method and Adaptations for Mating.* Oxford University Press, Oxford.

Russell, E.M. (1982). Patterns of parental care and parental investment in marsupials. *Biological Reviews of the Cambridge Philosophical Society*, **57**, 423–87.

Ross, C. (1987). The intrinsic rate of natural increase and reproductive effort in primates. *Journal of Zoology*, **214**, 199–220.

Sacher, G.A. (1959). Relationship of lifespan to brain weight and body weight in mammals. *C.I.B.A. Foundation Symposium on the Lifespan of Animals* (Ed. by G.E.W. Wolstenholme & M. Little), pp. 115–33. Brown, Boston, Massachusetts.

Sacher, G.A. & Staffeldt, E.F. (1974). Relationship of gestation time to brain weight for placental mammals. *American Naturalist*, **108**, 593–616.

Seim, E. & Saether, B.-E. (1983). On rethinking allometry: which regression model to use? *Journal of Theoretical Biology*, **104**, 161–8.

Stearns, S.C. (1977). The evolution of life history traits. *Annual Review of Ecology and Systematics*, **8**, 145–72.

Stearns, S.C. (1983). The influence of size and phylogeny of life history patterns. *Oikos*, **41**, 173–87.

Sutherland, W.J., Grafen, A. & Harvey, P.H. (1986). Life history correlations and demography. *Nature*, **320**, 88.

Turelli, M. & Petri, D. (1980). Density- and frequency-dependent selection in a random environment: an evolutionary process that can maintain stable population dynamics. *Proceedings of the National Academy of Sciences, USA*, **77**, 7501–5.

Western, D. (1979). Size, life history and ecology in mammals. *African Journal of Ecology*, **17**, 185–204.

Western D., & Ssemakula, J. (1982). Life history patterns in birds and mammals and their evolutionary significance. *Oecologia*, **54**, 281–90.

Section 4
Parental Investment:
Strategies for Offspring Survival

Among mammals, offspring survival depends on high levels of nutritional and protective care provided by the parents. For most species, the mother is the primary caretaker due to the constraints of lactation. However, other individuals such as fathers, friends, and family members can provide either care that substitutes for that of the mother — carnivores that regurgitate to young and thus are the nutritional providers — or care that complements that of the mother — active defence of a territory to exclude competitors for the mother's food, translating into milk for the infant.

This relatively high cost of offspring production among mammals is critical to their socioecology. Simple maximization of reproductive output or fertility rates is a strategy that will seldom occur, as the offspring will survive only under conditions of nurturing, and lost offspring may well impinge (through opportunity costs, time delays to next offspring) on the subsequent reproductive success of an individual and its kin.

Parenting is thus a major research focus, particularly so where humans are concerned. Humans have very altricial young, high levels of maternal investment, a long period of infant dependence and growth, and often a slow rate of reproduction, due to high inter-birth intervals. Overall reproductive success in both males and females is very dependent upon parental effort, not just upon mating opportunities or birth rate. The degree to which fathers participate in offspring care is of particular interest in studies of humans, but is also of relevance to other mammalian species where paternal effort can relieve the mother of some of the energy costs of reproduction, ensuring infant survival and increasing reproductive rates for both mothers and fathers.

Studies of parental investment and effort have focused on a series of important questions. The first of these is the problem of sex-biased investment. Since Fisher (1930) and Trivers (1972) developed theoretical models predicting the relative evolutionary 'worth' of males and females it has been clear that two aspects of reproduction are open to strategic manipulation (i.e. selection). The first is the sex ratio, and the second is the actual amount 'invested' in young dependent upon their sex. Fisher argued that there should normally be parity of birth: sex ratios of males and females when the reproductive success of each sex is balanced — sex, in other words, is a balanced polymorphism. Trivers, though, in his classic paper, showed that variance in male reproductive success exceeded that of females in most polygynously mating mammalian species. It followed from this that males and females provide very different evolutionary returns. From the parent's per-spective, either greater levels of effort can be directed towards sons than daughters (differential investment) or greater numbers of sons can be produced (sex ratio biasing).

Increased investment in sons has been observed in species such as elephant seals, red deer and African elephants — all species with competitive polygyny. Among some of the female kin-bonded primate species, on the other hand, investment appears to be biased in favour of daughters (see review in Clutton-Brock & Albon 1982). The important factor appears to be how the early differential parental effort has a long-term effect on the growth, survival or reproductive potential of each sex of offspring. Among those species where later competitive abilities of sons depends on early, lactationally mediated growth, early investment is male-biased. In those species where daughters attain their mothers' status, with its potential attendant reproductive advantages, mothers bias investment towards daughters. This example points to the complexity of issues in parental investment. The social context is a further consideration in whether or not to invest differentially.

The evidence for consistent sex ratio biasing is still ambiguous: few clear, statistically meaningful effects have been demonstrated for mammals (see Clutton-Brock & Albon 1982). Selective abortion or reabsorption of fetuses has been shown for coypus (Gosling 1986) depending on the energetic costs of males or females in relation to food availability. But sex ratio biasing on the basis of the potential variance in reproductive success is further complicated by competition between siblings of the same sex for local resources (local resource competition), or the beneficial effects of cooperation between siblings (local resource enhance-ment). The ecological and social contexts of the parental investment again require detailed investigation.

The chapters in this section examine some of the questions raised above. Lee's chapter examines the types of infant care that can be provided by individuals other than the mother, and the effects of this care on offspring survival, female energetics and fertility. The chapter concentrates on females living in groups composed of kin among two mammalian species — African elephants and vervet monkeys — with comparative information on general patterns of caretaking in humans. Lee suggests that caretaking strategies can be a positive selection for some kinds of sociality among females, as has been argued for birds and the callitrichid primates. One of the main conclusions to be drawn from Lee's chapter is that the same phenomena — allomothering — may in fact be a series of different responses to varying conditions (for example, coercive versus cooperative help in raising offspring).

Among humans, kin networks include post-reproductive females in some societies. Hawkes, O'Connell & Blurton Jones found that women past the age of childbearing in the Hadza hunter-gatherers spend more time and energy gathering food than do younger women, especially if they are grandmothers. Grandmothers lose more weight under the stress of foraging in the dry months than do the younger, reproductive women. Since food is shared amongst all women, they argue that grandmothers could be contributing towards reducing their daughter's foraging costs while maintaining the food intake of the grandchild.

This argument is developed further in the subsequent chapter by Blurton Jones, Hawkes & O'Connell. They examine the effort required to raise Hadza children, and compare this with effort among the !Kung. Hadza children are left in camp at a younger age, are weaned earlier and are nutritionally more independent when younger than are !Kung children. However, mothers and, as noted above, grandmothers continue to provide about half the food required for older children. They suggest that the early decrease in time and energy costs of maternal foraging for children's nutritional needs allows for a shortening of inter-birth intervals and ultimately for a higher reproductive rate among the Hadza in comparison to other hunter-gatherers.

What both these chapters highlight are the *costs* incurred in raising children, from both a proximate and an evolutionary perspective. In particular, we see with humans that costs can be variable within and between populations, depending upon the precise local conditions. Voland extends this idea by considering changes in child care within the same population through time. He presents historical data on the ecological background to investment in children by European parents. His analysis of patterns of infant mortality in the late seventeenth to nineteenth century Germany, shows that survival chances of newborns were affected by the investment strategies of their parents, which were influenced by the stability (or instability) of environmental conditions — famine, warfare, plagues. Sons received more attentive care, which he relates to their higher probability of marrying over that of daughters. But he also draws attention to the emotional expectations of parents, and how these influence investment decisions. If the expectation of survival is high, then different patterns of care, such as the number of godparents, were apparent. These expectations were facultative responses to environmental conditions, which have as their evolutionary roots the conditional parenting strategies widespread among mammals. Humans, as well as other mammals, seem to have the ability to assess the probabilities of the survival of their offspring and adjust their behaviour accordingly.

Borgerhoff Mulder focuses further on the problem of differential investment in sons and daughters by examining the inheritance practices among the pastoral Kipsigis of Kenya. Male-biased inheritance is common among many human societies, since wealth transmitted to sons is thought to confer greater reproductive advantages than that given to daughters. However, Borgerhoff Mulder demonstrates that both sons and daughters benefit reproductively from parental wealth — it is the timing of the transmission which varies. For daughters, wealthy fathers contribute towards their childhood health and an early age of menarche, which corresponds to a young age of marriage and a long reproductive lifespan. Among sons, inherited wealth is able to increase their ability to marry, or marry additional wives as adults. She stresses that it is essential to examine age-specific patterns of parental effort, within a cultural context of the types of resources that can be acquired and how these resources function for direct biological (nutritional) advantage or indirect advantages such as alliance formation and kinship ties.

It is perhaps no accident that within the context of this volume, the section with the largest number of direct studies of human rather than animal societies is that on parental investment. It is clear that the success of the human species is based on the solution of various demographic problems shared with other, non-human primates. Certainly studies of human evolution focus increasingly on aspects of growth and development, and strategies for the survival of offspring in relation to a variable resource base must have been critical to the emergence of the physical and behavioural characteristics of the species.

REFERENCES

Clutton-Brock, T.H. & Albon, S.D. (1982). Parental investment in male and female offspring in mammals. *Current Problems in Sociobiology* (Ed. by King's College Sociobiology Group), pp. 223–47. Cambridge University Press, Cambridge.

Fisher, R.A. (1930). *The Genetical Theory of Natural Selection*. Clarendon Press, Oxford.

Gosling, L.M. (1986). Selective abortion of entire litters in the coypu: adaptive control of offspring production in relation to quality and sex. *American Naturalist*, 127, 772–95.

Trivers, R.L. (1972). Parental investment and sexual selection. *Sexual Selection and the Descent of Man* (Ed. by B. Campbell.), pp. 136–79. Aldine, Chicago, Illinois.

Family structure, communal care and female reproductive effort

P.C. LEE

Sub-department of Animal Behaviour,
University of Cambridge, High Street, Madingley, Cambridge
CB3 8AA, UK

SUMMARY

1 Care given to infants from animals other than the mother potentially can enhance infant survival and reduce the energetic costs of maternal effort.

2 The behavioural components of caretaking are assessed for mammalian species, and communal care is defined by the contribution of the caretakers (alloparents) to survivorship and maternal energetics.

3 Specific examples from matrilineal family units of elephants and vervet monkeys, and from human societies, are compared to examine how the family structure affects the nature and extent of communal care.

4 It is suggested that high degrees of relatedness between females living in small, stable groups promote the expression of communal care when cooperation for defence and protection of infants is a major contribution to female lifetime reproductive success.

INTRODUCTION

Caring for other females' infants by members of a social group is a widespread phenomenon among mammalian species. In a recent review, Reidman (1982) lists over 120 species (free-ranging and captive) where such 'alloparenting' behaviour has been observed. A common theme amongst these species is that they live in small, stable groups of familiar individuals, and typically the groups are composed of kin of varying degrees of relatedness. Reproductive rates of females in alloparenting species are usually low and overlap between generations is frequent (Gittleman 1985).

I will explore the issue of alloparental care in terms of parental investment and the reproductive effort allocated to an individual offspring. In discussing parental investment, I shall concentrate on maternal strategies and focus on two major components of mammalian female reproductive effort, the energy costs of lactation and factors affecting infant survival. The specific questions addressed are (i) What types of behaviour can be considered to be alloparenting; (ii) Do alloparents contribute towards a mother's reproductive effort and hence her lifetime reproductive success; and (iii) How does the kin structure of the group affect the nature, frequency and outcome of alloparental care?

323

Among primates, a variety of interactions between group members and infants are subsumed under the term 'alloparenting', and as a result help and abuse are often confounded by a common terminology while differing in their functional outcome (see discussions in Hrdy 1976; Quiatt 1979; Reidman 1982; Gittleman 1985; Gray 1986). In order to minimize controversy in using the term alloparenting, I will define five specific functional components of alloparenting behaviour that will be used in this paper (see above references) and then discuss the effect of family structure on the expression of these behaviours.

The separate components of caretaking behaviour have been distinguished here specifically on the basis of their effects either on infant survival, or their contribution to maternal energetics. These components can then be further separated into those that involve little or no interaction between the members of a group, those that reflect dyadic, reciprocal relationships between individuals, and those where the potential costs of the behaviour and its beneficial effects are actively shared among group members.

Anti-predator behaviour

Providing protection for very young infants who are especially vulnerable to predation as a result of small body size, poor agility and coordination and lack of experience at escaping, is frequently cited as a form of alloparenting behaviour. Wildebeests passively dilute predation risks for their calves by synchronizing their breeding and 'swamping' predators with a superabundance of calves grouped together in large aggregations (Estes 1976). While active protection of infants by non-mothers against predation or infanticide is less common, the proximity of caretakers may reduce risks to infants.

Protection from environmental stresses

Most infants have reduced capacities for thermoregulation and infant survivorship may be affected by exposure to extremes of temperature. Communal nesting or denning may reduce the exposure of infants to environmental stresses such as cold, or to predators or infanticidal individuals. Here mothers passively 'cooperate' in the maintenance of a communal nesting or denning site. Again, direct assistance to infants in avoiding heat or cold stress is possible.

Reduction of maternal energy costs during foraging

Among species such as elephants and primates where milk energy yields in each suckling episode are relatively low (Martin 1984), the need for a mother to maintain close proximity for frequent feeding places constraints on maternal energy budgets. The costs of foraging for high quality food to sustain the metabolic requirements of lactation can be inflated by the costs of carrying a dependent infant (see, for example, Altmann 1980; Dunbar & Dunbar 1988), or by the costs of limiting foraging areas when accompanied by a metabolically inefficient infant (Dublin 1983). Some ungulate species creche calves (see Reidman 1982), leaving them together while the mothers move off to forage without the constraints of a slow, vulnerable infant, and return to suckle them at some energetically optimal interval. Thus the mother is able to maximize her own energy intake for lactation while minimizing the risks of predation on her isolated calf. Several mothers must be available before such a creche is possible.

Increasing milk availability

Among mammals, the duration of lactational investment depends on the rate of energy transfer through milk, which affects infant growth rates (Oftedal 1984; Harvey & Clutton-Brock 1985; Reiss 1985). Communal suckling of infants increases the milk available for each infant and should either increase its growth rate or make up for any deficit in maternal milk output. Indiscriminant suckling (e.g. milk theft) by young has been reported in bat species and northern elephant seals (reviewed by Reidman 1982). These species breed synchronously in large, high density colonies where individual recognition of young presumably is difficult. However, from the point of view of the mother, milk theft or communal suckling increases her energetic costs of lactation since she is supporting both her own offspring and the thieves. Thus communal suckling only reduces maternal effort when it is unequally spread across mothers and some mothers benefit from reduced energy costs of milk production at the expense of others.

Female reproductive success is affected by her inter-birth interval. The loss of female condition during lactation or the duration of hormonally medicated lactational anoestrus depends on suckling frequencies of the infant, which in turn influence the potential for reconception. The age at which high-quality, easily digested supplemental foods are introduced into the infant's diet has a major effect on probabilities of conception (reviewed by Short 1984). The introduction of supplemental foods interacts with age-specific changes in suckling frequencies to reduce infant demands while maintaining high growth rates, leading to lower suckling frequencies and an earlier resumption of oestrus cycling. Provisioning of the young with supplementary foods can be a major contribution both to a mother's lactational effort and to her reproductive success when inter-birth intervals are shortened.

Promoting infant independence and learning

Contacts between infants and other group members have been proposed as a way of acquiring social knowledge, for example learning of relative dominance ranks (see, for example, Cheney 1978), discriminating between kin and non-kin (Kurland 1977; Berman 1982), or of developing adult skills such as through play-mothering

(Lancaster 1971). With an increase in infant independence, weaning can be accelerated and mothers potentially can reconceive earlier (Altmann, Altmann & Hausfater 1978; Simpson *et al.* 1981).

Communal care

These separate components are often used to define alloparenting among social mammalian species. What, however, is relatively rare is the *co-occurence* of these different components. While many species exhibit one or perhaps two and only in their passive form, only a small number show all types where active participation is either reciprocal or shared. I will use the co-occurence of these different components, and specifically active participation, to define species where infant care is communally shared between the members of a social group.

Some interactions between 'alloparents' and infants can be classed as a counter-strategy employed to interfere in the reproductive success of other females rather than as communal care. Among baboons, it has been suggested that high-ranking females direct stressful aggressive attention to lower-ranking females thereby affecting their ability to gestate (see, for example, Dunbar 1980) and the survivorship of neonates (Wasser & Barash 1983). Among macaques, when the infants of low-ranking females are contacted by non-relatives, these contacts tend to be abusive (reviewed by Hrdy 1976; see Silk 1980).

A further important distinction as to the nature of caretaking should be made for species where caretaking initially appears to be communal using the components defined above. The social carnivores, often classed among the communal care species, suppress reproduction among subordinate females. In foxes, African hunting dogs, and wolves (reviewed by MacDonald & Moehlman 1982), often only a single female is able to breed in the group. The other group members assist with rearing the young by provisioning, protecting, and promoting independence through social interactions, yet these individuals are unable to reproduce. A similar form of social suppression of reproduction is seen among dwarf mongoose, who also exhibit high levels of infant care (Rood 1980). In common with helpers in many avian species, the caretakers have no options for breeding themselves when territories are saturated, mates are limited, or resources necessary for reproduction vary unpredictably (Emlen 1982).

Females in these groups are trading their immediate reproductive potential for survivorship through increased foraging success, and for delayed benefits such as inheriting a territory or acquiring higher dominance status. While some members of small family groups may be inhibited from breeding with close kin, the hormonal basis for infertility among supressed females (Abbott *et al.* 1986; Abbott, this volume; and see Wasser & Barash 1983) suggests that the subordinate females are not actively avoiding breeding with kin. Cooperative infant care combined with reproductive suppression I consider to be *coercive care* systems, and these are distinguished from the communal caring species.

We are left with a small number of species showing the full range of non-coercive communal care of infants. While many of the colobine species studied do show high levels of communal care (McKenna 1979), the limited number of studies in comparison to the cercopithicenes makes it difficult to categorize each species and they are therefore treated as a group. The family structure of the communal caring taxa has been defined by the nature of the kinship relationships within the group (cf. Wrangham 1980) and the kin available to mothers and infants within the different families are presented in Table 1.

In order to describe the nature of communal care in relation to family structure, I will review data on allomothering in two mammalian species, African elephants (*Loxodonta africana*) and vervet monkeys (*Cercopithecus aethiops*), and then discuss infant care among humans. Both the non-human species live in matrilineal kin groups, where males disperse from the group at or shortly after adolescence. Among vervets, as is typical for most cercopithecenes, a group can be composed of several lineages of related females with relatively permanently associated unrelated adult males. In contrast, elephant 'groups' typically consist of family units of about three or four related adult females and their immature offspring. Males are only temporarily and randomly associated with family units.

TABLE 1. Species showing communal care of infants. Family structure is defined from kin relationships within the group. Closed families consist of parents and recent offspring. Female kin-bonded groups are those with a small number of female lineages, and male kin-bonded are those where females transfer between groups. Communal care is defined by alloparents' contributions to maternal energetics, to infant survival, and to socialization. Relatedness of potential caretakers is also distinguished

	Family structure		
	Small, closed families	Female kin-bonded	Male kin-bonded
Kin available to mothers	Juvenile offspring	Juvenile offspring Adult/adolescent females	Juvenile offspring
Kin available to infants	Siblings	Siblings Adult/adolescent female	Siblings Adult/adolescent males Grandmothers
Taxa:	Jackals* Coyotes*	Coatis[†] Banded mongoose [‡] Lions ([‡]) Brown hyaena [§] Elephants [|] Vervets [|] Colobines [¶]	
			? Common chimpanzee**
	Humans	Humans	Humans

* MacDonald & Moehlman 1982. [†] Russell 1982. [‡] Reidman 1982, Gittleman 1985. [§] Stewart 1987. [\|] This paper. [¶] McKenna 1979. ** Nishida 1983.

Members of elephant family units associate with each other in fluid groups that range in size from 2 to over 500, with persisting associations between certain families. These associations of families or 'bond groups' (Moss & Poole 1983) are likely to be descended from a single lineage which has grown over time and fissioned into smaller but still associated family units (Moss, long-term data). Thus elephants tend to live in smaller, stable groups but associate with a larger number of more distantly related animals than do the closed female kin-bonded societies of primates.

The data reviewed are derived from field studies on free-ranging population of the two species in Amboseli National Park, Kenya. Details of methodology and analyses are presented in full elsewhere (Lee 1983, 1984, 1987a). Over 250 elephant calves, aged between birth and 5 years old, were observed between 1982 and 84. Scan and focal animal sampling techniques were used to record calf behaviour. Sixteen infant vervets from three adjacent groups were observed intensively in their first 3 months of life. The behaviour of a further fifty-two immature animals was sampled over a 2-year period in 1978–80. All observations were made using focal animal samples, but *ad libitum* data were used for some rarely occurring events. Rates of behaviour were calculated from focal samples only.

THE ELEPHANT EXAMPLE

African elephants have often been viewed as one of the more extreme examples of alloparenting species, possibly because 'communal suckling' has been reported (see, for example, Wilson 1975). While most 'communal suckling' probably is comfort suckling and involves non-lactating, nulliparous females (Lee 1987a), the importance of allomothers among elephants can be seen in the frequency with which infants receive assistance at avoiding environmental and social stresses and hazards. Survivorship of infants in the first 2 years of life (the most vulnerable period: Lee & Moss 1986) was positively related to the number of potential allomothers (i.e. immature females) within the family unit (Fig. 1 and see Lee 1987a).

Such an effect might result from differences in family quality, in that large families are large by virtue of having high calf survivorship. However, there was no relationship between calf mortality (averaged over 12 years within each family separately) and the size of that family ($r_s = -0.008$, $n = 53$, N.S.). Furthermore, mean calf mortality within separate families also declines with increasing numbers of potential allomothers (Fig. 1) Calf survival does appear to be related to the number of available allomothers, the juvenile and adolescent females, within the family unit.

Young, vulnerable calves typically have another family member close by and juvenile and adolescent females tend to be close to a calf when the mother has moved away to forage and feed (Fig. 2). These same animals also respond to distress vocalizations from the calves and will rush to defend the calf from other

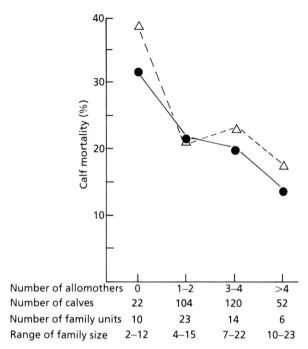

FIG. 1. The percentage of mortality in the first 24 months of life for elephant calves born to families with different numbers of juvenile and adolescent females (allomothers) (● — ●). The mean mortality per family unit averaged over 12 years for the median availability of allomothers is also presented (△ — △). The range of family sizes for each category is given. Data from C.J. Moss, unpublished long-term records.

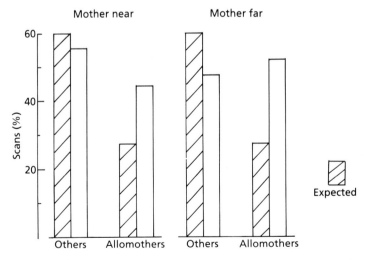

FIG. 2. The percentage of scan samples where the nearest neighbour to a young calf was either an allomother or another age−sex class of elephant (males, other calves, adult females) when mother was within 5 m (near) of the calf, and when she was >5 m (far). Expected proportions were calculated from the percentage of an average group in the different age−sex classes (see Lee 1987a).

elephants or from potential predators such as lions or humans, and to help it stand, swim or walk. Females, and especially those from the family unit, are most likely to assist or defend a calf when it is distressed (see Lee 1987a).

Elephant calves interact with other elephants in a variety of ways. While nutritive suckling tends to be restricted to mothers, comfort suckling with non-lactating females is most frequent with adolescents (Fig. 3). Interactions such as play are most frequent with peers, while those that involve greetings with another or investigating another's food, or comforting such as rubbing or dusting (a form of thermoregulation) tend to take place often with the juvenile and adolescent allomothers. These types of interactions are most frequent between siblings or other family females (Lee 1987a).

Elephant family units appear to act as a defensive unit for protecting and caring for young, risk-prone calves. Matrilineal families offer the potential for promoting calf survival when females cooperate to assist, defend and socialize calves. While large families containing more immature females tend to have higher calf survivorship, feeding competition between individuals may ultimately limit the advantages of having additional allomothers present (see Dublin 1983).

Since elephant calves are relatively slow-growing and have a period of lactational dependence of more than 2 years (Lee & Moss 1986), the reproductive effort of individual mothers within elephant families could be reduced when part of the costs of that care is taken up by the nulliparous females. Whether allomothering can reduce the inter-birth intervals of an elephant mother has yet to be determined. Since communal suckling is extremely rare, it would seem unlikely that the energy costs of lactation are met by females other than the mother, but the presence of caretakers might allow mothers to forage more efficiently. The main effect of the allomothers on reproductive effort appears to lie in their role in promoting calf survival.

NON-HUMAN PRIMATES: THE VERVET EXAMPLE

Among the cercopithecene primates, vervet monkeys have one of the most cooperative systems of communal infant care. While some of the components of communal care are seen in other species, among vervets contacts between infants and other group members are typically friendly and begin within a few hours of birth (reviewed in Lee 1983). Contacts between infant vervets and others are most common in the first month of life and decline in frequency throughout the first 3 months until the infant is weaned from ventral carrying (Lee 1983). Primate infants are most susceptible to environmental and social stresses in the first months of life, and Silk (1980) found that most harassments were directed towards macaque mothers during this period. For Amboseli vervets, very few such harassments were observed, and no infants were aggressively abused. While inefficient allomothering is a cause of infant mortality among captive vervets (Johnson *et al.* 1980; Fairbanks & McGuire 1984), this appears not to be the case in the wild (see Cheney *et al.* 1988).

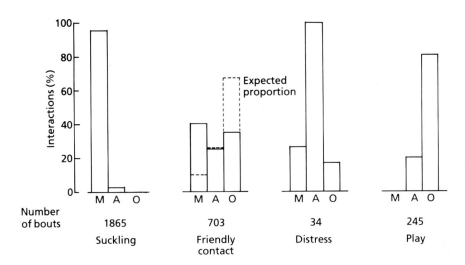

FIG. 3. The percentage of elephant calf social interactions that were with the mother (M), with allomothers (A) or with others (O). Expected proportions (dotted lines) were calculated only for friendly interactions. Since several individuals could participate in giving assistance during distress events, the totals are >100%.

The most frequent partners of infants are usually juvenile females, followed by adult females (Lee 1983). However, both juvenile and adult males occasionally carry, protect and 'babysit' infants. What, then, do these friendly contacts consist of? Infants were approached, sniffed, touched, cuddled, carried and groomed. When infants exhibited distress through 'whirring' vocalizations, another group member would typically respond to the distress call if the mother was unavailable. Infants who were frequently contacted were also frequently groomed by others ($r_s=0.576$, $n=16$, $P<0.05$). Grooming tends to become more frequent with increasing infant age, while contacts and ventral carrying decline.

High levels of proximity between an infant and older, experienced group members are maintained throughout the first 3 months of life, even though the types of interactions change in their relative importance. Mothers spend little time at distances of more than 2 m from infants in the first 3 months (range 0–30% of time; median=2%), and the time away was not significantly related to the rate of contacts received by infants ($r_s=0.301$, $n=16$, N.S.). However, when the mother was >2 m from the infant, another caretaker was almost always present (Fig. 4). Often, several infants would be left together with an allomother while the mothers would forage or interact at a distance. Whitten (1982) described reciprocity in allomothering such that vervet mothers who left their infants with another female would return to babysit that female's infant, and females with access to allo-mothers had higher food intakes than those without allomothers. Hrdy (1976) also pointed out that among langurs over 50% of infant transfers were followed by the mother going off to forage.

Family structure and reproductive effort

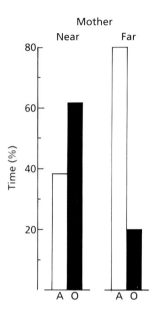

Fig. 4. The median percentage of time that an allomother was within 2 m of an infant vervet when the mother was near ($<$2 m) or far ($>$2 m) from the infant (n=16). (G Statistic = 124.1, d.f.=1, $P<$0.01.)

Unlike other cercopithecine species (e.g. macaques: Kurland 1977; Silk 1980; Berman 1982; baboons: Cheney 1978; Wasser & Barash 1983) and more like the colobines (Hrdy 1976; McKenna 1979), these friendly contacts were not limited to the closest relatives nor determined by the dyadic dominance relationships between mother and infant handler. The frequency of contacts was related to maternal dominance in that infants of high dominance mothers were contacted and cared for more frequently than those of lower dominance (Fig. 5), but the infant born first in the birth season was also frequently contacted, regardless of its maternal dominance. The attractiveness to allomothers of the infants with higher maternal dominance is atypical among primates, since it would be predicted that mothers of low rank should be less able to control the contacts received by their infants, and thus their infants would be handled more frequently and by a larger number of individuals. High ranking infants were both attractive (Cheney 1978; Seyfarth 1980), and available to allomothers, suggesting that allomothering takes place within relaxed contexts. The most frequent adult female allomother of an infant was equally likely to be of higher (n=11) or lower (n=8) rank than that infant's mother. Juvenile siblings (both sisters and, to a lesser degree, brothers) also contact and care for infants more frequently than do more distantly related group members (Lee 1987b). However, sibling preferences do not exclude other animals from contacting the infants. Of nineteen infants, only four of the eight with juvenile female siblings available were contacted most frequently by that sibling.

While it can be argued that the immature alloparents of vervet infants were gaining both experience for use with their own infants and protecting kin, it can

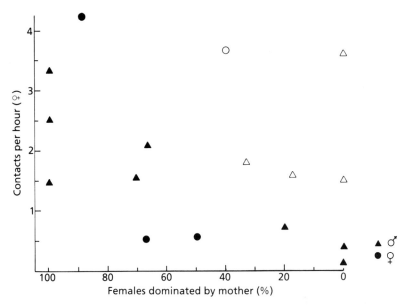

FIG. 5. The hourly rate of contacts to infant vervets by adult and non-sibling juvenile females plotted against relative maternal rank ($r_s=0.825$, $n=11$, $P<0.05$; first born infants excluded). Open symbols are first born infants. Circles are female infants, triangles are male infants.

also be asked if they had an effect on the reproductive effort of the mothers. Altmann, Altmann & Hausfater (1978) suggested that different 'styles' of inter-action between baboon mothers and infants could be related to the probability of subsequent conception (see also Simpson et al. 1981 for macaques). Mothers who were tolerant of their infants' attempts to leave proximity, or to contact other group members tended to resume cycling earlier than those who restrained their infants but also rejected them frequently from the nipple. While maternal styles have been described for captive vervets (Johnson et al. 1980), among free-ranging vervet mothers in Amboseli, such maternal styles could not be observed (Lee 1984). No mother consistently rejected her infant over time, nor did others consistently restrain their infants from leaving their proximity. All mothers, and especially those with access to allomothers, tended to be relaxed about their infants' contacts with others. They rejected infants from the nipple as a result of their ability to sustain milk production under nutritional constraints in a variable environment (Lee 1987c) and not as a function of dominance or ability to control their infants' contacts with others (see, for example, Altmann 1980). The prob-ability of reconception was strongly influenced by food availability and quality and appeared to be unrelated to maternal style. However, the lack of any consistent maternal styles among wild vervets may result from the presence of friendly and supportive allomothers who can buffer the mother against energy limitations during foraging and who provide a generally relaxed environment for maternal ınd infant social interactions.

Among the Amboseli vervets, the factor contributing most to female lifetime reproductive success was infant survival (Cheney *et al.* 1988). The contribution of allomothers to reproductive effort is potentially greatest when they provide direct protection for infants, and when they reduce maternal foraging costs, thus allowing mothers to maintain infant growth and again increasing the probability of survival.

DO HELPERS HELP?

The evidence for direct contributions by alloparents to maternal effort and reproductive success is relatively limited and tends to be correlational rather than causal. The effect of different types and levels of communal care can not yet be adequately demonstrated. However, the types of care are consistent, widespread and for some species (such as jackals: see MacDonald & Moehlman 1982) are related to survival. Among the species discussed here, elephants and vervets, caretakers do appear to assist the mother with at least some of the energetic burdens of reproduction. Allomothers appear to free the mother for more energetically efficient subsistence activities, and thus maintain infant growth rates and potentially reduce inter-birth intervals. Among other species in Table 1 such as hyaenas, where communal denning and provisioning of pups occurs, the same effects can be proposed but remain to be demonstrated. For elephants, vervets and possibly the colobines, where mortality is high before weaning as a result of predation, environmental stresses, or infanticide, defence of infants and assistance given to infants in confronting environmental stresses are major contributions to infant survivorship. Obviously, there may be advantages to the allomothers in terms of experience gained which will help with rearing their own offspring (see, for example, Reidman 1982). And among long-lived species, the potential for reciprocity in caretaking exists, especially between females in the matrilinial kin-bonded groups described (Whitten 1982; Lee 1987a).

How, then does the structure of the family affect the nature and extent of communal care? When few potential caretakers are available, as a result of dispersal, age at breeding, interbirth intervals and other life history variables, caretaking will be less common. In kin-structured groups, the possibility for cooperation between relatives increases the frequency and complexity of the types of caretaking observed. While larger families, and hence larger numbers of caretakers, appear to increase infant survivorship for jackals and elephants, resource limitations may constrain or introduce variance into the size of families without altering the kin structure of the group.

CARETAKING IN HUMAN SOCIETIES

Can we use the mammalian examples to interpret some aspects of caretaking among human societies? While not intended as a comprehensive review, what is of interest is whether contributions from alloparents to maternal energetics and infant survival can be distinguished, and whether the nature of the family has similar effects on the interactions between infants and other group members. Since human societies range from small, closed families, through matrilineal kin groups, to, most commonly, patrilineal and patrilocal groups (see, for example, van den Berghe 1979) — the male kin-bonded groups — we can predict differences between these families in the nature and extent of alloparenting.

Among humans, reports on alloparenting have tended to concentrate on the contribution of childminders to infant socialization and cognitive development. Cross-cultural analyses, however, suggest that in over 40% of societies, infants are cared for by non-mothers for significant proportions of time (Weisner & Gallimore 1977). Even in small, closed families typical of some industrialized countries, contacts with maternal and paternal kin can be maintained, and thus they are available as alloparents (Essock-Vitale & McGuire 1985).

Studies of alloparenting among humans have only rarely been designed to assess the effects of caretakers on infant survival and maternal energetics. Irons (1983) noted that a major way in which mothers could alter their reproductive effort was through the use of caretakers and Borgerhoff Mulder & Milton (1985) have suggested that polygynous marriages offer mothers the advantage of access to a greater number of caretakers. A relationship between household density (family size and number of potential caretakers) and childminding has been demonstrated for a number of societies (Munroe & Munroe 1971; Whiting & Whiting 1975; Chisholm 1983). Several consistent qualitative trends emerge. Firstly, kin of the infant tend to be involved in most caretaking events, and close kin (especially siblings) tend to be the most frequent childminders. While the degree of maternal tolerance of her baby is proportional to the number of adults in the household (see, for example, Munroe & Munroe 1971), the kin structure of the household should also predict maternal confidence in the quality of childminding. Among a matrilineal society such as the Navaho, female kin (sisters, aunts, maternal grandmothers, cousins) are especially important caretakers (Chisholm 1983). When Navaho women live in small nuclear families, they both maintain closer proximity to their infants and are less relaxed when household densities increase, possibly since infants exhibit more anxiety towards strangers. Among a patrilineal society such as the Kipsigis, full siblings are more attentive childminders than are children of co-wives (Borgerhoff Mulder & Milton 1985) but the number of available childminders correlates with a mother's ability to leave the household. Among the patrilineal societies, the potential for help from the paternal grand-mother is high (since she is likely to be the most available female kin) and more help from grandmothers could be predicted to occur.

Childminders among humans appear to play a major role in freeing the mother from some foraging constraints, although these are far more marked in agricultural societies than among gathering women (Nerlove 1974; Whiting & Whiting 1975). In agricultural societies, not only do families tend to be larger and thus more caretakers are available, but the location of a permanent cultivated plot may mean more predictable patterns of being near or far from the infant and childminders can contribute to infant care more reliably. When gathering distances are unpredictable, and the time to return to the child thus variable, mothers are more likely to carry the child in order to maintain suckling (Blurton-Jones 1972; Draper 1976; Konner 1976a). Furthermore, cultivated foods may be easier to process into breast milk supplements than gathered foods or meat. Thus one of the main contributions to female reproductive effort among agricultural societies comes through assisting with infant feeding and promoting infant growth while allowing the mother to maintain daily subsistance activies (Nerlove 1974). The consequent reproductive success of women whose inter-birth intervals are shortened with the use of supplemental infant foods potentially is enhanced. The Neolithic demographic transition to higher populations densities (Cohen 1980) may in part be explained by the increased availability of supplemental infant foods and reductions in reproductive effort leading to higher reproductive success among women.

Even when milk substitutes are less available, childminders provide protection for infants when mothers are not in close proximity, and are used to do so. Childhood traumatic illnesses can be a major source of mortality, and as in the case of elephants, extra eyes, ears and hands can protect infants from environmental stresses. Play groups of peers or mixed-aged children also provide social experience while keeping the infant in a relatively safe environment (Konner 1976b).

CONCLUSIONS: COMMUNAL CARE AND FEMALE KIN GROUPS

An interesting speculation can be posed by the examples examined above, relating to the question of female kin sociality. Discussions of sociality tend to revolve around the issues of the advantages of group defence of resources or alternatively, the avoidance of predation. While communal care of infants is generally mentioned as an advantage to group-living, it is seldom viewed as a causal factor for mammals (see Emlen 1982 for birds).

Among mammals, and especially primates, female bonding has been related to resource defense (see, for example, Wrangham 1980, and see van Schaik, this volume). If females need to cooperate to control access to food in the face of competition from other individuals, then typically they cooperate with kin. However, when infant survival contributes more to female lifetime reproductive success than does defence of resources, sociality of the sort that promotes infant survivorship will be at a selective premium. Relatively low quality food that is widely distributed (e.g. grass in the case of elephants, leaves in the case of the colobines), or food that is so patchy in distribution as to make the defence of those patches untenable (e.g. prey in the case of lions or hyaena) can lead to female sociality specifically related to infant survival. The most likely social grouping where members will communally protect and assist infants is that where the degree of relatedness between females is high, the probability of dispersing successfully to breed is low, and the fitness of relatives is unbiased (Vehrencamp 1985) by the actions of a dominant individual — in small, stable matrineal groups. Many of the communal caring species in Table 1 fall into this category.

Among the primates, infant care as a major factor in the evolution of sociality has been widely accepted for the New World monkeys (see, for example, Terborgh & Goldizen 1985). When the infant is a high proportion of the mother's body weight, grows rapidly and requires high energetic investment, then caretakers other than the mother will be exploited when possible, as exemplified by the polyandrous social systems among marmosets and tamarins where two males are required to assist a single female. Juvenile and adolescent helpers (both male and female) are also exploited, typically with reproductive suppression. While resource constraints keep the groups relatively small and the costs of dispersal are high, the advantages of several available individuals to assist with infant care promotes one of the more unusual primate systems of sociality.

High levels of communal care are reported for those primate species with small group sizes — the vervets and the colobines. The degree of relatedness between females in such groups can, potentially at least, be greater than in large groups composed of several matrilines such as in the macaques and baboons. Furthermore, as Silk & Boyd (1983) have argued, the degree of relatedness between females in a lineage can be increased with a single breeding male (e.g. in many colobine groups). With large group sizes and consequent differences in the degrees of relatedness between and within matrilines, infant care is most commonly seen within the matriline, while infant abuse tends to be seen across matrilines (Hrdy 1976; Silk 1980). Elephants, as well, live in relatively small matrilineal groups and exhibit high levels of communal care. When elephant families become large, fissioning tends to take place (Moss & Poole 1983; Moss, unpublished long-term data). High degrees of relatedness are thus maintained within the families.

The size of the group, and thus the degree of relatedness between and within matrilines, can ultimately be influenced by either resource defence, when large groups have an advantage in competition for access to resources, or by predation pressure, when large groups reduce predation risks. Furthermore group size may be constrained by resource availability and distribution, and consequently its defensibility. It is unexpected that vervets should show such high levels of communal care. Unlike most cercopithecines they are territorial — predicting a competitive advantage to larger group sizes. However, group size alone may not necessarily determine the outcome of territorial interactions. When groups are confined to small patches of seasonally variable resources, these can not be exploited by large numbers of individuals without increasing territory size beyond that which is defensible. If the advantages of large group size for resource defence or avoidance of predation are high, group size may increase and less communal care will be seen — each infant born in a distant lineage is a potential competitor for resources with a less related female's infant, especially when food is limited (see, for example, Dittus 1979). Even the frequency of intra-lineage infant care is expected to be reduced, since mothers would be less willing to allow other group members access to their vulnerable infants when some group members are likely to abuse infants. If, however, the inclusive fitness of females in the lineage is increased when high levels of infant care enhance infant survival, mothers should be willing to participate equally in caring for infants and allowing allomothers access to their infants. Trust in consanguity is evidenced by higher rates of caretaking.

Perhaps, among humans, where consanguity can be demonstrated not just through proximity and behaviour but also defined by language and obligations, trust that childminders will be non-abusive does not limit allomothering to small matrilineal groups. Even among humans, however, more childminding goes on between relatives, so kin relations are still important for defining and promoting access to childminders.

Explanations for the evolution of sociality obviously are not limited to the need for promoting infant survival through communal care. However, female kin-bonding within a group which results in communal care may indeed be a major advantage when the other costs or benefits of sociality are less influential.

ACKNOWLEDGEMENTS

I thank M. Borgerhoff Mulder, R. Foley, D. Kleiman, H. Kruuk, K. Lindsay, K. Stewart and F. White for stimulating discussions and critical comments. The vervet study was made possible by the assistance of D. Cheney and R. Seyfarth, and was funded by a L.S.B. Leakey Trust Studentship. The elephant study was funded by the National Geographic Society, and would not have been possible without the assistance, advice and support of C. Moss. R. Hinde and P. Bateson generously provided facilities for writing.

REFERENCES

Abbott, D.H., Keverne, E.B., Moore, G.F. & Yodyingyuad, U. (1986). Social suppression of reproduction in subordinate talapoin monkeys (*Miopithecus talapoin*). *Primate Ontogeny, Cognition and Social Behaviour* (Ed. by J.G. Else & P.C. Lee), pp. 329–41. Cambridge University Press, Cambridge.

Altmann, J. (1980). *Baboon Mothers and Infants*. Princeton University Press, Princeton, New Jersey.

Altmann, J., Altmann, S.A. & Hausfater, G. (1978). Primate infant's effects on mother's future reproduction. *Science*, **201**, 1028–31.

Berman, C. (1982). The ontogeny of social relationships with group companions among free-ranging infant rhesus monkeys. II. Differentiation and attractiveness. *Animal Behaviour*, **30**, 163–70.

Borgerhoff Mulder, M. & Milton, M. (1985). Factors affecting infant care in the Kipsigis. *Journal of Anthropological Research*, **41**, 231–61.

Blurton Jones, N. (1972). Comparative aspects of mother–child contact. *Ethological Studies of Child Behaviour* (Ed. by N. Blurton Jones), pp. 305–28. Cambridge University Press, Cambridge.

Cheney, D.L. (1978). Interactions of immature male and female baboons with adult females. *Animal Behaviour*, **26**, 389–408.

Cheney, D.L., Seyfarth, R.M., Andelman, S.J., & Lee, P.C. (1988). Reproductive success in vervet monkeys. *Reproductive Success* (Ed. by T.H. Clutton-Brock), pp. 384–402. University of Chicago Press, Chicago, Illinois.

Chisholm, J.S. (1983). *Navaho Infancy: An Ethological Study of Child Development*. Aldine, New York.

Cohen, M.N. (1980). Speculations on the evolution of density measurement and population regulation in *Homo sapiens*. *Biosocial Mechanisms of Population Regulation* (Ed. by M.N. Cohen, R.S. Malpass & H.G. Klein), pp. 275–303. Yale University Press, New Haven, Connecticut.

Dittus, W.J.P. (1979). The evolution of behaviour regulating density and age-specific sex ratios in a primate population. *Behaviour*, **69**, 265–302.

Draper, P. (1976). Social and economic constraints on child life among the !Kung. *Kalahari Hunter-Gatherers* (Ed. by R.B. Lee & I. DeVore), pp. 199–217. Harvard University Press, Cambridge, Massachusetts.

Dublin, H.T. (1983). Cooperation and reproductive competition among female African elephants. *Social Behaviour of Female Vertebrates* (Ed. by S.K. Wasser), pp. 291–313. Academic Press, New York.

Dunbar, R.I.M. (1980). Determinants and evolutionary consequences of dominance among female gelada baboons. *Behavioural Ecology and Sociobiology*, **7**, 253–65.

Dunbar, R.I.M. & Dunbar, P. (1988). Activity budgets and maternal energetics in gelada baboons. *Animal Behaviour*, **36**, 970–80.

Emlen, S.T. (1982). The evolution of helping. II. The role of behavioural conflict. *American Naturalist*, **119**, 40–53.

Essock-Vitale, S.M. & McGuire, M.T. (1985). Women's lives viewed from an evolutionary perspective. II. Patterns of helping. *Ethology and Sociobiology*, **6**, 155–73.

Estes, R.D. (1976). The significance of breeding synchrony in wildebeest. *East African Wildlife Journal*, **14**, 135–52.

Fairbanks, L.A. & McGuire, M.T. (1984). Determinants of fecundity and reproductive success in captive vervet monkeys. *American Journal of Primatology*, **7**, 27–38.

Gittleman, J.L. (1985). Functions of communal care in mammals. *Evolution: Essays in Honour of John Maynard Smith* (Ed. by P.J. Greenwood, P.H. Harvey & M. Slatkin), pp. 187–205. Cambridge University Press, Cambridge.

Gray, J.P. (1986). *Primate Sociobiology*. HRAF Press, New Haven, Connecticut.

Harvey, P.H. & Clutton-Brock, T.H. (1985). Life history variation in primates. *Evolution*, **39**, 559–81.

Hrdy, S.B. (1976). Care and exploitation of non-human primate infants by conspecifics other than the mother. *Advances in the Study of Behaviour*, **6**, 101–58.

Irons, W. (1983). Human female reproductive strategies. *Social Behaviour of Female Vertebrates* (Ed. by S.K. Wasser), pp. 169–213. Academic Press, New York.

Johnson, C.K., Koerner, C., Esrin, M. & Duoos, D. (1980). Alloparental care and kinship in captive social groups of vervet monkeys (*Cercopithecus aethiops sabaeus*). *Primates*, **21**, 406–15.

Konner, M.J. (1976a). Maternal care, infant behavior and development among the !Kung. *Kalahari Hunter-Gatherers* (Ed. by R.B. Lee & I. DeVore), pp. 218–45. Harvard University Press, Cambridge, Massachusetts.

Konner, M.J. (1976b). Relations among infants and juveniles in comparative perspective. *Social Science Information*, **15**, 371–402.

Kurland, J. (1977). *Kin Selection in Japanese Monkeys*. Karger, Basel.

Lancaster, J. (1971). Play-mothering: the relations between juvenile females and young infants among free-ranging vervet monkeys (*Cercopithecus aethiops*). *Folia Primatologica*, **15**, 161–82.

Lee, P.C. (1983). Caretaking of infants and mother-infant relationships. *Primate Social Relationships* (Ed. by R.A. Hinde), pp. 146–51. Blackwell Scientific Publications, Oxford.

Lee, P.C. (1984). Early infant development and maternal care in free-ranging vervet monkeys. *Primates*, **25**, 36–47.

Lee, P.C. (1987a). Allomothering among African elephants. *Animal Behaviour*, **35**, 278–91.

Lee, P.C. (1987b). Sibships: cooperation and competition among immature vervet monkeys. *Primates*, **28**, 47–59.

Lee, P.C. (1987c). Nutrition, fertility and maternal investment in primates. *Journal of Zoology*, **213**, 409–22.

Lee, P.C. & Moss, C.J. (1986). Early maternal investment in male and female African elephant calves. *Behavioural Ecology and Sociobiology*, **18**, 352–61.

MacDonald, D.W. & Moehlman, P.D. (1982). Cooperation, altruism and restraint in the reproduction of carivores. *Perspectives in Ethology*, **5**, *Ontogeny* (Ed. by P. Bateson & P.H. Klopfer), pp. 433–67. Plenum Press, New York.

McKenna, J.J. (1979). The evolution of allomothering behavior among colobine monkeys: function and opportunism in evolution. *American Anthropologist*, **81**, 818–40.

Martin, R.D. (1984). Scaling effects and adaptive strategies in mammalian lactation. *Physiological Strategies in Lactation* (Ed. by M. Peaker, R.G. Vernon & C.H. Knight). Symposium of the Zoological Society of London, 51, pp. 87–117. Academic Press, London.

Moss, C.J. & Poole, J.H. (1983). Relationships and social structure among African elephants. *Primate Social Relationships* (Ed. by R.A. Hinde), pp. 315–25. Blackwell Scientific Publications, Oxford.

Munroe, R.H. & Munroe, R.L. (1971). Household density and infant care in an East African society. *Journal of Social Psychology*, **83**, 3–13.

Nerlove, S.B. (1974). Women's workload and infant feeding practices: a relationship with demographic implications. *Ethnology*, **13**, 207–14.

Nishida, T. (1983). Alloparental behavior in wild chimpanzees of the Mahale Mountains, Tanzania. *Folia Primatologica*, **41**, 1–33.

Oftedal, O.T. (1984). Milk composition, milk yield and energy output at peak lactation: a comparative review. *Physiological Strategies in Lactation* (Ed. by M. Peaker, R.G. Vernon & C.H. Knight). *Symposium of the Zoological Society of London*, 51, pp. 33–85, Academic Press, London.

Quiatt, D. (1979). Aunts and mothers: implications of allomaternal behaviour of nonhuman primates. *American Anthropologist*, **81**, 310–19.

Reidman, M.L. (1982). The evolution of alloparental care and adoption in mammals and birds. *Quarterly Review of Biology*, **57**, 405–35.

Reiss, M.J. (1985). The allometry of reproduction: why larger species invest relatively less in their offspring. *Journal of Theoretical Biology*, **113**, 529–44.

Rood, J.P. (1980). Mating relationships and breeding suppression in the dwarf mongoose. *Animal Behaviour*, **28**, 143–50.

Russell, J.K. (1983). Altruism in coati bands: nepotism or reciprocity. *Social Behaviour of Female Vertebrates* (Ed. by S.K. Wasser), pp. 263–90. Academic Press, New York.

Seyfarth, R.M. (1980). The distribution of grooming and related behaviours among adult female vervet monkeys. *Animal Behaviour*, **28**, 789–813.

Short, R.V. (1984). Breast feeding. *Scientific American*, **250**, 35–41.

Silk, J.B. (1980). Kidnapping and female competition among captive bonnet macaques. *Primates*, **21**, 100–10.

Silk, J.B. & Boyd, R. (1983). Cooperation, competition and mate choice in matrilineal macaques groups. *Social Behavior of Female Vertebrates* (Ed. by S.K. Wasser), pp. 315–47. Academic Press, New York.

Simpson, M.J.A., Simpson, A.E., Hooley, J. & Zunz, M. (1981). Infant-related influences on birth intervals in rhesus monkeys. *Nature*, **290**, 49–51.

Stewart, K.J. (1987). Spotted hyaenas: the importance of being dominant. *Trends in Evolution and Ecology*, **2**, 88–9.

Terborgh, J. & Goldizen, A.W. (1985). On the mating system of the cooperatively breeding saddle-backed tamarin (*Saguinus fuscicollis*). *Behavioural Ecology and Sociobiology*, **16**, 293–300.

van den Berghe, P.L. (1979). *Human Family Systems: An Evolutionary Perspective*. Elsevier, New York.

Vehrencamp, S.L. (1983). A model for the evolution of despotic versus egalitarian societies. *Animal Behaviour*, **31**, 667–82.

Wasser, S.K. & Barash, D.P. (1983). Reproductive supression among female mammals: implications for biomedicine and sexual selection theory. *Quarterly Review of Biology*, **58**, 513–38.

Weisner, T.S. & Gallimore, R. (1977). My brother's keeper: child and sibling caretaking. *Current Anthropology*, **18**, 169–89.

Whiting, B., & Whiting, J.W.M. (1975). *Children of Six Cultures*. Harvard University Press, Cambridge, Massachusetts.

Whitten, P.L. (1982). *Female Reproductive Strategies Among Vervet Monkeys*. Ph.D. Thesis, Harvard University.

Wilson, E.O. (1975). *Sociobiology*. Belknap Press, Cambridge, Massachusetts.

Wrangham, R.W. (1980). An ecological model for the evolution of female bonded primate groups. *Behaviour*, **75**, 262–300.

Hardworking Hadza grandmothers

K. HAWKES*, J.F. O'CONNELL* AND
N.G. BLURTON JONES**

*Department of Anthropology, University of Utah, Salt Lake City, Utah 84112, USA, and
**Graduate School of Education, Department of Anthropology and Department of Psychiatry,
University of California, Los Angeles, CA 90024, USA

SUMMARY

Among Hadza hunter-gatherers in northern Tanzania, women spend a great deal of time in food acquisition, and those past child-bearing age spend more than younger women. We report this pattern of age-related work and consider explanations for it, concluding provisionally that it may be an optimal allocation of reproductive effort. We note that humans are distinguished from other primates by their capacity to extract resources with enough efficiency to regularly feed others as well as themselves. If female investments in resource acquisition compete with investments in fertility, older females may gain greater fitness by helping their adult daughters than by carrying additional, riskier pregnancies themselves. Among our hominid ancestors, this same capacity would have had profound consequences for patterns of sharing and life history strategies, including the evolution of menopause.

INTRODUCTION

Two developments in the recent literature on hunter-gatherers are of special interest here. The first is documentation of a wide range of variation in many aspects of hunter-gatherer behaviour, from sexual divisions of labour (Estioko-Griffin & Griffin 1985) through levels of work effort (Hill et al. 1985) to the complexity of political organization (Bean 1978; Marquardt 1986). The second has been the application of theory and models from evolutionary ecology to the study of this variation (see, for example, Winterhalder & Smith 1981; Hill & Kaplan 1988). These applications have been spurred by the successes gained in biology from conceptualizing patterns of behaviour as adaptations in the narrow sense (Williams 1966), that is, as strategies shaped by natural selection, and thus amenable to analysis and explanation in terms of their reproductive consequences for the individuals who display them, consequences which in turn depend on the characteristics of those individuals and their ecological context (see, for example, Krebs & Davies 1984; Stephens & Krebs 1986 for an extended discussion).

This paper is a contribution to both of these developments. More precisely, it is about women's work effort among the Hadza hunter-gatherers in northern Tanzania. In many human societies, the economic contribution of women declines

as they become grandmothers (Amoss & Harrell 1981). This is not so among the
Hadza, where women work harder and procure more food as they pass from
middle adulthood to their 'post-reproductive' years. Here we describe this pattern
of age-related work effort, and explore some hypotheses which might account for
it. We begin with a brief description of the study population, with special attention
to women's foraging, then outline our methods of data collection, and report some
quantitative results. Then we review some elementary components of life history
theory, and in this context, identify the factors which might underlie older women's
work. Evolutionary theory leads us to enquire whether the fitness-related costs and
benefits of food acquisition might be key determinants of Hadza women's behaviour,
and this prompts us to speculate about the role of these variables in human evolution.

THE HADZA

The Eastern Hadza are a group of 750 people who currently occupy a 2500 km^2
area south and east of Lake Eyasi. Much of this region is rough and hilly, and
covered with mixed savannah woodland. The climate is warm and dry. Annual
average rainfall is in the 300–600 mm range, most of it falling in the 6–7-month
wet season (November to May) (Schultz 1971).

At the time of first European contact, around the beginning of this century,
only the Hadza occupied this country (Woodburn 1964, and references therein).
They apparently lived entirely by hunting and gathering. Local incursions by non-
Hadza pastoral and agricultural groups are recorded in historic times as early as
the 1920s (Woodburn 1964, 1986; McDowell 1981). Archaeological evidence
suggests the periodic presence of farmers and pastoralists in the region for several
millennia, hunter-gatherers far longer (Mehlman, personal communication). Non-
Hadza settlement is now heaviest in the Mangola area, at the northern end of
traditional Hadza territory, and along the western edge of the Mbulu Plateau.

During the past 50 years, various segments of the Hadza population have been
subjected to a series of government and mission sponsored settlement schemes
designed to encourage them to abandon the foraging life in favour of full-time
farming (McDowell 1981; Ndagala 1986). None of these schemes has been successful; and in every case, most of the Hadza involved have returned to the bush,
usually within a few months. In each instance, some Hadza managed to avoid
settlement altogether, and continued to live as full-time hunter-gatherers.

Ethnographic data on the Eastern Hadza are available in a number of short
reports dating from the late nineteenth century to the mid-1960s (Baumann 1894;
Obst 1912; Reche 1914; Bagshawe 1925; Bleek 1931; Cooper 1949; Fosbrooke
1956; Tomita 1966). The first comprehensive description was provided by Kohl-
Larson (1958), based on field work in the 1930s. More recently, James Woodburn
(1964, 1968a, 1968b, 1970, 1972, 1979, 1986; Woodburn & Hudson 1966; see also
Bennett *et al.* 1970, 1975) has presented the results of several extended periods of
local research between 1958 and 1970. Topics addressed include demography,
kinship, social organization, subsistence and settlement patterns, and health and
nutrition. Several projects have also been undertaken within the last decade,

primarily concerned with ecological and ethnoarchaeological issues (Smith 1980, unpublished field notes; McDowell 1981; Vincent 1985a, 1985b; O'Connell, Hawkes & Blurton Jones 1988a, 1988b; H. Bunn & E. Kroll, personal communication).

Most of the data reported here were collected over a 15-month period in 1985−86 among Hadza living in the areas known locally as Tli'ika and Han!abi. These areas are situated astride a long, narrow (50 × 15 km), north-east to south-west trending ridge, flanked on the north-west and south-east by the Eyasi and Yaeda basins, respectively. The central portion of the ridge has an average elevation of 1200−1400 m; isolated peaks exceed 1500 m. Drainage is primarily to the north-west, toward Lake Eyasi. Nearly all streams are seasonally dry, although water can be found throughout the year in isolated pools and rockholes, and in subsurface catchments in stream channels. Lake Eyasi itself is a playa, which holds water only during the wet season and for a short period thereafter.

The crest and upper slopes of the ridge are covered by thick bush (*Acacia brevispica−Aloe−Dahlbergia*), in which giant baobab trees (*Adansonia digitata*) are a common and very striking element (Schultz 1971). The lower slopes and alluvial plains on the Eyasi side are dominated by *Acacia tortilis*−Commiphora woodland. The Yaeda Valley is a grassland, seasonally flooded at its south-west end. Medium and large herbivores, notably elephant, giraffe, buffalo, zebra, and several species of antelope are common throughout this region (Smith 1980). Large predators, including lion, leopard, and hyaena are also present.

Since the mid-1970s, the Tli'ika/Han!abi area has been occupied by about 200−300 Hadza, all of whom have lived primarily by hunting and gathering throughout most of that period (L.C. Smith, personal communication). Also present on occasion are members of a population of Datoga, a Nilotic pastoral group, whose range includes the lower slopes of ridge on both the Eyasi and Yaeda sides (see Klima 1985 for ethnographic background). Relationships between the Datoga and Hadza are strained at best, and have in the past been punctuated by violence (Mathiesson 1972). Contact between the two groups is limited, apart from early dry-season honey trade. Datoga livestock compete with other occupants of the region for access to water during the late dry season.

Many Tli'ika/Han!abi Hadza have relatives (mainly other Hadza, or Isanzu, with whom the Hadza have intermarried to some degree; see Woodburn 1986) living in or near the villages of Yaeda, Munguli, and Mangola, located 5−15 hours' walk to the south, south-west and north-east, respectively. Some of these people are full or part-time farmers. Bush-dwelling Hadza visit them occasionally for periods of up to a week in the dry season, usually to trade dried meat for tobacco, maize, or millet. Some may stay even longer during the late wet or very early dry season in order to benefit from the harvest.

WOMEN'S FORAGING

Hadza foraging patterns vary seasonally. During the 1985 dry season, the Hadza were concentrated in five, or sometimes six relatively large camps (population forty-five or sixty each) located near permanent water sources along the northern crest of the ridge overlooking Lake Eyasi. Camp composition was flexible with a notable tendency for married couples to live with the wife's mother (Woodburn 1968b describes this pattern as well). Parties of women foraged around these camps virtually every day. Their most common targets were tubers, most often //ekwa (*Vigna frutescens*). The variant called //ekwa hasa by Hadza in Mangola (Vincent 1985a) is the only type of //*ekwa* identified in the Tli'ika/Han!abi region. It was also taken in other seasons, usually in the same general way.

Women left camp early, sometimes by 07.00 hours, but usually between 08.00−09.00 hours. They were typically accompanied by infants under 2, children over about 6 or 7, and by armed teenage boys, or sometimes an adult man who provided protection against human strangers, most likely pastoralists who might be travelling through the area. After walking as a group from 4−60 minutes (mean = 25 ± 14 minutes for thirty-one trips), women reached their target patch for the day. They fanned out, tapping the ground around the base of the woody rootstalks with digging sticks to assess, they said, the probable presence and size of tubers in the rocky soil.

Having found a likely spot, each woman settled down to dig, usually just out of sight of other members of the party. Individuals kept track of each other with occasional calls. Periodically each woman loaded the tubers she had collected into a carrying cloth and moved on in search of another rootstalk. As she passed other women she might pause briefly to chat and check their progress. Sometimes more than one woman dug at the same rootstalk.

This was hard work. Excavations were sometimes more than a metre deep, and often involved moving large rocks through complex sequences of levering. Particularly intricate engineering problems drew the attention of several spectators. However, about three-quarters of the time spent in //*ekwa* patches was devoted to active digging. Women who had carried nursing infants sometimes enlisted older children to sit with them and hold the infant, interrupting their digging periodically to nurse.

Women usually dug to just past midday, when they gathered to cook the tubers on a common high flame fire. If the party was large, more than one fire was kindled. Tubers were rubbed free of dirt and grit, piled on the fire, often to a height of 30 cm or more, roasted and turned for 5−15 minutes, pulled off, stripped by hand of their tough outer skins, and eaten. After appetites were satisfied, or the tubers all eaten, a second bout of digging began and continued through the afternoon, followed by more cooking and eating. Women then loaded the remaining cooked tubers into their carrying cloaks and returned home. Just before reaching camp, those in the lead would pause so that any stragglers might catch up. The entire group then continued, arriving in camp together, often to the accompaniment of an excited chant by the children waiting there. Children and adults who had remained at home gathered around the women, and with varying degrees of insistence, claimed a share of tubers. Although acquisition loads were routinely recorded, the total remaining to carry home after consumption was not usually weighed during this period. Data on both acquisition totals and carry-home loads are available for a sample of 27 woman days. These indicate that 0.390 ±0.128 of the total weight of tubers acquired was consumed in the field by the collectors, other members of the party, or by passing hunters.

Although it was the norm, women did not always go collecting as a single party. Less often, they went accompanied only by their husband. One woman who did this frequently was a topic of censurious gossip among the other women for failing to follow the proper pattern and join the women's foraging party.

During the wet season, Hadza moved camp several times, first to the lower slopes of the ridge overlooking Lake Eyasi, then later back to the crest of the ridge, roughly along the north−south drainage line. Camps were somewhat smaller throughout this period, averaging about thirty to fifty occupants. Women continued to forage for roots (especially //ekwa), but during parts of this period they spent as many as half their foraging days taking berries. Species most commonly sought were undushi (Cordia spp.), which were found in large, dense patches, hundreds of metres across, on the lower alluvium south of Lake Eyasi; and kongoro (Grewia bicolor), which were encountered in large patches at the lower elevations and in much smaller patches throughout the uplands. Berry-collecting expeditions usually involved more than an hour of travel time each way (mean = 76±37 minutes, for seven kongoro trips), some as many as 2 hours, 10 minutes.

Women intending to take berries left camp as a group, usually between 08.00−09.00 hours. Party composition was as described for root-collecting trips. Party members often stopped to snack briefly on small patches of berries encountered as they walked. On arrival at the target patch, they typically spread out over an area 30−50 m in diameter and began to collect, alternately eating and filling their carrying cloaks or whatever other containers they had brought. Undushi are juicy enough to mash easily in the folds of a cloth and so must be carried in baskets or pots. Because the ripe fruit was often densely concentrated, it was not uncommon for several women to remain within a metre or two of one another. Women carrying nursing infants would sometimes set the infant down in the care of an older child, but often a mother picked with her infant held on her back in a carrying cloth. Much talking, joking, and often singing characterized these forays. When undushi were the target, women climbed into the trees to pick, leaving infants in the care of others below. They also directed older boys or the adult man accompanying them to cut down heavily laden branches for easier picking on the ground. By about midday, women had usually filled their containers (and themselves) with the juicy berries, and started the long, hot walk back to camp. When kongoro, a much drier fruit, was the target, women usually picked well into the afternoon, and continued snacking on any fruit encountered on the way home. As on tuber-collecting trips, the party typically regrouped just before arriving at camp. At sight of the women, children who had remained at home often set up a chant about the arrival of berries, and those present in camp gathered around the women to claim shares.

The other major resources acquired by women in 1985–86 were honey and baobab fruit. While the honey of several species of insects is taken by the Hadza, especially by boys, through most of the year, only *Apis mellifera* produces large amounts of honey and this quite seasonally, first in December and again from April through early June. This honey is taken by small parties of boys, mixed groups of adolescent boys and girls, and by nuclear families. Collecting tactics of men have been described by Woodburn (Woodburn & Hudson 1966). Baobab is usually one of the main resources taken by women in the dry season, but the 1985 crop was poor and women paid little attention to it. We defer discussion of both these resources to focus on women's work during October 1985, when roots, and March–April 1986, when roots and berries were the main resources taken by women.

METHODS

Foraging time is calculated from behavioural scans in study camps. These recorded the location and activity of each individual in camp at the instant they were first seen during the scan, and the general activity (when known) of individuals sleeping in the camp, but not present at the time of the scan. Scans were evenly distributed throughout the daylight hours, from 07.00 to 19.00 hours, and occurred no more frequently than once every half hour. The number of data points available for any individual varies depending on the number of days she slept in study camps. For the individuals considered here, the range is 13–122 observations per woman for the period 13 October to 1 November 1985, and 35–160 for the period 26 March to 19 April 1986.

These figures pertain to all women (but one) living in the camp under study and old enough to show marked breast development. (The exception is one very old woman resident in the 1985 dry season study camps who never joined the other women foraging and whose own foraging was not adequately monitored.) These women are divided into three age categories: (i) unmarried girls who have not yet had a pregnancy, (ii) women of child-bearing age (this includes two women who have been married at least 2 years but have not yet had a pregnancy), and (iii) women past child-bearing (none of these women has a child under about 15, all are said to be older than any of the women in category two). For the dry season of 1985, there were four girls in the first category with a total of 488 observations in camp scans, ten women in the second category with a total of 857 camp scans, and five women in the third category with a total of 295 camp scans. For the March–April 1986 season, there were only two girls in the first category with a total of 252 camp scans, seven women in the second category with a total of 871 camp scans, and five women in the third category with a total of 668 camp scans.

Time allocation during tuber-collecting trips is calculated from focal person follows, during which minute by minute activity records were kept of a single subject. A sample of 97 hours of follows provided the figures reported here. These include 69·7 hours on five women of child-bearing age, and 27·6 hours on two older women. Berry picking was not amenable to this method of observation because movement through and around berry bushes, or into trees made it impossible to accurately monitor picking rates, or to distinguish between picking to eat and picking to stash and carry for a single individual over a continuous period without disrupting her behaviour. Scans were used instead. At 5-minute intervals, the activity of each member of the party in sight (taken always in the same order) was described. Thirty-seven hours of berry trip scans supplied the data reported here. These covered only one girl in the first age category (175 observations total), six women in the second (1755 observations), and three women in the third (977 observations).

Return rates are reported in terms of kilograms of resource acquired per hour in patch, that is, excluding time spent walking to and from the areas of active picking or digging referred to here as 'resource patches'. For tubers, the total weight acquired was measured with a hanging spring scale for as many women as possible in a party before each cooking event. From the time women began to poke for tubers to their arrival at the cooking fire was counted as time in the patch. During the 97 hours of focal person follows tabulated here, activities other than active digging accounted for a mean of 27% of the time adult women spent in tuber patches. Travel to and from the patch and time spent cooking, eating and resting at the cooking site are counted as foraging time, but excluded from the calculation of return rates. Individual return rate averages are calculated from all occasions on which the time a woman spent in the patch and her total acquisition were measured. These are weighted by length of time in patch to calculate mean return rates for each individual as kilograms per hour in patch. The number of observed occasions from which an individual's average return rate was calculated varies from one to forty-four. The 1985 dry season means are calculated from all tuber foraging monitored from 10 September to 1 November. For the four girls in the first age category, this is a total of fifty occasions, for the ten women in the second 149 occasions and for the five women in the third 136 occasions. The March–April means are calculated from all tuber foraging monitored from 26 March to 19 April. Only one of the two girls in the first age category who was resident in the study camp was observed digging tubers during this period, and only on two occasions. Five of six women in the second category were observed on a total of twenty-eight occasions; five women in category three for thirty occasions.

Calculating acquisition rates for berries is less straightforward. Rates for the load carried home can be calculated in the same way as for roots, using weights of berries measured with a hanging spring scale divided by the time spent in the patch (Fig. 1a; Table 1, row A1). The points in Fig. 1a are calculated from nine *kongoro* picking occasions for the two girls in category I, 46 for the six women in category II, and 27 for the five women in category III. However, unlike tubers, berries are eaten continually during the time in patch. The load carried home is less than the total acquired. To correct for the amount eaten, we might assume an equivalent picking rate during the minutes spent eating in a patch. Since a mean of 47% as much time is spent eating as stashing, the in-patch returns might be

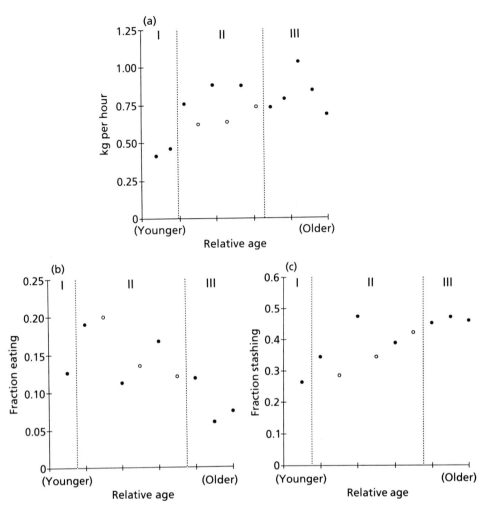

FIG. 1. Berry acquisition by relative age. Each point indicates an individual's mean carry-home rate. Individuals are ordered by relative age across the *x* axis from younger to older. Age categories are separated by vertical dotted lines and indicated by Roman numerals. In category I are unmarried girls with marked breast development who have not yet had a pregnancy. Category II contains child-bearing women, those who have young children or who are married and are younger than other women in this category. Category III contains women past child bearing who have no children under the age of about 15, and who are older than women in category II. Open circles represent those women who were carrying nursing infants when the observations were made. See text for description of methods. (a) Carry home rates for *kongoro* (kg/h in patch), 26 March–16 April 1986. (b) Time in the *kongoro* patch spent eating berries (fraction of in-patch scans). (c) Time in the *kongoro* patch spent stashing berries to carry home (fraction of in-patch scans).

TABLE 1. Age category means: women are assigned to three age categories. Age category mean values are calculated from the means for individuals in each of these categories (presented in Figs 1–4). In category I are unmarried girls with marked breast development who have not yet had a pregnancy. Category II contains child-bearing women, those who have young children or who are married and are younger than other women in this category. Category III contains women past child bearing who have no children under the age of about 15, and who are older than women in category II. See text for description of methods

	I Unmarried girls			II Child-bearing women			III Older women		
	Mean	S.D.	n	Mean	S.D.	n	Mean	S.D.	n
A Calculating *kongoro* rates									
1 Carry home rate for *kongoro* (kg/h in patch) 26 March–19 April	0·439	0·035	2	0·754	0·111	6	0·819	0·135	5
2 Eating *kongoro* (fraction of scans in patch) 37·3 h of scans at 5 min intervals	0·13		1	0·155	0·032	6	0·083	0·032	3
3 Stashing *kongoro* (fraction of scans in patch) 37·3 h of scans at 5 min intervals	0·27		1	0·377	0·064	6	0·460	0·010	3
B Total foraging time in two seasons									
1 Time foraging (min/day) 26 March–19 April	238	78	2	372	103	6	454	114	5
2 Time foraging (min/day) 13 October–1 November	141	59	4	201	88	10	389	170	5
C Return rate									
1 *//ekwa* acquisition rate (kg/h in patch) 8 September–1 November	0·841	0·320	4	1·859	0·558	10	1·966	0·512	5
2 *//ekwa* acquisition rate (kg/h in patch) 26 March–19 April	1·225		1	2·313	0·898	5	2·014	0·820	5
3 Estimated *kongoro* rate (kg/h in patch) 37·3 h of scans at 5 min intervals	0·51		1	1·082 (a) carrying infant	0·13	6	1·033	0·257	3
				0·970 (b) no infant	0·09	3			
				1·193	0·12	3			
D Time allocation to tubers and berries									
1 Time on tuber digging trips (min/day) 26 March–19 April	40	58	2	128	26	6	284	138	5
2 Time on berry picking trips (min/day) 26 March–19 April	198	134	2	177	81	7	170	56	5
3 Time on tuber digging trips (min/day) 13 October–1 November	135	667	4	166	91	10	392	170	5

estimated at 1·47 times the carry home rate. This correction is less satisfactory when return rates by age are the topic of interest, because the proportion of time in a patch spent eating is systematically related to age (Fig. 1b; Table 1, row A2). Older women spent less time eating than did child-bearing women (one tailed *t* test on difference in age category means, $P=0·013$), and (though this need not be so since half the time in patch is spent neither eating nor stashing) older women spend more time stashing than middle adults (Fig. 1c; Table 1, row A3; difference of means, $P=0·013$). For purposes of this discussion, we use the smaller set of return rates which are calculated from the berry patch scans where amounts eaten can be estimated directly. Acquisition rates are calculated by assuming an individual's picking rate for the minutes she spent eating was equivalent to her mean picking rate for the minutes she spent stashing, then calculating the average total acquisition rate for each individual.

Changes in women's body weights are used as an indirect index of investment. In the 1986 dry season, these weights were measured on an industrial electronic scale with read-out to hundredths of a kilo. Subjects were weighed barefoot, wearing light clothing. During this season, foraging time is measured as the fraction of 33 observation days on which an individual spent time in food acquisition.

Difference of means tests are used throughout to compare age categories. We used independent *t* tests, one tailed, variances pooled or separate depending on whether there is a significant difference ($P=0·05$) between the two variances.

THE QUANTITATIVE PICTURE OF WOMEN'S FORAGING

Hadza women spent a surprising amount of time in food acquisition during 1985−86: an average of 4 hours (±2 hours, 22 minutes) per day in the part of the dry season, 6 hours (±2 hours, 2 minutes) per day in the part of the wet season tallied here. When roots are the target, 48% of foraging time is spent travelling to and from the patch and cooking and eating the tubers, 38% in active digging, and the remaining 14% resting, walking between rootstalks, and packing tubers for transport, all within patch. For berries, the figures are 34% travelling, 26% picking to carry home, 9% eating, and about 23% walking among the bushes in patch. The other 8% is spent attending to infants, resting, drinking at water points, and (rarely) stopping for honey. This large time investment is quite surprising, especially given common generalizations about work effort among hunters, including the Hadza (Lee 1968; Sahlins 1968; Woodburn 1968a; cf. Hawkes & O'Connell 1981).

Even more interesting is the pattern of age-related variation in work effort (Fig. 2; Table 1, section B). Older, 'post-reproductive' women foraged an average of 6 hours, 29 minutes each day in the dry season sample, and 7 hours, 34 minutes each day in the wet season sample. These figures are 22−52% higher than the averages of child-bearing women, and 90−275% higher than those of unmarried girls. The differences in means for child-bearing and 'post-reproductive' women are significant for the dry season ($P=0·004$), but not for the wet ($P=0·122$). When

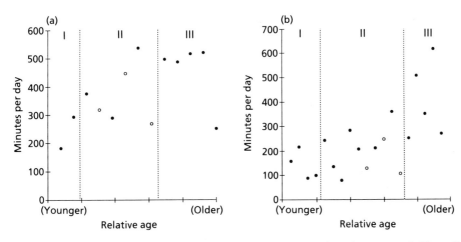

Fig. 2. Foraging time by relative age. Relative age, age categories, and nursing status as in Fig. 1. See text for description of methods. (a) Time spent foraging (min/day out of camp when the main activity was food acquisition), 26 March–19 April 1986. (b) Time spent foraging (min/day out of camp when the main activity was food acquisition), 13 October–1 November 1985.

data from the two seasons are combined, the figures are 2 hours, 53 minutes (±76 minutes) for girls, 4 hours, 24 minutes (±2 hours, 4 minutes) for child-bearing women, and 7 hours, 7 minutes (±2 hours, 13 minutes) for older women (for the first pair of means, $P=0.056$; for the second, 0.002).

Mean individual return rates for resources which comprise the bulk of women's acquisition during the periods in question are plotted by age category in Fig. 3. Age category means appear in Table 1, section C. The return rate distributions all show the same general pattern. There is an increase in efficiency up to the age of marriage (difference between means of category 1 and 2 for the 1985 dry season, Table 1, row C1, $P=0.002$). After reaching adulthood, rates remain fairly constant, although women carrying nursing infants may be less efficient foragers. The difference between these women and women in the same age category who are not carrying infants is significant for *kongoro* (Table 1, section C, rows 3a and b, $P=0.028$). However, even when the two seasons for //*ekwa* (for which means are not significantly different) are pooled, the difference between women carrying infants and other middle adult women does not reach significance. We are surprised at this lack of difference since nursing infants do seem to be an interference. Mothers carrying infants reduce this interference by enlisting older children or girls in age category one (whose own return rates are therefore further depressed) to sit beside them as baby sitters. Most berry picking is done standing with frequent moves among bushes. Although mothers do sometimes put their infants down to be tended by others they do this less often when berry picking than when digging tubers.

The older women, all past menopause, continue to show fairly high return rates. Because they spend more time foraging, they earn a foraging income higher

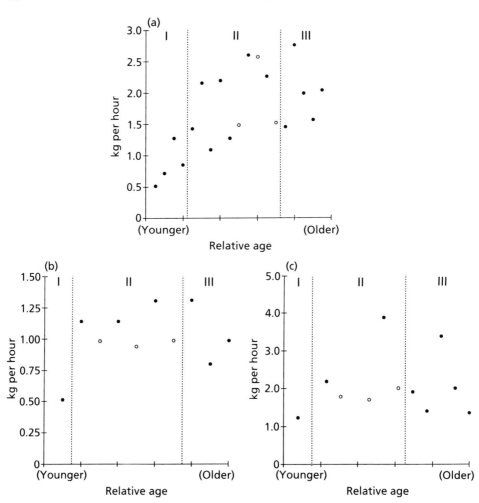

FIG. 3. Return rates by relative age. Relative age, age categories, and nursing status as in Fig. 1. See text for description of methods. (a) Return rates for *//ekwa* (kg/h in patch), 10 September–1 November 1985. (b) Return rates for *kongoro* (kg/h in patch) estimated from scans in *kongoro* patches plus amounts carried home on scan days. (c) Return rates for *//ekwa* (kg/h in patch), 26 March–19 April 1986.

than do younger women even though they have no young children of their own to feed. The pattern is especially interesting for its marked contrast with Lee's data for the !Kung. He estimated that during the dry season, women spent an average of less than 2 hours a day foraging. He also reports that women between the ages of 20 and 39 worked longer than those 40 to 59, who in turn worked longer than women over 60 (Lee 1979, p. 264, 1985). There are a few reports on older women's work effort among other foragers. The Cuiva of western Venezuela are more like the Hadza in that women over the age of 60 spend more time foraging than do younger women (Hurtado & Hill 1987), at least during the dry season, when the main resource taken was a tuber. Meehan (1982) reports yet another pattern for the Anbara of Arnhem Land, among whom women between 40 and 50 spent more time collecting shellfish than did women under 40 or over 60. Neither Cuiva nor Anbara women spend as much time foraging as do the Hadza. Guided by evolutionary theory, we draw on concepts and models about adaptive life history strategies in seeking to account for the Hadza patterns. We then briefly consider the differences between these and those reported for other modern hunter-gatherers, particularly the !Kung.

LIFE HISTORY THEORY

The alternative ways in which individuals schedule and adjust their allocation of the limited resources of a lifetime determine their probable contribution to the gene pools of future generations. Growth, maintenance, bearing offspring, and aiding the survival and reproduction of close kin are to a large extent competing expenses. The timing and degree of shifts in investment among them sets probable lifetime reproductive success. The fitness costs and benefits of alternative investment schedules depend on the present and probable future characteristics of the individual itself, of potential competitors, mates, and kin, and on other features of the environment including the food resources available and the costs and benefits of acquiring them.

Reproductive effort can be defined as effort made toward producing offspring, or toward augmenting the reproductive success of kin when that effort spent now reduces the effort which could be spent in the future. Reproductive value can be defined as the probable contribution an individual will *yet* make to the gene pool. Given those definitions, individuals would maximize their fitness by increasing their reproductive effort as their reproductive value declines (Fisher 1930; Williams 1966; Stearns 1976). As the probability of future reproduction decreases, there is less fitness reason to allocate resources to survival. How these expectations might apply to primates remains to be determined (Altman 1980). Clutton-Brock (1984) points out that it is difficult to tell whether the general prediction is widely met because the correlation of an array of features (e.g. size and experience) with age can give increasing reproductive success without requiring greater reproductive effort. In the same way, declining reproductive success with age could involve increased reproductive effort if senescence reduced efficiency in offspring production and care.

Menopause presents an interesting problem for life-history theory because child bearing terminates in advance of marked senescence (Williams 1957; Gaulin 1980; Mayer 1982). Some have argued that it is not an adaptation, but an artefact of increased longevity in post-industrial societies (see, for example, Weiss 1981). But life expectancy at reproductive age for hunting and gathering and horticultural societies allows most mothers to live past their middle forties (Howell 1979; Lancaster & King 1985). It is likely that long before the industrial revolution most women who lived to reproductive age also went through menopause.

In developing his general theory of senescence, Williams (1957) considered menopause and noted that natural selection could favour termination of offspring production in an organism for whom extended post-natal maternal investment was crucial for offspring survival. As a mother's probable survivorship began to decline, and her pregnancies became increasingly hazardous, she might increase her fitness more through additional investment in children already born, rather than by risking further pregnancies and the possible attendant loss of the dependent children she already had in order to bear another child which she would be unlikely to raise to maturity. She could trade off the probable small fitness gains from that additional pregnancy for the marginal gains from allocating the same investment to offspring (and other kin) already born. Williams emphasized that according to this theory of menopause, life beyond last childbirth is not 'post-reproductive'. Paradoxically, females could be increasing their effective reproductive effort as their reproductive value declined, even though they ceased bearing offspring.

Patterns of increasing probable mortality with age and extended maternal investment are common in primates. Hrdy (1981a) points out that variation in the behaviour of old females, both within and between some species of monkeys, varies in ways consistent with evolutionary theory: old females remaining dominant where a mother's rank can increase the reproductive success of her adult daughters, losing rank where it would not. Nevertheless, menopause, as seen in humans, is rare or non-existent in non-human primates. Sharp drops in fertility, if they occur at all, are associated with marked general senescence (Gould, Flint & Graham 1981; Graham, Kling & Steiner 1979; Dolhinow, McKenna & Vonder Haar Laws 1979). Perhaps the difference non-human mothers can make in the reproductive success of their maturing and adult offspring by abandoning child bearing is rarely great enough to outweigh the possible fitness benefits of continuing pregnancies, even if they are increasingly risky. Beyond nursing, non-human primate females rarely acquire food for their offspring (Silk 1978, 1979). The hominid capacity to feed others regularly in addition to themselves may allow older females to make a much bigger difference in the fertility of their maturing daughters.

We use this background to pose two questions. First, given that they are past child bearing, can the extra foraging of older Hadza women be understood as an optimal expenditure of reproductive effort? Second, can the fitness costs and benefits implied by the behaviour of these women contribute to explaining the evolution of the menopause itself?

THREE HYPOTHESES

The first matter is to consider explanations for the extra foraging of senior Hadza women. In addition to the implications of Williams's theory of senescence, other trade-offs may be different for older than for younger women. Since they are no longer bearing offspring, older women can have neither the same calls on their time for care of offspring, nor the same relationships with men. We pose three hypotheses, of which only the third draws on life history theory:

(a) The first hypothesis focuses on the opportunity costs of time spent foraging, i.e. the benefits which could be gained from activities foregone. The older women may give up less by spending time in food acquisition because they have less to gain from non-foraging activities. Since they have ceased bearing children, they might gain less from activities which compete with foraging in female time budgets. They have no young children of their own to tend, and if men provide resources in exchange for mating opportunities from women in child-bearing ages, women beyond those ages may be less likely to gain from time spent with men. Older women might then allocate more time to foraging because for them the cost of opportunities foregone when foraging is lower than it is for younger women.

(b) The second hypothesis has to do with possible differences in the value of the food acquired to women of different ages. The same amount of food may have higher value for older than for younger women because older women cannot trade sexual favours for resources acquired by men. Perhaps they trade food instead. If older women acquire vegetable foods to reciprocate for resources from men, the value of additional foraged resources would be greater for older women than for young ones who do not need vegetable food (or not so much) to acquire the same resources from men. Foraging longer, other things equal, would pay off for the senior women.

(c) The third hypothesis deals with the allocation of reproductive effort. Women past child-bearing age are not preparing for pregnancy and lactation. They may therefore get fitness gains by reducing the work load of their younger female kin who can thus allocate more of their own resources to pregnancy and lactation. This is an opportunity cost hypothesis like the first mentioned, but the opportunities at issue are not alternative allocations of time but of other resources. Some foraging activities appear to be very hard work. Digging //ekwa is the salient example. Adolescents might delay their first reproduction and child-bearing women lower their fertility by spending long hours at this energy-expensive task. If so, older women who are not investing in their own pregnancies or lactation might reduce the cost this foraging imposes on the reproduction of their daughters and younger female kin by increasing their own tuber acquisition, distributing the extra to their younger kin and so reducing the amount of digging that their younger kin do. Women of reproductive age with senior helpers would put less effort into tuber digging. Gains in reproductive success could follow from this reduced work. These gains might make the fitness earnings of the older women larger for reducing that work than for alternative allocations of the same investment.

These hypotheses are not mutually inconsistent. All three could be supported (or rejected). We can perform no decisive tests here, but report some additional aspects of the pattern of women's work which are relevant to these hypotheses.

COMPARISON OF BERRIES AND TUBERS

If senior women forage more because they have fewer other calls on their time, the pattern might be expected to hold across resources. General opportunity costs for time spent foraging can be gauged indirectly by comparing the time devoted to berry and tuber collecting by women of different ages during the March/April period when they take both resources (Fig. 4; Table 1, section D). If child-bearing women must pay the cost of reduced child care or reduced resource extraction from men when they forage, and so they forage less than senior women who would give up less, this might be expected to affect time allocation to all foraging. This would mean that older women spent more time foraging for all resources. Our data indicate that contrary to such an expectation from the first hypothesis, older women *do not* spend more time in all foraging tasks. Time spent on berry collecting is unrelated to age (difference of means, older women versus others, see Table 1, row D1, $P=0.0345$). The extra time older women spend foraging during this season is devoted to tubers (difference of means, older women versus others, see Table 1, row D2, $P=0.029$). If the opportunity costs for tuber-collecting forays are similiar to those for berry collecting trips, then the opportunity costs of foraging time cannot account for the extra tuber digging of the older women.

The extra time devoted to roots rather than berries is also inconsistent with the second hypothesis, that plant foods in general are more valuable for older women. The second hypothesis might also lead us to expect a relationship between extra foraging and hunting success. If older women were acquiring extra tubers to exchange for food with men, their efforts would correlate roughly with periods when men had food to trade. Hunters acquired large quantities of meat in the 1985 dry season, much less in the following wet season. Older women worked longer at tuber digging during both periods. From 13 October to 1 November, the dry season period sampled here (Fig. 4c; Table 1, row D3), hunters in the study camp killed eleven large animals and scavenged considerable portions of another four, earning a total of about 2110 kg of meat, or about 1.83 kg (live weight) per camp resident per day. During the part of the wet season sampled here (Fig. 4a; Table 1, row D1) 26 March to 19 April, hunters in the study camp killed no large game although a giraffe was killed by hunters in a neighbouring camp and some of it was transported to the study camp. This provided a maximum of about 0.2 kg of meat (live weight) per resident-day averaged through the period (although it all came in on one day). Through this season women, all women, did spend less time getting tubers than they had in the dry season. The berries were ripe and some of their foraging time was shifted from tubers to berries. In one sense then variations in time allocated to tuber digging *were* correlated with the amount of meat available. Yet older women allocated substantially more time to acquiring roots than did other women even when there was little meat to get (Fig. 4a and c; Table 1, rows D1 and 3: difference of means for age categories two and three, $P=0.021$ and 0.020, respectively). Senior women spent more than twice as much time digging tubers as their juniors whether or not there was meat in camp.

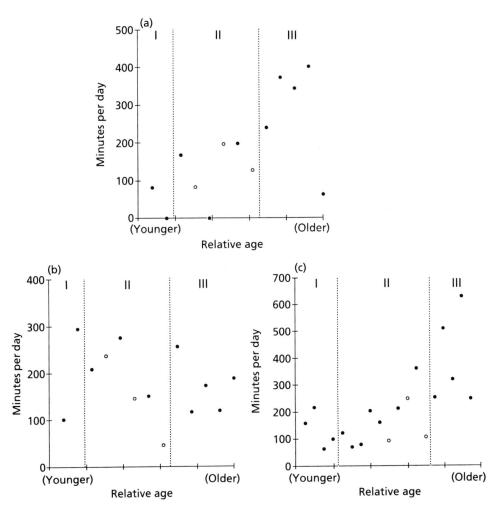

Fig. 4. Time spent on berries and tubers by relative age. Relative age, age categories, and nursing status as in Fig. 1. See text for description of methods. (a) Time spent on tuber-digging forays (min/day), 26 March−19 April 1986. (b) Time spent on berry-picking forays (min/day), 26 March−19 April 1986. (c) Time spent on tuber-digging forays (min/day), 13 October−1 November 1985.

The relationship between meat availability and extra tuber foraging by senior women merits further comment. It could be that longer-term relations of reciprocity play a role, so that women who did not provide tubers during periods of little meat risked penalties later. On the other hand, camp composition is flexible, so that co-residents at one time might later be living in different camps (Woodburn 1968b also reports this pattern). Even for those who remain co-resident throughout the year, actions over recent days or hours may be at least as important in reciprocal relationships as actions months past. More problematic is the fact that very small amounts of meat may be of very high value when there is little available.

A satisfactory account of the age difference in time allocated to berries and tubers would include a more general explanation of the pattern of alternating between these resources during periods when berries are ripe. This in turn demands an evaluation of return rates in nutritional terms. Such analysis is not yet complete. Here we can only note that switching between tubers and berries remains to be explained. Women do spend substantial amounts of time acquiring both. When both are available, they are usually taken on separate foraging trips, often in daily rotation.

We turn to the third hypothesis focused on reproductive effort. The apparent differences in the energetic expenditure associated with these two resource types are suggestive. For *kongoro* collecting, the main effort involves the walk to and from the patches and the load carried home. During the period covered by data reported above, the walk averaged 76 ($\pm$37) minutes each way. The cost of such walks in the sun cannot be negligible, but children over the age of 8 not only make these walks, but also collect enough to feed themselves. Even young girls always collect enough to carry some home. Releasing young women from some of this work may have little effect on their fertility, and so add little to the fitness of their mothers, grandmothers, and aunts. In contrast, tuber digging appears to be very hard work, and so may impose a fertility cost. In Blurton Jones, Hawkes & O'Connell (this volume) we argue that the low reproductive cost of older Hadza children implies that ecological constraints on child production fall mostly on pregnancy and lactation. Help with a young woman's energy budget may thus be especially productive of increased numbers of descendants.

Tests of this suggestion are simple in principle, but not in practice. We would expect differences in time spent digging tubers to make more difference in reproductive success than differences in time collecting berries. We would expect women with senior helpers to spend less time digging tubers, and to have higher reproductive success. The first of these predictions might be assessed through test of correlations between different amounts of foraging time spent in these activities and ovulatory regularity (Ellison, Peacock & Lager 1986). Satisfactory tests of the second are impeded by the small study population and the general problem of correlated features in individuals (Clutton-Brock 1984; Horn & Rubenstein 1984). For example, women who work harder may have higher reproductive success because women in very good condition do both, even though those same women may be lowering their potential fertility by their hard work. Still, we hope to perform such tests.

Data on body weights are potentially relevant. If older women are reducing the energy costs of their junior kin, this might show up as differential weight changes between the two categories of women. Over a given period of time, senior women spending more energy digging tubers might lose more weight (or gain less) than younger women, with differences between senior women and others varying in magnitude depending on the relative amounts of root digging.

Our data on weights during the two periods under discussion are insufficient to provide a critical test; but data from the 1986 dry season are quite striking. During this season, the study camp contained six older women, eight child-bearing women, and five girls with marked breast development who had not yet had a pregnancy. Percentage of time spent foraging by women of different age categories is shown in Table 2. As in earlier seasons, older women devoted more time to this activity than did child-bearing women and girls (difference in means for category 3 versus categories 1 and 2 combined, $P=0.00001$). Excluding one woman who bore a child, women lost an average 0.276 (± 1.84) kg during this period. The difference in weight lost by child-bearing women and unmarried girls is not significant. However, the older women lost significantly more than either child-bearing women or girls ($P=0.029$ and 0.0385, respectively).

DISCUSSION

The data show that all Hadza women spend a substantial amount of time foraging. This observation should help correct a tendency to see the small amount of foraging time reported in Lee's classic work diaries for the Dobe !Kung as typical of hunter-gatherers (see also Hawkes & O'Connell 1981; Hawkes et al. 1985; Hill et al. 1985; Hurtado et al. 1985; Hawkes 1986). The data on return rates show that among the Hadza, women's foraging efficiency increases from adolescence to a peak in middle adulthood, displays minor declines associated with the presence of nursing infants, and falls off only slightly after menopause. Hadza women generally work long hours foraging, grandmothers the longest. Higher foraging effort among older women is generally associated with tuber collecting. While we cannot decisively reject the possibility that differences in age-related opportunity costs of foraging or in age-related exchange values for tubers encourage extra work among older women, the contrast between time allocated to berries and roots and the persistence of extra work for older women in lean meat seasons are inconsistent with the simplest version of either hypothesis.

   The available data are most consistent with the proposition that senior women reduce the energy costs of foraging to their younger female kin by acquiring extra tubers for them. Although the pertinent data have not yet been analysed, it is our impression that the sharing of vegetable food is directed differentially toward close kin. The high tuber income of the senior women and the patterns of weight change among age ranks imply that much of what older women acquire is transferred to younger women. That older women lose more weight than younger ones could indicate that older women are not only investing reproductive effort, but that they are, paradoxically as 'post-reproductives', increasing their relative reproductive effort as their reproductive value declines. It is important to note that we have not analysed data on meat distributions, and the greater weight loss of the older women could also result from a lack of access to meat; but we think this is unlikely.

TABLE 2. Foraging time and weight changes, September–November 1986. Women are assigned to age categories as in Table 1. See text for description of methods

	Unmarried girls			Child-bearing women			Older women		
	Mean	S.D.	n	Mean	S.D.	n	Mean	S.D.	n
Fraction of 33 observation days spent out of camp foraging; September–November 1986	0·714	0·084	5	0·725	0·106	8	0·970	0·048	6
Weight change; September–November 1986	−0·63	1·96	5	−0·142	1·795	7	−0·958	0·539	6

If older Hadza women dig more tubers as a means of increasing their reproductive effort, why do senior women in some other foraging societies spend less time getting food? The answer may involve differences in the fitness-related costs and benefits of foraging. In the !Kung case, for example, two aspects of local ecology may be pertinent. First, the relatively harsh pressures imposed by seasonally high temperatures and lack of water in the Kalahari may make senior women's foraging return rates low, and their survivorship costs high (Blurton Jones & Sibly 1978). Draper (1975) describes a degree of exhaustion on return from a gathering trip for a woman with a 'grown daughter' unlike anything we observed among the Hadza. Second, for reasons yet unclear, babysitting seems to involve much more active supervision among the !Kung than it does among the Hadza. The amount of babysitting available plays a large role in the foraging income a !Kung woman can earn (Blurton Jones 1986, 1987; Blurton Jones, Hawkes & O'Connell, this volume), and so seniors may get the greatest increment in inclusive fitness by increased babysitting.

THE EVOLUTION OF MENOPAUSE

The argument developed above has interesting implications for the explanation of interspecific differences in female reproductive effort and their evolution. Females of some non-human primate species may cease sexual cycling before death (Graham, Kling & Steiner 1979), but such phenomena are rarely observed (Lancaster & King 1985). By contrast, human females who survive long enough to bear a child not only stop bearing early enough to see their last child reach sexual maturity, but can live on to see that child's children have children. The degree of senescence at menopause and the rate of senescence after it are remarkably low for women by non-human standards (Williams 1957; Lancaster & King 1985). As reviewed above, life history theory predicts that reproductive effort will increase as the probability of future reproduction declines (Fisher 1930). In the case of senior females, their lower probable survivorship lowers the probable fitness benefit of another pregnancy (Williams 1957). The magnitude of the fitness gains they might get from investing more in children already born, instead of bearing more, depends on the contribution they can make to the reproductive success of their younger close kin.

The data and interpretation presented in this paper suggest that the suite of resources exploited, and the age-specific costs and benefits of acquiring them may have a marked effect on age-specific reproductive effort. A major difference between humans and other primates is the difference in their extractive efficiency (Gaulin & Konner 1977). Early hominids may have been able to tap resources which were much higher quality than those exploited by their ancestors, allowing one individual to acquire enough to feed others as well as herself. Extraction of a high quality resource which is also energy expensive to acquire would provide an opportunity for females to make larger contributions to the reproductive success of their daughters, nieces, and granddaughters by foraging for them. Mechanisms such as critical fat (Frisch 1984) or direct work effort effects on fertility (Bentley 1985; Ellison, Peacock & Lager 1986) could be implicated. If older females could be productive enough, those who ceased child bearing and invested in their younger female kin may have earned greater fitness than those who did not.

By this line of argument, early termination of child bearing would be adaptive only where age-specific mortality and the importance of extended maternal investment reduced probable fitness gains from additional pregnancies, *and* where resource acquisition patterns could produce large enough marginal gains from assistance to younger kin. If females who stopped bearing infants earlier had higher fitness, it would enhance the fitness of their descendants to curtail the associated physiological preparation for pregnancy at an earlier age, thereby reducing personal maintenance costs, which in turn would allow them to allocate more resources to more effective 'reproductive effort'.

Perhaps menopause and a wider coincidence of interest among adult females are part of a common human, or even hominid pattern. This is not to suggest that women do not compete with each other. Conflicts of interest are expected between females because their fitness interests are rarely identical. It may be that ethnographers have failed to note the kinds of reproductive competition now being reported for non-human primates (Hrdy 1981b; Nicholson 1986) simply because they have not looked for them. On the other hand, one implication of the argument presented here is that greater cooperation means more in fitness terms for women than it does for other female primates.

Among modern hunter-gatherers, specific features of subsistence strategies and local ecology may determine the choice between childcare for grandchildren, resource acquisition for younger kin, or other alternatives including departure or death when an ageing senior might compete with younger kin for valuable resources. The costs and benefits of tuber exploitation in environments like that of the Hadza may provide an opportunity for seniors to raise the reproductive success of younger kin more by digging tubers than they could through other activities. By taking greater work costs themselves and so reducing these costs for their younger kin, allowing the latter to increase the number of offspring they bear, seniors may earn higher returns on their 'reproductive effort'.

Even if our argument about the relationship between Hadza women's foraging activities, reproductive effort, and features of their resource base is wrong, this inquiry should serve to make a very general point: many of the observations commonly made about the causes and consequences of divisions of labour and food sharing *between* the sexes (Lovejoy 1981; Lancaster & Lancaster 1983) should apply equally *within* the sexes, for example among individuals of different ages. Among other things, this suggests that divisions of labour and food sharing among adults could evolve under certain ecological circumstances as a result of adaptive adjustments in female life history strategies. We may come to see menopause as a consequence of foraging patterns among our female ancestors.

ACKNOWLEDGEMENTS

This work was supported financially by the National Science Foundation, the Swan Fund, Ms B. Bancroft, the University of Utah, and the University of California (Los Angeles). We thank Utafiti (Tanzanian National Research Council) for permission to pursue fieldwork, L.C. Smith for introducing us to the Hadza and for access to unpublished data, D. Bygott and J. Hanby for vital assistance in the field, and the Hadza themselves for tolerance and support. S. Beckerman, M. Borgerhoff Mulder, E. Charnov, P. Draper, K. Hill, M. Hurtado, S.B. Hrdy, P. Lee, and E.A. Smith read and commented on earlier drafts.

REFERENCES

Altmann, J. (1980). *Baboon Mothers and Infants*. Harvard University Press, Cambridge, Massachusetts.

Amoss, P.T. & Harrell, S. (1981). Introduction: an anthropological perspective on aging. *Other Ways of Growing Old: Anthropological Perspectives* (Ed. by P.T. Amoss & S. Harrell), pp. 1–24. Stanford University Press, Stanford, California.

Bagshawe, F. (1925). The peoples of the Happy Valley (East Africa); the aboriginal races of Kondoa Irangi. Part II, the Kanjegu. *Royal African Society Journal*, **24**, 117–30.

Baumann, O. (1894). *Durch Massailand zur Nilquelle*. Dietrich Reimer, Berlin.

Bean, L.J. (1978). Social Organization. *Handbook of North American Indians, Vol. 8, California* (Ed. by R.F. Heizer), pp. 673–82. Smithsonian Institution, Washington D.C.

Bennett, F., Barnicott, N., Woodburn, J., Pereira, M. & Henderson, B. (1975). Studies on viral, bacterial, rickettsial, and treponemal diseases of the Hadza of Tanzania, and a note on injuries. *Human Biology*, **2**, 61–8.

Bennett, F., Kagan, I., Barnicott, N. & Woodburn, J. (1970). Helminth and protozoal parasites of the Hadza of Tanzania. *Transactions of the Royal Society of Tropical Medicine and Hygiene*, **64**, 857–80.

Bentley, G. (1985). Hunter-gatherer energetics and fertility: a reassessment of the Kung San. *Human Ecology*, **13**, 79–109.

Bleek, D. (1931). Traces of former Bushman occupation in Tanganyika Territory. *South African Journal of Science*, **28**, 423–9.

Blurton Jones, N.G. (1986). Bushmen birth spacing: a test for optimal interbirth intervals. *Ethology and Sociobiology*, **7**, 91–105.

Blurton Jones, N.G. (1987). Bushmen birth spacing: direct tests of some simple predictions. *Ethology and Sociobiology*, **8**, 183–203.

Blurton Jones, N.G. & Sibly, R. (1978). Testing adaptiveness of culturally determined behavior: do Bushmen women maximize their reproductive success by spacing births widely and foraging seldom? *Human Behaviour and Adaptation* (Ed. by N. Blurton Jones & V. Reynolds), pp. 135–58. Society for Study of Human Biology Symposia, 18. Taylor & Francis, London.

Clutton-Brock, T.H. (1984). Reproductive effort and terminal investment in iteroparous animals. *American Naturalist*, **123**, 212–29.

Cooper, B. (1949). The Kindiga. *Tanganyika Notes and Records*, **27**, 8–15.

Dolhinow, P., McKenna, J. & Vonder Haar Laws, J. (1978). Rank and reproduction among female langur monkeys: aging and improvement (they're not just getting older, they're getting better). *Aggressive Behaviour*, **5**, 19–30.

Draper, P. (1975). Kung Women: contrasts in sexual egalitarianism in foraging and sedentary contexts. *Toward an Anthropology of Women* (Ed. by R. Reiter), pp. 77–109. Monthly Review Press, New York.

Ellison, P., Peacock, N. & Lager, K. (1986). Salivary progesterone and luteal function in two low-fertility populations of northeast Zaire. *Human Biology*, **58**, 473–86.

Estioko-Griffin, A., & Griffin, P. (1985). Women hunters: the implications for Pleistocene prehistory and contemporary ethnography. *Women in Asia and the Pacific: Towards an East-West Dialogue* (Ed. by M. Goodman), University of Hawaii Press, Honolulu.

Fisher, R. (1930). *The Genetical Theory of Natural Selection.* Dover, New York.

Fosbrooke, H. (1956). A stone age tribe in Tanganyika. *South African Archaeological Bulletin*, **11**, 41.

Frisch, R. (1984). Body fat, puberty, and fertility. *Science*, **199**, 22−30.

Gaulin, S. (1980). Sexual dimorphism in the human post-reproductive life-span: possible causes. *Journal of Human Evolution*, **9**, 227−32.

Gaulin, S. & Konner, M. (1977). On the natural diet of primates, including humans. *Nutrition and the Brain. Vol. 1* (Ed. by R. Wurtman & J. Wurtman), pp. 1−86. Raven Press, New York.

Gould, K., Flint, M. & Graham, C. (1981). Chimpanzee reproductive senescence: a possible model for the evolution of menopause. *Maturitas*, **3**, 157−66.

Graham, C., Kling, O. & Steiner, R. (1979). Reproductive senescence in female nonhuman primates. *Aging in Nonhuman Primates* (Ed. by D. Bowden), pp. 183−202. Van Nostrand Reinhold, New York.

Hawkes, K. (1986). How much food do foragers need? *Food and Evolution: Toward a Theory of Human Food Habits* (Ed. by M. Harris & E. Ross), pp. 341−56. Temple University Press, Philadelphia.

Hawkes, K., Hill, K., O'Connell, J. & Charnov, E. (1985). How much is enough: hunters and limited needs. *Ethology and Sociobiology*, **6**, 3−15.

Hawkes, K. & O'Connell, J. (1981). Affluent hunters? Some comments in light of the Alyawara case. *American Anthropologist*, **83**, 622−6.

Hill, K. & Kaplan, H. (1988). Tradeoffs in male and female reproductive strategies among the Ache. *Human Reproductive Effort* (Ed. by L. Betzig, M. Borgerhoff Mulder & P. Turke), pp. 277−89, 291−305. Cambridge University Press, Cambridge.

Hill, K., Kaplan, H., Hurtado, A. & Hawkes, K. (1985). Men's time allocation to subsistence work among the Ache of Eastern Paraguay. *Human Ecology*, **13**, 29−47.

Horn, H. & Rubenstein, D. (1984). Behavioral adaptations and life history. *Behavioral Ecology: an Evolutionary Approach* (Ed. by J.R. Krebs & N.B. Davies), pp. 279−300. Blackwell Scientific Publications, Oxford.

Howell, N. (1979). *Demography of the Dobe Kung.* Academic Press, New York.

Hrdy, S. (1981a). *The Woman that Never Evolved.* Harvard University Press, Cambridge, Massachusetts.

Hrdy, S. (1981b). 'Nepotists' and 'altruists': the behavior of old females among macaques and langur monkeys. *Other Ways of Growing Old: Anthropological Perspectives* (Ed. by P.T. Amoss & S. Harrell), pp. 59−76. Stanford University Press, Stanford, California.

Hurtado, A., Hawkes, K. Hill, K. & Kaplan, H. (1985). Female subsistence strategies among Ache hunter-gatherers of eastern Paraguay. *Human Ecology*, **13**, 1−28.

Hurtado, A. & Hill, K. (1987). Early dry season subsistence ecology of the Cuiva foragers of Venezuela. *Human Ecology*, **15**, 163−87.

Klima, G. (1985). *The Barabaig: East African Cattle Herders.* Waveland Press, Prospect Heights, Illinois.

Kohl-Larson, L. (1958). *Wildbeuter in Ost-Afrika: die Tindiga, ein Jäger- und Sammlervolk.* Dietrich Reimer, Berlin.

Krebs, J.R. & Davies, N.B. (Eds) (1984). *Behavioural Ecology: an Evolutionary Approach* (2nd edn.). Blackwell Scientific Publications, Oxford.

Lancaster, J. & King, B. (1985). An evolutionary perspective on menopause. *In Her Prime: A New View of Middle-Aged Women* (Ed. by J. Brown & V. Kerns), pp. 13−20. Bergen & Garvey Publishers, South Hadley, Massachusetts.

Lancaster, J. & Lancaster, C. (1983). Parental investment: the hominid adaptation. *How Humans Adapt: A Biocultural Odyssey* (Ed. by D. Ortner), pp. 33−66. Smithsonian Institution Press, Washington D.C.

Lee, R. (1968). What hunters do for a living, or, how to make out on scarce resources. *Man the Hunter* (Ed. by R. Lee & I. DeVore), pp. 30−48. Aldine, Chicago, Illinois.

Lee, R. (1979). *The Kung San: Men, Women, and Work in a Foraging Society.* Cambridge University Press, Cambridge.

Lee, R. (1985). Work, sexuality and aging among Kung women. *In Her Prime: A New View of Middle-Aged Women* (Ed. by J. Brown & V. Kerns), pp. 23–36. Bergen & Garvey Publishers, South Hadley, Massachusetts.

Lovejoy, C. (1981). The origin of man. *Science*, 211, 341–50.

Marquardt, W.H. (1986). Politics and production among the Calusa of South Florida. *Fourth International Conference on Hunting and Gathering Societies*. London School of Economics, 8–13 September 1986.

Matthiessen, P. (1972). *The Tree Where Man Was Born*. Collins, London.

Mayer, P. (1982). Evolutionary advantage of the menopause. *Human Ecology*, 10, 477–94.

McDowell, W. (1981). *A brief history of the Mangola Hadza*. Unpublished manuscript prepared for the Rift Valley Project, Ministry of Information and Culture, Division of Research, Dar es Salaam, Tanzania.

Meehan, B. (1982). *Shell Bed to Shell Midden*. Australian Institute of Aboriginal Studies, Canberra.

Mehlman, M. (in preparation). *Later Quaternary archaeological sequences in northern Tanzania*. Ph.D. thesis, University of Illinois, Urbana-Champaign.

Ndagala, D. (1986). *Free or doomed? Images of the Hadzabe hunters and gatherers of Tanzania*. Paper presented at the 4th International Conference on Hunting and Gathering Peoples. London School of Economics, London.

Nicholson, N. (1986). Infants, mothers, and other females. *Primate Societies* (Ed. by B. Smuts, D. Cheney, R. Seyfarth, R. Wrangham & T. Struhsaker), pp. 330–42. University of Chicago Press, Chicago, Illinois.

Obst, E. (1912). Von Mkalama ins Land der Wakindiga. *Mitteilungen der Geographischen Gesellschaft in Hamburg*, 26, 2–27.

O'Connell, J., Hawkes, K. & Blurton Jones, N. (1988a). Hadza scavenging: implications for Plio-Pleistocene hominid subsistence. *Current Anthropology*, 29, 356–63.

O'Connell, J., Hawkes, K. & Blurton Jones, N. (1988b). Hadza hunting, butchering, and bone transport, and their archaeological implications. *Journal of Anthropological Research*, 4, 113–61.

Reche, O. (1914). Zur Ethnographie des abflusslosen Gebietes Deutsch Ostafrikas auf Grund der Sammlung der Ostafrika-Expedition (Dr. E. Obst) der Geographischen Gesellschaft in Hamburg. *Abhandlungen des Hamburgischen Kolonialinstituts*, XVII.

Sahlins, M. (1968). Notes on the original affluent society. *Man the Hunter* (Ed. by R. Lee & I. De Vore), pp. 85–9. Aldine, Chicago, Illinois.

Schultz, J. (1971). *Agrarlandschaftliche Veränderungen in Tanzania (Mbulu/Hanang Districts)*. Weltform Verlag, Munich.

Silk, J. (1978). Patterns of food sharing among mother and infant chimpanzees at Gombe National Park, Tanzania. *Folia Primatologia*, 29, 129–41.

Silk, J. (1979). Feeding, foraging, and food sharing behavior in immature chimpanzees. *Folia Primatologia*, 31, 12–42.

Smith, L. (1980). *Resource survey of Hadza hunter-gatherers*. Unpublished manuscript on file, Department of Anthropology, University of Utah.

Stearns, S. (1976). Life-history tactics: a review of the ideas. *Quarterly Review of Biology*, 51, 3–47.

Stevens, D. & Krebs, J. (1986). *Foraging Theory*. Princeton University Press, Princeton, New Jersey.

Tomita, Y. (1966). Sources of food for the Hadzapi tribe: the life of a hunting and gathering tribe in East Africa. *Kyoto University African Studies*, 2, 157–71.

Vincent, A. (1985a). *Wild tubers as a harvestable resource in the East African savannas: ecological and ethnographic studies*. Ph.D. thesis, University of California, Berkeley.

Vincent, A. (1985b). Plant foods in savanna environments: a preliminary report of tubers eaten by the Hadza of northern Tanzania. *World Archaeology*, 17, 131–48.

Weiss, K. (1981). Evolutionary perspectives on human aging. *Other Ways of Growing Old: Anthropological Perspectives* (Ed. by P.T. Amoss & S. Harrell), pp. 25–58. Stanford University Press, Stanford, California.

Williams, G. (1957). Pleiotropy, natural selection, and the evolution of senescence. *Evolution*, 11, 398–411.

Williams, G. (1966). *Adaptation and Natural Selection*. Princeton University Press, Princeton, New Jersey.

Winterhalder, B. & Smith E. (Eds) (**1981**). *Hunter-Gatherer Foraging Strategies: Ethnographic and Archaelogical Interpretations*. University of Chicago Press, Chicago, Illinois.

Woodburn, J. (1964). *The social organization of the Hadza of North Tanganyika*. Ph.D. thesis, University of Cambridge.

Woodburn, J. (**1968a**). An introduction to Hadza ecology. *Man the Hunter* (Ed. by R. Lee & I. De Vore), pp. 49–55. Aldine, Chicago, Illinois.

Woodburn, J. (**1968b**). Stability and flexibility in Hadza residential groupings. *Man the Hunter* (Ed. by R. Lee & I. DeVore), pp. 103–10. Aldine, Chicago, Illinois.

Woodburn, J. (**1970**). *Hunters and Gatherers: Material Culture of the Nomadic Hadza*. The British Museum, London.

Woodburn, J. (**1972**). Ecology, nomadic movement and the composition of the local group among hunters and gatherers: an East African example and its implications. *Man, Settlement, and Urbanism* (Ed. by P. Ucko, R. Tringham & G. Dimpleby), pp. 193–206. Duckworth, London.

Woodburn, J. (**1979**). Minimal politics: the political organization of the Hadza of North Tanzania. *Politics in Leadership: a Comparative Perspective* (Ed. by W. Shack & P. Cohen), pp. 243–66. Clarendon Press, Oxford.

Woodburn, J. (**1986**). *African hunter-gatherer social organization: is it best understood as a product of encapsulation?* Paper presented at the 4th International Conference on Hunting and Gathering Peoples. London School of Economics, London.

Woodburn, J. & Hudson, S. (**1966**). *The Hadza: the Food Quest of an East African Hunting and Gathering Tribe* (16 mm film). London School of Economics, London.

Modelling and measuring costs of children in two foraging societies

N.G. BLURTON JONES*, K. HAWKES** AND
J.F. O'CONNELL**

*Graduate School of Education, Department of Anthropology and
Department of Psychiatry, University of California, Los Angeles, CA 90024, USA, and
**Department of Anthropology, University of Utah,
Salt Lake City, Utah 84112, USA

SUMMARY

1 We summarize Blurton Jones and Sibly's attempt to model costs of !Kung children. They aimed to explain the long (4-year) inter-birth intervals of !Kung women, and to show how these intervals relate to the circumstances of !Kung life. The model, when linked to the assumption that the reproductive system would maximize the rate of production of surviving offspring, successfully predicted several aspects of !Kung reproduction. Could it be generalized to another population of nomadic, savannah hunter-gatherers?

2 We describe our efforts to measure costs of raising children of nomadic Hadza in terms of the effects of a child upon its mother's acquisition of the resources that can be allocated to production and survivorship of children.

3 The costs of children appear to be lower among the Hadza than among the !Kung because:

(a) When Hadza mothers go out to forage they leave children behind in camp when the children are about 2½ years old, much younger than among the !Kung.

(b) Although Hadza babies continue to be suckled morning and evening beyond the age at which they are left at home, they are weaned sooner than !Kung children.

(c) Hadza children above 5 years old forage for themselves very effectively. Their return rates were such as to suggest that they could provide at least half their caloric requirements by their own efforts.

(d) Hadza children do a variety of tasks that aid adult subsistence work, from running errands to cutting down baobab fruit from high trees.

(e) Hadza children receive resources from grandmothers and other female relatives, who work harder and as productively as do the mothers of the children.

4 Mobile, nomadic hunter-gatherers are often held to have universally long inter-birth intervals (IBI). But if schedules of reproduction vary adaptively, these lower costs of Hadza children imply that we should expect higher fertility and shorter IBI in Hadza women than in !Kung women. Efforts to test this expectation are underway.

367

5 We suggest that since older children receive resources from their own efforts and those of their female relatives, the fitness benefit to mother from putting increased effort into feeding her older children (over the age of about 5) is small. Conversely the cost to mother from reduced provisioning of these children will be much less — since they will feed themselves and scrounge harder from their grandmother, than it would be for !Kung mothers (whose children we suppose to be unable to forage productively under normal circumstances).

6 We suggest that this implies that for Hadza women fitness returns are greatest from increased effort at production and care of babies and children under 5.

7 Hadza women's ability to acquire resources may not be limited by backload because:

(a) Although Hadza mothers work longer hours, they travel much less far to find food than do !Kung mothers.

(b) The Hadza environment provides less heat stress than the !Kung environment in the dry season. Coupled with the short journeys and similar loads this argues against heat stress and the ability to carry heavy loads being a factor that limits the ability of Hadza women to acquire more food. Although Hadza women are unquestionably mobile and nomadic, their reproduction does not appear to be limited by load carrying or other aspects of this mobility.

8 Hadza women's ability to acquire resources may be limited by time. Thus time may be a better measure of the cost of Hadza children than backload. Nomadic hunter-gatherers have been stereotyped as leading lives of leisure, under no pressure of time. We suspect that this characterization will not fit the Hadza. Further predictions must await completion of a model of the time costs of raising Hadza children, and collection of data on children's foraging during the wet seasons.

INTRODUCTION

A well-constructed model of costs of children would enable us to see how altered circumstances change the way in which a parent who maximizes its reproductive success should behave. Costs and reproductive success are the subjects of Trivers's (1972) concept of parental investment: 'any investment by the parent in an individual offspring that increases the offspring's chance of surviving (and hence reproductive success) at the cost of the parent's ability to invest in other offspring'. Trivers further states 'I measure the size of a parental investment by reference to its negative effect on the parent's ability to invest in other offspring: a large parental investment is one that strongly decreases the parent's ability to produce other offspring'. But to render Trivers's abstract statement predictive of differences between populations we need to relate it to the practical circumstances that we believe determine the combination of behaviours that form an optimal strategy.

   The concept of economic costs of children that has long interested demographers, economists, and anthropologists (see, for example, Nag, White & Pett (1978) may help us bridge the gap between parental investment theory and real life circumstances. In some societies, children impair mothers' subsistence work and economic production, in others they increase the family income or wealth, for instance by herding and other chores. The economic costs can easily be expressed in a common currency (e.g. the dollar equivalent of family income), the currency is regarded as being in short, or limited supply; and as being offset against reproduction: when children create wealth people are expected to have more children (e.g. peasant farmers); when children decrease wealth people are expected to have fewer (e.g. mobile nomadic hunter-gatherers). We suggest that a promising combination of the evolutionist and the economist concept of cost of children would be: the effect of the child upon the mother's ability to acquire resources that could be allocated to other children.

Resources that are needed to raise one offspring will be similar to resources needed to raise another offspring. It should be possible to assess the resources needed by an offspring (that give it a good chance of growth and survival), and what is entailed for the mother in acquiring these resources. It should also be possible to assess the effect of other aspects of child care upon the mother's ability to acquire resources. It may be possible to relate these effects to each other in a common currency. If in addition such a model successfully predicts features of reproduction then it may indeed have correctly captured the relationships between circumstances, behaviour and reproduction. Blurton Jones & Sibly's (1978) 'backload model' (henceforth 'BJ&S') was an attempt at such a model, applied to the reproduction of !Kung women. It made a number of successful predictions. It shows very strongly how a number of features of !Kung life and circumstances would influence an individual woman's reproductive success (number of offspring raised to maturity). We wished to emulate BJ&S with another population of mobile hunter-gatherers, the Hadza, and in doing so, to explore the nature and generality of such models. BJ&S was very specific to !Kung circumstances. Can it tell us what to expect of the Hadza, or do we have to start afresh with each population? Is load carrying and mobility a universal cost of reproduction for nomadic hunter-gatherers?

The !Kung are well known for their long inter-birth intervals (IBI) (average about 4 years; Howell 1979), which is an interesting challenge to the evolutionists' expectation that reproductive systems (whether physiological, behavioural, or cultural) maximize the number of descendants. !Kung women wean children late, and carry them with them when they go gathering, sometimes until the child is 4 years old. As a result of their gathering efforts women carry home 58% or more of the calories reaching camp. Women who adopt IBI shorter than 4 years have more children to feed, and sometimes have to carry two children at once (unless they can get more baby sitting done for them). Thus if a woman adopts shorter IBI but wishes to continue to feed her children adequately, she has to carry higher average weights of food and of child home from her gathering excursions (we refer to these weights as 'backload').

Lee (1972) discussed this effect and linked it to the low !Kung fertility. In their 'backload model', Blurton Jones & Sibly (1978) quantified the effect and showed that it is strong and sharp. They calculated the weight of plant food needed to provide the mother's share of the daily allowance of calories for a child (using the same calorie allowances as used by Lee), and added this to the average weight of child carried (again using figures from Lee) (Fig. 1). When these weights are accumulated for the families resulting from adopting various IBI it was found that the weights are not much affected by changes in IBI above 4 years but increase steeply as IBI shortens below 4 years (Fig. 2). This effect is due to food require-ments as much as due to weight of child. BJ&S argued that carrying significant weights on the long journeys between sources of staple plant foods and camps at permanent water in the late dry season involved serious risks of heat stress. Thus a woman who adopts short IBI faces the possibility of being unable to safely bring home enough food for her children. Doubtless she might explore ways to change the situation, such as getting her husband to provide more food (if he is not already doing as much as suits him), and enlisting more help from female relatives (if they are not already helping as much as suits them). But if these things also have some limit we can expect shorter IBI to at some point lead to shortage of food for the children.

BJ&S argued that the shortage of food might lead to deaths of infants and children, and that therefore mortality should be expected to increase as IBI decreases, in the same way that backload increases. BJ&S suggested that this mortality might offset the advantage of a greater rate of births, such that the IBI that maximizes number of surviving offspring (the optimum IBI) would be at the observed mean IBI of 4 years. These predictions were tested by Blurton Jones

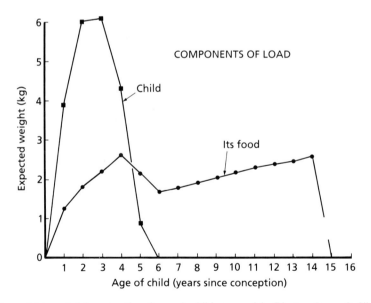

FIG. 1. The weight carried by a mother for each child, as used in Blurton Jones & Sibly's (1978) 'backload' model for !Kung women.

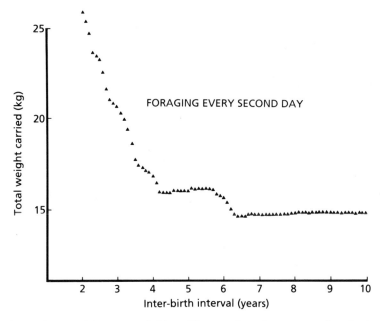

FIG. 2. The maximum weight ever carried by a !Kung mother who goes to collect food every second day, in relation to different birth spacings that she might adopt. (From Blurton Jones & Sibly 1978.)

(1986, 1987) using Howell's data on reproductive careers of individual women. The optimum IBI is just over 4 years. Many other predictions from the 'backload model' were also confirmed by Howell's data. This model accounted for much of the variance in IBI and in mortality.

The model shows the importance for reproductive success of many features of !Kung life: location of camp, baby-sitting, cost of processing mongongo nuts, father's provisioning, time of day for foraging, the costliness of carrying babies and feeding teenagers, would all make a difference to the quantities in the model. The model lets us see why, as Lee already argued, !Kung women at cattle posts (who escape many of the constraints of the model) can raise children at a faster rate, and why the survivorship of these women's children shows little relationship to inter-birth interval, and it enables us to understand why !Kung now use donkeys when collecting mongongo nuts, as BJ&S long ago expected they would. However, the !Kung backload model makes it clear that the most important properties of the model arise from a geomorphological accident — the distance between permanent water (near which dry season camps are placed) and mongongo nut groves (on fossil sand dunes far from the water courses). It is this distance which makes heavy loads economic, and a limit from heat stress plausible, and ultimately leads us to expect long inter-birth intervals. Yet long IBI have been held to be a characteristic of all mobile hunter-gatherers. Is there a contradiction here? If we tried to repeat BJ&S in another population what would happen? Would we still find it useful to measure costs by backload?

Thus our questions are:

(a) How does the effect of a Hadza child upon its mother's work compare with the effect of a !Kung child on its mothers work? What parameters would we enter into a BJ&S 'backload' model of Hadza child care?

(b) What does this suggest about Hadza reproduction? Do we expect different IBI from the !Kung?

(c) Is backload an appropriate measure of costs for the Hadza? What general principles can guide us in constructing models of costs of children that relate observable behaviour and circumstances to the evolutionary cost of a child (the effect of its production and survivorship upon the production and survivorship of siblings)?

We report data that was gathered to answer these questions. While not all the required information is available much of it is already more direct, and better quantified than that available to BJ&S for the !Kung.

THE HADZA

The contemporary situation and recent history of the Hadza are summarized in Hawkes, O'Connell & Blurton Jones, this volume. The research reported here comes from some of the 200−250 Hadza who currently live a predominantly hunting and gathering life.

The Hadza are a particularly interesting group to compare with the !Kung. Some of both groups are full-time hunters and gatherers, in the savannah of sub-Saharan Africa, foraging with traditional technology among much the same fauna and flora. Because of our adaptationist viewpoint we are attracted by differences between the Hadza and the !Kung. Woodburn's (1964, 1968a, 1968b) accounts of the Hadza in the 1950s first indicated differences in behaviour and social organization between the Hadza and the !Kung. While mostly non-quantitative, Woodburn's accounts accord with many of our impressions and data on the Hadza in the 1980s. Particularly relevant for this paper are the reports that Hadza children (Jelliffe *et al.* 1962; Woodburn & Hudson 1966) forage for themselves at an early age while !Kung children do not forage for themselves until mid to late teens (Draper 1976; Lee 1979).

METHODS

Sample sizes

The data reported here is derived from observations between late August and early November 1986, with additions from October 1985 as shown in the tables and text.

We worked in one Hadza camp at a time, so the numbers of people upon whom we collected data is limited. In September to November 1986 the data were collected in a camp of up to around seventy people. Regular inhabitants were ten adult men, six boys 0–10 years old, eleven boys 10–20, four girls 0–10, eight girls 10–20, ten women of child-bearing age and six older women.

Estimates of children's ages

Most Hadza do not know their own age, and there are only very few official records (just a handful of Hadza who live in villages have birth certificates). Age was estimated by ranking the children, partly by our inspection of the children, partly by asking the mothers and other informants. Presence or absence of the child in the camp lists from our pilot visit in 1984, and from Blurton Jones & Smith's census in 1985 were used to determine which children were born before August 1984, and before June 1985. Thus our 1-year old and 2-year old age marks are rather reliable. Older children were also ranked and their ages estimated in relation to children known to have been babies in Smith's 1977 census (9–10 years old in October 1986). Some young adults also have ages known from their birth during the visits of previous fieldworkers. Fitting a polynomial regression of age-rank upon known age allowed ages between these points to be estimated, within the limits due to errors in age ranking of any particular individual.

Data collection procedures

We tried to measure the same features of raising and providing for children as seemed important in the BJ&S model: amount children are carried when mother forages, age of weaning, amount of child's food provided by mother, distance travelled by mother, weight carried and heat environment. We also paid special attention to trying to quantify the ethnographic comments about Hadza children's foraging. How productive is it? How much of their food do they get for themselves? Some of our observations were aimed more widely at capturing other unanticipated costs of Hadza children.

The time living in a Hadza camp during September to November 1986 was divided into three types of work day:

(a) Anthropometry days ($n = 3$), usually the first day back in camp after a break, since it disrupted all other kinds of data collection. All people were weighed using a Weylux load cell digital industrial weigher model 850, which provides a very rapid readout and extreme accuracy. Upper arm circumference was measured following the instructions in Morley & Woodland (1979). Stature was measured with extreme crudity, using a steel tape measure fixed to the side of the Land Rover, and a spirit level as a head board. Each person was paid with a mug of maize flour, which combined to give each family a satisfying meal, and was used up by the following day.

(b) Days in camp (21 days). Several kinds of data were collected on each of these days.

(i) A schedule of 1-hour follows of children aged 0–10 years. Brief interruptions of up to 10 minutes were allowed, and the observation resumed and continued to complete an hour of observation time. Longer interruptions, such as when the women returned from gathering with food to be weighed, or when meat was brought into camp and needed to be weighed led to the observation being discarded.

(ii) The departure of the women from camp (almost always as a group leaving together, and heralded by noisy sharpening of digging sticks) was noted, and the identity of children who departed with them. The time of their return was noted and the food that they brought home was weighed. Care was needed during the weighing to remove shoes, knives, and newly-found hammer stones from the woman's load. Food weights were taken with hanging spring scales.

(iii) Any meat or other food brought into camp by men or children was weighed.

(iv) Children spending the day in camp were scanned, and the time of leaving camp and returning with food (roots, baobab, birds, sometimes honey) was noted. The food was weighed on their return. Baobab processing was observed to the exclusion of other kinds of observation. The aim was to find out how long children took to process and eat the baobab that they brought in. Much data was lost because two or more children unevenly pooled their resources and their labour. (Eighty-eight follows of a child collecting and processing baobab were recorded, from twenty-one children, it is possible to record this data from several children simultaneously). Children were also followed and observed in detail digging roots (fifty-two follows on seven boys and eight girls). The roots were weighed to give records of weight of root extracted per minute of digging. Children also often cooked their roots.

(c) Days following people in the bush (10 days, up to sixteen women and eight teenage girls, various numbers of boys and one to three men).

(i) The observer left camp with the women and boys in the morning and stayed with them until they returned to camp in the afternoon. Their times of departure and return were noted, and the identities of the women, men, and children accompanying them. All roots gathered were weighed before cooking, and after the midday meal in the bush, food acquired during the afternoon was weighed before cooking, all food taken back to camp was weighed.

(ii) During the time that the women were digging the observer divided his time between 1-hour follows of babies and their mothers, and of boys who were almost always searching for honey (15 hours on seven boys), and a few 1-hour follows of the 10–18 year old girls.

RESULTS

How much do Hadza mothers carry their babies?

We computed the proportion of days that babies of a given age left camp with mother when she went gathering for September to October 1986. For October

1985 the data comes from almost daily follows of women, linked to repeated scans of people in camp. Unfortunately there were few toddler-age children in the October '85 camp. Hadza toddlers out with mother ride on her most of the time except when she is digging or cooking, or occasionally when entertained by an older child or when carried by grandmother.

The data are presented in Table 1, with the weights of the children recorded in the month of observation, and their estimated ages. There is a rather striking 'all or none' effect, babies up to a year old always go with mother. Children over 3 never go. Children seem to stop accompanying mother suddenly, rather soon after they are 2 years old. The departure of Hadza mothers without their 2-year olds is also striking to the ears — the children cry and scream and totter after their mother, only to be carried back again or gathered up and scolded by an older child.

Hadza children stop being carried much earlier than !Kung children. They are left at home under minimal supervision. Often they are in the company only of older children, sometimes in the seemingly inattentive presence of whichever men (if any) are at home, either in camp or at a 'men's place' from 5 to 30 metres outside the area of the houses.

Age at weaning

Data on Hadza suckling are, so far, very limited. In Table 1, we list which of the babies and toddlers were still observed to suckle at all. Toddlers that are left at home are suckled before their mother departs and as soon as she returns. No child over a probable age of 3 was seen to be suckled. Thus Hadza children are weaned perhaps as much as a year earlier than !Kung children

Proportion of children's food provided by Hadza mothers

We have yet to calculate proportions of food acquired by Hadza men and women. We know that there is extreme seasonal variation in the amount of meat arriving in a Hadza camp, both as a result of hunting and of scavenging (O'Connell, Hawkes & Blurton Jones 1988). Thus we know that there would be great variation in the amount of a child's food that comes from the work of men. We also have data yet to be analysed which will show how much food children receive from different people. Rather than present a premature and misleading account of the proportions of food reaching a child from men, mothers and other female relatives, we concentrate on reporting the more striking feature of Hadza children's subsistence: the foraging that they do for themselves.

Foraging by children

In the BJ&S model it was assumed that !Kung children provided no food for themselves until the age of 15, at which point they became completely independent. The ethnographies suggested (see, for example, Lee 1968), and the data reported

TABLE 1. Frequency with which Hadza children accompany mother when she goes gathering (if not carried, these children stay in camp, rather older ones begin to accompany mother on foot). Based on 31 days' data collection

Child ID	Age (estimated)	Suckle	Weight (kg)	No. of times child taken with mother	No. of times mother observed foraging	% days child taken foraging
PAKSB	1 month	+	3·30	5	5	100
HISELFS Oct. 85	8 months	+	5·5	17	17	100
PLAIDSB	9 months	+	8·2	2	2	100
SUSA Oct. 85	9 months	+	7·0	15	22	73
THRUSHB	>1	+	7·05	11	11	100
TBB	>1	+	8·10	20	20	100
SUSA Oct. 86	1¾	+	8·95	0	12	0
FATS Oct. 85	<2	+	11·0	9	10	90
WOE	<2	+	9·0	0	26	0
BMAR (Sick)	>2	+	9·40	7	17	41
FBB	2½	+	11·15	3	18	17
SABENA1	2½			0	4	0
KOJOJ	2½	−	14·00	0	23	0
UMT	2½	−	14·00	0	18	0
ELIZAS	3			0	10	0
MCS	3	−	11·45	0	18	0
RED Oct. 85	3½		12·5	0	21	0
13B Oct. 85	3½		12·0	0	16	0
RED Oct. 86	4½	−	13·60	0	17	0
GREEN Oct. 85	4½	−		0	22	0

by Draper (1976), show that !Kung children seldom go out foraging, and that children are dependent on their parents late into their teens.

As Woodburn reported in his 1968 papers, Hadza children foraged for themselves from an early age. Children dug for roots, collected and processed baobab fruit, chopped trees for honey, and shot at birds and small mammals. It proved easier to measure their rates of returns (grams per hour) from these activities than to measure total time spent in them. We will thus arrive at a slightly indirect estimate of the amount of food that children of various ages gathered. They did not consume all this food themselves; even quite small children gave food to others.

Tubers

Children most often collected a root called Makarita (*Eminia antenullifera*) that was found just below ground all around the camp, and less often take //ekwa (*Vigna frutescens*), the main target for adult women. Table 2 shows average grams per hour of digging excursion for Makarita and //ekwa for children of a variety of ages (52 'follows'). Two girls of about 2, and a boy of about 2½ years were the youngest observed trying to dig and each obtained a small root just once. Their returns were almost unmeasurably small but they knew what to do. Smaller

Table 2. Children and their foraging returns (in grams per hour)

ID	Age	Height	Weight	Makalita	//ekwa	Honey	Baobab
SUSA	1¾		8·95	23			Tried
KOJO	2½	89	14·0				Tried
UMT	2½	84	14·0				351
MCS	3	93	11·45				661
RED	4½	93	13·6				247
GREEN	5½	97	13·5	205			451
NAKED	7	87	17·0	160			712
ADAM	8	67	16·7	221			741
LOST	8	105	15·45				721
FQ	9	129	16·9	392			415
IYA	10	131	17·75	414			465
HAMN	10	95	19·3	348			405
MOGI	10	130	18·4	409	208		412
SIMA	11	128	19·2				681
DUTU	12	125	24·2			54	726
MOGE	13	140	22·8				901
JP	13	128	28·1				842
GUDO	13	132	29·2			102	1015
GID	14	131	27·2	2986	936	112	583
EC#2	14	131	26·6			152	833
FLYC	14	135	31·0			159	968
GIDI	15	133	29·2		589		
MINJ	15	135	28·2	1022			807
LILS	14	128	26·8	1418			
SHARP	15	145	35·8	797			1357
OYA	17					144	1481
MANA	18					190	1481
2WOMEN	>20						1934

children dig and scrape at the ground anywhere and anytime 'in play'. Children of 9 and over (two were known to be small babies in Smith's 1977 census, thus 9–10 years old) sometimes accompanied the women on their foraging trips, sometimes stayed in camp and foraged nearer camp. Their returns seemed to be higher in camp. Perhaps they accompanied the women mainly for the lunchtime share out, or for the opportunity to pursue honey (see below).

Vincent (1985) gives a caloric value of Makarita of 530 cal (2226 J)/kg (although she does not state whether this applies to fresh or to cooked and peeled roots). If we take Vincent's figure the children of 5–10 years, who acquire about 307 g/h, may acquire around 163 cal (684·6 J)/h by digging for Makarita, a very low rate of return. At 2·7 cal (11·34 J)/min this barely reaches the caloric cost of walking for an adult (Durnin & Passmore 1967) but might exceed that for a 30 kg child (McArdle, Katch & Katch 1986), and is more than twice the RDA (converted to cal (J)/min).

Baobab

Any child that could walk could get baobab pods (the fruit of *Adansonia digitata*) during the dry season of 1986. Collecting baobab was easier after the camp moved in mid-October. This made us wonder whether children deplete the supply of fallen baobabs on the ground near camp and thus might show differences in returns before and after a camp move. (Indeed the possibility should be considered that camp moves could be provoked by children's declining returns from foraging, making their demands on adults more noticeable.) Collecting times varied, prob- ably depending on how much children play while out collecting, how much they wait for each other, and with small children, how skilfully they tie their bundle of pods and how often they stop to pick up pods they drop and to retie their bundle. The distances covered are normally very small indeed, from 0 to 200 metres out of camp.

Baobab fruits are easy to collect but they are hard to process in the dry season. Two year olds try but mostly fail to crack the outer shell. Children over 3 break them quite easily with hammerstones, or by beating the pod onto a rock. Older children, like adults, can break them by these methods, or with a stick, or by stamping on them. The pith is then tipped out on the rock and hammered to pulp with a round stone. These spherical stones (diameter 5–10 cm) are collected and carried home by adult women who use them for pounding baobab and leave them unattended on the rock where they process baobab.

There is a wet process and a dry process for preparing and eating baobab. Younger children mostly use the wet process — they pour a small amount of water on the pith as they hammer, pick up a handful and squeeze and suck. The process is repeated with the residue, eventually the seeds are cracked and their contents gradually eaten by sucking and chewing, pieces of shell being spat out. Older children sometimes use the dry process. Water is not added; when the nuts are hammered the pile is winnowed on a small leather mat. Eventually a fine white powder is produced, which is then eaten raw with water, and sometimes cooked with water, although not normally by children. Both these processes are time-consuming. Children were timed collecting and processing weighed amounts of pod and pith on eighty-eight occasions (Table 2).

The amount of flour gained from a known weight of pods was calculated separately using data collected by KH. A 1 kg pod gives 0·6 kg pith which gives 0·3 kg of edible flour. Wehmeyer, Lee & Whiting (1969) give a caloric value of 388 cal (1629·6 J)/100 g of baobab as eaten. If we apply this figure to the returns in g/hour shown in Table 2 we find that children aged 5–10 years collect and process about 540 g of raw pods per hour (629 cal (2641·8 J)/h). Children aged 10–15 get about 1014 cal (4258·8 J)/h. This makes baobab the most rewarding food for children and it is the one most taken by the younger children.

Hunting by children

Woodburn's film (Woodburn & Hudson 1966), his British Museum guide (1970) and Jelliffe *et al.* (1962) could be read as implying that Hadza boys of 10 or more feed themselves by shooting birds and small animals. Our data suggest that they entertain themselves by trying to shoot birds and small animals but feed themselves with baobab, honey, payments for errands, scrounging, and are also fed by mother and grandmother.

During the 31 days' observation in the dry season of 1986, thirty passerine birds were presented to NBJ for weighing; they weighed on average 13 g each. None was eaten by the boy who shot them. Two were taken by young women and eaten. Most were given to toddlers who carried them about and sometimes eventually cooked and ate them. Eight larger birds (mean weight 172 g), one (210 g) bush baby (*Galago* sp.), and one 96 g elephant shrew (Macroscelididae, probably *Nasilio brachyrhynchus*) were shot by boys during this period, all at dusk. It is not known who ate many of these. When the boys were searching for honey they seldom pursued birds. When they pursued birds they usually missed. Boys of this age can and do shoot things just as in the Woodburn film, but this was not a significant part of their diet. We only followed children during the dry season of 1986 so that these comments may have to be revised if there is seasonal variation in boys hunting. A teenage girl captured a 1·1 kg tortoise and disappeared into the bush with it and two girl friends, apparently to eat it.

Boys in villages at the edge of Hadza country trap birds successfully. We did not observe Hadza boys doing this.

We twice observed 'scavenging' by children. On one occasion several 12−15 year old boys found a dead spotted hyaena (*Crocuta crocuta*) near camp and cooked and ate many parts of it, bringing some of the cooked parts into camp. On another occasion three children aged 4−8 cooked the discarded, plucked wings of a vulture (brought home by an adult man for the feathers). Their meal was interrupted from time to time by adults telling them that vultures were not food.

Honey

Adult search for honey is strongly seasonal, peaking in June and concentrating on *Apis mellifera*. However, boys and youths aged between 10 and 20 spent much time in the 1986 dry season collecting honey, mostly from other species. Any boy who could get the use of an axe spent a considerable amount of time searching for honey. Boys with no axe sometimes accompanied and spotted for boys with an axe. Indeed it was my strong impression that the boys of 12−15 were very much better at finding the nests of a small 'sweat bee' (called *kanoa* in Hadza) than were older boys and men. However, older boys wield an axe faster and get higher returns.

Fifteen hours of follows were conducted. Returns were measured in two ways: (i) weighing the honey and comb taken from a nest, (ii) recording the estimated

volume of portions eaten. The weighing showed *kanoa* nests to average 45 g
($n=27$) with relatively little variation. This was used to calibrate follows on which
weighing was not done (getting sticky as well as very hot appeals a good deal
more to Hadza boys than to researchers). A find that led to substantial portions
being eaten was counted as a find of 45 g of honey.

Six boys aged 10–15 acquired on average 111·5 g/h (Table 2) (averaging just
over two finds per hour). At 304 cal/100 g, they acquired about 339 cal (1424 J)/h.
Boys who went into the bush with the women tended to do at least 2 and more
often 3–4 hours of honey hacking. Thus they could acquire 650–1350 cal (2730–
5670 J) of honey each day. The variation seemed to be correlated with age, skill,
and luck. Two older boys got returns of 438 cal (1839·6J)/h and 578 cal (2427·6J)/h.
Honey from *Apis mellifera* whose nests are much larger, often containing more
than 1 kg of honey, was seen taken by boys on two occasions.

An estimate of total food acquired by children

We do not yet have good data on the amount of time that Hadza children spend
foraging for themselves. It was difficult to collect such data alongside a good
sample of follows of particular foraging trips, and the observations of food sharing
that were the top priority in the 1986 dry season. The food brought into camp by
children and weighed amounted to 614 calories (2578·8 joules) per child per day.
This can only be taken as a rough minimum figure; by no means all the food that
children acquired was weighed on every day spent observing in camp. Children
appear to eat a small proportion of the food they gather while they are still out of
camp.

Our impression is that boys and girls who are in camp for the day go on at
least one baobab excursion. Some often went on two or more in a day. Boys in
the bush pursue honey for several hours. Girls in the bush dig for several hours
but with interruptions (fewer as they get older) for playtime and baby-entertaining.
They get very low returns (we suspect they go for their share of the meal, and are
encouraged by older women as babysitters and as learners). Teenage girls,
though often followed by teenage boys, continue energetic digging while they
chat.

If children spent just 2 hours per day foraging for baobab, or (for older boys
with access to an axe) honey, they would acquire around 800–1000 calories,
almost half the calories they need. It seems realistic, from our observations, to
suppose that many children do spend this much time foraging each day during
the dry season. Children's returns from their own foraging increase with age,
probably outstripping the increase in food requirements as they get older. Thus it
is quite possible that Hadza children from the age of about 5 years old provide
half of their food by their own efforts, and older children much more.

The amount of food to be gathered by a mother (to ensure a basic caloric
requirement for her children) is thus a good deal less than for the !Kung, and it

probably declines with a child's age rather than increasing. If we take into account the greater contributions of men and old women (see Hawkes, O'Connell & Blurton Jones, this volume), the Hadza woman's contribution to her child's diet may be even more reduced. Furthermore, the effect of her behaviour upon the child's survivorship will also be weakened. If she neglects it, it can simply obtain more food for itself, or curry favour with its grandmother or other female kin.

Economic benefits of Hadza children

Unlike !Kung children, Hadza children help adults in many ways. In camp children run errands, fetching and carrying food or utensils, and carry messages. They take messages, food, water, and utensils between camp and the 'men's place'. They often fetch water from the waterhole at the command of an adult. They entertain and occasionally feed toddlers in camp. They feed toddlers with baobab, sometimes with roots, and on rare occasions with tiny birds. Older girls who go into the bush with the women bring food home. The amounts of these things vary, and the circumstances of this variation will be interesting to establish.

In addition to most of the above (including feeding and occupying toddlers) boys aged 10–20 accompany women into the bush, playing two important roles. They act as guards against Datoga pastoralists and wild animals. They climb baobab trees and cut and knock down vast quantities of fruit that the women would otherwise not obtain.

The work of Hadza mothers — is backload an appropriate measure?

BJ&S followed Lee's lead and treated weights to be carried by !Kung mothers as a crucial measure of the cost entailed in raising children. There are four reasons to think this was appropriate for the !Kung case. First it turned out to give correct predictions about birth spacing. Second, there was an astonishingly close correspondence between required backload and child mortality. Third, it seemed likely that the energy expenditure of carrying heavy loads (15–20 kg) over long distances of 6+ miles, in a severe climate was a limiting factor. Fourth, it proved possible to translate many consequences of child rearing into the common currency of 'backload'. We can begin to estimate the backloads required for Hadza women to raise a child. But is backload at all likely to limit Hadza women in their acquisition of resources?

Hadza backloads and journey distances

The average load carried from digging place to the midday meal site was 13·6 kg for mothers who had a baby with them (maximum 16·6 kg). Weights carried back to camp were always considerably less (61% of the weight gathered was taken home: Hawkes, O'Connell & Blurton Jones, this volume). Distances could not be measured accurately but were between 0·5 and 2 miles most of the time. Journeys to root patches varied between 4 and 60 min, mean 25 min, and were walked at a moderate pace.

The heat environment

In the Kalahari Lee (1976) reported September to October temperatures of over 93° F (30° C) every day, and Wyndham *et al.* (1964) report globe thermometer readings of up to 140° F. To assess heat stress one needs to know wet bulb (WB) and globe thermometer (GT) readings. In 1981 Dr Ben Nieuwoudt, the medical doctor at Tsumkwe in Namibian Ngamiland collected GT and WB data for us, in what he was told was an unusually cool year. Globe thermometer temperatures for this period, and for the 1986 dry season in Hadza country are given in Table 3. The Hadza figures are much lower than the Ngamiland figures. Only 14% of days in Hadza country had highest (noon to 5 pm) GT readings above 115° F, while 44% of days in Ngamiland had noon GT readings above 115° F. Wet bulb readings are low in both environments. Though stressful for the researcher, and the subject of occasional complaint by the Hadza, the Hadza conditions are not particularly dangerous for a journey of only 25 minutes carrying 15 kg.

In Hadza country it seems most unlikely that heat stress is a problem for Hadza women carrying babies and food. Whether heat stress limits time and energy spent digging is another question. Complaints by men and women about the power of the sun, and that Hadza women often sought medicine for headaches at the end of specially hot days suggests that the working Hadza woman may experience some heat stress in the late dry season.

Time and efficiency costs of Hadza children

Hadza women may run out of time for acquiring and processing food. Hawkes, O'Connell & Blurton Jones, this volume, present data that shows that in all seasons Hadza women work longer hours foraging out of camp than do !Kung women. Time spent processing food and in other tasks has to be added to this. Often women were seen still hammering baobabs at the rock as darkness fell. If we measure costs of children not in weight carried nor in energy expended, but in time used, the cost of feeding Hadza babies increases considerably. Babies could be also costed by their nuisance value, that is, the time taken to feed and quiet them, and thus lost to food acquisition, and by the detrimental effect they have on the mother's skill, speed and strength at digging.

The root-digging returns of women carrying babies were substantially lower in the 1986 dry season than those of women of child-bearing age who were not carrying babies. Three women who carried babies, and one pregnant woman in the month before and the month after she gave birth averaged 0·997 kg/h (Table 4). (Their returns were 66% of those of women of comparable age who were not carrying babies.) These four had lower returns than all but one of the mature

Table 3. Radiant heat measures from dry season in Hadza and !Kung country (GT readings collected in 1981 by Dr B. Nieuwoudt at Tsumkwe, by NBJ in 1986 in the Tliika region of Hadza country)

GT (°F)	Hadza (Tliika) No. days	!Kung (Tsumkwe) No. days
< 91	0	5
91– 95	0	1
96–100	1	5
101–105	2	5
106–110	7	16
111–115	8	17
116–120	2	21
121–125	1	14
>125	0	3
Total	21	87

(completed breast development) women. Four women of child-bearing age who did not carry babies averaged 1·536 kg/h. The women with babies were also lower than the five teenage girls and childless young women who averaged 1·171 kg/h. This lowering of returns was assumed to be due to time lost to baby care, exhaustion, and clumsiness at digging with a baby strapped to one's back. However, the number of women studied is small, and although a similar impression is given by the data from dry season 1985 and wet season 1986, that difference is not statistically significant (Hawkes, O'Connell & Blurton Jones, this volume). Not only is more data needed, but we should take into account differences in the ages of the babies, and in the availability and interest of child 'caretakers'.

Thus babies who still ride on mother may turn out to impose a time cost that lowers the mother's rate of acquisition of food for all her other children and 'dependants'. There may be some very interesting consequences of using time as a measure of a mother's costs.

It would be possible, though difficult, to make the same comparison for baobab processing. Babies and toddlers interrupt this work quite considerably. Mothers try to suckle while they pound the baobab into flour but it seldom works. Children stop mothers' work by suckling, by getting too near — their fingers could be crushed by the mothers' hammering — by walking on other people's work, by grizzling and by falling about. The time taken by these interruptions looks significant and should be quantified. Some of the toddlers who stay at home in the day also interfere strongly with mothers' baobab processing, particularly if they try to process baobab soon after returning home, when the toddler is eager to be suckled.

Completing a model that assesses cost of children in time needed to provide for them, or in energy expended, will have to wait until later. But it is an important issue for the generality of the BJ&S model, and for generalizations about mobile hunter-gatherers that weight seems so inappropriate for the Hadza case.

TABLE 4. Weights and work of five categories of women: teenage girls, post-pubertal young women yet to give birth, mothers still of child-bearing age but not carrying babies and women past child-bearing age. September to November 1986. ('B' = gave birth between weighing dates)

ID	Height	Weight 1	Weight 2	Change	Days in	Days out	kg/day out	h/day out	kg/h
IYA (10)	131	18·65	17·75	−	18	6	0·58	3·75	0·155
MOGI (10)	130	18·25	18·40	+	18	3	0·53	5·35	0·096
KEJA (14)	137	31·10	30·55	−	3	26	2·81	5·37	0·525
BUD2 (15)	153	37·55	38·15	+	8	19	2·47	5·46	0·452
SHARP (15)	145	32·40	35·85	+	7	20	4·11	5·57	0·738
RIPO (16)	138	36·40	33·70	−	6	22	5·94	5·64	1·054
CF (17)	145	44·85	42·90	−	6	22	7·43	5·28	1·205
GONGS (18)	156	46·25	48·45	+	7	20	6·94	5·85	1·186
SOKI (21)	141	45·80	44·70	−	9	17	7·07	5·75	1·230
PEND (23)	142	43·85	44·25	+	9	14	6·74	5·70	1·182
THRM (19)	141	34·60	35·45	+	10	12	4·27	5·06	0·844
PAK	140	43·75	41·05	B	7	23	5·24	5·61	0·935
KRS	146	44·10	45·50	+	7	22	6·36	5·70	1·116
TB	156	50·85	50·90	+	4	25	7·08	5·93	1·195
FB	160	60·05	58·50	−	10	20	7·17	5·67	1·265
WELE	158	43·55	46·35	+	8	14	11·36	5·35	2·123
TNB	167	61·30	65·35	+	8	19	8·07	5·69	1·419
SHEL	154	51·90	50·65	−	4	23	7·80	5·83	1·337
ISAM	143	38·75	37·60	−	0	30	10·26	5·75	1·785
OTB	154	45·50	45·35	−	0	29	9·10	5·81	1·566
EARS	139	37·45	37·00	−	2	27	9·35	6·12	1·526
NC	156	53·70	52·15	−	0	27	9·20	5·87	1·566
TWRD	140	40·15	38·85	−	0	29	10·13	6·04	1·667
OT	154	45·30	44·15	−	3	25	6·04	5·50	1·098

DISCUSSION

Differences between costs of Hadza and !Kung children

The analysis of the Hadza data is at a very early stage. It is important to emphasize that the data comes only from the dry season of 1986, with some additions and comparisons with the dry season of 1985. No data is presented on the wet season, and we do not know whether the wet season is more difficult or less difficult for Hadza child rearing.

How do costs of Hadza child rearing compare with costs of !Kung child rearing as discussed by BJ&S?

(a) Hadza children accompany their mother when she goes gathering until the child is about 2 years old. In the bush the child is primarily carried by its mother. According to Lee (1972), !Kung children accompany their mother to a greater age, the proportion of time they are carried by mother declines gradually until at around the age of 4 the child always stays at home or walks. Hadza children thus take up much less of the mother's backload than do !Kung children. Hadza children aged 2−4 must be at some increased risk when left unattended in camp but the risks are not obvious. Are they measurable? Are they any greater for !Kung children?

(b) Hadza children are weaned around 2½ years old, 6−9 months earlier than !Kung children. This difference is also puzzling. Perhaps tubers, and, in the dry season, baobab, provide the Hadza with better weaning foods. But in both cultures children eat a variety of adult foods long before they are weaned.

(c) Hadza children forage for themselves quite effectively and quite often. In contrast !Kung children are described as very seldom foraging (Draper 1976; Lee 1979). As well as feeding themselves, Hadza children sometimes feed their younger siblings or other close relatives. This appears to be a most important difference between the Hadza and the !Kung. It is essential for us to get an understanding of why this difference exists.

(d) Hadza children perform many small and not so small services for adults. This is a strong contrast with the !Kung (Draper 1976). Hadza children may fetch water because it is so near (0·25−0·5 miles (0·4−0·8 km)) whereas !Kung camp about 1 mile (1·7 km) from water in the late dry season (Lee 1972, 1979).

(e) Hadza children receive food from older siblings, grandmothers, other people, as well as from their own efforts and from their mothers, and indirectly from men. We expect to be able to quantify the roles of these sources.

Comparable data is not available for the !Kung. But plant foods are held to be shared much less than meat. BJ&S assumed that the !Kung mother collected most of the plant foods eaten by her children, and followed Lee in expecting this to amount to 58% of the child's calorie intake. BJ&S assumed in effect either that no one but mother and father fed the child, or that food received from others was balanced by food given by mother and father to others (in other words that sharing was very even, as suggested by their reading of the ethnographies). This is an important assumption. A flow of food from grandmother to children that was not reciprocated by the mother would ease the mother's burden. We believe such a flow to be an important feature of Hadza life (see Hawkes, O'Connell & Blurton Jones, this volume).

(f) Although they forage more often, Hadza women go much less far to get food than do !Kung women. This seems to be a simple consequence of the location of water and plant resources in the two areas.

(g) The climate in Hadza country in the late dry season is less harsh than in !Kung country in the late dry season.

(h) Hadza women with children sometimes carry weights similar to those carried by !Kung women. But as the journeys are very short, and the climate easier, we do not think that heat stress imposes any limit on loads that can be carried by Hadza women. Consequently we do not think that backload would be as good a measure of costs of Hadza children as it seemed to be for !Kung children. We do not think backload, or similar results of mobility, are universal constraints on the reproduction of nomadic hunter-gatherers.

(i) Though the effect is not statistically significant, Hadza women who carry small children with them seem to acquire less roots per unit time than do similar-aged women who are not carrying babies. Lactating women work significantly less time than non-lactating women of child-bearing age (K. Hawkes, J.F. O'Connell & N.G. Blurton Jones, unpublished data), while the grandmothers work more time than anyone. Small Hadza children interfere with baobab processing. We think these imply that time required to acquire and process the food needed by a child would be an interesting measure of the costs of a Hadza child. We do not think that Hadza women will fit the contemporary view of hunter-gatherers as people of leisure, unconstrained by time.

Expected differences between Hadza and !Kung rates of birth

From these differences we are tempted to conclude that the work involved in raising a Hadza child is considerably less than the work involved in raising a !Kung child. If we apply the same model as in the !Kung research, we should expect shorter IBI among the Hadza. However, if the best measure of costs is not backload but something else, such as time, then we cannot at this stage judge what length of IBI we expect. Completing a model of Hadza child costs will require further data (for instance on nutritional value of plant food, proportion of food supplied by women and men, etc.). We also need to decide upon the most appropriate measure of costs for the Hadza model (see below). It remains likely that, as biological common sense should have told us anyway, when opportunity arises, hunter-gatherers can and do reproduce faster than !Kung. Our data suggests that such an opportunity may exist in the Hadza's patch of savannah–woodland.

Which is the best measure of costs of children?

We have argued that 'backload' was a successful measure of costs for the !Kung but that it seems less appropriate for the Hadza. How should we choose a measure of cost?

For the evolutionary ecologist, the ultimate measure of costs is the effect of a child (or any features of raising it) upon the number and survivorship of its siblings. However, not only is this very hard indeed to measure, but it leaves us uninformed about the reasons for the effect. It gives us no clues as to what conditions or changes in conditions might modify this effect. At the other extreme, simply to document consequences of children, what they make mother do, leaves us unable to relate these consequences to each other or to the reproductive strategy. Thus a promising intermediate measure of cost seemed to us to be the effect of the child upon the mother's ability to acquire the resources needed to produce and raise children.

It seems clear that we need a common currency into which we can translate consequences of raising children. 'Backload' was a useful common currency for the !Kung because so many features of raising and providing for !Kung children could be seen to have consequences for the weight of food and child to be carried by the mother. But other common currencies are possible, and may be just as good. Thus we can translate the calories needed to produce milk for the baby into the weight of food to be gathered to provide these calories, or the time needed to gather this food. We can translate many consequences of child care into calories or time. For example, children's interruptions of baobab processing could be measured in time, and in calories of food that could be provided if that time had been spent processing. The weight of food unprocessed during the interruption could also be calculated. But is this a helpful measure of the cost of the interruption? The interruption would not be redressed by bringing back more baobab.

We should use a cost of which the mother can only afford to incur a limited amount. If she can bear infinite costs then she can raise infinite numbers of offspring. BJ&S felt they had some reason to believe that backload would be limited to something not far above 15 kg. There is a limit to the amount of time in a day, though we might debate whether Hadza women should be expected to process food after dark! (sometimes they do!). There must be some limit to the number of calories a woman can expend in a day, and it is conceivable that her efficiency at many tasks deteriorates as she gets exhausted.

We cannot offer a complete solution to the question of how to measure costs of children. Those of us who wish to exploit Trivers's stimulating concept of parental investment to its fullest, still have some way to go. It is clear that the combination of the traditional anthropological interest in economic costs of children with the concept of parental investment offers promising directions.

Relative costs of lactation and pregnancy and of older children

One of the most interesting results of beginning to construct a model of costs of Hadza children seems to be the emphasis that falls on the costs of children under 2 years old. This was seen to be particularly important in the analysis of the activity of post-reproductive women (Hawkes, O'Connell & Blurton Jones, this volume).

Hadza children from 5 years old seem able to easily acquire a substantial proportion of the food that they need. This is likely to mean that any neglect of the child by its mother will have little effect on its survivorship, surely less than it would have on the survivorship of a !Kung child. We might expect the neglected Hadza child simply to forage more. Conversely, the fitness returns for kin investing in an older Hadza child will be much less than for a !Kung relative investing in an older child. It also follows that returns from investing in younger children and in the production of children by the mother would give greater returns.

The lower costs of reduced care for older children may permit Hadza parents both to be more rejecting and punitive to children, and to make use of children's labour to an extent that the !Kung cannot. The Hadza parent has less to lose by compelling the child to expend energy on the parent's behalf.

This argument is largely dependent on the assumption that !Kung children would be unable to acquire sufficient resources if forced to forage. This assumption is, as stated above, untested. Lee (1979) reports an instance of successful foraging by !Kung boys aged 10–14, interpreting it as arising from the large proportion of dependents in their camp. This episode took place in July, a dry cool month, but Lee gives no additional information, such as the distance between camp and the foods that were gathered and how this compares to other camps in which children were observed not to gather.

Could the !Kung raise children as cheaply as the Hadza?

The differences between consequences of raising Hadza and !Kung children draws attention to the many features of !Kung life that seem to increase the cost of children. BJ&S already raised the question: could a !Kung woman rearrange her life in such a way as to enable her to raise children at a faster rate? Seeing the way Hadza produce children we have to return to this question even more forcefully. Specifically we must wonder, for the proposedly limiting late dry season:
(a) Why don't !Kung women leave children at home at an earlier age? Hadza women do not seem to need baby-sitters, why do !Kung? Is it merely a question of camp size? Have Hadza passed some demographic threshold which ensures a supply of teenage baby-sitters? This does not seem to be the reason, for most of those baby-sitters are out in the bush with the women and not at home with the small children! Is the Hadza environment with its apparently greater density of wildlife but cooler climate really so much safer than the !Kung environment?
(b) Why don't !Kung children forage for themselves more often and from an earlier age? Would they get lost in their flat environment when Hadza children can use slopes, kopjies, baobabs, the view across the rift valley, as navigation aids? Is there anything rewarding for !Kung children to forage on within safe range of camp?

Why don't elderly !Kung women work harder to feed their grandchildren?

While these weaken the model's claim that !Kung women are reproducing at the optimal rate (unless we investigate the explanations suggested above), they illustrate the strength of the model as a way of integrating many aspects of !Kung life and circumstances and showing how they would influence reproductive success. They can be regarded as entry points to extending the models to allow us to assess the fitness consequences of more and more aspects of !Kung or Hadza life.

ACKNOWLEDGEMENTS

This work was supported financially by the National Science Foundation, the Swan Fund, Ms B. Bancroft, the University of California, Los Angeles and the University of Utah. We thank Utafiti (Tanzania National Scientific Research Council) for permission to conduct field work in Tanzania. We wish to thank for invaluable help at many stages of this research Mrs Adie Lyaruu, Mrs Flora Matemu, Enock Maranga, D. Bygott and J. Hanby, Michael Leach, Nick and Bridget Evans. We are grateful to many Hadza for their patience, protection and companionship. The research could never have started without the generous and capable help of Lars C. Smith who introduced us to the Hadza and to Tanzania.

REFERENCES

Blurton Jones, N.G. (1986). Bushman birth spacing: a test for optimal interbirth intervals. *Ethology and Sociobiology*, 7, 91–105.

Blurton Jones, N.G. (1987). Bushman birth spacing: direct tests of some simple predictions. *Ethology and Sociobiology*, 8, 183–203.

Blurton Jones, N.G. & Sibly, R.M. (1978). Testing adaptiveness of culturally determined behaviour: Do bushman women maximise their reproductive success by spacing births widely and foraging seldom? *Human Behaviour and Adaptation* (Ed. by N. Blurton Jones & V. Reynolds), pp. 135–58. Society for Study of Human Biology Symposia, 18. Taylor & Francis, London.

Draper, P. (1976). Social and economic constraints on child life among the !Kung. *Kalahari Hunter-gatherers* (Ed. by R.B. Lee & I. DeVore) Harvard University Press, Cambridge, Massachusetts.

Durnin, J.V.G.A. & Passmore, R. (1967). *Energy, Work and Leisure*. Heinmann Educational Books, London.

Howell, N. (1979). *Demography of the Dobe !Kung*. Academic Press, New York.

Jelliffe, D.B., Woodburn, J., Bennett, M.B. & Jelliffe, E.F.P. (1962). The children of the Hadza hunters. *Tropical Paediatrics*, 60, 907–13.

Lee, R.B. (1968). What hunters do for a living, or, how to make out on scarce resources. *Man the Hunter* (Ed. by R.B. Lee & I. DeVore), pp. 30–48. Aldine, Chicago, Illinois.

Lee, R.B. (1972). Population growth and the beginnings of sedentary life among the !Kung bushmen. *Population Growth: Anthropological Implications* (Ed. by B. Spooner). M.I.T. Press, Cambridge, Massachusetts.

Lee, R.B. (1979). *The !Kung San*. Cambridge University Press, Cambridge.

McArdle, W.D., Katch, F.I. & Katch, V.L. (1986). *Exercise Physiology*. Lea & Febiger, Philadelphia.

Morley, D. & Woodland, M. (1979). *See How They Grow*. Macmillan, London.

Nag, M., White, B.N.F. & Peet, R.C. (1978). An anthropological approach to the study of the economic value of children in Java and Nepal. *Current Antropology*, 19, 293–306.

O'Connell, J.F., Hawkes, K., & Blurton Jones, N.G. (1988). Hadza scavenging: implications for plio-pleistocene hominid subsistence. *Current Anthropology*, 29, 356–63.

Trivers, R.L. (1972). Parental investment and sexual selection. *Sexual Selection and the Descent of Man* (Ed. by B. Campbell), pp. 136–79. Aldine, Chicago, Illinois.

Wehmeyer, A.S., Lee, R.B. & Whiting, M. (1969). The nutrient composition and dietary importance of some vegetable foods eaten by the !Kung Bushmen. *South African Medical Journal*, **95**, 1529–30.

Woodburn, J.C. (1964). *Social organisation of the Hadza of northern Tanganyika*. Unpublished Ph.D. dissertation, Cambridge University.

Woodburn, J.C. (1968a). An introduction to Hadza ecology. *Man the Hunter* (Ed. by R. Lee & I. DeVore), pp. 49–55. Aldine, Chicago, Illinois.

Woodburn, J.C. (1968b). Stability and flexibility in Hadza residential groupings. *Man the Hunter* (Ed. by R. Lee & I. DeVore), pp. 103–10. Aldine, Chicago, Illinois.

Woodburn, J.C. (1970). *Hunters and Gatherers: The Material Culture of the Nomadic Hadza.* The British Museum, London.

Woodburn, J.C. & Hudson, S. (1966). *The Hadza: The Food Quest of a Hunting and Gathering Tribe of Tanzania* (16 mm film). London School of Economics, London.

Wyndham, C.H., Strydom, N.B., Ward, J.S., Morrison, J.F., Williams, C.G., Bredell, G.A.G., Von Rahden, M.J.E., Holdsworth, L.D., Van Graan, C.H., Van Rensburg, A.J. & Munro, A. (1964). Physiological reactions to heat of Bushmen and of unacclimatised and acclimatised Bantu. *Journal of Applied Physiology*, **19**, 885–8.

Vincent, A. (1985). Plant foods in savanna environments: a preliminary report of tubers eaten by the Hadza of northern Tanzania. *World Archaeology*, **17**, 131–48.

Yellen, J.E. & Lee, R.B. (1976). The Dobe − /Du/da environment: background to a hunting and gathering way of life. *Kalahari Hunter-Gatherers* (Ed. by R.B. Lee & I. DeVore), pp. 27–46. Harvard University Press, Cambridge, Massachusetts.

Differential parental investment: some ideas on the contact area of European social history and evolutionary biology

E. VOLAND

Institut für Anthropologie der Universität,
Bürgerstrasse 50, D-3400 Göttingen, FRG

SUMMARY

1 In European history the variation of infant and child mortality is based, among many other factors, on parental attitudes towards the life and death of offspring.
2 Such differences are the result of specific ecological and historical experiences of the populations: The degree of fatalism or responsibility with which parents regard their newborn children depends on the amount of stress to which the population has been exposed (such as epidemics, wars, subsistence crises).
3 When comparing populations, differential parental effort could be interpreted as a biological adaptation to the more or less severely fluctuating environments.
4 Within populations, part of the variation in infant and child mortality is attributable to discriminatory parental behaviour.
5 The desire for children and the love and affection bestowed upon them are based on the reproductive value of both parents and children. Accordingly, differences in infant and child mortality have as their ultimate causes different adaptive investment strategies.
6 Proximately, investment decisions are guided by the prevailing traditional attitudes and values of a population. Therefore, cultural norms and models for reproduction and investment have an eminently biological function: on average, they support parents in their striving for fitness maximization.

INTRODUCTION

Numerous causes for the high rate of infant and child mortality in the past have been suggested. Poor hygienic conditions coincided with unsatisfactory medical care, infant nutrition was frequently artificial, mothers had extreme work loads, economic fluctuations resulted in subsistence crises, and many other circumstances have been held responsible for the detrimental and frequently life-threatening living conditions of infants.

Social historians are of the opinion that these popular explanations for the frequently horrendous mortality figures only skim the surface, because they are unable to explain a large part of the observed statistical variance between families. Instead, there is a second level underneath these causes. Infant and child mortality not only reflects the external ecological and economic pressures, but also depends on the nature of parental resource allocation to their offspring.

Figure 1 illustrates the relation between rye prices and the number of abandoned children in the French community of Limoges (1726–90). The figure shows children were increasingly abandoned during periods of food shortage. Under such conditions parents made decisions concerning the allocation of investment amongst their offspring. It does not require much to imagine that parents had a certain amount of leeway for exerting a strategically differentiated influence on the 'God-given' fate of their children. Hence, parental reproductive interests stood between the socioecological pressure and the fate of the children as a filter, a buffer and perhaps at times even as a reinforcer. The history of differential infant and child mortality is also a history of differential parental behaviour.

The purpose of this chapter is to examine some of the factors likely to affect parental investment decision making.

PARENTAL BEHAVIOUR AND INFANT MORTALITY: RELATIONS AT THE POPULATION LEVEL

Ecological experience and parental care

At the end of the nineteenth century infant mortality averaged over 30% in most regions of southern Germany while at the same time it remained under 15% in

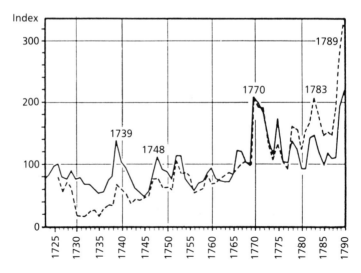

FIG. 1. The relation between rye prices and the number of abandoned children in Limoges (1726–90). —— Rye prices; ---- Foundlings in the Hôpital général. (From Peyronnet 1976.)

north-western Germany. Imhof (1984) suggests that this extreme difference could be attributed to collective psychological factors, because both populations underwent completely different ecological and historical experiences. Whereas southern Germany was repeatedly afflicted by catastrophic events, such as the frequent and occasionally long-lasting wars of the seventeenth and eighteenth centuries (by the Thirty Years War, in particular), as well as by the disastrous plagues of the late Middle Ages and early modern times, north-western Germany remained practically untouched by these threats. These dissimilarities in ecological experience were so pronounced that they led to population differences in mentality and hence to attitude differences towards children, and moreover, towards life strategies in general, even up to the end of the nineteenth century, i.e. to a time in which the overall ecological stress has lost most of its unpredictability and menace (Imhof 1984).

Stress which cannot be controlled (e.g. constantly recurring epidemics, disastrous military campaigns and oppressive subsistence crises) shape, in the long run, the belief in what can be achieved on earth. If repeated heavy losses of population due to war, illness or hunger lead to the destruction of the means for existence again and again, then this experience will bury itself deeply and permanently into the memory of whole generations. It seems fair to assume that in those populations that have been heavily 'traumatized' (Imhof 1984) there would be a gradual development of a more or less fatalistic attitude towards worldly goods, towards property and poverty but also towards life and death, compared with regions not exposed to the same traumatization (Imhof 1984). One consequence of this is that even the bearing and the loss of children is perceived from a view that has proved its value as a practical guide to interpreting and understanding the world. At the same time, the world view provides a rationalization and a justification for one's own beliefs and behaviour — including parental investment. It is this pattern of a population's psychohistorical constitution which co-determines to what degree parents strive to shield their offspring from ecological and economic pressures.

Demographic adjustment and biological adaptation

Clearly the interpretation of mortality differences by social historians discussed above, raises intriguing questions for evolutionary biologists.

In the first place, it is evident that ecological influences on human reproductive behaviour also have important historical constraints. The perception of the environment accumulates a certain body of experience over generations, which influences individual reproductive decisions via psychological and cultural mechanisms. Anthropologists have for the most part ignored the role of historical factors when they search for the ecological constraints for specific investment patterns. Their research generally focuses on the current adaptive value of, for example, wide birth spacing (Blurton Jones & Sibly 1978) or female infanticide (among the Eipo: Schiefenhövel 1984, or among the Netsilingmiut Eskimos: Irwin, in press). Caro & Borgerhoff Mulder (1987) rightly emphasize the epistemological difficulty to conclude such investment patterns being products of natural selection, because a long enough social and ecological stability can hardly be proven for traditional societies. There are no *a priori* reasons to assume a correlation between a lack of historical knowledge and a stability of the population's living conditions in the past (Caro & Borgerhoff Mulder 1987). As long as it is not clear how enduring the collective memory is under changed living conditions, there will always remain a certain amount of doubt with respect to the designation and description of behavioural adaptations. This is more marked when populations pass through a strong sociocultural change. For example, the above-average fear that Greek immigrants to Australia have of losing a child, and their correspondingly high fertility, is best explained by the knowledge of the high rate of infant mortality in rural Greece rather than experience actually acquired in their new home (Callan 1980).

Secondly, you are immediately reminded of the *r*- and *K*-strategies of the behavioural ecology models. Does not the higher 'turnover' of life, i.e. the higher fertility and the lower per capita investment in the historically more severely traumatized regions of southern Germany, correspond to the evolutionarily successful oportunistic *r*-strategy reaction to unpredictable environmental fluctuations, while the relatively stable socioecological situation in Germany's north-west favoured the *K*-strategy reduction of the number of children in conjunction with an increase in the per capita investment? Is not perhaps even the demographic transition to be understood as a biological reaction to the transformation from the uncertain living conditions of the later Middle Ages and early modern times to the increasingly safer living conditions of the nineteenth and twentieth centuries? Three aspects of the demographic transition support this assumption:

(a) Contrary to previously held views, it has become increasingly evident that human fertility, even in pre-modern societies, was adapted to the prevailing socioecological living conditions. 'Natural fertility' was not a question of fate, but a deliberate, quasi-'rational' measure for mastering life. Accordingly, the demographic transition is to be understood as a behavioural adjustment process to changed socioecological living conditions, and not as an effect of increasing family planning competence (Carlsson 1966; Andorka 1978; Easterlin 1978; Wrigley 1978; Freedman 1979; Lee & Bulatao 1983; Lindert 1983; McLaren 1984). Certainly, pre- and post-transitory reproductive strategies equally reflect the 'unconscious rationality' (Wrigley 1978) to strive for an adjusted balance of people and resources at the family level. This point fits in very well with anthropological conceptions of human reproductive behaviour (LeVine & Scrimshaw 1983), and there is no identifiable reason why Alland's (1970) conclusion that humans have always controlled their reproductive patterns to a certain degree should not also apply to pre-modern Europe.

(b) The demographic transition begins with a certain amount of delay after a radical change in the chances for survival. The extreme fluctuations in mortality during the late Middle Ages and early modern times due to pestilence, hunger and wars, that is, to Malthus' three positive checks, gradually decrease and stabilize at an initially quite high, but relatively invariable and therefore calculable level (Flinn 1981). One of the direct primary causes for this change lies in the increasing replacement of epidemic by endemic causes of death. Against this background of decreasing environmental fluctuations, an advantageous rise in 'quality' of offspring can become an increasingly realistic and promising option in reproductive strategies. Reynolds (1986, p. 112) argues similarly: The fertility decline 'occurs when people perceive the environment in which they live as stable and trustworthy as opposed to being hazardous and likely to collapse around them'.

(c) Together with the demographic transition, there was indeed an average rise in 'quality' of offspring, that is, an increase in per capita costs (Andorka 1978; Birdsall 1983; Lindert 1983). Historically, the leading social groups (the bourgeoisie or the nobility, respectively) were those who initiated the demographic transition (Livi-Bacci 1986; see, for example, the situation in France: Lévy & Henry 1960; particularly in Rouen: Bardet 1983). The 'trendsetters' were, therefore, those social groups with socially strategic reasons for aiming at a rise in 'quality' of their offspring and the latter's advantageous social placement.

All of these observations can be reconciled without any contradictions with the assumption that the demographic transition is the result of an increasing upgrading and efficiency of K-strategy options. In my opinion this suggests an evolutionary influence in human reproductive performance even in our most recent past. Of course, to interpret the demographic transition not only as a change in psychological, social and economic, but also in biologically adaptive strategies, is only of value as a hypothesis until there is empirical evidence that in the transitory and in the post-transitory societies the reduction in the number of offspring has led to increased inclusive fitness.

Doubts have been legitimately raised as to whether social success obtained at the expense of the number of offspring is really translated into increased fitness (Lande 1978; Barkow & Burley 1980; Hill 1984; Vining 1986). In any case, more detailed studies than those currently available are necessary in order to make valid statements about the biological adaptivity or maladaptivity of the modern fertility decline (see Kaplan & Hill 1986).

Let us summarize briefly at this point: Differential parental investment reflects, among many other things, differences in the psychohistorical constitution of populations. The collective experience with the predictability of living circumstances has shaped parental attitudes towards life and death of their children and has, in this way, influenced the quality and extent of parental primary care. There is some evidence that the phenotypic adjustment of parental behaviour to existing living conditions is guided by an adaptive reproductive strategy competence.

A second aspect of differential parental investment has less to do with reproductive performance at the population level than with parental strategies within populations.

INTRA-FAMILIAL DIFFERENCES IN
PARENTAL INVESTMENT

The psychosocial desirability of children and their reproductive values

If sociobiologists are right in claiming that human reproductive behaviour is optimized for fitness maximization by natural selection, then it should be possible — even within populations — to find patterns of differential parental investment that serve adaptive reproductive strategies. It should make sense also to investigate the variations of infant and child mortality from a perspective that is aimed at the possible ultimate causes for these variations rather than the proximate ones.

Evolutionary theory predicts that the amount of parental investment will be determined by the reproductive values of both parents and children. The probability of subsequent participation in genetic reproduction of a population by parents or their children, respectively, should therefore be the decisive factor for the amount of parental investment in any single child. Reproductive value is influenced by several biogenic and sociogenic factors. These are age and sex, the constellation of the family, the prevailing demographic pattern of the population, and finally, in socially stratified societies, the sociocultural position of the respective family as a direct determinant of the access to resources and to living chances in general.

With this point in mind, a working team, established some years ago, has attempted to reconstitute as many complete family histories for as many generations as possible on the basis of the vital and social data found in entries taken from parish registers, tax rolls and other historical sources. By means of historical demographic methods we intend to describe the patterns of genetic and social reproduction and to examine them for possible adaptive consequences. Our data stem from a region known as Ostfriesland, which is situated on the coast of north-western Germany, and cover the time span from the end of the seventeenth century to the end of the nineteenth century.

Table 1 contains some of our initial findings on the correlates of infant and child mortality in the form of a brief synoptic summary. In reference to specific independent variables, our results indicate that infant and child mortality reveal correlates that were predicted from evolutionary biology considerations. For each of the effects found, a cost−benefit oriented behavioural strategy can be named that should lead to fitness maximization on the average. This especially applies to the obvious underinvestment young widows make in their only sons or daughters. Please see the reference given for further details.

Boone (1986, 1988) reports adaptive patterns of sex-biased parental investment in fifteenth to sixteenth century Portuguese élite. Among the high nobility males outreproduced females, whereas among the low nobility females outreproduced males. Correspondingly, the tendency to concentrate investment in sons increased with status. This reproductive strategy aimed at maximizing the potential for lineage survival and posterity generation after generation.

TABLE 1. Differential infant and child mortality in one-parent families (= mother or father died within first year of last-born child, $n = 870$) (Ostfriesland, late seventeenth to nineteenth century). (From Voland 1988)

Independent variable	Dependent effect: higher mortality	Statistical significance		Possible adaptive behavioural strategy
		Mother-only households, $P<$	Father-only households, $P<$	
Age of parent	— in families with young fathers or mothers	0·05	0·05	Increase of parental investment with declining reproductive potential
Sex of infant	— of sons	0·10	—	Daughter-biased investment with economic stress
Birth rank of child	— of first born	0·001	N.S.	Resumption of the woman's reproductive activity in a second marriage: Childless women have increased remarriage chances (does not apply to men)
Remarriage	— with remarriage	0·10	N.S.	(a) as above (b) investment in one's biological children at the expense of one's stepchildren

Figure 2 illustrates a similar example of adaptive investment strategies within a population. Infant mortality of first-born children in relation to sex of child and social class of parents in Leezen (Germany, 1720–1869) is shown. These data reveal that the survival chances of newborns were manipulated by differential parental investment. The more attentative care for sons and the greater neglect of daughters by the land-owning farmers seem to be mainly caused by the prevailing inheritance practices favouring male descendants (Fig. 3). Other reasons for different investments made in boys and girls by the various social groups are related to one's chances of marriage. Sons of farmers had the greatest prospect, while sons of labourers had the least prospect of marrying in their native parish (Voland 1984); daughters of farmers ran the risk of lowering their social status by marrying, whereas daughters of smallholders had a certain chance of hypergamy (Fig. 4). I think that these investment patterns are in close agreement with the Trivers–Willard (1973) hypothesis.

From these illustrations it becomes evident that in the seventeenth to nineteenth centuries, death did not threaten all infants and children equally, but that there was a certain amount of parental discrimination against some offspring. The

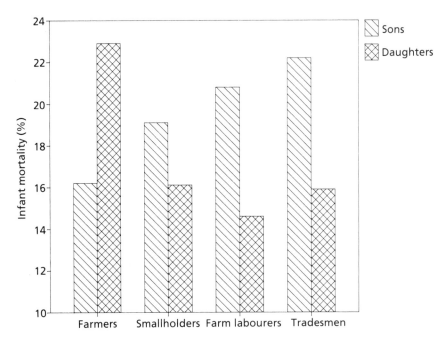

Fig. 2. Infant mortality of first borns (excluding stillbirths), according to sex and social class in Leezen (1720–1869) (n=3197). (From Gehrmann 1984; Voland 1984.)

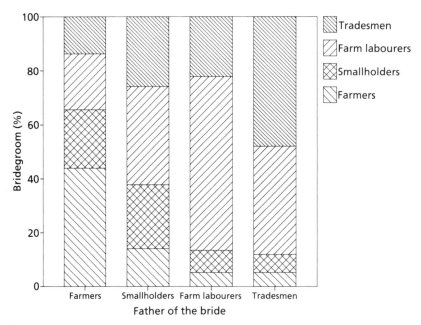

Fig. 3. Social background of spouses in Leezen (1720–1869). 100%=all the brides' fathers in the corresponding social class (n=579). (From Gehrmann 1984; Voland 1984.)

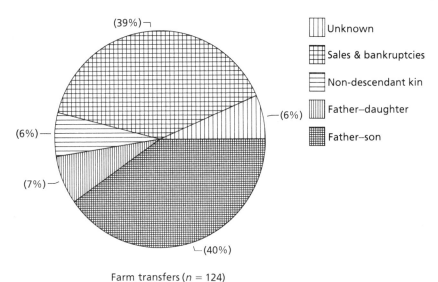

Farm transfers ($n = 124$)

FIG. 4. Farm transfers in Leezen (1720–1869) ($n=124$) (From Gehrmann 1984; Voland 1984.)

Christian 'respect for life' was obviously not a comprehensive, egalitarian life-protecting attitude. On the contrary, only particular children were able to benefit, whereas others were subjected *a priori* to an increased psychosocially induced probability of dying. A historical, even an intra-familial coexistence of sincere parental love on one side and of the psychological and physical rejection and neglect of children on the other, illustrates the latent ambivalence of parental behaviour. It is probably due to this close temporal and psychological contiguity that the psychohistorians of the history of childhood have come to such extremely differing conclusions about children's living reality in the past (cf. Pollock's 1983 re-analysis of the classic views of Ariès 1960; DeMause 1974; Shorter 1975; Stone 1977 and others). Empirical evidence that parental manipulation indeed contributed to genetic fitness maximization is still lacking. However, the observed patterns of selective discrimination against certain groups of children correspond, by and large, to the *theoretical* assumptions of evolutionary biology, according to which parents do everything in order to manage successfully their limited investment resources, so that with a certain probability this will lead to the maximally achievable reproductive success in their lifetimes (cf. Dickemann 1979, 1984; Hammel, Johansson & Ginsberg 1983; Scrimshaw 1984; Voland 1988).

Viewed from this perspective, the psychological desirability of children and the amount of parental love and affection bestowed upon them as proximate determinants of parental care are based on other, ultimate variables, namely the respective reproductive values of the children involved. The emotions, the social and, if you will, the material interests that differently bind parents to their children do not appear to be unaffected by the pressures of natural selection. On the contrary, they are relevant components of biological adaptation since they belong to the proximate mechanisms with which humans sensitively react to the various culturally induced selection pressures.

Popular wisdom and reproductive decisions

One might ask though, how parents actually 'know' what will serve maximum lifetime reproduction. Little is known about the cognitive processes involved, other than that humans (and other higher primates) have the capacity for 'rational preselection', which allows the optimal behavioural pattern to be found for the respective given circumstances, without having to engage in the blind 'trial-and-error' game of evolution (Boehm 1978). This is achieved by anticipating the varying effects of alternative behaviours on the fundamental biological problems of maintenance and reproduction. This cognitive mechanism also provides for an evaluation of a child's reproductive value, at least partially, at a very early stage in the child's life, namely prior to the baptismal ceremony. The following will illustrate this point:

For the parish of Reepsholt there is an almost complete register of the godparents for the second half of the eighteenth century. Otte (1985) has taken advantage of this to check upon an old local adage which says that 'nice children have many names' (*'Liebe Kinder haben viele Namen'*). Originally, this proverb expressed the popular wisdom that 'beloved children have many godparents'.

Table 2 illustrates the average number of godparents for those children who survived their first birthdays and for those who did not. Though just falling short of statistical significance, the figures reveal a rather curious trend ($0.1 > P > 0.05$): The surviving children usually had more godparents! This holds true for the farmers and smallholders. Interestingly enough, the children of labourers and tradesmen do not appear to have been exposed to what appears to be parental manipulation.

Of course, the fate of a newborn does not directly depend on the number of godparents, but both factors — parental care, which influences a child's chances of survival, and the number of godparents — may have their common origin in the parental motivation to invest care in that particular child. Obviously, parents already 'know' prior to a child's baptism whether this is going to be a beloved child or not. These *a priori* differential estimations and expectations of newborns have a lasting influence on the willingness to invest in any one child, and this seems to be primarily independent of the child's individual character and personality. Thus, the differential motivation for investment relativizes parental love to a quantifiable human trait, and there were a lot of children for whom intrafamilial variability of parental affection was fatal.

This may appear to be trivial at first, because it matches our daily experience for the most part, but it contains a far-reaching implication. Namely, if 'parental calculations' lead to adaptive results on the average, as our findings seem to indicate, then the prevailing cultural models for reproduction and investment

TABLE 2. Average number of godparents for children who survived their first birthday (*) or not (†), as well as the differences (D), by social status and sex of the godchildren, Reepsholt (Ostfriesland, 1645–1700). (From Otte 1985)

	[Boys]				[Girls]			
	n	*	†	D	*n*	*	†	D
Farmers	209	2·58	2·50	*0·08*	216	2·53	2·38	*0·15*
Smallholders	263	2·26	2·10	*0·16*	249	2·21	2·07	*0·14*
Labourers and tradesmen	300	2·31	2·32	*−0·01*	296	2·08	2·08	*0·00*
Total	772	2·37	2·30	*0·07*	761	2·25	2·18	*0·09*

have an eminently biological function: they serve biological reproduction by supporting parents in their striving for fitness maximization and are therefore components of human biological adaptation.

CONCLUSION

It should now be obvious that for analysing the variability of parental investment and the reasons for its historical change, close cooperation between social historians and evolutionary biologists is necessary. The history of parental behaviour has a biological dimension that has previously received little attention, namely the striving for fitness maximization as the ultimate motive behind one's daily behaviour. This is why an explanation of historical facts remains incomplete if the biological nature of humans is not taken into consideration.

At the same time adaptive parental strategies have a direct historical dimension, because they are also based on the cognitive processing of collective socioecological experience that is passed from generation to generation. The analysis of human adaptability also remains incomplete for as long as there is no consideration of history, its interpretation and its cognitive processing into a cultural perception that profoundly affects human reproductive behaviour.

ACKNOWLEDGEMENTS

Financially support was provided by the Deutsche Forschungsgemeinschaft. I thank Monique Borgerhoff Mulder, Valerie Standen, Christian Vogel and Paul Winkler for their valuable and helpful comments on an earlier draft of this paper.

REFERENCES

Alland, A. (1970). *Adaptation in Cultural Evolution: An Approach to Medical Anthropology.* Human Science Press, New York.
Andorka, R. (1978). *Determinants of Fertility in Advanced Societies.* Methuen, London.
Ariès, P. (1960). *L'enfant et la vie familiale sous l'ancien régime.* Plon, Paris.
Bardet, J.-P. (1983). *Rouen aux XVIIe et XVIIIe siècles: Les mutations d'un espace social.* S.E.D.E.S., Paris.

Barkow, J.H. & Burley, N. (1980). Human fertility, evolutionary biology, and the demographic transition. *Ethology and Sociobiology*, **1**, 163–80.

Birdsall, N. (1983). Fertility and economic change in eighteenth and nineteenth century Europe: a comment. *Population and Development Review*, **9**, 111–23.

Blurton Jones, N.G. & Sibly, R.M. (1978). Testing adaptiveness of culturally determined behaviour: do Bushmen women maximize their reproductive success by spacing births widely and foraging seldom? *Human Behaviour and Adaptation* (Ed. by N. Blurton Jones & V. Reynolds), pp. 135–58. Society for Study of Human Biology Symposia, 18. Taylor & Francis, London.

Boehm, C. (1978). Rational preselection from Hamadryas to *Homo sapiens*: The place of decision in adaptive process. *American Anthropologist*, **80**, 265–96.

Boone III, J.L. (1986). Parental investment and elite family structure in preindustrial states: A case study of late medieval–early modern Portuguese genealogies. *American Anthropologists*, **88**, 859–78.

Boone III, J.L. (1988). Parental investment, social subordination and population processes among the 15th and 16th century Portuguese nobility. *Human Reproductive Behaviour: A Darwinian Perspective* (Ed. by L.L. Betzig, M. Borgerhoff Mulder & P. Turke), pp. 201–19. Cambridge University Press, Cambridge.

Callan, V.J. (1980). *The Value of Children to Australian, Greek, and Italian Parents in Sydney*. Paper No. 60-c. East–West Population Institute, Honululu. (Quoted from Hollerbach 1983).

Carlsson, G. (1966). The decline of fertility: innovation or adjustment process. *Population Studies*, **20**, 149–74.

Caro, T.M. & Borgerhoff Mulder, M. (1987). The problem of adaptation in the study of human behavior. *Ethology and Sociobiology*, **8**, 61–72.

DeMause, L. (Ed) (1974). *The History of Childhood*. The Psychohistory Press, New York.

Dickemann, M. (1979). Female infanticide, reproductive strategies, and social stratification: a preliminary model. *Evolutionary Biology and Human Social Behavior — An Anthropological Perspective* (Ed. by N.A. Chagnon & W. Irons), pp. 321–67. Duxbury, North Scituate, Massachusetts.

Dickemann, M. (1984). Concepts and classification in the study of human infanticide: Sectional introduction and some cautionary notes. *Infanticide — Comparative and Evolutionary Perspectives* (Ed. by G. Hausfater & S. Blaffer Hrdy), pp. 427–37. Aldine, New York.

Easterlin, R.A. (1978). The economics and sociology of fertility: A synthesis. *Historical Studies of Changing Fertility* (Ed. by C. Tilly), pp. 57–133. Princeton University Press, Princeton, New Jersey.

Flinn, M.W. (1981). *The European Demographic System, 1500–1820*. The Johns Hopkins University Press, Baltimore.

Freedman, R. (1979). Theories of fertility decline: A reappraisal. *World Population and Development — Challenges and Prospects* (Ed. by P.M. Hauser), pp. 63–79. Syracuse University Press, Syracuse.

Gehrmann, R. (1984). *Leezen 1720–1870 — Ein historisch–demographischer Beitrag zur Sozialgeschichte des ländlichen Schleswig-Holstein*. Wachholtz, Neumünster.

Hammel, E.A., Johansson, S.R. & Ginsberg, C.A. (1983). The value of children during industrialization: Sex ratios in childhood in nineteenth-century America. *Journal of Family History*, **8**, 346–66.

Hill, J. (1984). Prestige and reproductive success in man. *Ethology and Sociobiology*, **5**, 77–95.

Hollerbach, P.E. (1983). Fertility decision-making processes: A critical essay. *Determinants of Fertility in Developing Countries Vol. 2: Fertility Regulation and Institutional Influences* (Ed. by R.A. Bulatao & R.D. Lee), pp. 340–80. Academic Press, New York.

Imhof, A.E. (1984). The amazing simultaneousness of big differences and the boom in the 19th century — some facts and hypotheses about infant and maternal mortality in Germany, 18th to 20th century. *Pre-industrial Population Change — The Mortality Decline and Short-term Population Movements* (Ed. by T. Bengtsson, G. Fridlizius & R. Ohlsson), pp. 191–222. Almquist & Wiksell, Stockholm.

Irwin, C. (in press). The sociocultural biology of Netsilinguit female infanticide. *The Sociobiology of Sexual and Reproductive Strategies* (Ed. by A.E. Rasa, C. Vogel & E. Voland), Chapman & Hall, London.

Kaplan, H. & Hill, K. (1986). Sexual strategies and social-class differences in fitness in modern industrial societies. *The Behavioral and Brain Sciences*, **9**, 198–201.

Lande, R. (1978). Are humans maximizing reproductive success? *Behavioral Ecology and Sociobiology*, **3**, 95–8.

Lee, R.D. & Bulatao, R.A. (1983). The demand for children: A critical essay. *Determinants of Fertility in Developing Countries. Vol. 1: Supply and Demand for Children* (Ed. by R.A. Bulatao & R.D. Lee), pp. 233–87. Academic Press, New York.

LeVine, R.A. & Scrimshaw, S.C.M. (1983). Effects of culture on fertility: anthropological contributions. *Determinants of Fertility in Developing Countries, Vol. 2: Fertility Regulation and Institutional Influences* (Ed. by R.A. Bulatao & R.D. Lee), pp. 660–95. Academic Press, New York.

Lévy, C. & Henry, L. (1960). Ducs et pairs sous l'ancien régime — Charactéristiques démographiques d'une caste. *Population*, **15**, 807–30.

Lindert, P.H. (1983). The changing economic costs and benefits of having children. *Determinants of Fertility in Developing Countries. Vol. 1: Supply and Demand for Children* (Ed. by R.A. Bulatao & R.D. Lee), pp. 494–516. Academic Press, New York.

Livi-Bacci, M. (1986). Social group forerunners of fertility control in Europe. *The Decline of Fertility in Europe* (Ed. by A.J. Coale & S.C. Watkins), pp. 182–200. Princeton University Press, Princeton, New Jersey.

McLaren, A. (1984). *Reproductive Rituals: The Perception of Fertility in England from the Sixteenth Century to the Nineteenth Century.* Methuen, London & New York.

Otte, R. (1985). *Soziologische Beziehungen zwischen Kindern und ihren Paten in einer historischen Population (Reepsholt, Ostfriesland 1645–1700) als Ausdruck elterlicher Investmentstrategien.* Unpublished Masters Thesis, University of Göttingen.

Peyronnet, J.C. (1976). Les enfants abandonnés et leurs nourrices à Limoges au XVIII-siècle. *Revue d'Histoire Moderne et Contemporaine*, **23**, 418–41.

Pollock, L. (1983). *Forgotten Children — Parent–Child Relations from 1500 to 1900.* Cambridge University Press, Cambridge.

Reynolds, V. (1986). Religious rules and reproductive strategies. *Essays in Human Sociobiology, Vol. 2.* (Ed. by J. Wind & V. Reynolds), pp. 105–17. Free University of Brussels, Brussels.

Schiefenhövel, W. (1984). Preferential female infanticide and other mechanisms regulating population size among the Eipo. *Population and Biology — Bridge between Disciplines* (Ed. by N. Keyfitz), pp. 169–92. Ordina édition, Liège.

Scrimshaw, S.C.M. (1984). Infanticide in human populations: Societal and individual concerns. *Infanticide — Comparative and Evolutionary Perspectives* (Ed. by G. Hausfater & S. Blaffer Hrdy), pp. 439–62. Aldine, New York.

Shorter, E. (1975). *The Making of the Modern Family.* Basic Books, New York.

Stone, L. (1977). *The Family, Sex and Marriage in England 1500–1800.* London.

Trivers, R.L. & Willard, D.E. (1973). Natural selection of parental ability to vary the sex ratio of offspring. *Science*, **179**, 90–2.

Vining, D.R. (1986). Social versus reproductive success: The central theoretical problem of human sociobiology. *The Behavioral and Brain Sciences*, **9**, 167–216.

Voland, E. (1984). Human sex-ratio manipulation: Historical data from a German parish. *Journal of Human Evolution*, **13**, 99–107.

Voland, E. (1988). Differential infant and child mortality in evolutionary perspective: Data from late 17th to 19th century Ostfriesland (Germany). *Human Reproductive Behaviour: A Darwinian Perspective* (Ed. by L.L. Betzig, M. Borgerhoff Mulder & P. Turke), pp. 253–61. Cambridge University Press, Cambridge.

Wrigley, E.A. (1978). Fertility strategy for the individual and the group. *Historical Studies of Changing Fertility* (Ed. by C. Tilly), pp. 135–54. Princeton University Press, Princeton, New Jersey.

Reproductive consequences of sex-biased inheritance for the Kipsigis

M. BORGERHOFF MULDER

Evolution and Human Behavior Program,
University of Michigan, Ann Arbor,
Michigan 48109–1070, USA

SUMMARY

1 Most human societies are characterized by the patrilineal inheritance of resources. Hartung (1982) predicted that parents will preferentially select males as heirs where males can use resources to acquire a large number of mates. He provided support for his hypothesis by demonstrating strong associations between polygyny, bridewealth and male-biased inheritance across a world-wide sample of societies. His argument rested on three assumptions: first, that variance in male reproductive success can substantially exceed variance in female reproductive success; second, that polygyny is contingent on wealth differences among men; and third, that inherited resources confer stronger reproductive benefits on males than females.
2 While there is considerable support for the first two assumptions, the third remains untested in traditional societies.
3 Among the polygynous Kipsigis of Kenya inheritance is passed exclusively to sons. I use demographic and socioeconomic data to examine the effects of parental resources on the reproductive success of Kipsigis sons and daughters.
4 Results show that parental cattle ownership confers significantly greater reproductive benefits on sons than on daughters, but parental land ownership is an equally strong predictor of the reproductive success of both sexes.
6 These results partially support Hartung's hypothesis, and raise important issues concerning the constraints on inheritance decisions that are contingent on the nature of locally critical resources, the flexibility of inheritance customs in novel environmental conditions and the validity of equating inheritance with parental investment.

INTRODUCTION

In human cultures property is generally inherited by males. For example, of 411 societies coded in the *Ethnographic Atlas* (Murdock 1967), 84% exhibit highly male-biased inheritance (Hartung 1982). Societies in which inheritance is shared, either equally or unequally, among children of both sexes are relatively rare in the *Ethnographic Atlas* and there are no societies where only daughters inherit.

Surprisingly few explanations have been proposed for why sex-specific inheritance rules differ between societies, and why women are so rarely beneficiaries;

indeed a recent review of cross-cultural differences in the life-cycle events of women barely touches on the issue (Brown 1981). The principal explanation remains that offered by Goody (1969a, see also 1973a), who suggested that in labour-intensive social systems sons are the sole heirs, whereas in capital-intensive systems (that are often highly stratified) daughters receive inheritance to gain, or at least retain, social status at marriage. Dickemann (1979, 1982a) viewed this link between social mobility and inheritance by daughters as evidence of adaptive variation in sex-specific parental investment (cf. Trivers & Willard 1973), thereby providing the rudiments of an ultimate explanation (*viz.* Tinbergen 1963) for inheritance by women in stratified capital-intensive societies.

Hartung (1982, see also 1976) used cross-cultural data to test the more general evolutionary biological proposition that parents can maximize their fitness by passing a greater proportion of their wealth to individual offspring of the sex upon whom this wealth has stronger reproductive effects (Trivers & Willard 1973; Clutton-Brock, Albon & Guinness 1984). The prediction was made that in polygynous societies, where variance in male reproductive success exceeds that of females and where a man's chances of becoming polygynous are enhanced by wealth, preferential inheritance by males is expected. This was confirmed with strong statistical evidence from a world-wide sample of societies. In polygynyous populations such as the East African pastoral Nandi, the Trobriand horticulturalists of the South Pacific and Aranda hunter-gatherers of central Australia, inheritance tends to be strongly male-biased, whereas in largely monogamous groups such as the Marquesian islanders and the Central American Zapotec Indians inheritance is divided more equally among the sexes.

Objections to Hartung's argument stemmed from two assumptions in his model: first, that polygyny is a good indicator of variance in male reproductive success (Fix 1976) and second, that differential resource control (cf. Emlen & Oring 1977) is a principal cause of polygyny in humans (Dickemann 1982b). These assumptions can be supported: (i) it is a fundamental characteristic of most non-human sexual species that variance in male reproductive success is commonly associated with variance in mate number (Bateman 1948); (ii) differential access to resources is a prime determinant of number of wives in societies ranging from hunter-gatherers to highly stratified states (as suggested by Hartung's [1982] correlations of bridewealth and polygyny, and reviewed in Betzig 1986; Borgerhoff Mulder 1987a); this does not imply that polygyny is necessarily based on differential resource access in all cases (Chagnon 1979; Borgerhoff Mulder, in press a).

There is, however, a third assumption in Hartung's model, that inherited resources are *more strongly* associated with the reproductive success of sons than daughters. This issue has not yet been examined because detailed data on parental factors affecting offspring fitness have not been available, at least for traditional societies.

In this paper, I test this third assumption by examining the effects of parental resources on the reproductive success of men and women among the Kipsigis of

Kenya. The test is particularly apt because Kipsigis are culturally and linguistically very close to the Nandi (see Orchardson 1961; Toweett 1979), who are included in Hartung's sample.

My findings raise two sets of questions for discussion. First, what is the adaptive significance of current inheritance practices among Kipsigis, do the data support Hartung's original hypothesis, and what factors might constrain inheritance decisions? Second, how can we explain the paradox of preferential inheritance by sons throughout so many human societies (including Kipsigis) in the light of sex allocation theory (Fisher 1958; Charnov 1982). This latter issue prompts important questions about the degree to which societal rules governing the inter-generational transfer of property can be equated with parental investment (*sensu* Trivers 1972).

More generally, in focusing on the reproductive consequences of sex-biased inheritance, this chapter explores the possible adaptive significance of cultural variability in inheritance practices; the strengths and weaknesses of such investigations in humans are discussed in detail elsewhere (Borgerhoff Mulder 1987).

ETHNOGRAPHIC BACKGROUND

History and subsistence

The Kipsigis are a Nilo-Hamitic people of the Kalenjin language family living in Kericho District, south-western Kenya. Until the 1930s, Kipsigis were largely pastoralists, raising herds of cattle and goats over large communally-owned areas and tending small millet gardens (Peristiany 1939). With the arrival of the British in the area, Kipsigis were evicted from their land, settling either on specially designated 'Reserves' or moving to areas traditionally used only by Masai (Manners 1967). During this period of settlement maize was increasingly adopted as a subsistence (and subsequently cash) crop, a development that led to the emergence of individual land tenure (Manners 1967; Saltman 1977).

Land is now the primary productive resource and, with increasing land shortages that commenced in the late 1930s, is a highly valued asset (Manners 1967). While all families own some land, large differences in land ownership exist, principally as a result of differences in the initial areas occupied and subsequently fenced for cultivation and grazing by each family; indeed, little land has changed hands since the period of settlement.

Despite Kipsigis' adoption of cultivation, livestock (primarily cattle) are still highly valued as capital investment, a symbol of prestige, currency for procuring wives and a source of dairy and meat produce, as among all Kalenjin (see Schneider 1979). Agricultural surpluses are generally converted into livestock.

Inheritance

Like livestock (Peristiany 1939), land is inherited from fathers to sons (patrilineally), with each son receiving an equal portion. In polygynous households, livestock and land are divided equally among co-wives (Orchardson 1961; Borgerhoff Mulder 1987c), and then shared equally among the sons of each wife (Peristiany 1939). A man receives an initial share of his patrimony at, or even before, his marriage, and this capital he calls his own; he does not inherit his full share of land and cattle until his father is very old (often over 60) or dies unexpectedly. While inheritance is a gradual and ill-defined process, (as among many pastoralists, see, for example, Gulliver 1963 for the Arusha), effective patterns of resource *use* are less ambiguous: the sons of a particular wife will jointly use the plot and herd allocated to their mother even though they and their father may dispute, or more commonly tactfully deny, exclusive ownership.

Women do not receive an inheritance. Prior to marriage, a woman is entirely dependent on her parents. At marriage she leaves her natal home, bringing nothing of her own other than clothing and a few household items, for example gourds for culturing sour milk and baskets for serving food. She becomes economically dependent on her husband and makes a major labour contribution, cultivating his fields, milking his cows and marketing locally any surplus produce (see Borgerhoff Mulder 1987c). She will visit her natal family, particularly if they live nearby, both to help her mother and, if she is neglected in her marital home, to seek economic and emotional support (Borgerhoff Mulder 1988a).

Reproductive life history

Men are circumcised at between 11 and 21 years of age into distinct named age sets (*ipinwek*, singular *ipinda*) and women undergo a clitoridectomy operation immediately on reaching menarche (12–19 years in this population). Reproductive careers for both sexes start almost invariably at marriage, with men and women marrying at 15–47 and 12–20 years of age respectively. Marriage requires the payment of bridewealth to the bride's father, approximately six cows, six goats (or sheep) and a cash payment approximating £40 in 1983 (Borgerhoff Mulder 1988a). A man's first marriage is financed by his father, whereas he himself is responsible for the bridewealth for subsequent wives.

METHODS

Study site

The study area (35 km²) comprises three adjacent areas with different settlement histories on the borders of Kericho and Narok Districts: Kipkelat 'Reserve', settled in the early 1930s, Masailand, first settled by squatters in the later 1930s and the Gelegele Settlement Scheme, established on a European farm at Independence in 1963. The terrain is varied and hilly, including poorly drained heavy loam soils with scrubby vegetation (suitable for grazing), easily cultivated gently sloping hillsides and stony thicketted ridges (see Pilgrim 1961 for details). The area has received little administrative attention from local authorities, partly because it is the least productive region of Kericho District and partly because of the continuing land disputes with Masai. Primary school has been attended by growing proportions of boys and girls since the 1960s.

Reproductive measures

Reproductive questionaires were administered in Kipsigis to all individuals of reproductive age in each household in the study area between June 1982 and December 1983; date of birth, circumcision and marriage, the number of surviving offspring and, where possible, the number of live births were determined. Detailed cross-checks for reliability were made, and past life history events were dated to the year by cross-referencing with male circumcision ceremonies, severe droughts and other events of known date. Years of education, occupation (for men) and location of natal home (for women) were recorded.

The number of surviving offspring produced per married year was the principal measure of reproductive success used in these analyses. This was calculated as the number of surviving offspring divided by the length of the reproductive career to date (the interval between marriage and 1982–83), thus controlling for differences among individuals in the sampled length of their reproductive careers. Males and females married less than 5 years were excluded from the analyses of reproductive success because unusually high variability characterizes reproductive rates measured over short periods.

Reproductive performance was broken down into constituent components (see notes to each Table and Borgerhoff Mulder 1988b for details).

Measures of wealth and inheritance

Wealth was measured in terms of the ownership of land (in acres) and cattle (including calves). This information was gathered through interviews with household heads, direct enumeration of livestock, and examination of government land allocation registers, where available (Borgerhoff Mulder 1987a).

Because the aim of this chapter is to examine the effects of parental wealth on offspring fitness, a measure of how an individual's parents' wealth contributes to the resources he or she has available for reproduction was developed, using the following procedures that differ for males and females.

The entire amount of land that a man receives from his parents, subsequently called parental acres (PA_m) was calculated as $PA_m = OA_m + MA_m$, where OA_m (own acres) = the acres a man receives at marriage, and MA_m ('mother's acres') = the number of his father's acres divided by the number of his father's wives, a legitimate procedure given the equal division of resources among Kipsigis cowives (see above). The cattle a man receives from his parents, termed parental cattle

(PC_m) was calculated in identical manner: $PC_m = OC_m$ (own cattle received at marriage) + MC_m ('mother's cattle').[*, †, ‡]

Parental acres and parental cattle could only be calculated for younger men whose fathers were still alive (or died during the study) and effectively in control of family resources.

A woman does not bring any of her parent's property to her marital household. Parental acres for a woman (PA_f) were measured as the number of her father's acres divided by the number of his wives; similarly parental cattle (PC_f) were measured as a woman's father's cattle divided by the number of his wives. Sample sizes of married women for whom parental wealth was known were smaller than those for males, because Kipsigis women move at marriage; if their natal homes lay outside of the study area I was unable to assess parental wealth.

Analysis

Analyses were conducted within cohorts in order to control for changing conditions of reproduction and wealth accumulation over the last 50 years. Men were allocated to cohorts according to their *ipinda* (age set) affiliation: women do not belong to named age-sets and were allocated to arbitrarily defined 'cohorts', according to the year of their marriage (Borgerhoff Mulder 1988b).

Parental wealth and reproductive measures were available for four cohorts of men: eleven individuals in *Chuma II* (circumcised 1939–45), fifteen in *Sawe I* (1946–56), twenty-five in *Sawe II* (1957–63) and sixty-nine in *Korongoro* (1964–78); parental wealth measures were only available for two cohorts of women: forty-four individuals in Cohort II (married 1964–73) and thirty-five in Cohort III (married 1974–83). *Chuma II* and *Sawe I* men were combined in order to enhance the sample size. In order to assess the reliability of reproductive data on younger individuals, comparisons were drawn with two older cohorts of men and women, for whom measures of lifetime reproductive rates were available (cf. Borgerhoff Mulder 1988b): thirty-six *Maina* men (circumcised 1922–30), forty-one *Chuma I* men (1931–38), fifty-five '*Maina*' women (married 1930–39) and 201 '*Chuma*' women (1940–53).

Regression coefficients were calculated using stepwise and forced entry models

---

* Competition between full brothers over the wealth of their mother can be incorporated into the equation by dividing the mother's wealth by the number of her sons (e.g. $PA_m = OA_m + MA_m$/sons). Analyses using this terms produced a very similar pattern of results to those presented here and, because of problems in assessing at what age a brother can be considered as a competitor, the simpler term was used. Similarly, no measure of competition among daughters was incorporated into the measure of parental wealth for women.
† The terms 'mother's acres' and 'mother's cattle' are used only to indicate the availability of resources to offspring; mothers do not *own* such capital assets.
‡ Parental wealth measures were not calculated for men and women who live in Gelegele (settled in the 1960s) because it was not possible to determine accurately the land available to these families prior to settlement, land on which the offspring may have been raised and to which the family may still retain some rights.

(SPSSx Users' Manual 1983), and their gradients compared against a Z distribution (Bailey 1981), quoting d-values with two-tailed statistical probabilities. Significance levels for directional predictions tested with Pearson correlation coefficients are one-tailed. Statistical trends (T 0·05<P<0·10) are noted if consistent in direction with significant results from other cohorts.

RESULTS

Parental resources and reproductive success

The effects of parental cattle (PC_m) on reproductive success of males are shown in Fig. 1. For the three cohorts with information on parental wealth, the Pearson correlation coefficients between parental cattle and the number of surviving off-spring produced per married year were positive and significant (*Chuma II/Sawe I*: r=0·56, n=26, P<0·001; *Sawe II*: r=0·58, n=25, P<0·001; *Korongoro*: r=0·31, n=68, P<0·01). Turning to the components of reproductive success, parental cattle wealth was positively associated with the number of a man's wives (polygyny) and the average number of surviving offspring produced per year by a man's wives in all three cohorts; it was not consistently associated with age at marriage (Table 1).

Parental acres (PA_m) were also positively associated with the number of surviving offspring produced per married year (*Chuma II/Sawe I*: r=0·42, n=26, P<0·01; *Sawe II*: r=0·46, n=26, P<0·001; *Korongoro*: r=0·41, n=69, P<0·001,

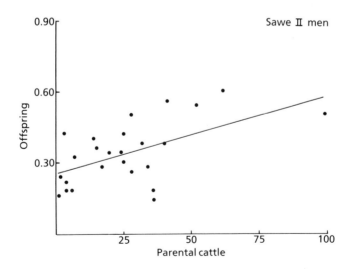

FIG. 1. Surviving offspring per married year plotted against parental cattle (PC_m) for twenty-five men in the *Sawe* II cohort: y=0·2519 + 0·0033 (x), $F_{1, 23}$=11·59, P<0·01 (*Chuma II/Sawe I*: y=0·2384 +0·0036 (x), $F_{1, 24}$=11·04, P<0·01; *Korongoro*: y=·3200 + 0·0024 (x), $F_{1, 66}$=6·90, P<0·01, — not illustrated).

TABLE 1. Pearson correlations between parental cattle and components of reproductive success in men

	Average offspring per wife[1]	Polygyny[2]	Age at marriage[3]
Chuma II/Sawe I[4]	0·34 T	0·57**	−0·35
	(25)	(25)	(12)
Sawe II	0·44**	0·35*	0·14
	(25)	(25)	(16)
Korongoro	0·25*	0·20*	0·20*
	(68)	(68)	(78)

[1] Average number of offspring per wife calculated as number of surviving offspring born to a man divided by his total number of reproductive wife-years (see Borgerhoff Mulder 1988b).
[2] Polygyny calculated as mean number of wives of reproductive age per married year (see Borgerhoff Mulder 1988b).
[3] For the analysis of the effects of parental wealth on age at marriage, men married for less than 5 years were not excluded.
[4] Bracketted numbers denote sample sizes; and in Tables below.
* $P<0.05$. ** $P<0.01$. T $0.05<P<0.10$.

see Fig. 2). As regards the components of reproductive success, parental acreage was consistently associated with the number of surviving offspring produced per year by a man's wives and, in one cohort, with polygyny (Table 2). In the *Chuma II/Sawe I* cohort sons of land-rich men married younger than sons of poorer men.

For the two cohorts of women for whom adequate-sized samples of individuals with known parental wealth were available, parental cattle holding (PC_f) was not significantly correlated with the number of surviving offspring produced per married year by women in either cohort (Cohort II: $r=0.01$, $n=39$, N.S.; III: $r=-0.05$, $n=33$, N.S.; Fig. 3). However turning to land-based wealth, parental acres (PA_f) and number of surviving offspring produced per married year were significantly correlated (Cohort II: $r=0.41$, $n=44$, $P<0.001$; III: $r=0.32$, $n=35$, $P<0.01$, see Fig. 4). Daughters of parents with large farms had significantly earlier ages of menarche and higher offspring survivorship (Table 3).

Confounding variables

To what extent are the effects of parental wealth independent of other factors shown to be important determinants of reproductive success in men and women?

In men, off-farm employment (other than casual labouring) and length of education are associated with moderate increases in reproductive success (Borgerhoff Mulder 1987a). The effect of parental cattle (PC_m) on the number of surviving offspring produced per married year was nevertheless independent of employment (*Chuma II/Sawe I*: $F_{1,23}=8.70$, $P<0.01$; *Sawe II*: $F_{1,22}=11.06$, $P<0.01$; *Korongoro*: $F_{1,65}=5.57$, $P<0.05$; similarly for the effects of parental acres (PA_m) $F_{1,23}=1.78$, N.S.; $F_{1,23}=8.30$, $P<0.01$; $F_{1,65}=11.21$, $P<0.001$). Parental wealth effects were also largely independent of the number of years education a man had followed (parental cattle: $F_{1,23}=9.14$, $P<0.01$; $F_{1,22}=10.51$,

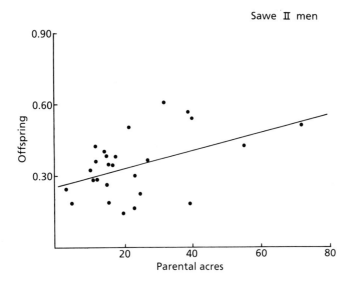

FIG. 2. Surviving offspring per married year plotted against parental acres (PA_m) for twenty-six men in the *Sawe II* cohort: $y=0.2519 + 0.0038$ (x), $F_{1,24}=6.55$, $P<0.05$ (*Chuma II/Sawe I*: $y=0.2582 + 0.0031$ (x), $F_{1,24}=5.23$, $P<0.05$; *Korongoro*: $y=0.2906 + 0.0030$ (x), $F_{1,67}=13.34$, $P<0.001$ — not illustrated).

TABLE 2. Pearson correlations between parental acres and components of reproductive success in men

	Average offspring per wife[1]	Polygyny[2]	Age at marriage[3]
Chuma II/Sawe I[4]	0.39*	0.19	−0.56*
	(26)	(26)	(12)
Sawe II	0.56***	0.07	0.20
	(26)	(26)	(17)
Korongoro	0.31***	0.37***	−0.04
	(69)	(69)	(51)

[1-4] As for Table I.
* $P<0.05$. *** $P<0.001$.

$P<0.01$; $F_{1,65}=6.68$, $P<0.01$; parental acres: $F_{1,23}=1.67$, N.S.; $F_{1,23}=7.74$, $P<0.01$; $F_{1,66}=12.19$, $P<0.001$).

In women, to what extent are the effects of parental acres (PA_f) independent of the wealth of their husbands and their own education (measured in years), both important determinants of women reproductive success (Borgerhoff Mulder 1987c) Parental acreage was still a significant determinant of the number of surviving offspring born per married year after husband's acres were entered into a multiple regression (Cohort II: $F_{1,41}=6.26$, $P<0.05$; III: $F_{1,30}=3.18$, $0.05<P<0.10$); parental and husband's acres were weakly positively correlated in both cohorts

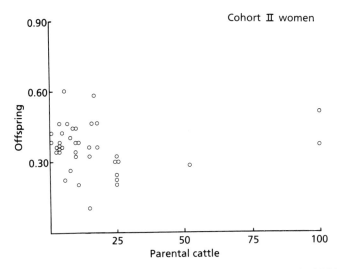

FIG. 3. Surviving offspring per married year plotted against parental cattle (PC_f) for thirty-nine women in Cohort II: $y=0\cdot3600 + 0\cdot0000$ (x), $F_{1,37}=0\cdot96$, N.S. (Cohort III: $y=0\cdot3504 - 0\cdot0008$ (x), $F_{1,31}=0\cdot08$, N.S. — not illustrated).

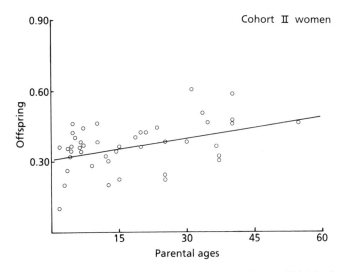

FIG. 4. Surviving offspring per married year plotted against parental acres (PA_f) for forty-four women in Cohort II: $y=0\cdot3098 + 0\cdot0030$ (x), $F_{1,42}=8\cdot70$, $P<0\cdot01$ (Cohort III: $y=0\cdot2793 + 0\cdot0038$ (x), $F_{1,33}=3\cdot83$, $P<0\cdot05$ — not illustrated).

(Cohort II: $r=0\cdot23$; $n=44$, $0\cdot05<P<0\cdot10$; III: $r=0\cdot10$, $n=33$, N.S.), indicating only slight intermarriage among the rich (Borgerhoff Mulder 1988a). Similarly, parental acres made a significant contribution to the regression model after the effects of woman's education were taken out (Cohort II: $F_{1,41}=8\cdot50$, $P<0\cdot01$; III: $F_{1,31}=2\cdot83$, $0\cdot05<P<0\cdot10$); a woman's education and her parents' acreage were

Table 3. Pearson correlation coefficients between parental acres and components of reproductive success in women

	Age at menarche	Fertility[1]	Probability offspring survive[2]
Cohort II	−0·35**	0·20 T	0·28*
	(38)	(44)	(44)
Cohort III[3]	−0·24*	0·07	0·25 T
	(48)	(34)	(34)

[1] Fertility was calculated as the number of live births divided by length of reproductive career to date (see Borgerhoff Mulder 1987c).
[2] The probability that offspring survive was calculated as the number of surviving offspring divided by number of live births (see Borgerhoff Mulder 1987c).
[3] For the analysis of the effects of parental acres on age at menarche, women married less than 5 years were not excluded.
* $P<0·05$. ** $P<0·01$. T $0·05<P<0·10$.

not significantly intercorrelated (Cohort II: $r=0·10$, $n=44$, N.S.; III: $r=0·16$, $n=34$, N.S. respectively).

   Finally, women with parents living nearby were greatly over-represented (mean distance between natal and marital households was 1·2 km for sampled women and 29·6 for non-sampled women of Cohorts II and III). To what extent does this sample bias affect the results? In so far as the distance between natal and marital homes did not affect reproductive success (Cohorts II: $r=−0·07$, $n=43$, N.S.; III: $r=0·14$, $n=34$, N.S.), parental acres was still a significant predictor of reproductive success after marital distance was entered into the regression ($F_{1,40}=8·25$, $P<0·01$; $F_{1,31}=3·08$, $0·05<P<0·10$ respectively).

Sex differences in the effects of parental wealth

Sex differences in the gradient of the slope of reproductive success on parental resources were tested by comparing the regression coefficients for similarly-aged cohorts. Turning first to cattle wealth, the effects of parental cattle on the reproductive success of men and women showed a significant difference in one of the two comparisons (*Sawe II* men and Cohort II women: $d=2·68$, $P<0·01$); in the second comparison (*Korongoro* men and Cohort III women) the slope was steeper for males but the difference was non-significant ($d=1·09$, N.S.). In contrast, there were no significant differences in the effects of parental acres on the reproductive success of men and women in either cohort (*Sawe II* men and Cohort II women: $d=0·47$, N.S.; *Korongoro* men and Cohort III women: $d=0·36$ N.S.). The effects of the different types of parental wealth on the reproductive success of men and women are summarized in Table 4 and diagrammed in Figure 5. In short, parental cattle enhance the fitness of sons but not daughters, whereas parental acres enhance the fitness of both sexes.

TABLE 4. The effects of parental wealth on the reproductive success of men and women[1]

| | Men | | | Women | |
	Chuma II/Sawe I	Sawe II	Korongoro	Cohort II	Cohort III
Parental wealth[2]					
Cows	0·56	0·58	0·31	0·01	−0·05
	(26)	(25)	(68)	(39) N.S.	(33) N.S.
	***	***	**		
Acres	0·42	0·46	0·41	0·41	0·32
	(26)	(26)	(69)	(4)	(35)
	**	***	***	***	**

[1] Reproductive success measured as the number of surviving offspring produced per married year.
[2] Parental cattle were measured for men as PC_m and for women as PC_f; and parental acres as PA_m and PA_f, see Methods.
** $P<0·01$. *** $P<0·001$

Supplementary analyses

There are at least two obvious problems with these analyses.
(a) Measures of the effects of parental resources on the reproductive success of men and women are by necessity (see Methods) taken from younger individuals who have not yet terminated reproduction. Reproductive careers of men are longer and less age-dependent than those of women (Borgerhoff Mulder, in press a), and potentially more subject to biases resulting from sample truncation: for example, after 20 years of marriage the correlation coefficient between age-specific and terminal reproductive success is only $r_s=0·60$, (it exceeds $r_s=0·95$ after 45 years). Such truncation of male reproductive careers may produce misleading results if the effects of wealth on male reproductive success are stronger in the later rather than the earlier reproductive years.

To counter this problem, the effects of resources on reproductive success of *Maina* and *Chuma I* men were examined; these men are over 50 years of age and have already completed, or almost completed, their reproductive careers (see Borgerhoff Mulder 1988b for details). Because measures of *parental* wealth were not available, the regression coefficients of numbers of surviving offspring produced per year on own acres (OA_m) and own cattle (OC_m) were calculated. In so far as men over 50 years of age have generally received their full inheritances, these measures incorporate both inherited and personally-accrued wealth.

The effects of wealth on the number of surviving offspring produced per married year by *Maina* and *Chuma I* men are shown in Table 5. The regression coefficients were very similar to those presented in the legends to Figs 1 and 2, suggesting that the effects of wealth on male reproductive success do not increase markedly with age, and are not therefore more pronounced among older men. Regression coefficients of reproductive success on parental cattle and acres of like-aged women were not available, but can perhaps be assumed to be similar to

(a)

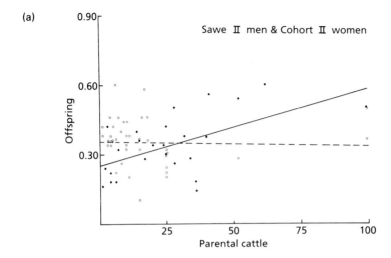

(b)

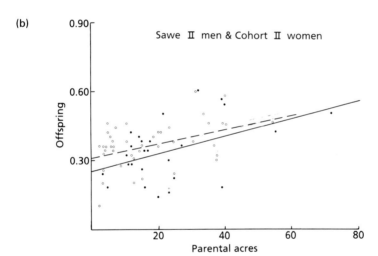

FIG. 5. (a) The effects of parental cattle on the number of surviving offspring produced per married year by men (solid circles and solid line) and by women (plain circles and dashed line); the difference in slope is significant, see text. Parental cattle measured for men and women as PC_m and PC_f respectively. (b) The effects of parental acres on the number of surviving offspring produced per married year by men and women; the differences are not significant, see text. Parental acres measured for men and women as PA_m and PA_f respectively.

those found in Cohort II (PC_f 0·0000; PA_f 0·0030, see legends to Figs 3 and 4). Comparisons of the coefficients for *Maina* and *Chuma I* men with those for Cohort II women show marked sex differences in the effects of cattle on reproductive success (*Maina* men and Cohort II women: $d=4·38$, $P<0·001$; *Chuma I* and Cohort II women: $d=2·74$, $P<0·01$), but no significant sex differences in the effects of acres on reproductive success ($d=0·39$; $d=-0·80$, N.S. respectively).

TABLE 5. Regression coefficients of surviving offspring produced per married year on the number of cows and acres owned by a man, for men with information on lifetime reproductive success

	Cattle			Acres		
	r^1	b^2	F-value	r	b	F-value
Maina	0·83	0·0040	$F_{1,\,28}=61·47^{***}$	0·87	0·0034	$F_{1,\,34}=102·53^{***}$
Chuma I	0·50	0·0036	$F_{1,\,34}=11·56^{***}$	0·36	0·0014	$F_{1,\,39}=5·83^{*}$

[1] Regression coefficient.
[2] Regression slope.
* $P<0·05$. ** $P<0·01$. *** $P<0·001$

These results indicate that the sex-specific effects of parental resources on reproductive rates found in the younger cohorts of men (cf. Table 4) are very similar to sex differences in the effects of resources on a man's entire reproductive output, and can therefore perhaps be considered as adequate estimates.

(b) A more appropriate measure of the potential reproductive effects of passing resources to daughters might be the slope of women's reproductive success on *husband's* wealth. Because a measure of parents' wealth is not required for this analysis, data can be used for older women who have completed their reproductive careers. The effects of husbands' cattle and acres on the reproductive success of older women (Table 6) are very similar in pattern to those of parents' wealth (cf. Figs 3 and 4), in so far as land is positively associated with the number of surviving offspring produced per married year whereas cattle are not. Comparison of these slopes with those of like-aged men (cf. Table 5) shows, again, that there is a significant sex difference in the reproductive consequences of cattle ownership (*Maina*: $d=4·87$, $P<0·001$; *Chuma*: $d=0·64$, $0·05<P<0·10$) but not land ownership (*Maina*: $d=0·39$, N.S.; *Chuma*: $d=-1·04$, N.S.). This analysis suggests that were parents to bequeath land to a daughter this would enhance her fitness; cattle given to a daughter, by contrast, would not have marked fitness consequences.

Sex ratio effects

Parents can modify investment by varying the relative numbers of each sex raised, so sex ratios were examined. No significant biases from unity were found in sex ratios either at birth or among offspring surviving beyond 5 years (Table 7).

To determine whether wealthy parents produce more male than female offspring, the sex ratio of a woman's progeny was examined in relation to the wealth available to her through her husband (husband's acres, or cattle, divided by the number of co-wives, see Borgerhoff Mulder 1987c). No consistent biases towards male offspring, either at birth or reaching age 5, were found among women who fell above the mean wealth value of acres or cows (Table 7).

TABLE 6. Regression coefficients of surviving offspring produced per married year on the number of cows and acres available to a woman, for women with information on lifetime reproductive success

	Cattle			Acres		
	r^1	b^2	F-value	r	b	F-value
Maina	0·06	0·0006	$F_{1, 41}=0·15$ N.S.	0·24	0·0028	$F_{1, 53}=3·18$ T
Chuma	0·03	0·0003	$F_{1, 185}=0·18$ N.S.	0·24	0·0023	$F_{1, 199}=12·65$***

[1] Regression coefficient.
[2] Regression slope.
T $0·05<P<0·10$
*** $P<0·001$

TABLE 7. The effects of wealth on the sex ratio of offspring born and surviving to women of three cohorts

	All sample[1]	Acres		t-test[2]	Cattle		t-test
		$>\bar{x}$	$<\bar{x}$		$>\bar{x}$	$<\bar{x}$	
Live births							
Maina	1·16	1·18	1·14	0·34	1.00	0·97	0·15
	(46)	(14)	(26)		(13)	(17)	
Chuma	1·03	1·02	1·04	−0·41	1·06	0·97	1·60
	(171)	(59)	(112)		(62)	(96)	
Cohort I	1·02	1·02	1·03	−0·25	1·04	1·02	0·33
	(168)	(51)	(117)		(52)	(104)	
Surviving offspring[4]							
Maina	1·17	1·11	1·17	−0·34	1·08	1·06	0·11
	(64)	(18)	(35)		(17)	(14)	
Chuma	0·99	1·02	0·97	0·69	1·04	0·95	1·32
	(204)	(73)	(125)		(74)	(106)	
Cohort I	1·02	0·98	1·03	−0·78	1·05	1·01	0·67
	(183)	(51)	(130)		(51)	(118)	

[1] Samples for wealth comparisons are lower than full samples, due to limited availability of information on cows and acres.
[2] Women were compared according to whether they fell above (or at) or below the mean acres or cows score for that cohort. All student's t-tests were non-significant.
[3] Maina, Chuma and Cohort I women were married between 1930−39, 1940−53 and 1954−1963 respectively.
[4] Surviving to 5 years old; mortality past age 5 was low (approximately 4% compared to 25% prior to age 5, see Borgerhoff Mulder 1988b). Younger cohorts were dropped from these analyses because small proportions of their children had reached age 5.

Summary of results

These findings show that parental cattle enhance the reproductive success of Kipsigis men but not women. Conversely, the size of the parental plot is an equally strong predictor of reproductive success in women as it is in men. These results remain robust whether parental resources are measured as those to which a young man has direct access, or as the full capital held by a man towards the end of his life. The results are replicated across cohorts, and are independent of measured confounding variables. No significant sex ratio effects were observed.

DISCUSSION

I first discuss the adaptive significance of Kipsigis inheritance patterns, examining whether these data support the hypothesis that parents select as heirs offspring of the sex on whom resources have greater reproductive effects; this leads to a review of the constraints on inheritance decisions. I then turn to the broader issue of how male-biased inheritance can be interpreted in the light of sex ratio theory, raising the question of whether inheritance patterns provide a good measure of parental investment.

Resources, sons and daughters in the Kipsigis

Preferential inheritance by sons is an adaptive strategy where variance in male reproductive success exceeds that of females, where polygyny is contingent on resource access, and resources affect the reproductive success of sons more than daughters. The first two conditions are met among the Kipsigis (Borgerhoff Mulder 1988b). What evidence is there of the third?

Turning first to livestock, the more traditional form of wealth among Kipsigis, parental cattle wealth holdings increase a man's reproductive success, largely through facilitating polygynous marriage. A man's first marriage is financed by his parents, and is therefore directly contingent on the availability of parental wealth; payments for subsequent wives are taken from the stock a man received at his marriage, their progeny or animals acquired independently (through raiding, trade and negotiation, see Borgerhoff Mulder 1987a). So although polygyny in the Kipsigis is, at least in part, contingent on differences in personally accrued rather than inherited wealth, (cf. Pospisil 1982 for Kapauku Papuans of New Guinea), parental cattle holding is clearly an important determinant.

By contrast, parental cattle holding bore no relationship at all to the reproductive success of women. Differences in parental cattle ownership are therefore more strongly associated with the fitness of sons than of daughters, supporting a hitherto untested assumption of Hartung's (1982) model. With the present data it is still not possible to conclude that parents *cannot* enhance their daughters' reproductive success through giving them cows, in so far as no Kipsigis woman actually receives cows from her father at her marriage. However, the fact that *husband's* cattle are not associated with a woman's reproductive success suggests that access to a large herd does not have marked effects on a woman's reproductive success. Current evidence therefore suggests that parents do pass their cattle to the sex on whom cattle confer greater reproductive benefits, as the original model predicts.

Parental holdings of land, the newer source of wealth among Kipsigis, show a different pattern of reproductive effects from that of parental cattle. The amount of land owned by parents is positively associated with the reproductive success of both sons and daughters. The effects of parental acres on daughters' reproductive success can perhaps best be explained as a consequence of developmental differences, in so far as girls raised on larger farms have an earlier age of menarche than girls raised on smaller farms. This is consistent with arguments that nutritional status influences timing of menarche in humans and other mammals (Frisch & McArthur 1974; Garn 1979, Garn & La Velle 1983; Bronson & Rissman 1986; discussed in Borgerhoff Mulder 1987c) and with previous findings that early-maturing Kipsigis women have higher reproductive success due to longer reproductive lifespans and higher fertility per year married (Borgerhoff Mulder, in press b). The finding that age at menarche is affected by parental acreage and not parental cattle is not understood; it could perhaps reflect the greater nutritional advantages of living on a large farm (Borgerhoff Mulder 1987c), where a maize surplus (stored throughout the year) is available.

Again there is a problem in determining the potential reproductive effects of passing land to daughters, in so far as it is sons and not daughters who obtain rights to their parents land. However, women whose husbands have large farms benefit in terms of health, nutrition and reproduction (Borgerhoff Mulder 1987c) and it therefore seems reasonable to conclude that parents *could* improve the reproductive chances of their daughters by giving them land at marriage. As regards the inheritance of land, Kipsigis parents appear to deviate from a behavioural strategy that could earn them more grandoffspring.

These results raise a very important question. Have Kipsigis failed to adapt their inheritance practices to the new circumstances they face as settled cultivators, or should we incorporate further considerations into the model for explaining the adaptive significance of inheritance patterns? I turn initially to the second alternative, and examine what ecological and social factors may constrain inheritance practices.

First, land is not a resource that can be moved; rights of use and ownership restrict residence decisions. If land were to be transmitted to Kipsigis daughters and daughters continued to co-reside with their husbands, this would lead to a dissolution of the corporate land-owning group, as anthropologists have long recognized. Peters (1978) points out that the Bedouin of Cyrenaica were reluctant to apply islamic inheritance rules (prescribing inheritance to sons *and* daughters) to the inter-generational transmission of land and livestock for fear of breaking up their unilineal corporate property-holding groups. Such considerations may also constrain Kipsigis' inheritance decisions. Because of the potential importance of post-marital residence rules in constraining inheritance practices, some investigators have chosen to focus primarily on residence patterns (see, for example, Irons 1979, see also Murdock 1949). This may in the end prove to be a more fruitful approach to the study of inheritance.

Second, land is a non-renewable and inherently divisible resource. Partition of land among Kipsigis sons and daughters would lead to an even more rapid reduction in the size of plots than is already occurring (Manners 1967). Solutions to such problems arising from the bilateral transmission of land (Goody 1969a, 1973b; Fox 1978) are found in well-established land tenure systems such as those of ancient Egypt (Middleton 1962) and Ceylon (Leach 1961a; Yalman 1967) where cross-cousin (and even sibling) marriages are arranged so as to prevent progressive fragmentation of the estate. For Kipsigis, for whom the inter-generational management of land holdings is a new challenge, a more attractive option may be to continue to exclude daughters from inheritance (now of land) and attempt to assist them in other ways (see below).

Returning now to the first alternative, Kipsigis may not yet have had time to adapt their inheritance practices to the novel environmental conditions of today. A time lag in the appearance of an adaptive response may well be expected, given the conservative force of custom. Indeed, contemporary Kipsigis parents can barely countenance the thought of daughters as heirs. Clearly evolutionary biological anthropologists' belief that the human phenotypic response is practically infinitely flexible requires serious investigation (Caro & Borgerhoff Mulder 1986).

In sum, while there is evidence that Kipsigis transmit cattle between generations in a way that maximizes the production of grandoffspring, land would appear to be inherited in a way that is not perhaps optimal in terms of maximizing fitness. Two suggestions for this latter result were offered. First, inheritance decisions are likely to be affected by the nature of the resource (whether it is movable, partible and renewable), such that the model must be respecified. Second, inheritance may be constrained by the conservative force of custom, and therefore less flexible than adaptationist models have assumed.

Equal investment in the sexes?

Fisher's (1958) sex ratio theory (see Charnov 1982) states that the average investment of parents in their sons and daughters should be equal; higher average investment in one sex is expected to be offset by higher mortality in that sex. Evidence from humans, particularly societies (such as the Kipsigis) practising exclusive formal inheritance by males, suggests that total investment in males and females is not equal, even though sex ratios after the period of peak mortality are not biased to females. Is Fisher's theory inappropriate under certain conditions (Frank 1987), or does this finding simply reflect the problems in measuring parental investment (Trivers 1972) and its effects in humans? I now turn to this latter point.

First, higher investment in one sex may be offset by subsequent benefits received from offspring of that sex, thereby lowering the *net* investment in that sex (cf. Clark 1978, Clutton-Brock & Albon 1982). Thus exclusive inheritance by Kipsigis sons may be explicable in terms of their continuing coresidence with their parents. As Kipsigis men and women age, their sons take on an increasing proportion of the animal husbandry, farm work and homestead maintenance.

Furthermore, if they find off-farm employment, they invariably make cash contributions to their fathers. By contrast, daughters (because they move away) make rarer labour contributions to their parents (Borgerhoff Mulder 1988a), and are unable, except in unusual circumstances, to provide economic assistance. These observations indicate that despite exclusive inheritance to Kipsigis sons, an overall sex difference in net parental investment may not exist.

Second, the sexes may benefit from different sorts of investment at different periods in their lifetimes. Kipsigis data suggest that women on large farms benefit principally in terms of advantages accrued during development that promote early menarche, whereas men reap the advantages of inherited parental wealth later in life when they are considering marriage and remarriage. Gross comparisons of the effects of parental resources on the reproductive success of sons and daughters as measured by total reproductive output will obscure the adaptive significance of varying investment according to such immediate factors as the extent to which an offspring can benefit at that particular age (Horrocks & Hunte 1983; see also Borgerhoff Mulder 1985). There was no evidence of sex differences in care awarded Kipsigis boys and girls during late infancy (Borgerhoff Mulder & Milton 1985). In short, it may be important to look at the age-specific effects of investment in the two sexes to understand the adaptive significance of inheritance practices, particularly when inheritance occurs at advanced age.

Third, sex allocation could differ according to socioeconomic status, with, for example, high-ranking parents investing preferentially in sons and low-ranking parents in daughters, through varying either the sex ratio (Trivers & Willard 1973) or postnatal investment (Reiter, Stinson & Le Boeuf 1978); in this way, overall investment (as averaged across parents) in the sexes may be equal, despite strong sex biases in different social strata. Such societal splits in investment have been the focus of a number of recent studies that look at class differences in sex-specific mortality (Johansson 1984; Voland 1984; this volume), marriage chances (Dickemann 1979, 1982a; Boone 1988) and association patterns (Betzig & Turke 1986). In contrast to these studies, the Kipsigis data showed no effects of socio-economic status on sex ratios of surviving offspring; nor was there any evidence that the pattern of exclusive inheritance to males is reversed in poorer families, at least in so far as no Kipsigis women receive an inheritance. Furthermore, there was no evidence that resources were more strongly associated with daughters' reproductive success in poorer families or with sons' in the richer families (as seen from the distribution of the data in Figs 1−4).

Fourth, to what extent does the normative description of inheritance practices reflect, or accurately gauge, parental investment (in Trivers' sense)? Categorizing societies according to the conventional inheritance rule (matrilineal, patrilineal, etc.) not only obscures intra-societal variability in the transference of wealth, as anthropologists have long recognized (Leach 1961b; Needham 1962; and above) but also detracts attention from other more subtle forms of investment.

So, what hidden forms of investment in daughters may occur among Kipsigis?

Married daughters sometimes ask for maize (and other food) from their parents when they have insufficient in their own barn. Of seventy-nine overnight visits that women made to their parental homes (among a sample of 118 women studied over a year) fifteen (19%) were expressly for this purpose, although the amounts of maize involved were small.

Kipsigis women can also rely on their parents in other ways: if they are poorly treated by their husbands, they can 'run away' to their parental homes (Borgerhoff Mulder 1988a), staying for months, years or even permanently. There were forty-six 'runaway' women (4% of the total sample, $n=1226$) who were at the time of my survey entirely economically dependent on their parents.

Another form of hidden investment in daughters may be parents' effort in arranging advantageous marriages for their daughters. In some societies (although not in the Kipsigis, Borgerhoff Mulder 1988a) this may be at considerable cost, namely through lowering bridewealth demands (West African LoWiili: Goody 1969b; South African Swazi: Kuper 1978; New Guinea Melpa: Strathern 1980) in order to attract husbands of high socioeconomic rank (see also Flinn 1988, where fathers risk physical injury in defence of their daughters' marriage chances). Parents may even compete to attract high-ranking husbands for their daughters through providing sumptuous dowries (Dickemann 1982a). While dowries are often considered as a form of inheritance (Goody 1973b), reduced bridewealth demands are not, even though parents may suffer economically as a result and have fewer resources available for the marriage payments of their other sons; this underscores the arbitrariness of equating simple inheritance conventions with investment.

For several reasons then, preferential investment in sons may not be as extreme as the anthropological label 'patriliny' might suggest: parents may accrue benefits from coresident adult sons, daughters may benefit from parental resources long before inheritance takes place, intra-societal splits in inheritance may occur, and forms of investment other than inheritance in its strict sense are likely to be important.

CONCLUSIONS

Sex-biased inheritance can be broadly predicted by the differential effects of resources on reproductive success of sons and daughters as first demonstrated by Hartung (1982) and substantiated with Kipsigis data. The association between the nature of the resource base and the effects of resources on the reproductive success of sons and daughters has not yet been determined; but differences in inherited resources, whether they are mobile, partible and renewable, are likely to influence and place constraints on inheritance decisions.

Parents' decisions concerning the inter-generational transmission of novel resources appear to be constrained by customary norms that are less flexible in the face of novel circumstances than evolutionary biological anthropologists anticipated.

Patterns of sex-specific investment may bear little relationship to nominal inheritance systems, as classified by conventional anthropological nomenclature, in so far as investment may take many forms other than formal inheritance and intra-societal variations in investment may occur. Neither the net investment of parents in sons and daughters nor the effective extent of offspring dependence on their parents are yet known for any society.

By focusing on the age-specific effects of parental investment in the two sexes rather than the correlations between resources and reproductive success, it may be possible to throw further light on cultural differences in investment decisions, residential patterns and inheritance practices.

ACKNOWLEDGEMENTS

I am thankful to all the Kipsigis families who put up with my endless questions and offered me such friendship and hospitality; also to the Office of the President for permission to conduct research in Kenya. Laura Betzig, Tim Caro, Sarah Durant, Steve Frank and Paul Turke all offered suggestions concerning the analysis of these data; Tim Caro made particularly helpful comments on many drafts; John Hartung, Tim Ingold, Joan Silk, Meg Symington McFarland and an anonymous referee identified further problems; thanks to you all. Field work was supported by the National Geographic Society.

REFERENCES

Bailey, N.T.J. (1981). *Statistical Methods in Biology*, 2nd edn. Wiley, New York.
Bateman, A.J. (1948). Intrasexual selection in *Drosophila*. *Heredity*, 2, 349–68.
Betzig, L.L. (1986). *Despotism and Differential Reproduction: A Darwinian View of History*. Aldine, New York.
Betzig, L.L. & Turke, P.W. (1986). Parental investment by sex on Ifaluk. *Ethology and Sociobiology*, 7, 29–37.
Boone, J.L. III. (1988). Parental investment, social subordination, and population processes among 15th and 16th century Portuguese nobility. *Human Reproductive Behaviour: A Darwinian Perspective* (Ed. by L. Betzig, M. Borgerhoff Mulder & P. Turke), pp. 201–19. Cambridge University Press, Cambridge.
Borgerhoff Mulder, M. (1985). Resource certainty or paternity uncertainty? Comment on Hartung. *Brain and Behavioral Sciences*, 8, 677–8.
Borgerhoff Mulder, M. (1987a). On cultural and reproductive success: Kipsigis evidence. *American Anthropologist*, 87, 617–34.
Borgerhoff Mulder, M. (1987b). Adaptation and evolutionary approaches to anthropology. *Man*, 22, 25–41.
Borgerhoff Mulder, M. (1987c). Resources and reproductive success in women with an example from the Kipsigis of Kenya. *Journal of Zoology*, 213, 485–505.
Borgerhoff Mulder, M. (1988a). Kipsigis bridewealth payments. *Human Reproductive Behaviour: A Darwinian Perspective* (Ed. by L. Betzig, M. Borgerhoff Mulder & P. Turke), pp. 65–82. Cambridge University Press, Cambridge.
Borgerhoff Mulder, M. (1988b). Reproductive success in three Kipsigis cohorts. *Reproductive Success: Studies of Individual Variation* (Ed. by T.H. Clutton-Brock), pp. 419–35. Chicago University Press, Chicago, Illinois.

Borgerhoff Mulder, M. (in press a). Is the polygyny threshold model relevant to human societies? Kipsigis evidence. *Mating Patterns* (Ed. by C.G.N. Mascie Taylor & A.J. Boyce) Cambridge University Press, Cambridge.

Borgerhoff Mulder, M. (in press b). Early maturing Kipsigis women have higher reproductive success than later maturing women, and cost more to marry. *Behavioural Ecology & Sociobiology.*

Borgerhoff Mulder, M. & Milton, M. (1985). Factors affecting infant care in the Kipsigis. *Journal of Anthropological Research*, **41**, 231–62.

Bronson, F.H. & Rissman, E.F. (1986). The biology of puberty. *Biological Reviews*, **61**, 157–95.

Brown, J.K. (1981). Cross-cultural perspectives on the female life cycle. *Handbook of Cross-cultural Human Development* (Ed. by R.H. Munroe, R.L. Munroe & B.B. Whiting), pp. 581–610. Gartland STPB Press, New York.

Caro, T.M. & Borgerhoff Mulder, M. (1987). The problem of adaptation in the study of human behaviour. *Ethology and Sociobiology*, **8**, 61–72.

Chagnon, N. (1979). Is reproductive success equal in egalitarian societies? *Evolutionary Biology and Human Social Behavior: An Anthropological Perspective* (Ed. by N.A. Chagnon & W. Irons), pp. 374–401. Duxbury Press, North Scituate, Massachusetts.

Charnov, E.L. (1982). *The Theory of Sex Allocation.* Princeton University Press, Princeton, New Jersey.

Clark, A.B. (1978). Sex ratio and local resource competition in a prosimian primate. *Science*, **201**, 163–5.

Clutton-Brock, T.H. & Albon, S.D. (1982). Parental investment in male and female offspring in mammals. *Current Problems in Sociobiology* (Ed. by King's College Sociobiology Group), pp. 223–47. Cambridge University Press, Cambridge.

Clutton-Brock, T.H., Albon, S.D. & Guinness, F.E. (1984). Maternal dominance, breeding sucess and birth sex ratios in red deer. *Nature*, **309**, 358–60.

Dickemann, M. (1979). Female infanticide, reproductive strategies and social stratification: a preliminary model. *Evolutionary Biology and Human Social Behavior: An Anthropological Perspective* (Ed. by N.A. Chagnon & W. Irons), pp. 321–67. Duxbury Press, North Scituate, Massachusetts.

Dickemann, M. (1982a). Paternal confidence and dowry competition: A biocultural analysis of purdah. *Natural Selection and Human Behavior* (Ed. by R.D. Alexander & D.W. Tinkle), pp. 417–38. Chiron Press, New York.

Dickemann, M. (1982b). Comment on Hartung. *Current Anthropology*, **23**, 8–9.

Emlen, S.T. & Oring, L.W. (1977). Ecology, sexual selection and the evolution of mating systems. *Science*, **197**, 215–23.

Fisher, R.A. (1958). *The Genetical Theory of Natural Selection*, Oxford University Press, Oxford.

Fix, A.G. (1976). Comment on Hartung. *Current Anthropology*, **17**, 616–17.

Flinn, M.J. (1988). Parent-offspring interactions in a Trinidadian Village: Daughter guarding. *Human Reproductive Behaviour: A Darwinian Perspective.* (Ed. by. L. Betzig, M. Borgerhoff Mulder & P. Turke), pp. 189–200. Cambridge University Press, Cambridge.

Fox, H. (1978). *The Tory Islanders: A People of the Celtic Fringe.* Cambridge University Press, Cambridge.

Frank, S.A. (1987). Individual and population sex allocation patterns. *Theoretical Population Biology*, **31**, 47–74.

Frisch, R.E. & McArthur, J.W. (1974). Menstrual cycles: fatness as a determinant of minimum weight for height necessary for the maintenance and onset. *Science*, **185**, 949–51.

Garn, S.M. (1979). Continuities and change in maturational timing. *Constancy and Change in Human Development* (Ed. by J. Kagan & O.G. Brim), pp. 113–162. Harvard University Press, Cambridge, Massachusetts.

Garn, S.M. & La Velle, M. (1983). Reproductive histories of low weight girls and women. *American Journal of Clinical Nutrition*, **37**, 862–6.

Goody, J. (1969a). Inheritance, property and marriage in Africa and Eurasia. *Sociology*, **3**, 55–76.

Goody, J. (1969b). "Normative", "recollected" and "actual" marriage payments among the LoWiili, 1951–1968. *Africa*, **39**, 54–61.

Goody, J. (1973a). Strategies of heirship. *Comparative Studies in Society and History*, **15**, 3–20.

Goody, J. (1973b). Bridewealth and dowry in Africa and Eurasia. *Bridewealth and Dowry* (Ed. by J. Goody & S.J. Tambiah), pp. 1–58. Cambridge University Press, Cambridge.

Gulliver, P.H. (1963). *Social Control in an African Society*. New York University Press, New York.

Hartung, J. (1976). On natural selection and the inheritance of wealth. *Current Anthropology*, **17**, 607–13.

Hartung, J. (1982). Polygyny and inheritance of wealth. *Current Anthropology*, **23**, 1–12.

Horrocks, J. & Hunte, W. (1983). Rank relations in vervet sisters: A critique of the role of reproductive value. *American Naturalist*, **122**, 417–21.

Irons, W. (1979). Investment in primary social dyads. *Evolutionary Biology and Human Social Behavior: An Anthropological Perspective* (Ed. by N.A. Chagnon & W. Irons), pp. 181–213. Duxbury Press, North Scituate, Massachusetts.

Johansson, S.R. (1984). Deferred infanticide: Excess female mortality during childhood. *Infanticide: Comparative and Evolutionary Perspectives* (Ed. by G. Hausfater & S.B. Hrdy), pp. 463–85. Aldine, New York.

Kuper, A. (1978). Rank and preferential marriage in southern Africa: the Swazi. *Man*, **13**, 567–79.

Leach, E.R. (1961a). *Pul Eliya: A Village in Ceylon: A Study of Land Tenure and Kinship*. Cambridge University Press, Cambridge.

Leach, E.R. (1961b). *Rethinking Anthropology and Other Essays*. London, Athlone Press, New York.

Manners, R.A. (1967). The Kipsigis of Kenya: Cultural change in a "model" East African tribe. *Contemporary Change in Traditional Societies, Volume 1: Introduction and African Tribes* (Ed. by J. Steward), pp. 207–359. University of Illinois Press, Urbana.

Middleton, R. (1962). Brother–sister and father–daughter marriage in Ancient Egypt. *American Sociological Review*, **27**, 603–11.

Murdock, G.P. (1949). *Social Structure*. MacMillan, New York.

Murdock, G.P. (1967). *Ethnographic Atlas*. University of Pittsburgh Press, Pittsburgh, Philadelphia.

Needham, R. (1962). Remarks on the analysis of kinship and marriage. *Rethinking Kinship and Marriage* (Ed. by R. Needham), pp. 1–34. Tavistock, London.

Orchardson, I.Q. (1961). *The Kipsigis*. Kenya Literature Bureau, Nairobi.

Peristiany, J.G. (1939). *The Social Institutions of the Kipsigis*. Routledge & Kegan Paul, London.

Peters, E.L. (1978). The status of women in 4 Middle East communities. *Women in the Muslim World* (Ed. by L. Beck & N. Keddie), pp. 311–50. Harvard University Press, Cambridge.

Pilgrim, J. (1961). *Social and Economic Consequences of Land Enclosure in the Kipsigis Reserve*. East African Institute of Social Research, Kampala.

Pospisil, L. (1982). Comment on Hartung. *Current Anthropology*, **23**, 9–10.

Reiter, J., Stinson, N.L. & Le Boeuf, B.J. (1978). Northern elephant seal development: the transition from weaning to nutritional development. *Behavioral Ecology and Sociobiology*, **3**, 337–67.

Saltman, M. (1977). *The Kipsigis: A Case Study in Changing Customary Law*. Shehkman Publishing Company, Massachusetts.

Schneider, H.K. (1979). *Livestock and Equality in East Africa*. Indiana University Press, Bloomington, Indiana.

SPSSx Users' Guide (1983). McGraw-Hill, Chicago, Illinois.

Strathern, A.J. (1980). The central and the contingent: bridewealth among the Melpa and Wiru. *The Meaning of Marriage Payments* (Ed. by J.L. Comaroff), pp. 49–66. Academic Press, New York.

Tinbergen, N. (1963). On aims and methods of ethology. *Zeitschrift für Tierpsychologie*, **20**, 410–63.

Toweett, T. (1979). *Oral Traditional History of the Kipsigis*. Kenya Literature Bureau, Nairobi.

Trivers, R.L. (1972). Parental investment and sexual selection. *Sexual Selection and the Descent of Man* (Ed. by B. Campbell), pp. 139–79. Aldine, Chicago, Illinois.

Trivers, R.L. & Willard, D.E. (1973). Natural selection and parental ability to vary sex ratio of offspring. *Science*, **179**, 190–2.

Voland, E. (1984). Human sex-ratio manipulation: historical data from a German parish. *Journal of Human Evolution*, **13**, 99–107.

Yalman, N. (1971). *Under the Bo Tree*. University of California, California.

.

Section 5
Hominoids, Hominids and Humans

Humans present perhaps the most problematic species for socioecology — although as the papers by Dyson-Hudson; Hawkes, O'Connell & Blurton Jones; Blurton Jones, Hawkes and O'Connell; Voland and Borgerhoff Mulder show, the approach is possible and yields both empirical results and extensions of various models developed for non-human animals. The question remains, though, does socioecology have the power to avoid the pitfalls into which other broadly evolutionary approaches to anthropology fell?

The most recent of these was sociobiology as espoused by E.O. Wilson in his book *Sociobiology: the New Synthesis* (1975). Apart from the criticisms levelled at this and subsequent books (Wilson 1975; Lumsden & Wilson 1981) for their supposed political motivation and justification of certain right wing political systems, the 'Wilsonian paradigm of sociobiology' was also subject to a more directly biological critique: that it was overly genetically deterministic, looking for simple genetic controls of complex behavioural patterns; that it was species-specific, demanding uniformity of behaviour across all equivalent members of a population; and that it assumed that the behaviour of cultures could be treated as equivalent to that of species of non-human animals. The sociobiological model presented in the 1970s was one in which both humans and animals were perceived as inflexible in a way that seemed at variance with many empirical observations.

The shift of emphasis away from genetics towards the more immediate behavioural response, and towards more precise specification of the environment, as illustrated in the chapters of this volume, is the hallmark of socioecology, and its distinction from the earlier formulations of sociobiology. The stress on flexibility of response and within-species variability, in contrast to fixed genetic determinism, would seem to be entirely appropriate for humans, the variability of whose foraging and social adaptations across time and space are immense. The question remains, though, whether socioecology can deal not just with the superficial similarities of humans to the rest of the natural world, but also with the actual mechanisms by which humans arrive at a relationship to their resources and their environments.

A large number of criteria have been used by anthropologists to divorce humans from the rest of the biological world, and hence the applicability of evolutionary principles to their behaviour. Some, such as the exclusivity of technology as a human trait (see McGrew, this volume), have been overtaken by the advance of knowledge concerning the African apes. Most others, though, lie firmly in that no man's land in which non-human primates possess these features in rudimentary form relative to modern humans, or have analogous characteristics leaving it unclear whether humans are unique or the end of a continuous and natural procession. Central to this debate are human consciousness and motivation in behaviour, the importance of reproductive maximization as a behavioural goal, the subservience of individual goals to those of the group, the plasticity of human behaviour, and ultimately, whether what is referred to as cultural behaviour can be reduced to an equivalence with the acceptably Darwinian behaviour of animals.

The papers in this section are concerned with these issues from two angles — from the point of view of the theoretical constructs of social anthropology and the extent to which they are compatible with those of socioecology; and in terms of the direct comparative study of humans and hominoids. The chapters by Layton and Ingold fall into the former category. Layton's is essentially conciliatory, looking for ways of integrating classic sociobiology and social anthropology. His central point is that much of the clash between the two disciplines is a result of different levels of explanation — proximate and ultimate. Most theory within social anthropology is concerned with explanations within a cultural context; that of sociobiology with explanations independent of specific historical trajectories. Such a difference is reminiscent of Tinbergen's identification of the four basic questions in ethology — and the difference between explanations of function or selective value and explanations of direct causal mechanism. The distinction is an important one. Animals adapt to the problems posed by their environment; many of these problems are shared, but different animals will respond variably depending upon context — ecological context, phylogenetic context, etc. For this same reason different human societies may find alternative responses depending upon their cultural repertoire, rules and history. As Layton argues, in these terms the types of explanation preferred by socioecologists and social anthropologists may be more complementary than competitive. The particular significance of this view is that it combines the generalizing principles of socioecology with the fact of human variability.

Ingold, in contrast, is less sanguine about the integration of human and animal sociality. He argues that the essence of human sociality lies in the existence of *rules* that govern relationships. This separates human relationships from those of animals which are based on interactions. As these rules are derived from the institutions and cultural frameworks of a particular society, the relationships themselves are further distanced from the factors determining non-human patterns of social relationship and structure. Ingold's argument is a powerful one, focusing sharply onto the characteristics that are most distinctively human — our ability to make rules and actively to construct both social and ecological situations. According to Ingold, the onus would be on evolutionary biologists to incorporate these features more fully into their models. The onus must also lie, though, on social anthropologists to take into account the developments that there have been within socioecology — the shift away from a crude resource-based and physical notion of the environment and the emphasis on the individual, not the society, as the source of behaviour. In Layton's terms, rule making and suchlike are the specifically human way of solving adaptive problems, and therefore belong to the realm of proximate mechanisms. The development of the comparative approach proposed here will rest to a considerable extent on whether 'man the rulemaker' can be shown to be making (and indeed subsequently breaking) rules according to the expectations of the general principles of behavioural ecology.

The papers by McGrew and Foley do not address the problem of human uniqueness in this way. Rather, they simply attempt to make comparisons across species to see whether patterns may be discerned. Both papers emphasize the same point — that ecological and behavioural differences must be rooted in phylogenetic relationships, for the phylogeny provides the initial conditions for any evolutionary development. Here must lie one of the potential fusion points for the biological and social sciences, for phylogeny is to a biologists what a cultural tradition is for an anthropologist — the framework in which any adaptation, and any evolutionary novelty must develop. McGrew's analysis of tool making in extant apes is unable to find simple 'environmental' correlations or evolutionary progressions. What is central, is his argument that this is not necessarily surprising as the variables will compound each other — physical environment versus phylogeny — and further, as argued earlier, that the scale of the significant environment is unknown, for example 'captive' or 'wild' may, in some circumstances, be more significant than species or habitat.

Ultimately, the reason for expecting an evolutionary explanation of human behaviour lies in the generally accepted fact that humans *have* evolved in other ways. An evolutionary approach should therefore ideally be dynamic, considering change through time more directly. Traditionally comparative socioecology is unable to do this, being forced instead to compare across extant species as if they represented different points in the evolutionary time–space continuum. The direct consideration of evolutionary processes and patterns through time can only be made through palaeobiology and the study of fossils, renowned for their intractability. However, even a superficial examination is of value, for the hominid fossil record provides us with an observation that is central to the project of comparative socioecology. The gaps between species (including that between humans and other primates), viewed in the present day, are not absolute, but are a function of differential extinction. The endless gradations of evolutionary change are the ultimate justification for a comparative approach to socioecology, for differences between individuals will become differences between populations, and then between species and higher taxa. Socioecology is one of the ways in which small-scale processes, so important for day to day survival of individuals, can be built up into the grand patterns that we hope to reveal in the evolution of organic diversity.

REFERENCES

Lumsden, C. & Wilson, E.O. (1981). *Genes, Mind and Culture*. Harvard/Belknap, Cambridge, Massachusetts.
Wilson, E.O. (1975). *Sociobiology: the New Synthesis*. Harvard/Belknap, Cambridge, Massachusetts.

Are sociobiology and social anthropology compatible? The significance of sociocultural resources in human evolution

R. H. LAYTON

Department of Anthropology, University of Durham, 43 Old Elvet, Durham DH1 3HN, UK

SUMMARY

1 Theoretical differences between sociobiologists and social anthropologists are discussed and resolved in the form of general hypotheses:

2 In human evolution patterns of cooperation and reciprocal altruism facilitated the exploitation of natural resources. These same patterns also favoured interactive behavioural adaptations which exploited social resources in the social environment.

3 Social kinship developed to identify close kin and facilitate kin-selected altruism but also as a means to create potential for reciprocal altruism beyond the limits of close biological kinship.

4 As the success of strategies depended on cooperation or reciprocal altruism, the exchange of information of actors' intentions would be favoured, and hence communities tend to develop codes for transmitting such information.

5 While sociobiologists are concerned to locate the ultimate causes of behaviour in natural selection, much of the variability in human behaviour studied by social anthropologists represents the proximate content of adaptations which are themselves historically derived and may vary randomly. The adaptive consequences of utilizing a particular set of historically-derived strategies may be different to those arising from utilizing a different set, independently derived, in a similar environment. Sociobiologists cannot therefore disregard questions about proximate causation.

6 Two case studies are presented which exemplify the general propositions.

7 Opportunities for successful cooperation between sociobiologists and social anthropologists exist. Sociobiologists must (i) pay more attention to reciprocal altruism, (ii) examine the significance of convention and code for the origin of adaptive strategies, and (iii) explore the implications of the lateral transmission of learned behaviour. Social anthropologists must (i) learn to think of social evolution in adaptive rather the progressive terms, (ii) consider whether the search for power and wealth are proximate goals which ultimately contribute to reproductive success, and (iii) explore the possibility that some aspects of cultural variability can be interpreted within a Darwinian framework.

INTRODUCTION

Much of the debate between sociobiologists and sociocultural anthropologists has been acrimonious. The aim of this paper is to identify some of the sources of this disagreement and suggest how the issues may be resolved. There appear to be four major points of theory or method obstructing the reconciliation of sociobiology and social anthropology. Two lie with sociobiology, two with social anthropology. Sociobiology often seems to minimize the significance of variation between human cultures and to underestimate the emergent properties of human social interaction which may condition the adaptive value of interactive strategies. Social anthropology either pays relatively little attention to social process as opposed to social structure, or is prone to conceive of evolution as progress rather than adaptation. These problems will be discussed in the first half of the paper. The second half of the paper presents some case studies, through which an approach to reconciling the two disciplines is suggested.

CRITICISMS OF SOCIOBIOLOGY

Sociobiology and cultural variation

Sahlin's (1976) criticism of sociobiology is based largely on sociobiology's use of kin selection as an explanatory model. A simple mathematical formula predicts that the selective consequence for kin-selected altruism varies according to the degree of genetic relatedness between the actor and the recipient of his behaviour. But Sahlins points out that human interaction based on socially-recognized kinship is remarkable for its capacity to extend well beyond genealogical second cousins (normally accepted as the limit for effective kin selection), and that behaviour toward socially recognized kin often varies independently of the degree of biological relatedness. Such behaviour is therefore said by Sahlins and others to be interpretable only in terms of the conventional logic of the cultural system itself and not referrable to general biological principles. In kinship systems based on regular alliances between lineages, for example, cross-cousins are ideal marriage partners whereas parallel cousins, although genealogically equally related, are prohibited as marriage partners (Lévi-Strauss, 1969).

A useful example of the social anthropological position on this issue is provided by Goody (1956), who pointed out an interesting contrast between the range of socially-defined 'incest' prohibitions among the Tallensi and Ashanti of Ghana. The Ashanti have local matrilineages. These are classed into eight dispersed matriclans. Any man who has sexual relations with a woman born into his matrilineage or matriclan is deemed to have committed incest and is executed. Yet the clan is a social construct whose members are dispersed throughout the territory of the Ashanti kingdom. The rule cannot be interpreted as a dispersal mechanism whose function is merely to restrict breeding between siblings, or parent and offspring. To have sexual relations with the wife of another man in one's own matrilineage among the Ashanti is merely to commit adultery. The offender is fined. Among the Tallensi the converse is the case. The Tallensi have patrilineages. Sexual relations with the wife of a man in one's patrilineage are regarded with 'horror'. It is, however, permissible to have an extra-marital affair with a woman born into one's own lineage, providing she belongs to a genealogically distant segment. Goody argues that the worse offence is in both cases deemed to be that against the women whose children perpetuate the lineage. He regards the rules as devices for regulating political relations within the lineage.

The question why one society recognizes patrilineal descent and the other matrilineal, is not regarded as one social anthropology is competent to answer (contrast Irons 1979, 1983 who has offered a sociobiological explanation). A further difficulty with the social anthropologist's critique is that, just as Sahlins appears to discount the possibility of biological kin selection operating in human society, Goody (1956) refuses to treat relations within the nuclear family as a special case. He regards ego's own mother among the Tallensi as merely one of those women married into the lineage. Kaberry (1969) has criticized him for this, pointing out that the nuclear family is differently regarded. Tallensi consider incest with one's own mother such a monstrous iniquity that only the mentally deranged could commit it. This implies a different regulative mechanism is at work among biologically close kin.

Although I see some force in the anthropologists' position I do not wish to argue that kin selection or biologically based mating avoidance patterns are inoperative in human communities, merely that to confine investigation to this phenomenon is to disregard two of the most interesting features of human social interaction: its extension beyond the range of close kin, and its inter-cultural variability. I will argue in this paper that sociobiologists have not, generally, fully acknowledged the force of the social anthropological position on this score. It is contended here that reciprocal altruism offers a better explanation for much human interaction conducted in the idiom of kinship.

Among the French villagers with whom I conducted my Ph.D. research, the tendency for close kin to cooperate preferentially was well-recognized. But although this was accepted as legitimate and normal where it concerned joint purchase of agricultural equipment by two or three households it was discouraged in the management of community-wide affairs. Both the commune and many village agricultural cooperatives forbad father and son, or brothers, seeking election to the same council. Participants recognized that it was *in their best interests* to prevent management of shared resources falling into the hands of a kin-based faction, and the majority therefore policed attempts by minorities to monopolize resources. In his analyses of Yanomamo behaviour, Chagnon is explicit in his intention to explore the explanatory power of a kin-selecting model (1982) but on the strength of his data a case may be put that reciprocal altruism also plays a part in structuring Yanomamo behaviour toward socially recognized kin. Chagnon recognizes that culturally-based distinctions between parallel and cross-cousins, and generational divisions, are crucial to marriage exchange. Further, 40% of marriages are between people culturally classified as cross-cousins, but not in fact first cousins (Chagnon 1982 Figs 14.12–13; 1979). Even more significantly, competition for spouses is facilitated by coalitions between individuals who are members of different kin groups, particularly different lineages, who are willing to yield women to each other as marriageable partners (Chagnon 1979). I suggest that the classificatory kinship terminology, according to which a man calls women of his own lineage 'sister', and those of an allied lineage 'wife', can be regarded as a signal of commitment to continued reciprocal, altruistic exchange between the two groups. When such a classification no longer reflects political expediency, the leader of a lineage may take the initiative in signalling lineage fission. He does so by reclassifying distant 'sisters' (distant parallel cousins) as 'wives' (Chagnon 1979). But to conclude that man is therefore more a rule breaker than a rule maker (Chagnon 1982) leaves open the question of who then makes the rules.

The particular virtue of repeated cross-cousin marriage as practised by the Yanomano is that it generates a convergence between reciprocal and kin-selecting altruism: one increasingly becomes genetically related to one's allies (Chagnon 1979).

Such examples suggest that reciprocal altruism, as well as kin-selected altruism, has played an important part in the evolution of human social behaviour. Arguably, this has been recognized in social anthropology since Mauss's work on the gift (1954, originally published in 1925). Plausible contexts for the early practice of reciprocal altruism include meat sharing (see below) and child care (see Lee, Harcourt and others, this volume, and Irons 1983).

Not all aspects of sociobiology suffer from the limitations of the simple kin-selecting altruism model. The Dyson-Hudson & Smith model of human territoriality (1978) for instance, was impressively successful at identifying which of the case studies it considered, would display perimeter defence; only those in which resources were dense and predictable. Different resource distributions will make other exploitative strategies more appropriate. Other recent sociobiological studies have presented evidence that the appropriate kin-selected strategy will also vary with social context. Irons (1979, 1983) argues that a mother is faced with two alternative strategies in seeking help with the care of her children: either to invest in social relations with her own kin (sisters) or her husband's kin (brothers in law). He concludes that the culturally-dominant strategy varies according to which sex contributes most to subsistence (see note 1). Borgerhoff Mulder (1987) argues that Kipsigis do not cultivate relations with affines (e.g. by paying higher bride-wealth to secure brides from wealthy families) because the social milieu does not render affines important allies; a contrast with, for instance, the Yanomamo (Chagnon 1979) or the Rajputs of northern India (Parry 1979).

The content and adaptive consequences of culture

A second aspect of sociobiology's minimization of cultural variability lies in its tendency to underrate the distinction between the conditions to which behaviour is adapted (the ultimate explanation for variable reproductive success), and the content of the adaptation (the proximate explanation for the evolution of a particular mechanism). To take an anatomical example, a bird's wing, bat's wing and insect's wing are all adapted to aerial locomotion; but the form of the adaptation must be understood in terms of evolutionary history (cf. Foley 1985). Which anatomical features provided the source of the adaptation, and how were elements of the species' physiology modified as adaptation proceeded? The same point applies to behavioural adaptations. Cultural strategies have their origins in the distinctive historical trajectory of the community. Desert Aboriginal and Kalahari San patterns of territoriality, for instance, probably constitute alternative adaptation to a semi-arid environment, each with a distinct cultural content and particular political consequences (Layton 1986a and see note 1). Alternative cultural traditions among humans may therefore change in a fashion analogous to the phenomenon of convergence noted between different genera in biological evolution. That the *proximate* source of human adaptations is often to be found in learned, cultural strategies rather than in specific physiological traits is clearly an important element in understanding the course of human evolution. What are the evolutionary consequences of adopting one learned strategy rather than another? Although Borgerhoff Mulder rightly argues that it is not relevant to the application of a Darwinian model to establish whether traits contributing to differential fitness are due to genetic or learned differences (Borgerhoff Mulder 1987), this issue cannot be disregarded when investigating the reasons why learned differences have to a degree replaced physiological ones in the course of hominid evolution (the fact that they do not depend on vertical transmission from parent to child may be crucial). Myers has recently criticized the tendency of some writers on the San ('bushmen' of southern Africa) to minimize the cultural form of their behaviour: 'the human being as culture-user is largely ignored ... and the San are now seen as the archetypal hunter-gatherers without all that "cultural clutter"' (Myers 1986a, p. 175). Yet the source of behavioural strategies may have a bearing on their adaptive significance. Morphy and I compared case studies of rapid social change in Aboriginal Australia and the French Jura, and argued that, under such circumstances of uncertainty, neighbouring but politically autonomous units (clans or villages) could be seen to adopt alternative strategies, from within the cultural repertoire, as solutions to common economic 'problems'. Such alternatives may prove to differ in their success (Morphy & Layton 1981). We argued that cultural strategies are, up to a point, reversible, allowing some units to discard their strategy and adopt the alternative which proved more viable.

Learned strategies are certainly not incompatible with sociobiological explanations. Dyson-Hudson & Smith (1978), for instance, draw attention to the use of culturally-specific devices to maintain territorial boundaries. Reciprocal altruism and territoriality are organizational strategies, and there is no reason to assume they are subject even in non-human species to such specific genetic control as, for example, sickle-cell anaemia. Learned strategies could enhance adaptive behaviour without supplanting genetic dispositions. For this reason I prefer Durham's model of cultural evolution in which natural selection favours phenotypes which are capable of modification through learned behaviour (Durham 1979), to that of Rindos, who supposes specific learning skills replace specific genetic determinations (Rindos 1985). Alexander cogently argues that to ask whether a trait is the product of natural selection (genes) or the environment (culture) is to pose a false dichotomy (Alexander 1979). If learned strategies facilitate reproductive success, natural selection will favour learning skills.

Sociobiology and the evolutionary advantages of behaviour for groups

Trivers argues that if the Wynn-Edwards hypothesis of group selection is rejected 'whole worlds of sociology, anthropology and political science come crashing to the ground' (Trivers 1985). This is an over-statement. Marx (1930, translated from the 1867 original) and Durkheim's (1938, translated from the 1895 original) approach to sociology rested not so much on a concept of group selection but on the premise that, for two reasons, human social behaviour could not be explained in terms of the inherent properties of the individual participants. Patterns of interaction within social groups have *emergent qualities*, and different communities are governed by different cultural values — laws, currency, principles of exchange — which individuals must learn, and submit to.

What is meant by 'emergent qualities'? A simple example would be that of cooperative hunting: the nature of the strategy emerges through interaction. The sharing of individual hunters' kills as a form of risk-minimizing reciprocal altruism is a more complex example, since it depends on continued association of participants beyond single hunting events. Both these phenomena are well-known to sociobiology. More complex still are instances where information is shared about the location of food resources or territorial allegiance. Much human communication takes place through complex codes based on learned, culturally-specific conventions (language, art and ritual). An analysis of culture in Darwinian terms would hypothesize that learned traits enhanced reproductive success: that the exchange of information about territorial allegiance through body decoration, for instance, aided reciprocal altruism, or that the coding of topographic information in symbolic form on maps, ground paintings, etc., facilitated foraging strategies and thus, indirectly, reproductive success. This would not necessitate disregarding the source of such adaptive strategies in a learned, shared tradition (cf. Durham 1979). On the contrary, it would help to assess the significance of such traditions in human adaptation (cf. Alexander 1979). No hypothesis of group selection is entailed.

In comparison with the social environment of other species, the human social environment contains extremely rich and complex resources (Smith 1983). To take

two cases cited by Trivers (1985), the quantity and complexity of information exchanged in human interaction vastly exceeds that contained in the warning of
approaching predators transmitted between Belding's ground squirrels (*Spernophilus
beldingi*). The quantity and complexity of material resources exchanged in human
communities greatly exceeds the meals of blood transferred between vampire bats
(*Desmodus rotundus*). Adaptation to resources generated *within* the social environment is likely to have been as important, in human evolution, as adaptation
to natural resources (cf. Durham 1979). Dwyer's recent case-study of a hunt in
New Guinea exemplifies the proposition that human organizational strategies may
be directed as much toward social ends such as the maintenance of relationships
with affines, as with foraging for the natural resources which will yield most
calories for least effort (Dwyer 1985). Dwyer's points do not attack the use of an
optimizing approach in general, but show that the value of goods as tokens of
social relationships must be considered alongside their immediate value as items
of consumption. The argument is reminiscent of the debate between the formalists
and substantivists in economic anthropology 20 years ago (Firth 1967).

CRITICISMS OF SOCIAL ANTHROPOLOGY

Of the two problems inherent in the method of social anthropology, the most
fundamental is perhaps the tendency to analyse social behaviour in terms of
structural types rather than interactive process. Of the three classic theories
available (functionalism, structuralism and Marxism) only one, Marxism, deals
explicitly with social *process* and this theoretical orientation often suffers from the
second shortcoming, to be mentioned below.

Social structure and social process

Radcliffe-Brown saw himself as a Linneus of social anthropology writing, for
instance, 'the task I have set myself since 1910 is that of determining and classifying
these varieties (of Australian Aboriginal kinship), since I believe that science
begins with classification' (1951, p. 55). Radcliffe-Brown's method consisted of
abstracting constant patterns of interaction from observed behaviour in any community in order to construct a typology of societies. He argued that 'the actual
relations of Tom, Dick and Harry may go down in our field notebooks and may
provide illustrations for a general description. But what we need for scientific
purposes is an account of the form of the structure' (1952, p. 197). Lévi-Strauss
(1969) investigated the cultural logic behind some of the types of society identified
by Radcliffe-Brown, but he too emphasized structure at the expense of process.
In his analysis of myth, for instance, he disregarded variation in performances and
looked for a constant structure of opposed images recurring in all versions of an
abstract type, 'the myth'.

    Both Radcliffe-Brown and Lévi-Strauss may have regarded their techniques as
scientific on the grounds that they separated constant relationships (in human

thought or interaction), that were relatively predictable, from the contingent circumstances in which those relationships were manifested. The consequence, however, is that both treated societies, or cultures, as ideal types and disregarded the variability of behaviour among participants in a community which is as crucial to an integrated theory of anthropology as is variability within biological populations to the Darwinian theory of evolution.

A second consequence of the anthropological approach is the tendency to attribute too concrete a reality to the structures inferred, and to write as if society's interests override those of its individual members.

Other social anthropologists have attempted to find ways of reintroducing actual behaviour into accounts of social structure. Some, like Barth's analysis of Pathan politics in northern Pakistan (Barth 1959), resemble Chagnon's approach in regarding rules as something to be circumvented, and have been criticized for their failure to take the distinctive qualities of local cultural strategies into account (see Asad 1972 & Ahmed 1976). A recent paper of interest is Verdon's (1982) reanalysis of Nuer society. Whereas Evans-Pritchard saw descent and affinity as structural principles which imposed themselves on individual Nuer, Verdon regards them as relationships constituted by social transactions. Verdon develops an approach earlier proposed by Glickman (1971). Glickman highlighted the incongruity in Evans-Pritchard's claim 'it would seem it may be partly just because the agnatic (patrilineal) principle is unchallenged in Nuer society that the tracing of descent through women is so prominent' (Evans-Pritchard 1951, p. 28). Glickman contended that patrilineal descent was manifest primarily through the assistance a father gave his son in obtaining the cattle required for bridewealth. Verdon similarly argues that *Buth* (patrilineal kinship) describes social relationships through which cattle are inherited, *mar* (cognatic kinship) describes relationships created by the transfer of cattle as bridewealth. Evans-Pritchard saw certain relationships as characteristic of patrilineages: joint ownership of land, mutual support in disputes, exogamy. Verdon points out that Evans-Pritchard described the unity of the hamlet in similar terms, yet showed the hamlet to be inhabited by a variety of cognatic kin. Verdon argues that actual relationships of mutual dependence brought about by common residence are constantly recast in a patrilineal idiom as an affirmation of shared rights and obligations, while actual patrikin who live elsewhere are forgotten (Verdon 1982; cf. Sahlins 1976). Such an approach offers insight into how cultural rules are sustained by social interaction.

Recent adherents of structuralism have also tended to emphasize actual discourse rather than the abstract systems posited by Saussure (with regard to language) or Lévi-Strauss (with regard to cosmology). A processual model of cultural behaviour focuses on the continual construction of meaningful interaction within the idiom of a culture. 'Vocabulary' (units of action) and 'grammar' (rules for combining such units) can be distinguished from the ongoing process of choosing elements and sequences which constitute actual behaviour (Goodenough 1965). As a consequence of usage change will occur over time in both the vocabulary and grammar of a cultural tradition. Some recent theories are summarized by Giddens (1979).

Social evolution: progress or adaptation?

It is Spencer rather than Darwin who, retrospectively, is regarded as the exponent of evolution as progress (Freeman 1974; Rindos 1985). Trivers puts the neo-Darwinian position: 'since traits are only defined as useful by reference to a given environment, the concept of natural selection provides no objective support for terms like 'higher' and 'lower'. Darwin made notes to himself to stop referring to higher and lower animals . . .' (Trivers 1985, pp. 31−2). There is, however, a trend in cultural anthropology to regard parallel sequences of culture change, especially in the origins of agriculture and State formation, as evidence that simpler forms of social interaction tend inevitably to be replaced by more complex forms because complex forms provide in a general sense 'better adaptations'. While Marvin Harris (1968) rightly argues that the recurrent nature of such trends would not justify the conclusion that they are 'morally good and proper' (p. 653), he is opposed to the 'lingering acceptance of the possibility that endless diversity may still turn out to be the best model of cultural evolution'. He adapts White's proposition that 'culture evolves as the amount of energy harnessed per capita per year is increased, or as the efficiency of the means of putting the energy to work is increased'. On these criteria, it is argued, certain social *systems* can be described objectively as 'better adapted' than others because they organize human labour more effectively. Human cultures will tend to 'evolve' along esssentially similar paths when confronted with similar 'techno-environmental situations'. Despite the group-selectionist implication, Harris finds support from Darwin for this view (Harris 1968) (see note 2).

If Harris's argument is true, then it would appear that social evolution has a unique quality: that social strategies follow a relatively predictable sequence, generating increasingly complex forms of social interaction. But can it be assumed that such a sequence inevitably confers increasingly greater reproductive success on the participants? In some environments a hundred-fold increase in population resulted from the transition from hunting and gathering to farming. But in environments to which agriculture is maladapted, hunting and gathering continued to provide the most effective set of behavioural strategies (Irons 1983). Moreover, the general category 'hunter-gatherers' embraces very diverse behavioural adaptations (Eskimo, Australian Aboriginal, etc.), all far removed from earlier hominid subsistence behaviour and clearly related to contemporary environmental pressures.

Bender's comparison of prehistoric social systems in Brittany and the North American mid continent casts doubt on the existence of a correlation between increased social complexity and the commencement of agriculture which Harris' model seems to predict (Bender 1985). In Brittany differentiation into social classes seems to have declined by the end of the Neolithic. In the North American mid-continent, hunting-gathering communities went through a cycle of increasing social stratification and ramifying trade networks, which might be construed as evidence of increased energy harnessed per capita, yet this broke down prior to the onset of maize cultivation. Rather than accept increasing elaboration of social interaction as inevitable, it may be more helpful to return to Marx's original questions concerning social change and ask which types of social system are inherently stable, and why; which systems tend to promote increasing change, and which tend to disintegrate (Marx 1965). For Marx the driving force was the practical consequences of particular concepts of property and exchange. One possibility compatible with Darwinian theory is that if strategies for social interaction generate more social resources, adaptations will increasingly be directed to exploitation of the social as well as the natural environment. Thus the form and complexity of hunter-gatherer tools and weapons can be largely explained in terms of adaptation to the natural environment (Torrence 1983). Much hunter-gatherer technology can be used by individuals and does not depend on cooperation. The Industrial Revolution is better understood, however, as the modification of social strategies such as mercantile exchange to exploit a social resource, the surplus of human labour produced by enclosures of common land. From this resulted the industrial technology which depended on organizing a large labour force. The character of the physical environment plays relatively little part in the explanation.

The competing theoretical orientations which interpret change as adaptation or progress can be seen behind recent work on the Middle-Upper Palaeolithic transition in Europe. It is arguable that recognizably modern human societies appear in Europe with the advent of art, diversifying foraging strategies and periodic aggregations at special habitation sites (Mellars 1973; White 1982). But what kind of process is here represented? Clark & Strauss's view that social change was a response to population growth (1983) seems implausible, since all hunter-gatherer populations appear to be capable of more rapid growth than in fact takes place (Bates & Lees 1979). Was it, then, in some general sense 'progress' in human adaptation? Binford implies such a generalized ladder of progress when he compares Mousterian tools to the opportunistic tool manufacture of Australian Aborigines, and Upper Palaeolithic tools to the curated tool manufacture of Eskimo (Binford 1979). Later, however, Binford is careful to explain contemporary variation in hunter-gatherer subsistence in terms of adaptation to specific environments (see, for example, Binford 1980). Can Binford's former model be justified on the grounds that the physical environment changed during the transition from Middle to to Upper Palaeolithic? Gamble points out that glaciations occurred during the Middle Palaeolithic which forced Mousterian populations to withdraw, whereas under similar conditions Upper Palaeolithic populations remained *in situ* (1982). Gamble (1982) proposes that the appearance of 'Venus' figurines is an expression of inter-group alliances which gave individuals in different bands reciprocal access to territories and hence provided an insurance against local fluctuations in food resources. But Gamble also treads a dangerous path: 'We must expect that the course of social evolution has taken the path of ever-increasing elaboration and complexity in terms of alliances, as the principles of group organization passed from the sociobiological to the social' (p. 101). What is the unspecified vital force at work here? Is it more satisfactory to argue that Upper Palaeolithic populations had specifically discovered strategies which enabled a more effective exploitation of the social environment offered by band society? Unlike Binford, Gamble draws parallels between the Upper Palaeolithic and modern Aboriginal behaviour (1982). The issue of variation between modern hunter-gatherer cultures will be taken up again in the second half of the paper.

Some general propositions

The implications of the preceding discussion may be summarized in the form of some general hypotheses which should (I hope) take account of arguments advanced both in sociobiology and social anthropology:

(a) In the course of human evolution patterns of cooperation and reciprocal altruism generated or facilitated exploitation of a variety of resources. This development favoured interactive behavioural adaptations that took advantage of resources available in the social environment.

(b) Social kinship developed, not only to identify close kin and hence facilitate kin-selecting altruism, but also as a means of creating the potential for reciprocal altruism beyond the limits of close biological kinship (compare Dunbar, this volume). Where similar strategies for exploiting the potential of social reciprocity are found in different cultural traditions it is therefore worth asking whether they represent adaptations to common patterns of resource dispersion.

(c) If behavioural strategies became more complex, behaviour would become less predictable. To the extent that the success of strategies depended on cooperation or reciprocal altruism, the exchange of information about actors' intentions would be favoured. Hence communities would also tend to develop codes for transmitting such information (cf de Waal, this volume).

(d) Much of the variability in human behaviour studied by sociocultural anthropology represents the proximate, historically derived content of adaptation, while sociobiologists are more concerned to locate the ultimate causes of behaviour in natural selection. It is, for instance, well-established that the association of sounds and meanings in spoken language is entirely 'arbitrary' or conventional; and an explanation of *why* English rather than Urdu came to be spoken in sixteenth Century Britain would probably lie beyond the scope of sociobiology. To a large extent this may be true of classificatory kinship terminologies, the form of interpersonal greeting rituals, etc. What we term 'cultures' have come about because different communities within the same (human) species have, by consent or negotiation, accumulated distinctive sets of conventional strategies in the organization of behaviour. The cultural content of strategies might be predicted to vary randomly where their contributions to reproductive success is equivalent.

The practical consequences of utilizing a historically-derived set of strategies may, however, have adaptive consequences different to those arising from utilizing a different set, independently derived, in a similar environment (Layton 1986a). Sociobiologists cannot therefore entirely disregard questions about proximate causation.

CASE STUDIES

The following examples are intended to exemplify these general propositions. The first case study is drawn from my field work on social change in the French Jura, a region of rural France near the Swiss border. The second draws on a number of studies of social interaction among hunter-gatherers which seek to demonstrate the adaptive advantages that derive from participation in social networks and social groups.

To illustrate the importance of reciprocal altruism in human behaviour, I outline the importance of greeting rituals and reciprocal aid in the French case, and evidence for the material and symbolic value of hunter-gatherer exchange networks. With regard to the conventional nature of social strategies, I argue that cultural definitions of legitimate heirship and the principles governing the exploitation of human labour must be taken into account in understanding the way in which nineteenth century French farmers coped with shortage of land.

The significance of human kinship as a mode of creating relationships based on reciprocal altruism has been discussed above, with reference to the Yanomamo. The French case illustrates the value of reciprocal exchange between farming households. The circumstances under which open networks or closed groups tend to form among hunter-gatherers throws further light on the adaptive functions of social relationships which are often structured on the basis of classificatory kinship.

The use of rituals or tokens to express continued commitment to reciprocal relations suggests that the exchange of information as well as material transactions plays a significant role in human interaction.

Relations within and between households in the French Jura

Settlement in the high plateaux of the French Jura is nucleated: farmers live in a village, with their fields scattered around the community. At the time of my fieldwork in 1969 and 1972 farming households depended on each others' labour at times, especially on unpredictable occasions such as crises following an accident. Actors therefore invested in inter-personal relationships in the expectation that reciprocal altruism will guarantee return help on future occasions. But such relationships may be fragile. People repeatedly relied on greeting rituals which signal others' intentions: are they to be construed as generous, proud or foolish? Relationships were made and broken as much on the information obtained through these rituals as in transactions involving labour or equipment (Layton 1971, 1981).

Over the last century the structure of social procedure has changed considerably. Until the early twentieth century French agriculture was labour intensive. Farming households were larger than those occupied by craftsmen. The appropriate strategy for a farming couple should have been to bear many children, in order to create a large labour force. But the Jura was overpopulated 100 years ago, and the rule of partible inheritance meant that to have a large number of children threatened the farm's future viability.

Nineteenth century household censuses suggest that two strategies were employed to overcome this problem. One was to enforce life-long celibacy on some members of the farming family, so that the household head had adult, unmarried brothers and sisters living with him (Hajnal 1965 described this as 'the European Pattern' for farming households in the time leading up to the Industrial Revolution, although Lazlett 1972 shows the pattern across Europe to have been more variable). A second strategy was to substitute the labour of unrelated persons for that of family members by retaining resident agricultural labourers (Table 1; cf. Lazlett 1972).

There are two points to be highlighted with regard to the paper's theme. One is that to avoid dividing the farm land it was legitimate heirs, not just biological offspring, that had to be limited. All but one of the eight households containing illegitimate children 100 years ago were those of agricultural labourers who would have owned little or no land (total = 307 households in three villages, 1872 census). I have no direct evidence of who the fathers of these children were, but Viazzo (1986) has examined data from some other areas of Europe in which the European pattern obtained. Viazzo shows that late marriage and adult celibacy did not inevitably result in high levels of illegitimacy. Where this *was* the case, Viazzo proposes two explanations on the basis of historical evidence. Either the fathers were also members of a poverty-stricken sector of society: day labourers, miners or itinerant farmworkers, or farmers adhered to low nuptiality to avoid dividing the land but, by fathering bastards, reproduced a labour force with no claim on inherited resources. The latter argument assumes that genetic paternity was less important than the social entitlement, which resulted from specific cultural conventions, determining the transmission of property.

The second point relevant to the paper's theme is that, for the purpose of operating a farm, the labour power of genetically unrelated employees was as productive as that of one's own children. Farmers could, hypothetically, enhance their own children's chances of survival and successful reproduction by tapping the labour of genetically unrelated others (cf. Borgerhoff Mulder 1987). Individuals' 'life chances' are therefore conditioned by their position within the community. Nineteenth century census data shows that many households containing only agricultural labourers and their children lacked a married couple. The household head was often either unmarried or widowed. It may have been impossible for incomplete families to maintain an independent unit of production, and hence they were obliged to contribute their labour to other households' success (Table 2; cf. Irons 1979; Foley 1985).

Circumstances changed dramatically with the growth of French industry. There was massive migration to towns, particularly among agricultural labourers.

TABLE 1. Household size in three villages

1872	Average size of household	Total
Farmers		
Without resident employees	4·5	(136)
With resident employees		
Family size	3·8 ⎫	
Total household size	5.1 ⎬	(51)
Craftsmen		
Without resident employees	3·8	(47)
Without resident employees		
Family size	2·7 ⎫	
Total household size	3·7 ⎬	(7)
Day labourers	2·6	(31)
1972 (no households have resident employees)		
Farmers	4·6	(85)
Craftsmen and factory workers	3·5	(105)

TABLE 2. Marital status and occupation in three villages (individuals aged 21 or older)

	Single(%)	Married(%)	Widowed(%)	Total
Farmers 1872	34	59	7	(485)
Farmers 1972	17	81·5	1·5	(130)
Labourers and servants 1872	61	13	26	(87)
Factory workers 1972	25	73·5	1·5	(61)

For a while agriculture seemed doomed, but then an alternative method of production using farm machinery presented a solution. Instead of households increasing in size to compensate for the loss of labourers, machinery replaced hired manpower. Partible inheritance, however, remained a constant threat to the viability of farms.

With the growth of urban employment farming came to occupy progressively fewer households in the surrounding villages. Village institutions such as the dairy cooperatives and management of common land were restructured through political initiatives, to cope with changing circumstances (Morphy & Layton 1981). Mechanization did not reduce the inter-dependence of farming households. On the contrary, farmers relied greatly on each other for information about innovations during a period of rapid change as well as for continuing mutual assistance. Failure to conform to the currently 'normal' state of mechanization is interpreted as withdrawal from the moral community of reciprocal altruism. Both radical innovators and traditionalists risk ridicule and isolation (Layton 1973). This phenomenon illustrates how important the lateral transmission of cultural strategies may be in human adaptation. The skills required to learn new techniques may be genetically conditioned, but the techniques themselves are by no means solely transmitted from parent to child.

Social strategies and adaptation in hunting and gathering communities

The second case study summarizes some recent research into hunting and gathering societies. In examining interaction based on an idiom of kinship in such societies two opposed patterns can be identified: open networks based on inter-personal links, and closed, corporate groups whose members share exclusive rights. Woodburn (1980) regards the sharing behaviour characteristic of open networks as the product of an autonomous political philosophy of 'immediate return', while the vesting of property in closed lineages is the product of the opposed philosophy of 'delayed return'. It will here be argued that the recurrence of such strategies in certain ecological settings, or with regard to certain resources, argues (contrary to Woodburn's contention) in favour of an adaptive function, even though the specific form of the strategies can only be explained in terms of the culture's history (cf. Smith 1983). Under what circumstances are networks generated, and when do groups emerge? To what kind of resources is access facilitated: natural resources such as food and water or socially-generated ones such as information or lasting patterns of debt and credit?

Networks

Winterhalder (1986) has recently devised a mathematical model to predict the consequences of hunter-gatherer food sharing. If a set of individuals hunt and gather independently, pooling their catches will not increase the average yield, but will compensate for variation in each actor's personal catch. If all foragers do equally well each time, there would be no benefit from sharing, but if hunting success is unpredictable, and if the hunter's catch (when he is successful) is more than he and his immediate kin can consume, it will benefit each hunter to share his catch when he is lucky, and to receive through reciprocal exchange when others are successful. At some point the advantage of pooling will be offset by an increase in band size depleting local prey. Winterhalder (1986) finds that six foragers sharing will reduce variation by 80% under any circumstances, although he stops short of hypothesizing that this is an optimal number for exchange partners. Winterhalder points out that storage offers, in theory, an alternative strategy, but that in an environment where hunting success is random it might not offset the effect of a long run of bad luck. It should be added that food storage is generally impracticable in low latitudes. Storage is appropriate when variation in hunting success affects all participants simultaneously, such as is brought about by the seasonal variation of temperate and arctic climates (Winterhalder 1986); climates in which food preservation is also more feasible. Thus Nunamuit eskimo households, and the coalitions of households that constitute north-west coast Indian lineages, both practiced storage of seasonally available resources (Gubser 1965; Garfield & Wingert 1966; Binford 1979) a strategy Woodburn would class as 'delayed return'.

An alternative hypothesis for sharing networks among hunter-gatherers to that of Winterhalder is proposed by Myers (1986b). According to this argument, the intrinsic value of the goods exchanged is less important than their symbolic value as tokens of friendly intent. Myers contends that the immediate use-value of the tools, clothing and food exchanged among the Pintupi of central Australia is not great and they are readily obtained, constructed and replaced. More significant is the capacity of exchange to 'provide a moral basis for continued and ongoing co-residence and cooperation among members of a band.' The function of material exchanges would correspond to that attributed above to the greetings exchanged between French farmers.

These alternative hypotheses can to some extent be tested by looking at the intrinsic value of resources that are shared. In some cases meat from game animals forms the substance of exchange, in other cases it is personal possessions. Altman & Peterson (1986) describe sharing in an Aboriginal community in Arnhem Land. Here large game (macropod, emu, etc.) is referred to as *maih nagimuk*, and distinguished from *maih yawut*, small game (fish, lizard, bird). People are normally expected to bring large game back to camp and distribute it among resident households. 'The rules relating to sharing are not directed at the small game essential to provisioning the household but to occasions when there is an immediate surplus to households needs' (Altmann & Peterson 1986, p. 4). The crux of the distinction is that other households cannot legitimately claim a share of small game.

Among the Nyae-Nyae !Kung, meat sharing is similarly exercised with respect to big game animals weighing hundreds of pounds (Marshall 1976).

Kaplan & Hill (1985) examined Ache food sharing in the light of three hypotheses (kin selected and reciprocal altruism, tolerated theft). They concluded that kin-based food sharing was confined to the nuclear family and that reciprocal altruism offered the most likely explanation for most sharing. Meat was shared most often, followed by honey. The benefits were not, however, short term. Sharing was directed through the whole band and good hunters ate significantly less of their own kills than did others. Kaplan & Hill (1985) suggest the long-term pay-offs for the good hunter stem from other band members' wish to keep him in the group. His children may be better protected from the threat of infanticide, his close relatives more likely to be cared for when sick, and other hunters may trade sexual access to their wives for his allegiance to the band. All these cases support the hypothesis that sharing is a risk-reducing strategy related to the intrinsic food value of the goods shared.

Meat, however, is not the only substance of exchange among hunter-gatherers. The exchange of personal possessions corresponds best to Myers' hypothesis. In addition to Myers' data, relevant research has been conducted by Wiessner and

Marshall on the !Kung. Marshall (1976) writes that almost everything a !Kung person has may have been given to him, and may be passed on to others in time. 'Instead of keeping things, they use them as gifts to express generosity and friendly intent, and to put people under obligation to make return tokens of friendship'. Wiessner (1982) interprets reciprocal gift exchange networks among the !Kung as a means of reducing risk. She argues that the network has a specific cultural form, it is known as *hxaro* and that its consequences 'cannot be predicted from environmental variables alone'. Gifts exchanged through *hxaro* can be any *non-food items* and in fact constitute the bulk of a !Kung's material wealth. However, there is a connection with meat sharing: any *hxaro* partners in camp are likely to be in 'the first wave' of meat sharing by the owner of the kill, and will be included in daily reciprocal exchanges in camp.

Hxaro partners extend well beyond one's own band; indeed the aim is to have them in as many independent bands as possible, bands whose territories, or *N!ore*, contain as great a variety of resources as possible. Where partners live far apart, a continual balanced flow of gifts is important, to let each partner know the relationship is intact. In this way, the risk of resources failing in one's own *N!ore* is offset by guaranteeing access to other bands. Winterhalder has suggested that risk reduction may provide an explanation of the ultimate causation of such networks, while the symbolic dimension provides the proximate mechanism by which egalitarian interactions are constructed and sustained (personal communication). Whereas greeting rituals offer an appropriate medium for communication when participants regularly meet face-to-face, material gifts such as those exchanged in *hxaro* are more appropriate when participants live many miles apart. Hewlett, van der Koppel & Cavalli-Sforza (1982) calculated that Aka pygmies of the Central African Republic made over twice as many expeditions to visit relatives as to forage for food (see their Tables 6 and 7), but this is to some extent an artefact of the way their data were collected (see note 3).

Both San and Australian Aborigines possess customs for classifying non-kin as if they were kin: 'sections' and 'sub-sections' among Aborigines, namesakes among San. Despite their different political implications (see Layton 1986a) both customs facilitate elaborate networks of social interaction. The resources which flow through such networks are not just natural goods such as meat, but artefacts, information, spouses and reciprocal rights of access to foraging ranges. It was the initiation of such strategies which Gamble (1982) postulated as the distinguishing feature of Upper Palaeolithic social interaction in Europe.

Lineages represent closed coalitions rather than open networks. They are corporate groups recruited on (putative) common descent and share some form of property to the exclusion of outsiders. The premise of Dyson-Hudson & Smith's paper (1978) is that human and animal territoriality is not genetically fixed but (like food sharing) an organizational strategy to be expected when 'the costs of exclusive use and defence of an area are outweighed by the benefits' (Dyson-Hudson & Smith 1978, p. 23). A fact perhaps not sufficiently examined in Dyson-Hudson and Smith's paper is that human territoriality is exercised by groups rather than individuals (but see MacDonald & Carr, this volume). Probably the best case among hunter-gatherers of boundary defence of collective territories is that of the North-west Coast Indians of North America. Lineages defended areas of land and sea containing a diversity of patchy but dense productive resources. Those which were superabundant were not defended (Layton 1986a; Richardson 1986). Presumably territoriality is here interacting with the optimization of re-sources which are best exploited, through cooperation within the group (cf. Smith 1983; Foley 1985). The Western Shoshoni of California contrasted markedly with the North-west Coast pattern. Living in a harsh environment, Western Shoshoni families subsisted on wild grass seeds and solitary game during summer months. Because the game did not form herds there was no advantage in cooperative hunting and families were scattered in search of unpredictable resources. In winter, Piñon nuts were the principal food source. Families coalesced at Piñon nut groves. However, each grove only fruited once every 3−4 years. 'The very erratic pattern of yields brought different families together at different places each fall ..., *no consistent group of families* could amalgamate and establish either band or family ownership of piñon groves' (Dyson-Hudson & Smith 1978, p. 28, my emphasis).

Although the contrast between Shoshoni and North-west Coast organizational strategies seems clearly related to the distribution of natural resources, Australian Aboriginal descent groups are not so obviously recruited around the exploitation and defence of material resources. Here it seems more important to consider the intervention of culture, because their principal property is a body of shared religious knowledge, physically manifested in 'sacred sites' created by ancestral heroes. Unauthorized access to these sites is punished, but members of neighbouring clans allow each other reciprocal access to subsistence resources in the 'estates' on which their sacred sites lie. Conception, birth or residence on an estate may be grounds for adoption into the clan controlling it. Such adoption normally requires instruction into the clan's body of knowledge and is signalled by participation in the clan's ritual or control of its sacred objects (Myers 1986a; Williams 1986). Such knowledge has real value, and is not mere superstition (Layton 1986b). Cashdan (1983) argued that Australian Aborigines and Kalahari San defend access to the social group exploiting resources over a particular range through greeting ceremonies to mark the acceptance of visitors, because the area is too large effectively to defend its perimeter. This not only permits effective control over natural resources but allows the exchange of information about their location. It also requires visitors to acknowledge an obligation to extend reciprocal access to their hosts on future occasions. E.A. Smith accepts this argument in a recent paper (1986) and interprets 'social boundary defence' as the logical transformation of perimeter defence where there is a lack of synchrony in the availability of resources among neighbouring bands. It is, in other words, an extension of the meat-sharing hypothesis to an inter-band level, but with the 'pay-offs' including information sharing as well as reciprocal debt between members of bands in contact with one another. Winterhalder's model for effective sharing predicts that

80% of potential risk reduction from pooling and division can be gained by only six foragers (Winterhalder 1986). Birdsell's model, derived from a study of tribal (dialect) boundaries in Australia predicted that each local group will have six others contiguous with it (Birdsell 1958; cf. Gamble 1982), providing members of any one band with the option of joining six others.

Peterson (1975), Cashdan and I have argued that in Aboriginal Australia the need to care for sacred sites, and the desire to die in one's own estate, draw people back to their own country and hence, although culturally defined, have a spacing function which counteracts residential movement between bands. There is, however, a temptation to go a step further than this in proposing an ethological explanation. The geographical distribution of sacred sites closely follows the distribution of surface water. Surface water is probably the most dense and predictable, albeit patchy, of subsistence resources (see Layton 1986b, Fig. 2). Water is therefore the resource most likely to be subject to boundary defence, according to the predictions of the Dyson-Hudson & Smith (1978) model. Is the defence of access to sacred sites covertly the defence of access to water? While the coincidence of site and water distributions cannot be accidental, the answer is not so simple. Water can usually be obtained without trespass on sacred sites, which are small or located a short distance away. Myers considers whether the Pintupi do in fact defend access to water. He concludes that this is not directly the case; people are likely to return to their home area during drought because there they are most likely to find the congenial relations of close kin, and be sure of support in disputes at a time when movement away would not be practicable, rather than because they would be physically excluded from other base camps (Myers 1986a). They are not defending the water itself so much as a privileged set of social relationships. He concludes that Pintupi society 'does not merely reify "ecological necessities" but ... has taken on its own emergent values' (p. 192). Sociobiologists will, however, note that the social relationships being defended are those obtaining between close kin: despite the elaborate cultural form there is perhaps an element of simple, kin-selecting altruism at work when drought afflicts all groups in the region. Despite the elaborate cultural forms of Aboriginal social behaviour there are therefore a number of points at which plausible sociobiological hypotheses can be proposed to explain the adaptive significance of such behaviour for participants. I conclude that despite variation in the cultural content of hunter-gatherer social relationships, open networks through which resources are shared (Woodburn's immediate return systems) tend to develop in relation to resources whose spatial distribution is scarce or unpredictable while closed coalitions defending exclusive access to resources (Woodburn's delayed return) develop in relation to resources whose spatial distribution is dense and predictable.

CONCLUSION

Perhaps the research interests of sociobiologists and sociocultural anthropologists will never completely coincide, but it has been argued here that they overlap sufficiently to yield useful grounds for collaboration. The opportunities for successful cooperation would be enhanced if both parties made certain concessions to the other's insights. Sociobiologists might find it profitable to examine further the significance of reciprocal altruism, rather than kin-selecting altruism, in human social behaviour; to examine more fully the origin of human adaptive strategies in conventional, cultural codes, and the evolutionary implications of the lateral transmission of learned behaviour. Sociocultural anthropologists, on the other hand, must rid themselves of the tendency to reify groups, and learn to think of social evolution in adaptive rather than progressive terms. Instead of regarding the search for power or wealth as sufficient goals for human competition, they might consider whether as Borgerhoff Mulder argues (1987), they constitute proximate goals which ultimately contribute to reproductive success. Rather than rest content with the premise that cultural forms are purely conventional, they should consider whether some aspects of cultural variability can be objectively interpreted within a Darwinian framework.

NOTES

1 Promising though Irons' hypotheses appear, I suspect that they require further refinement. He cites the Tiwi of northern Australia as a case where women's labour is more critical than men's, because the bulk of the diet is vegetable foods gathered by women. Tiwi do practice sororal polygyny: men marry later than women, and many men have several wives. The wives are likely to be close kin to one another, constituting a cooperative group. It will benefit each wife's inclusive fitness to help care for the other wives' children. Neat though this appears, the implication seems to be that the more important the vegetable component in hunter-gatherer diet, the greater the incidence of sororal polygyny. In Australia it should then be most prevalent in the central desert. However, the opposite is the case: it occurs most often on the north coast, among the Tiwi and Yolngu. It is much rarer among the Yolngu's neighbours, the coastal Gidjingali, who have a different classificatory kinship system which gives the oldest of a set of brothers less opportunity to secure all the appropriate marriage partners (Keen 1982). Almost certainly there are other ecological factors to be taken into account, but the Gidjingali demonstrate that two historically divergent, cultural configurations offer viable patterns of behaviour in a given environment.

2 Both Trivers (1985, p. 22) and Harris (1968, p. 118) cite Darwin's vision of the 'grandeur' of his evolutionary perspective, but only Harris includes the preceding remark that natural selection results in 'the most exalted object of which we are capable of conceiving, namely the production of the higher animals'. Harris further quotes from Darwin's *Descent of Man* the view that 'civilized' have supplanted 'barbarous' nations as a consequence of their perfection of the intellect (*Descent of Man*, p. 154, cited Harris 1968, p. 121).

3 The sample of named sites visited was predetermined by the researchers and includes many lying outside the foraging range. All cases where five or more visits were made to a site are regarded as of equal value.

ACKNOWLEDGEMENTS

This paper would probably not have been written but for the stimulus of working in a department which combines social and biological anthropology, and I am grateful to my colleagues, especially Malcolm Smith, Rob Foley, Michael Carrithers and Gil Manley, for the encouragement they provided. I have been particularly fortunate in receiving detailed comments on the original version of the paper from Eric Alden Smith, Monique Borgerhoff Mulder, Malcolm Smith, Val Standen, Ekhart Voland and Bruce Winterhalder, and have done my best to do justice to their insights in revising the paper for publication.

REFERENCES

Ahmed, A.S. (1976). *Millenium and Charisma among Pathans*. Routledge, London.

Alexander, R. (1979). Evolution and culture. *Evolutionary Biology and Human Social Behaviour* (Ed. by N. Chagnon & W. Irons), pp. 59–78. Duxbury, North Scituate, Massachusetts.

Altman, J. & Peterson, N. (1986). *Rights to Game and Rights to Cash among Contemporary Australian Hunter-gatherers*. Paper presented to the 4th International Conference on Hunting and Gathering Societies, London.

Asad, T. (1972). Market model, class-structure and consent. *Man* (N.S.), 7, 74–94.

Barth, F. (1959). Segmentary opposition and the theory of games. *Journal of the Royal Anthropological Institute*, 89, 5–21.

Bates, D.G. & Lees, S.H. (1979). The myth of population regulation. *Evolutionary Biology and Human Social Behaviour* (Ed. by N. Chagnon & W. Irons), pp. 273–89. Duxbury, North Scituate, Massachusetts.

Bender, B. (1985). Prehistoric developments in the American midcontinent and in Brittany, Northwest France. *Prehistoric Hunter-gatherers: the Emergence of Cultural Complexity* (Ed. by T.D. Price & J.A. Brown), pp. 21–57. Academic Press, New York.

Binford, L.R. (1979). Organization and formation processes: looking at curated technologies. *Journal of Anthropological Research*, 35, 255–73.

Binford, L.R. (1980). Willow smoke and dogs' tails: hunter-gatherer settlement systems and archaeological site formation. *American Antiquity*, 45, 4–20.

Birdsell, J.B. (1958). On population structure in generalized hunting and gathering populations. *Evolution*, 12, 189–205.

Borgerhoff Mulder, M. (1987). Adaptation and evolutionary approaches to anthropology. *Man* (N.S.), 22, 25–41.

Cashdan, E. (1983). Territoriality among human foragers: ecological models and an application to four Bushmen groups. *Current Anthropology*, 24, 47–66.

Chagnon, N. (1979). Mate competition, favouring close kin, and village fissioning among the Yanomamo Indians. *Evolutionary Biology and Human Social Behaviour* (Ed. by N. Chagnon & W. Irons), pp. 86–132. Duxbury, North Scituate, Massachusetts.

Chagnon, N. (1982). Sociodemographic attributes of nepotism in tribal populations: man the rule-breaker. *Current Problems in Sociobiology* (Ed. by King's College Sociobiology Group), pp. 291–318. Cambridge University Press, Cambridge.

Clark, G.A. & Strauss, L.G. (1983). Late Pleistocene hunter-gatherer adaptations in cantabrian Spain. *Hunter-gatherer Economy in Prehistory* (Ed. by G. Bailey), pp. 131–48. Cambridge University Press, Cambridge.

Durham, W.H. (1979). Towards a co-evolutionary theory of human biology and culture. *Evolutionary Biology and Human Social Behaviour* (Ed. by N. Chagnon & W. Irons), pp. 39–59. Duxbury, North Scituate, Massachusetts.

Durkheim, E. (1938). *The Rules of Sociological Method* (Translated by S.A. Solovay & J.H. Mueller). Free Press, New York.

Dwyer, P.D. (1985). A hunt in New Guinea: some difficulties for optimal foraging theory. *Man* (N.S.), 20, 243–53.

Dyson-Hudson, R. and Smith, E.A. (1978). Human territoriality: an ecological reassessment. *American Anthropologist*, **80**, 21–41.

Evans-Pritchard, E.E. (1951). *Kinship and Marriage among the Nuer*. Clarendon Press, Oxford.

Firth, R. (Ed.) (1967). *Themes in Economic Anthropology*. Tavistock, London.

Foley, R. (1985). Optimality theory in anthropology. *Man* (N.S.), **20**, 222–42.

Freeman, D. (1974). The evolutionary theories of Charles Darwin and Herbert Spencer. *Current Anthropology*, **15**, 211–37.

Gamble, C. (1982). Interaction and alliance in palaeolithic society. *Man* (N.S.), **17**, 92–107.

Garfield, V. & Wingert, P.S. (1966). *The Tsimshian Indians and their Arts*. University of Washington Press, Seattle.

Giddens, A. (1979). *Central Problems in Social Theory: Action, Structure and Contradiction in Social Analysis*. Macmillan, London.

Glickman, M. (1971). Kinship and credit among the Nuer. *Africa*, **41**, 306–19.

Goodenough, W.H. (1965). Rethinking status and role. *The Relevance of Models for Social Anthropology* (Ed. by M. Banton), pp. 1–24. Tavistock, London.

Goody, J. (1956). A comparative approach to incest and adultery. *British Journal of Sociology*, **7**, 286–305.

Gubser, N.J. (1965). The Nunamiut Eskimo: Hunters of Caribou. Yale University Press, New Haven, Connecticut.

Hajnal, J. (1965). European marriage patterns in perspective. *Population in History: Essays in Historical Demography* (Ed. by D.V. Glass & D.E.C. Eversley), pp. 101–43. Arnold, London.

Harris, M. (1968). *The Rise of Anthropological Theory: a History of Theories of Culture*. Routledge, London.

Hewlett, B., van der Koppel, J.M.H. & Cavalli-Sforza, L.L. (1982). Exploration ranges of Aka pygmies of the Central African Republic. *Man* (N.S.), **17**, 418–30.

Irons, W. (1979). Investment and primary social dyads. *Evolutionary Biology and Human Social Behaviour* (Ed. by N. Chagnon & W. Irons), pp. 181–213. Duxbury, North Scituate, Massachusetts.

Irons, W. (1983). Human female reproductive strategies. *Social Behaviour of Female Vertebrates* (Ed. by S.K. Wasser), pp. 169–213. Academic Press, New York.

Kaberry, P. (1969). Witchcraft of the sun: incest in Nso. *Man in Africa* (Ed. by M. Douglas & P. Kaberry), pp. 175–95. Tavistock, London.

Kaplan, H. & Hill, K. (1985). Food sharing among Ache foragers: tests of explanatory hypotheses. *Current Anthropology*, **26**, 223–46.

Keen, I. (1982). How some Nurngin men marry ten wives. *Man* (N.S.), **17**, 620–42.

Layton, R. (1971). Patterns of informal interaction in Pellaport. *Gifts and Poison, the Politics of Reputation* (Ed. by F.G. Bailey), pp. 97–118. B.H. Blackwell, Oxford.

Layton, R. (1973). Pellaport. *Debate and Compromise, the Politics of Innovation* (Ed. by F.G. Bailey), pp. 48–74. B.H. Blackwell, Oxford.

Layton, R. (1981). Communication and ritual: two examples and their social context. *Canberra Anthropology*, **4**, 110–24.

Layton, R. (1986a). Political and territorial structures among hunter-gatherers. *Man* (N.S.), **21**, 18–33.

Layton, R. (1986b). *Uluru: an Aboriginal History of Ayers Rock*. Australian Institute of Aboriginal Studies, Canberra.

Lazlett, P. (1972). Introduction. *Household and Family in Past Time* (Ed. by P. Lazlett & R. Wall), pp. 1–73. Cambridge University Press, Cambridge.

Lévi-Strauss, C. (1969). *The Elementary Structures of Kinship* (Translated by J.H. Bell & J.R. von Sturmer). Eyre & Spottiswoode, London.

Marshall, L. (1976). Sharing, talking and giving: relief of social tensions among the !Kung. *Kalahari Hunter-gatherers: Studies of the !Kung San and Their Neighbours* (Ed. by R.B. Lee & I. deVore), pp. 350–71. Harvard University Press, Cambridge, Massachusetts.

Marx, K. (1867 [1930]), *Capital. Vol. 1* (Translated by E. & C. Paul). Dent, London.

Marx, K. (1965). *Pre-capitalist Economic Formations* (Ed. by E.J. Hobsbawm). International Press, New York.

Mauss, M. (1954). *The Gift: Forms and Functions of Exchange in Archaic Societies* (Translated by I. Cunnison). Cohen & West, London.

Mellars, P. (1973). The character of the middle-upper Palaeolithic transition in south-west France. *The Explanation of Culture Change* (Ed. by C. Renfrew), pp. 255–76. Duckworth, London.

Morphy, H. & Layton, R. (1981). Choosing among alternatives: cultural transformations and social change in Aboriginal Australia and French Jura. *Mankind*, 13, 56–73.

Myers, F. (1986a). Always ask: resource use and land ownership among Pintupi Aborigines of the Australian Western Desert. *Resource Managers: North American and Australian Hunter-gatherers* (Ed. by N. Williams & E. Hunn), pp. 173–95. Australian Institute of Aboriginal Studies, Canberra.

Myers, F. (1986b). *Burning the Truck and Holding the Country: Property, Time and the Negotiation of Identity among Pintupi Aborigines.* Paper presented to the 4th International Conference on Hunting and Gathering Societies, London.

Parry, J.P. (1979). *Caste and Kinship in Kangara*. Routledge, London.

Peterson, N. (1975). Hunter-gatherer territoriality: the perspective from Australia. *American Anthropologist*, 77, 53–68.

Radcliffe-Brown, A.R. (1951). Murngin social organization. *American Anthropologist*, 53, 37–55.

Radcliffe-Brown, A.R. (1952). *Structure and Function in Primitive Society*. Cohen & West, London.

Richardson, A. (1986). The control of productive resources on the northwest coast of North America. *Resource managers: North American and Australian Hunter-gatherers* (Ed. by N. Williams & E. Hunn), pp. 93–112. Australian Institute of Aboriginal Studies, Canberra.

Rindos, D. (1985). Darwinian selection, symbolic variation and the evolution of culture. *Current Anthropology*, 26, 65–88.

Sahlins, M. (1976). *The Use and Abuse of Biology: an Anthropological Critique of Sociobiology.* University of Michigan Press, Ann Arbor, Michigan.

Smith, E.A. (1983). Anthropological applications of optimal foraging theory: a critical review. *Current Anthropology*, 24, 625–51.

Smith, E.A. (1986). *Environmental Uncertainty and Ecological Risk: Implications for Hunter-gatherer Settlement and Social Organisation.* Paper presented to the 4th International Conference on Hunting and Gathering Societies, London.

Torrence, R. (1983). Time budgeting and hunter-gatherer technology. *Hunter-gatherer Economy in Prehistory* (Ed. by G. Bailey), pp. 11–22. Cambridge University Press, Cambridge.

Trivers, R. (1985). *Social Evolution*. Cummings, Menlo Park, California.

Verdon, M. (1982). Where have all the lineages gone? Cattle and descent among the Nuer. *American Anthropologist*, 84, 566–79.

Viazzo, P.P. (1986). Illegitimacy and the European marriage pattern. *The World we have Gained: Histories of Population and Social Structure* (Ed. by L. Bonfield, R.M. Smith & K. Wrightson), pp. 100–21. B.H. Blackwell, Oxford.

Wiessner, P. (1982). Risk, reciprocity and social influences on !Kung san economics. *Politics and History in Band Societies*, (Ed. by E. Leacock & R. Lee), pp. 61–84. Cambridge University Press, Cambridge.

White, R. (1982). Rethinking the middle/upper Palaeolithic transition. *Current Anthropology*, 23, 169–92.

Williams (1986). *The Yolngu and their Land*. Institute of Aboriginal studies, Canberra.

Winterhalder, B. (1986). Diet choice, risk and food sharing in a stochastic environment. *Journal of Anthropological Archaeology*, 5, 369–92.

Wolf, E.R. (1982). *Europe and the People without History*. University of California Press, Berkeley.

Woodburn, J. (1980). Hunters and gatherers today and reconstruction of the past. *Soviet and Western Anthropology* (Ed. by E. Gellner), pp. 95–177. Duckworth, London.

Why is ape tool use so confusing?

W.C. McGREW

Department of Psychology, University of Stirling,
Stirling FK9 4LA, Scotland, UK

SUMMARY

1 The extent and nature of tools used by living apes (Hominoidea) are compared. Data from six types are contrasted: chimpanzee (*Pan troglodytes*), bonobo (*Pan paniscus*), highland gorilla (*Gorilla gorilla beringei*, *G. g. graueri*), lowland gorilla (*G. g. gorilla*), orang-utan (*Pongo pygmaeus*), and gibbon (*Hylobates* spp.).

2 Tool-use is classified as occuring in four contexts, two in free-ranging settings, i.e. pristine versus human-influenced, and two in captivity, i.e. spontaneous versus human-induced. Disappointingly, there are gaps in our knowledge, and there is little consistency across contexts.

3 The ape types are classified on phylogenetic grounds in terms of their last common ancestor in relation to humans, and on socioecological grounds in terms of terrestriality, plant and animal portions of the diet, and social structure.

4 When the types of ape are ranked qualitatively by their extent of tool use, and these rankings are compared with the other variables, only one apparent relationship emerges: tool use seems to be positively correlated with faunivory. Phylogeny and terrestriality seem unrelated, and plant foods and sociability cannot be meaningfully compared on such crude, uni-dimensional criteria.

5 Two types of ancestral hominoid are hypothesized: One is an older, conservative stem-form of pongid with highly variable tool use performance according to environmental pressures. The other is a younger, African proto-hominid with cultural capacities and convergence with hominids.

INTRODUCTION

This paper has three aims:
(a) To compare the extent of tool use across living apes;
(b) To relate variation in tool use by (i) homology to phylogenetic relationships, and (ii) analogy to socioecological variation;
(c) To synthesize these findings to infer aspects of tool use among ancestral hominoids which led to the origins of material culture.

The first aim entails updating the evidence, as new findings on tools used by apes continue to mount. More and more, it is clear that context is important: How an organism behaves in captivity may or may not reflect its actions in nature. Here, the exercise shows not only the extent of current knowledge, but also its persisting gaps. Many questions remain about about the tool use of even these well-studied mammals.

The two parts of the second aim present different problems. The fossil record for apes is sketchy. Molecular anthropology now provides a wealth of data for tackling phylogeny, but the conclusions do not always agree. Phylogenetic relationships among the African Pongidae and Hominidae remain obscure, although a consensus seems to be emerging (see Foley 1987). In arguing by analogy in terms of socioecology, the difficulty is in choosing variables to the right degree of specificity. Given limited space, this paper addresses this set of questions only crudely.

In undertaking the third aim, this paper starts from the simple premise that each branching in the phylogenetic tree represents an ancestral hominoid. Each of these is fair game for what Tooby & DeVore (1987) call 'strategic modelling', that is, the construction of conceptual models of human and non-human primate behaviour based on current understanding of evolutionary theory. Here, such a conceptual model is referentially based on living apes, with all of the limitations that this entails.

METHODS

To survey all primary sources would be exhausting, so this paper relies upon Beck's (1980) and Tuttle's (1986) comprehensive syntheses. Other material comes from original reports (e.g. Whitten 1982), relevant reviews (e.g. Lethmate 1982), or comprehensive anthologies (e.g. Susman 1984). Despite this, there are surprising gaps in knowledge: a rigorous, overall assessment of intelligence in great apes is overdue, having not been done since Rumbaugh (1970); a systematic, empirical attempt to induce tool use in captive lowland gorillas remains to be tried.

To contrast the living apes, this paper splits them into six types. These are taxonomically messy, but ecologically revealing. All three geographical races of *Pan troglodytes* are lumped, as are Bornean and Sumatran *Pongo pygmaeus*. *Gorilla* is split into lowland (*G. g. gorilla*) and highland (*G. g. beringei* and *G. g. graueri*) on ecological grounds. All forms of gibbons, including siamang, are lumped into *Hylobates*. *Pan paniscus* retains specific status. This scheme may offend partisans, but it seems sensible here.

It is possible to fit five of the six forms into a phylogenetic tree, indicating relative degrees of closeness and thus a sequence of common ancestry. (Comparative data on lowland and highland gorillas are not yet available.) Sibley & Ahlquist (1984) have done this on the basis of DNA–DNA hybridization–dissociation tests. Calibrating the 'molecular clock' to obtain absolute timing is more contentious (see Foley 1987 for essentials). Sibley & Ahlquist have done so on the basis of divergence of the African and Asian apes at 16 million years ago (see Fig. 1). This yields a basis for comparison of tool use across hominoids on grounds of homology.

The six types of apes are then compared in two subject areas: use of tools and socioecology. The resulting classifications are crude and are useful only for qualitative comparisons.

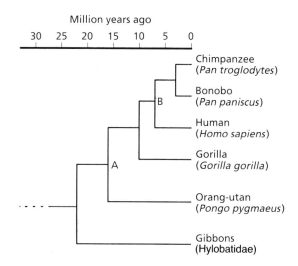

Million years ago

30   25   20   15   10   5   0

Chimpanzee
(*Pan troglodytes*)

Bonobo
(*Pan paniscus*)

Human
(*Homo sapiens*)

Gorilla
(*Gorilla gorilla*)

Orang-utan
(*Pongo pygmaeus*)

Gibbons
(Hylobatidae)

Fɪɢ. ɪ. Phylogeny of living apes and humans (Hominoidea). (From Sibley & Ahlquist 1984.)

For tool use (see Table 1), *captive* means apes living in confinement where both the means and ends of tool use are largely absent, while *free-ranging* is the opposite, where there is, for example, access to natural vegetation. In captivity, tool use is termed either *spontaneous*, that is, unprompted by humans, or *induced* by human intervention, often in a structured, experimental way. In free ranging, the sub-division is between *pristine*, which means natural conditions, and *human-influenced*, in which wild apes are provisioned or previously captive apes are released. Figures 2 and 3 show these four types of context for chimpanzees.

This analysis gives a matrix of twenty-four cells, each of which has a forced-choice coding of one of five types, as given in Table 1: '++' means that tool use is well-known from several individuals in several populations, as recorded by several investigators; '+' means that tool use has been convincingly noted, at least once, somewhere; '− −' means that tool use is notably absent from long-term studies of several populations; '−' means that tool use has not been seen, but studies have been few, or short, or limited to a few subjects; '?' means that this combination of ape and context has yet to be properly studied. Overall, presence is probably more accurate than absence, given the whimsy of negative evidence.

For socioecology (see Table 2), the matrix also has twenty-four cells: the six types of ape plotted against four types of social or environmental categorization. *Vertical distribution* refers to the extent to which apes spend more of their waking hours on the ground (*Ter*) or above it in vegetation (*Arb*). The vegetative portion of the diet is classified by the predominant part of the *plant* eaten, either fruit (*Frug*), or foliage (*Fol*). For the *animal* portion of the diet, the types of prey are given, for example mammal (*M*). For social organization, Wrangham's (1987) criteria for membership of parties are followed whenever possible: *stable* means that composition of the temporary grouping stays constant over months, while *variable* means that it changes over weeks, days or hours.

TABLE 1. Use of tools by living apes in four settings

Type	Captive		Free-ranging	
	Spontaneous	Induced	Human-influenced	Pristine
Chimpanzee	++	++	++	++
Bonobo	++	?	−	−
Highland gorilla	+	+	?	− −
Lowland gorilla	+	?	?	−
Orang-utan	++	++	++	− −
Gibbon	+	?	+	− −

++ = Well-known.
 + = Seen, at least anecdotally.
− − = Consistently absent.
 − = Not (yet?) seen.
 ? = Unstudied.

TABLE 2. Selected aspects of socioecology of living apes

Type	Vertical distribution	Diet		Parties[§]
		Plant	Animal	
Chimpanzee	Ter/Arb*	Frug[†]	M,B,E,I[‡]	Variable
Bonobo	Arb/Ter	Frug/Fol	M,E,I	Variable
Highland gorilla	Ter	Fol	(I)	Stable
Lowland gorilla	Ter/Arb	Fol/Frug	I	Stable(?)
Orang-utan	Arb	Frug	B,I	Solitary
Gibbon	Arb	Frug	I	Stable

* Ter=terrestrial, Arb=arboreal.
[†] Frug=frugivorous, Fol=folivorous
[‡] M=mammal, B=bird, E=egg, I=insect.
[§] Modified from Wrangham (1987).

PATTERNS OF TOOL USE

Chimpanzee

Chimpanzees in all settings use tools regularly (Beck 1980; Tuttle 1986). In the wild, they use a variety of tools made from a variety of materials to accomplish a variety of tasks. This is true of the eastern, central, and western geographical races, and is known in habitats ranging from savannah to primary forest. Well-known types of tools include probes of vegetation to obtain social insects (Fig.

2a), hammers of stone to crack open nuts, sponges of leaves to soak up fluids, and weapons of woody branches to deter or dominate opponents. Wild chimpanzees also *make* tools and show flexibility in doing so. They use a variety of raw materials to make the same tool, for example, twig, vine, or bark to fashion a probe for termite fishing. The same raw material may be used to make various tools: for example, a leaf may be modified to be a sponge, napkin, probe, or billet-doux (Nishida 1980). However, marked contrasts occur across populations, and some of the differences seem to be cultural, that is they result from social traditions and reflect more than just environmental affordances (McGrew, Tutin & Baldwin 1979). Finally, some groups of wild chimpanzees seem to have more impressive tool kits than others, but it is not yet clear whether these contrasts are real or are artefacts of differing observational conditions or techniques.

A similar range of tool use is shown by free-ranging chimpanzees influenced by varying degrees of human contact. Chimpanzees released after years of confinement onto forested islands used hammers to crack open nuts (Hannah & McGrew 1987; Fig. 2b). Crop-raiding chimpanzees in a relict Guinean population living near human settlement showed similar use of hammer-stones (Sugiyama & Koman 1979). Both before and after provisioning, chimpanzees at Gombe showed the same kind of tool use directed to natural prey, but after provisioning with bananas, they also added new tool use to their repertoire, for example, levers to prise open metal boxes supplying bananas (Goodall 1968).

In captivity, chimpanzees spontaneously show every mode of tool use seen in the wild (Beck 1980), with the degree and range of expression being largely a function of opportunity. For example, given an artificial 'termite mound' containing prized food, Edinburgh Zoo chimpanzees made tools from their bedding branches to probe for it (Nash 1982). Further, captive chimpanzees invented new types of tools to solve new problems, for example, poles used as ladders to escape from enclosures (Menzel 1973). The most extensive tool users are chimpanzees reared in human homes, who learn by imitation to use many household implements, from keys to vacuum cleaners (Temerlin 1975).

Studies of induced tool use in captive chimpanzees (Figs 3a & b) began at least 70 years ago, with the early efforts of Kohts and Kohler. Kohler (1925) set standards with tasks which are still used: a rake to obtain out-of-reach incentive, boxes stacked to obtain incentive suspended overhead, etc. Circumstances which induce tool use vary from merely providing materials in a structured setting, such as crayons and paper for drawing, to carefully demonstrating how to solve problems (Hayes & Hayes 1954). More recently, tool use has been linked to other intellectual tasks, e.g. one chimpanzee using a symbol-system to ask for a tool from another, with the recipient using the tool to obtain a food item which both then share (Savage-Rumbaugh, Rumbugh & Boysen 1978).

All in all, chimpanzees show impressive performance of tool-making and tool use in all settings. Even their apparent limitations are consistent: no one has yet recorded a chimpanzee using a tool to make another tool. This reassuring uniformity does not apply to other apes, however.

(a)

Bonobo

Only recently have field studies of the bonobo reached the stage at which good behavioural observations have accumulated (see articles in Susman 1984). Primatologists at both sites in north-central Zaire, Lomako and Wamba, have not yet reported evidence of habitual tool use, though it has been keenly sought. A similar absence at both sites has been noted although the methods of study differ: at Wamba the apes are heavily provisioned at an artificially cleared feeding site, while at Lomako no provisioning is done.

In captivity, no studies of induced tool use seem to have been tried, perhaps because few bonobos are in captivity and almost all of these are not in laboratories, but in zoological gardens, where testing is not so easy. Jordan (1982) reported observations of groups at zoos in Antwerp, Frankfurt and Stuttgart. Using Beck's

(b)

Fig. 2. Chimpanzee tool use in free ranging. (a) Wild chimpanzee use of a stick to dip for driver ants. (Photograph by C.E.G. Tutin.) (b) Rehabilitated chimpanzee use of a hammer-stone and anvil to crack open palm nuts. (Photograph by A.C. Hannah.)

(1980) modes of tool use, she reported a range of spontaneous behavioural patterns indistinguishable from that of chimpanzees.

Orang-utan

The orang-utan presents a puzzle. Long-term field studies, both at Tanjung Puting in Borneo (Galdikas 1982) and at Ketambe in Sumatra (Rijksen 1978) have revealed no evidence of tool-use. Apart from dropping twigs and branches from the canopy, wild orang-utans present negative evidence. Most conspicuously absent are cases in feeding; only one isolated case is recorded (Rijksen 1978, p. 84). These findings after years of careful study confirm those from shorter studies at other sites (for example, Rodman 1977).

There seem to have been no studies of wild orang-utans being provisioned by scientists. However, both Tanjung Puting and Ketambe double as rehabilitation centres as well as field sites. Conveniently for comparison, captive apes, such as those confiscated by wildlife officials, are released into the same habitats in which their wild counterparts live. These released orang-utans show a rich array of tool use, some of it remarkably inventive, for example, use of floating objects to raft across a river. Much of their tool use involves artificial objects and tasks, but other usages are naturalistic, such as trying to open spiny fruits with a stick.

(a)

Despite being studied largely in zoological parks and not in laboratories, orang-utans in captivity are unparalleled tool users (Lethmate 1982). For example, one orang-utan wound 'wood-wool' around a cracked stick to mend it for use as a tool. Lethmate compared orang-utans in detail with chimpanzees on twenty-three categories of tool use and five of tool making. There were few differences. These accomplishments apply equally to spontaneous and to induced tool use. Perhaps the most striking example of the latter is that of the making and using of flaked stone tools (Wright 1972). After only a few hours of human demonstration, the ape used a quartzite hammer to knap flint flakes, the sharp edges of which he used to cut a cord, allowing access to food in a box.

Gorilla

Highland gorillas also pose problems. In the wild, the negative evidence is overwhelming. Many years of study of *G. g. beringei* in the Virunga Volcanoes of

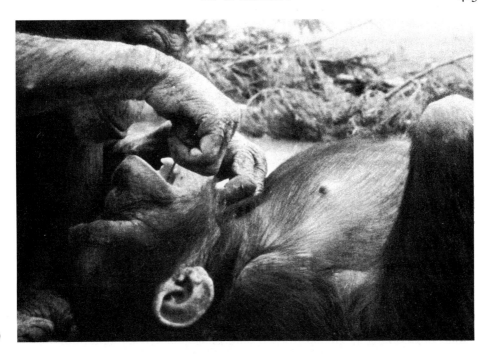

(b)

Fɪɢ. 3. Chimpanzee tool use in captivity. (a) Induced use of a twig as a utensil to get sticky food from a jar. (Photograph by C.E.G. Tutin.) (b) Spontaneous tool use in dental grooming to remove loose milk teeth.

Rwanda and Zaire have produced no signs of tool use (Fossey & Harcourt 1977). This holds despite excellent observations of completely relaxed subjects at a few metres' distance on the ground. Shorter periods of study on *G. g. graueri* at Kahuzi-Biega in Zaire also have been negative (Goodall 1979). Highland gorillas have been studied only in pristine conditions, without rehabilitation, provisioning, or crop raiding, so nothing can yet be said about human-influenced tool use during free ranging.

In captivity, there are now no highland gorillas to be studied, but there were two early investigations. Yerkes (1927) thoroughly tested and re-tested a young female on a battery of Köhler-like tasks. Her performance on these tests improved somewhat with age, but he remained disappointed in her poor showing. Carpenter (1937) watched a pair of males in the San Diego Zoo for about 6 weeks. His observations were casual, but he saw spontaneous use of containers for drinking.

Lowland gorillas are even more problematical. In the wild, they have scarcely been studied behaviourally. Further, no indirect data such as discarded tools have been found, even in areas where sympatric chimpanzees leave such circumstantial evidence (Jones & Sabater Pi 1971). Possible human influence on free-ranging lowland gorillas remains to be assessed. No wild population has yet been provisioned nor has rehabilitation into the wild of captive gorillas been tried.

In captivity, there seems to be only one spontaneous case of tool use in print: Wood (1984) reported that gorillas in a large captive colony in England modified branches into sticks in order to rake in food lying beyond reach outside their cage. The habit is well established in this captive colony which provides the most stimulating social and physical environment yet devised. Surprisingly, no research on induced tool use seems to have been tried in lowland gorillas, despite there being equivalent opportunities in zoological gardens to that of studying orang-utans. The nearest approximation has been objects manipulated in intelligence testing, such as patterned string problems (see, for example, Fischer & Kitchener 1965) or Piagetian research (see, for example, Redshaw 1978), but these seem to have been limited to youngsters.

Gibbon

Field studies of gibbons pre-date those of all other apes, having been done for over 50 years in various parts of South-East Asia (Whitten 1982). None of these studies has reported tool use, apart from the occasional dropping of branches onto observers below. However, all the findings have suffered from the limiting conditions of observers on the ground watching subjects high in the trees. Several studies of free-ranging gibbons loosed onto islands have been done, but most of these were short or intermittent, and superficial. Only one case of tool use seems to have been seen: Baldwin & Teleki (1976) saw a young female repeatedly use a leaf as a sponge to dip water from a pool. A release of captive gibbons into the wild in Thailand yielded few behavioural data and no mention of tool use (Tingpalapong *et al.* 1981). Provisioning of wild gibbons remains to be done.

In captivity, only one gibbon's spontaneous tool use has been described. Rumbaugh (1970) reported how a young female used a cloth as a sponge and a rope to make a swing. Surprisingly, there seem to have been no systematic attempts to elicit tool use from captive gibbons. Perhaps investigators have been deterred by the unpromising anatomy of the hylobatid hand as a manipulatory organ. However, Beck (1967) showed that although gibbons fail a string pulling task when it is presented on a flat surface, they quickly pass if the strings are elevated so that they can be easily grasped.

PATTERNS OF SOCIOECOLOGY

The six types of apes can be ranked in terms of degree of terrestriality (Tuttle 1986). This seems worth doing, as almost all tool use in nature takes place on the ground. (Only the use of a leafy sponge to sop up drinking water from tree-holes seems to be obligatorily elevated. Perhaps even more precarious is the arboreal use of hammer-stones to open nuts, in which a bough serves as an anvil.) Table 2 shows the apes ranked from least to most time during waking hours spent on the ground. The range is wide, from gibbons who apparently never descend to the ground, to adult male highland gorillas who may never leave it.

All apes are primarily frugivorous, except for the high-altitude gorillas who have become secondarily folivorous in a largely fruitless environment (Tuttle 1986). However, ranking the six types is difficult, as some differences within species are greater than some differences between species. More to the point, the plant portions of the diets of the apes cannot readily be ranked on a single variable, for example, extent of nutrition. Even using difficulty of processing is problematical, for example, chimpanzees in the Ivory Coast use hammers to crack open *Detarium* nuts in order to eat the kernels (Boesch & Boesch 1983), but lowland gorillas in Gabon apparently break them open with their teeth (Williamson, personal communication). There are derived ways of ranking diet, such as in terms of quality, but that is beyond the scope of this paper.

Animal matter in the diet, or faunivory, can be more easily ranked across the six types of ape, from most to least (Tuttle 1986). Again, the range is wide. At one end, chimpanzees in all types of habitat prey on mammals, birds and their eggs, and social insects. They use a variety of tactics geared to the vulnerability of the prey and often out-compete other sympatric predators; for example, baboons are limited to grabbing the winged emergents of termites, but chimpanzees use tools to extract the underground castes. Lowland gorillas in north-eastern Gabon regularly eat termites too, but these are obtained by destroying their mounds and picking up the prey by hand (Tutin & Fernandez 1983). At the other end of the scale, highland gorillas eat animal matter only inadvertently in the form of invertebrates living in the plants eaten (Harcourt & Harcourt 1984).

Social organization varies more widely across apes than the single category in Table 3 indicates (see, for example, Wrangham 1986, 1987). The problem here is how to rank sensibly a multi-dimensional phenomenon. Chimpanzee parties may be less stable than those of bonobos, but how does one compare either of these with the constancy of the harem in gorillas or of the nuclear family in gibbons? What does one do with a solitary type like an adult male orang-utan — is its social life infinitely stable or individually variable over time? Finally, the composition and durability of parties of lowland gorillas is simply unknown.

Overall, the apes present a diverse radiation on a variety of socioecological criteria. This is clear even from the small set of features dealt with here, the number of which could easily be doubled or trebled, or made more specific. No univariate analysis will do justice to this variety, and multivariate analysis is not yet possible.

DISCUSSION

Several general points emerge from the survey. First, the data remain incomplete. Only two of the six types of apes (chimpanzee and orang-utan) are sufficiently studied in all four settings to even begin to draw conclusions. Second, there is a disappointing lack of agreement across settings, that is, for a given type of ape, the occurrence and extent of tool use in one setting is not necessarily found in another. The best set of data comes from free-ranging apes in pristine settings,

TABLE 3. Living apes ranked in terms of tool use, relation to *Homo*, terrestriality and faunivory

	Tool-use	Related to human	Terrestriality	Faunivory
Most	Chimpanzee	⌐ Chimpanzee ? ⌊ Bonobo	Highland gorilla	Chimpanzee
	⌐ Orang-utan ? ⌊ Bonobo		Lowland gorilla	Bonobo
		⌐ Highland gorilla ? ⌊ Lowland gorilla	Chimpanzee	Orang-utan
	⌐ Highland gorilla		Bonobo	Gibbon
	? ⊢ Lowland gorilla	Orang-utan	Orang-utan	Lowland gorilla
Least	⌊ Gibbon	Gibbon	Gibbon	Highland gorilla

but it is also the most negative. The next best is of spontaneous tool use in captivity, and it is the most positive! Adding data from the other two, sparser settings adds little to the picture. So, what to make of the contrasts, especially the striking discrepancy between orang-utans in nature and their counterparts in all other settings, or between captive and free-ranging bonobos?

Several caveats need stating. First, spontaneous tool use is not the same as being unaffected by human influence. Spontaneous only means *untaught*, and even that is not always clear. Reinforcement which shapes learning can be given unintentionally. Second, many cases of tool use by apes, especially young ones, may be direct imitation of foster, human caretakers. This is often an unintended by-product of emotional attachment. Third, older captive apes have plenty of 'free' time in settings which may present a wealth of objects, or conversely, focus their attention on a few. Yet, however artificial and contrived the captive environment, one must still explain the abilities manifest there, especially as these contrast so markedly with performance in nature.

The most parsimonious interpretation for Table 1 may be that *Pan troglodytes* is the only true tool user, given its consistency across all four settings. All tool use by other apes can then be written off as freak accidents or as somehow prompted by contact with humans. If this is so, then the question marks in the non-pristine settings (in Table 1) can be expected to change to double-plusses as more studies are done and as data build up.

For purposes of present comparison, however, Table 3 gives cautious rankings of apes in terms of tool use, based on Table 1. So, how does this reflect homology and analogy?

For homology, no correlation is apparent. Whereas one African ape, the chimpanzee, is both closest to humans phylogenetically and technologically, another closely related type, the gorilla, shows virtually no tool use. In fact, the tool use of gorillas is indistinguishable from that of the only distantly related gibbons. Arguably the most impressive tool-user in captivity, the orang-utan, is very distantly related to the hominids, having diverged in the early Miocene (Sibley & Ahlquist 1984). Analysis of tool use by phylogeny seems hopeless.

Socioecologically, the results range from the puzzling to the perverse. For vertical distribution, the least technological apes are at the extremes: arboreal gibbon and terrestrial gorilla. The most frequest tool user, the chimpanzee, falls in the middle, as having daytime activity most balanced between life in the trees and on the ground.

For diet, there is a marked contrast between herbivory and faunivory, when tool use is considered. Plant foods do not lend themselves to easy ranking, at least in terms of the crude criteria given here. There is no obvious relation between, for example, degree of frugivory and frequency of tool use. A possibility worth exploring is that of food quality in optimal foraging terms, for example, net energy gain after taking into account searching time, handling time, etc. But, but for use of animal foods, a striking correlation appears: The more types of animals eaten, the more tool use shown. The link between vertebrate prey and frequent tool use is especially intriguing. These two variables do *not* reiterate each other, as only chimpanzees use tools to get animal prey.

With no uni-dimensional ranking of social structure obvious, all that can be said here is that these data suggest no simple relation between tool use and sociability. Perhaps more detailed comparisons along the lines of Wrangham's (1987) set of fourteen socioecological variables would yield useful results. Another approach, even more fine-grained, would be to compare technological and socio-ecological variation *within* a species, as enough data become available for enough populations studied similarly enough.

The results of the second aim can be summarized in a sentence: The only relationship apparent from these simple comparisons by homology and analogy is that the more similar the animal portion of the diet in two types of apes, the more alike also is their known tool use.

ANCESTRAL HOMINOIDS

If living hominoids are a muddle, what can possibly be said about extinct ones? Complications abound. Caution is advisable, for several reasons: First, current opinion repeatedly stresses the pitfalls of referential modelling based on a single species (Tooby & DeVore 1987). Second, by definition, living species cannot be ancestral. In chimpanzees at least, cultural evolution is likely to have occurred in parallel with and inseparably from organic evolution for millions of years. Third, the palaeo-archaeological and palaeontological records are biased against perishable tools and soft food items.

Given these and other obstacles, a conservative model for testing hypotheses about the evolutionary origins of tool use would be a Miocene hominoid descended from a dryopithecine (hylobatid-like) ape. (See point A in Fig. 1). This ancestral proto-pongid could have had the intellectual capacities of living great apes, and its tool use might have been more or less developed according to some local combination of socioecological forces. Sometimes this ape may have acted like a chimpanzee, sometimes like a gorilla, sometimes like an orang-utan. In this line of reasoning, living wild orang-utans are non-tool users because they have 'given up' its phenotypic expression by making a latter-day commitment to arboreality (cf. Galdikas 1982; Lethmate 1982). Highland gorillas have similarly traded off tool use for life in the 'salad-bowl', that is, technology has been 'sacrificed' for a dependable and abundant, but low-quality folivorous diet. Paradoxically, bonobos have ended up having it both ways (or neither?), as a sort of ecological hybrid between arboreal, fruit-eating orang-utan and terrestrial, foliage-eating gorilla.

This model is attractive because it accounts phylogenetically and socioecologically for the contrast between tool use shown by apes in captivity and in nature. All great apes are smart enough to use tools but they only do so in useful circumstances.

Another model hominoid would be an African Pliocene form ancestral to living *Pan* and *Homo* (see point B in Fig 1). This later-living model is perhaps more radical, as such a form need not have been a tool user. In this scenario, the bonobo remains least changed of all living apes, with its greater reliance on terrestrial herbaceous vegetation and lack of competition from sympatric apes (and maybe even humans until recently) (see Wrangham 1986, for the source of this argument). Sometime later in antiquity, proto-chimpanzees and proto-hominids convergently evolved the beginnings of tool use in their respectively less and more open habitats. Even as recently as 1–2 million years ago, long after the divergence in brain size, their archaeological records could have been indistinguishable. (However, in this model it is difficult to explain the rich but latent tool using talents of orang-utans, as Asian apes long separated from their African cousins, other than as a weak assertion of likely evolutionary convergences. This does not seem satisfying.)

What can be said in summary about the paper's third aim? It seems likely that ancestral hominoids made and used tools no less complicated than those used by living chimpanzees. This means that material culture (and the term is used advisedly) long predates the first lithic artefacts recognizable in the archaeological record. If the tool kits of some apes differ little from those of some human beings (McGrew 1987), then the same is likely to be true of a common ancestor. As Foley (1987) has recently stressed, we will only be able to begin to infer further the nature and extent of ancestral hominoid life by doing more palaeosocioecology. Comparisons across living apes remain confusing but fascinating.

ACKNOWLEDGEMENTS

I thank A.C. Hannah and C.E.G. Tutin for photographs, K. Valley for figures, and A. Bowes, R. Layton, J. Lethmate, V. Standen, E. Williamson, and an anonymous reviewer for helpful comments on the manuscript.

REFERENCES

Baldwin, L.A. & Teleki, G. (1976). Patterns of gibbon behavior on Hall's Island, Bermuda. A preliminary ethogram for *Hylobates lar*. *Gibbon and Siamang, Vol. 4* (Ed. by D.M. Rumbaugh), pp. 21–105. Karger, Basel.

Beck, B.B. (1967). A study of problem solving by gibbons. *Behaviour*, **28**, 95–109.

Beck, B.B. (1980). *Animal Tool Behavior*. Garland STPM, New York.

Boesch, C. & Boesch, H. (1983). Optimisation of nut-cracking with natural hammers by wild chimpanzees. *Behaviour*, **83**, 265–86.

Carpenter, C.R. (1937). An observational study of two captive mountain gorillas (*Gorilla beringei*). *Human Biology*, **9**, 175–96.

Fischer, G.J. & Kitchener, S.L. (1965). Comparative learning in young gorillas and orang-utans. *Journal of Genetic Psychology*, **107**, 337–48.

Foley, R. (1987). *Another Unique Species*. Longman, London.

Fossey, D. & Harcourt, A.H. (1977). Feeding ecology of free-ranging mountain gorilla (*Gorilla gorilla beringei*). *Primate Ecology* (Ed. by T.H. Clutton-Brock), pp. 415–47. Academic Press, London.

Galdikas, B.M.F. (1982). Orang-utan tool-use at Tanjung Puting Reserve, Central Indonesian Borneo (Kalimantan Tengah). *Journal of Human Evolution*, **10**, 19–33.

Goodall, A. (1979). *The Wandering Gorillas*. Collins, London.

Goodall, J.v.L. (1968). The behaviour of free-living chimpanzees in the Gombe Stream Reserve. *Animal Behaviour Monographs*, **1**, 161–311.

Hannah, A.C. & McGrew W.C. (1987). Chimpanzees using stones to crack open palm nuts in Liberia. *Primates*, **2**, 31–46.

Harcourt, A.H. & Harcourt, S.A. (1984). Insectivory by gorillas. *Folia primatologica*, **43**, 229–33.

Hayes, K.J. & Hayes, C. (1954). The cultural capacity of chimpanzees. *Human Biology*, **26**, 288–303.

Jones, C. & Sabater, Pi, J. (1971). Comparative ecology of *Gorilla gorilla* (Savage and Wyman) and *Pan troglodytes* (Blumenbach) in Rio Muni, West Africa. *Bibliotheca Primatologica*, **13**, 1–96.

Jordan, C. (1982). Object manipulation and tool-use in captive pygmy chimpanzees (*Pan paniscus*). *Journal of Human Evolution*, **11**, 35–9.

Kohler, W. (1925). *The Mentality of Apes*. Routledge & Kegan Paul, London.

Lethmate, J. (1982). Tool-using skills of orang-utans. *Journal of Human Evolution*, **11**, 49–64.

McGrew, W.C. (1987). Tools to get food: The subsistants of Tasmanian aborigines and Tanzanian chimpanzees compared. *Journal of Anthropological Research*, **43**, 247–58.

McGrew, W.C., Tutin, C.E.G. & Baldwin, P.J. (1979). Chimpanzees, tools, and termites: cross-cultural comparisons of Senegal, Tanzania, and Rio Muni. *Man N.S.*, **14**, 185–214.

Menzel, E.W. (1973). Further observations on the use of ladders in a group of young chimpanzees. *Folia Primatologica*, **19**, 450–7.

Nash, V.J. (1982). Tool use by captive chimpanzees at an artificial termite mound. *Zoo Biology*, **1**, 211–21.

Nishida, T. (1980). The leaf-clipping display: A newly-discovered expressive gesture in wild chimpanzees. *Journal of Human Evolution*, **9**, 117–28.

Redshaw, M. (1978). Cognitive development in human and gorilla infants. *Journal of Human Evolution*, **7**, 133–41.

Rijksen, H.D. (1978). *A Fieldstudy on Sumatran Orang Utans (Pongo pygmaeus abelii Lesson 1827)*. H. Veenman & Zonen B.V., Wageningen.

Rodman, P.S. (1977). Feeding behaviour of orang-utans of the Kutai Nature Reserve, East Kalimantan. *Primate Ecology* (Ed. by T.H. Clutton-Brock), pp. 383–413. Academic Press, London.

Rumbaugh, D.M. (1970). Learning skills of anthropoids. *Primate Behavior. Developments in Field and Laboratory Research* (Ed. by L.A. Rosenblum), pp. 1–70. Academic Press, New York.

Savage-Rumbaugh, E.S., Rumbaugh, D.M. & Boysen, S. (1978). Linguistically-mediated tool use and exchange by chimpanzees (*Pan troglodytes*). *Behavioral and Brain Sciences*, **1**, 539–54.

Sibley, C.G. & Ahlquist, J.E. (1984). The phylogeny of hominoid primates, as indicated by DNA–DNA hybridization. *Journal of Molecular Evolution*, **20**, 2–15.

Sugiyama, Y. & Koman, J. (1979). Tool-using and -making behavior in wild chimpanzees at Bossou, Guinea. *Primates*, **20**, 513–24.

Susman, R.L. (Ed.) (1984). *The Pygmy Chimpanzee*. Plenum, New York.

Temerlin, M.K. (1975). *Lucy: Growing Up Human*. Science and Behavior Books, Palo Alto.

Tingpalapong, M., Watson, W.T., Whitmire, R.E., Chapple, F.E. & Marshall, J.T. (1981). Reactions of captive gibbons to natural habitat and wild conspecifics after release. *Natural History Bulletin of the Siam Society*, **29**, 31–40.

Tooby, J. & DeVore, I. (1987). The reconstruction of hominid behavioral evolution through strategic modeling. *The Evolution of Human Behavior: Primate Models* (Ed. by W.G. Kinzey), pp. 183–237. State University of New York Press, Albany.

Tutin, C.E.G. & Fernandez, M. (1983). Gorillas feeding on termites in Gabon, West Africa. *Journal of Mammalogy*, **64**, 530–1.

Tuttle, R.H. (1986). *Apes of the World*. Noyes, Park Ridge.

Whitten, A.J. (1982). Diet and feeding behaviour of Kloss gibbons on Siberut Island, Indonesia. *Folia Primatologica*, **37**, 177–208.

Wood, R.J. (1984). Spontaneous use of sticks by gorillas at Howletts Zoo Park, England. *International Zoo News*, **31(3)**, 13–18.

Wrangham, R.W. (1986). Ecology and social relationships in two species of chimpanzee. *Ecological Aspects of Social Evolution. Birds and Mammals* (Ed. by D.I. Rubenstein & R.W. Wrangham), pp. 352–78. Princeton University Press, Princeton, New Jersey.

Wrangham, R.W. (1987). The significance of African apes for reconstructing human evolution. *The Evolution of Human Behavior: Primate Models* (Ed. by W.G. Kinzey), pp. 28–47. State University of New York Press, Albany.

Wright, R.V.S. (1972). Imitative learning of a flaked stone technology — the case of an orangutan. *Mankind*, **8**, 296–306.

Yerkes, R.M. (1927). The mind of a gorilla. Parts I & II. *Genetic Psychology Monographs*, **2**, 1–193, 337–551.

The evolution of hominid social behaviour

R.A. FOLEY

Department of Biological Anthropology, University of Cambridge, Downing Street, Cambridge CB2 3DZ, UK

SUMMARY

1 As the only means of determining past evolutionary states, and because living species represent only a very small fraction of total biological variability, palaeobiology is a central element of evolutionary biology. Reconstructions of evolutionary patterns based solely on living organisms cannot address the question of how such patterns evolved.

2 The apparent gap between the socioecology of humans and other animals is a function of the number of intermediate animals (apes and hominids) that have become extinct during the course of human evolution.

3 The direction of evolutionary change is determined by the conjunction of current selective pressures and the nature of the existing biological material (phylogenetic heritage).

4 The direct environmental selective pressures operating on the early hominids appear to result from their occupation of drier and more seasonal tropical environments in Africa some 5–2 million years ago.

5 The phylogenetic heritage of the hominids stems directly from the African apes, and so the socioecology of these apes is central to understanding the direction taken by human evolution.

6 Examination of comparative socioecological models in the context of hominid phylogeny can provide a means for determining the character of the early hominid social systems. Such a cladistic analysis suggests that the earliest African hominoids had a system approximating to that of the gorilla, while that of the earliest hominids most closely resembled chimpanzees.

7 The most appropriate model for social evolution in hominids is one based on the development of kin-based male–male alliances and the increase of paternal care, both occurring in the context of increased group size. These characteristics were a response to the open environments in which they were living.

8 A model of the socioecology of the robust australopithecines suggests a markedly different pattern of behaviour from that of the early members of the genus *Homo*.

PALAEOBIOLOGY AND THE EVOLUTION OF HUMAN SOCIOECOLOGY

Evolution — the dynamic product of natural selection — occurs through time. Neoecologists can gain access to evolution either by detailed comparative studies that relate adaptive features to the environments in which they are found, or through longitudinal studies of the life history strategies of particular populations, providing direct access to biological processes occurring over short periods of time. And yet it is over very long time-spans — what Gould (1977) has referred to as 'deep time' — that major changes in adaptive strategy occur. The importance of palaeobiology — the direct study of biological systems in the past, through geology, palaeontology and archaeology — lies in the fact that it provides information about past evolutionary states. These past evolutionary states are of great significance, for they represent the starting point for selective processes that lead to extant and observable evolutionary and ecological situations. Furthermore, palaeobiology has a particular significance for the study of human socioecology, deriving from the relative diversity of living and extinct hominoids and the extent of the morphological and behavioural gap between modern humans and other animals.

The relative diversity of extinct and extant species

By far the majority of species are extinct. Extant species, although numbering in their millions, are only a small proportion of total evolutionary diversity. Primates and hominids are no exception to this (Table 1). Owing to the problems associated with the fossil record it is not possible to calculate exactly the number of known fossil species of primates, but according to Szalay & Delson (1979) a figure of 253 is a reasonable approximation. Even this may be a serious underestimate of the number of primate species that have ever lived. Assuming a species longevity of approximately 1 million years (Stanley 1979), Martin (1984) has calculated that the some 6000 species of primate of modern aspect (i.e. excluding primitive Palaeocene forms) have existed since the Eocene. In other words, living primates, of which there are 183 species, constitute only 3% of total primate variability. Turning to the Hominoidea, where the palaeontological evidence is relatively good, some forty-six species are known from the fossil record; but this figure should be compared with that of 180 — the number of hominoid species, using Martin's method, that may be expected to have evolved. The implications of this become clear when we consider humans. Assuming they have evolved like any other species, then, given an origin some 6–8 million years ago (Sibley &

TABLE 1. Relative diversity of extinct and living primates, hominoids and hominids. See text for a discussion of the method for calculating these figures

Number of:	All primates	Hominoids	Hominids
Living species	183	12	1
Fossil species	250 (62)*	46	9 (max.)
Total 'expected' species	6000	180	16 ?

* Number of 'substantially known' species.

Ahlquist 1984), the Hominidae should include between fourteen and sixteen species, of which at maximum nine are currently known from the fossil record (Bilsborough 1986; Foley 1987). This compares strikingly with the fact that only one species is extant: *Homo sapiens*. The comparative socioecology, or indeed any aspect of their biology, of hominids based on living species is only a very partial representation of the context in which modern humans evolved.

The effect of differential extinction

Comparisons between modern humans and other animals have always been a source of great controversy within anthropology. It is clear that humans differ in many ways from other animals. The degree and nature of these differences has led some anthropologists to argue that the gap between humans and other animals is one of kind, not degree. Many aspects of human behaviour, according to this view (see Ingold 1986; this volume) cannot be accounted for by the same principles used to explain non-human animal behaviour.

However, the number of extinct hominids has a direct bearing on the magnitude of the gap between humans and the rest of the biological world. As Darwin himself explained (1871), the gap is a function of the rate of extinction. For example, Fig. 1a shows the evolutionary or adaptive distance between modern humans and their closest living relatives, *Pan*, the chimpanzees, assuming constant evolutionary divergence. However, at least part of that distance is accounted for by the fact that most species of hominid known in the fossil record have become extinct. Had they survived to coexist with modern humans, then the distance in morphology, behaviour and life history strategy between modern humans and 'other animals' would not be so great (Fig. 1b). Furthermore, the fossil record is by no means complete. In particular we know nothing of the character of extinct Pliocene and Pleistocene African apes. Some of these may have been more similar to the Hominidae than living forms, and so would further close the gap between humans and other animals (Fig. 1b). Overall, the behavioural and morphological distance between humans and other primates, specifically African hominoids, is in no sense a 'real' biological gap, but is a function of the number of species that have become extinct during the course of human evolution. This rate of extinction is in turn a function of competitive and environmental factors (i.e. of specific historical factors). Differential extinction is the principal determinant of how greatly humans differ from other animals.

DETERMINANTS OF EVOLUTIONARY TRENDS

A comparative framework for human socioecology, therefore, depends upon investigating and understanding the evolutionary character of the extinct hominids, for they bridge the gap between humans and the rest of the animal world. How can this be achieved?

The fossil record must lie at the heart of any answer. It is, however, notorious

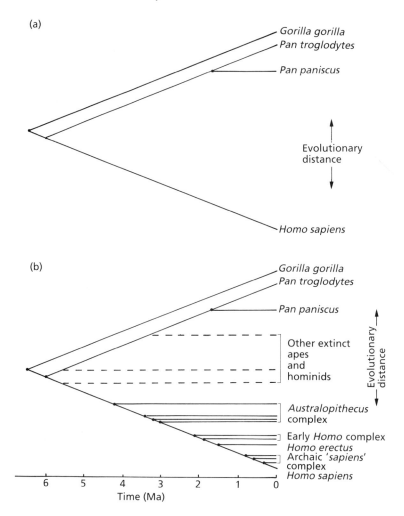

Fig. 1. The evolutionary distance between humans and other animals. (a) The distance based on living species is that between modern humans and Pan; (b) the distance when known fossil hominid species are taken into account and hominoids that are not currently known in the fossil record are included. The implication is that the apparent gap between humans and other species is based on the rate of extinction, and is not an intrinsic characteristic. (Ma: millions of years ago.)

for its incompleteness and limited evidence, and on its own is unlikely to yield up a very full picture of the evolution of hominoid social behaviour. The fossil record must be considered in relation to the fuller knowledge of living species, and within the theoretical framework of the factors that determine the direction the evolution of a species will take. Integrating the two sources of information through the principles of evolutionary ecology enables us to construct models of the socioecology of extinct species. What follows is an exercise in such model building.

Selection and phylogeny

Evolutionary trajectories are determined by the conjunction of two factors. The first of these is the central mechanism of natural selection, through which animals adapt according to the demands and constraints of the environments in which they live. Their evolutionary trajectory is therefore shaped principally by the current selective pressures operating on them. Adaptations, the outcomes of evolutionary changes, will reflect the specific problems of survival faced by organisms, and may to some extent be viewed as 'optimal' solutions to those problems.

However, as has been observed many times (see for example, Maynard Smith 1978) these adaptations do not arise *de novo*, but are always modifications of existing biological characteristics. Natural selection can only operate on the raw material available. This may be referred to as the effect of phylogenetic heritage. To a very large extent the means by which an organism responds to a novel or changing selective pressure depends upon the nature of existing adaptations and characteristics. For example, many catarrhines have adopted a more terrestrial way of life during the Late Tertiary. Those belonging to the Cercopithecoidea have adopted a palmigrade form of quadrupedalism, whereas the hominoids are more upright and use their forelimbs to a lesser extent and in a digitigrade manner. These different responses to the same problem — terrestrial locomotion — may be best understood in terms of their phylogenetic heritage. The ancestors of the cercopithecoids appear to have been above branch quadrupeds, where the ancestors of the living hominoids employed more suspensory behaviour in their arboreal locomotion (Aiello 1981; Andrews & Aiello 1984). These alternative 'starting points' meant that the cercopithecoids could simply maintain their quadrupedalism in a terrestrial context, whereas the hominoids required further modification — knuckle walking and bipedalism (Foley 1987).

To reconstruct the evolution of human socioecology we need to determine both the environments in which early hominids were living (current selection) and the character of the hominoids as they began to adapt to this environment (phylogenetic heritage).

HOMINID PALAEOENVIRONMENTS AND PHYLOGENY

The early hominid environment

The colonization of more open and drier tropical African environments was a major factor in the evolution of the early hominids. The preferred habitats of Late Miocene African hominoids are poorly known, but the earliest hominids occurring during the Pliocene ($5\cdot0-1\cdot8$ Ma) have been recovered from relatively dry African environments (Vrba 1976; Raup & Vondra 1981; Isaac 1984). Palaeo-environmental reconstruction of both the East and South African hominid local-ities suggests that all localities were dominated by what might broadly be referred to as 'savannah' vegetation — that is, environments where grass forms the pre-dominant understory, although there may often be other elements to the flora (bush and light woodland). These environments are today characterized by low rainfall, high levels of seasonality, and frequently a large biomass of large herbi-vores and their associated carnivores. Some variability in habitat (Taung in South Africa is relatively wet and the most heavily wooded, Peninj in East Africa the most open and dry), is evident, as is a slight trend towards habitats becoming more open with time. This dry, seasonal and open tropical environment that expanded considerably during the Late Tertiary undoubtedly imposed some major and novel selective pressures on the hominids.

Hominid phylogeny

Selection for characteristics that would enhance survival in this environment would operate on the specific phylogenetically derived attributes of the hominids. These attributes are gradually becoming known.

Consensus is emerging, based on the studies of molecular evolution, that hominids share a closer phylogenetic affinity with the African apes — *Pan troglodytes*, *Pan paniscus*, and *Gorilla gorilla* — than any of these do with the Asian great ape, the orang-utan (*Pongo pygmaeus*) (Sarich 1983). Furthermore, the separation of the African apes and hominids occurred relatively recently, probably between 5 and 8 million years ago (see Andrews 1986 for a review). Molecular evidence has until recently suggested a trichotemy at this stage — that is, a three-way split between the gorilla, chimpanzee and hominids, but recent work by Sibley & Alquist (1985) using DNA–DNA hybridization techniques indicates that the gorilla–chimpanzee split was earlier, making chimpanzees and hominids monophyletic (Fig. 2). The early hominids were phylogenetically very close to ancestral chimpanzees. This is now having an impact on interpretations of the behaviour and anatomy of the earliest hominids (*Australopithecus afarensis*). Stern & Susman (1983) have suggested a far greater degree of arboreal behaviour in *A. afarensis* than has previously been thought, while Kimbel, White & Johanson (1983) have pointed out some marked similarities in cranial features between chimpanzees and *A. afarensis*. From the socioecological perspective, this close relationship enables us to model the phylogenetically inherited characteristics of the earliest hominids.

COMPARATIVE HOMINOID SOCIOECOLOGY

Such modelling depends greatly upon data drawn from modern hominoids. These data are essential to determining the social characteristics of the last common ancestor of African apes and hominids and that of the earliest hominids. Studies of living hominoids and other species have shown that habitat, resource-base, diet, body size, life history parameters and social relationships are all tightly integrated.

The basic ecological variables which affect social structure are the size, quality and structure of food patches, and the ways in which these patches can be either exploited singly or defended (Wrangham 1980). Food patch size and distribution is consequently one of the main factors compared between the hominoid species. Since female reproduction is critically related to energy and nutrient intake, the effect of these food patch variables will be most marked for females. Males, on the other hand, distribute themselves so as to maintain access to females, and can solely or jointly defend either the females themselves, or the resources required by these females, against male competitors. The final factor which influences patterns of social organization is the growth, survivorship and period of dependence of offspring, which can require differential types and levels of parental care from the two sexes (Krebs & Davies 1981, 1984).

While I shall not attempt a comprehensive review, a brief description of hominoid socioecology (see Table 2 for summary and references) is essential to the arguments developed in the remainder of this paper.

Pongo pygmaeus

The orang-utan is currently confined to the forests of Borneo and Sumatra. The most striking feature of its social organization, relative to most other haplorhine primates, is that it is largely solitary. Lone males occupy and defend a fairly large home range. Females are solitary within a smaller core area within these larger home ranges. Associations between individuals are either temporary sexual liasons between males and females, or more prolonged mother−offspring relationships. This pattern of social organization reflects the extremely dispersed distribution of food. Orang-utans eat fruits that are relatively rare and have constantly shifting distributions. Patches are of a size that can only support a single individual, hence the lack of permanent sociality and the absence of group defence. Home range sizes are smaller for females than males and sexual dimorphism is marked (Table 2).

Gorilla gorilla

Gorillas, the largest living primates, are to be found in the rain forests of Central and Western Africa. Although there are two distinct sub-species — the highland gorilla (G. g. beringei) and the lowland gorilla (G. g. gorilla) — their socioecology appears to be broadly similar. Gorillas live in relatively small groups, the structure of which is centred upon a single older male, the silverback, to whom several typically unrelated females are bonded. Offspring and occasionally unrelated younger males are attached to the group. Social relationships tend to focus upon the silverback, with relatively relaxed and uncompetitive interactions between the females. This 'harem' system involves marked sexual dimorphism, male−male competition and female defence territoriality. The principal food supply is leaves, which are relatively continuously distributed in the environment.

TABLE 2. Comparative socioecology of larger hominoids. Sources used to compile this information include: *P. pygmaeus*: MacKinnon (1974); Galdikas & Teleki (1981); Rodman (1984); Rodman & Mitani (1987). *G. gorilla*: Fossey & Harcourt (1977); Harcourt (1977); Dixson (1981); Tutin & Fernandez (1985); Stewart & Harcourt (1987). *P. troglodytes*: Wrangham (1977); Nishida (1979); Wrangham & Smuts (1980); Ghiglieri (1984); Goodall (1986). *P. paniscus*: Susman (1984); White (1986); Wrangham (1986); Nishida & Hiraiwa Hasegawa (1987)

	Pongo	*Gorilla*	*Pan troglodytes*	*Pan paniscus*	*Homo sapiens*
Body size (weight, kg)	M 84 F 38	M 160 F 89	M 40 F 30	M 43 F 35	M 70 F 63
Sexual dimorphism	F/M 45%	F/M 48%	F/M 75%	F/M 80%	F/M 90%
Diet	Frugivorous (folivorous)	Folivorous (frugivorous)	Frugivorous (folivorous)	Frugivorous (folivorous)	Eclectic (high % meat)
Habitat	Forest (continuous large patches)	Forest (dispersed)	Forest (dispersed small patches)	Forest (dispersed large patches)	'Tropical woodland'
Male–male relationships	Solitary/ competitive	Competitive	M kin alliance	Unknown (possibly competitive)	M kin alliances and lineages
Male–female relationships	Temporary/ sexual	Strong; M central	Sexual consorts	Sexual liaisons	Strong bonds
Female–female relationships	None	Variable	Undifferentiated	Strong ('sexual')	Strong
Group size	1	12–15	26	30 ?	25++
Transfer	Both M & F	Both M & F	F	F M?	F
Parental care	F only	Strong M role preventing infanticide	Predominantly F (short term)	Predominantly F (short term)	M and F (very long term)
Age at first birth	7–14 years	5 years (min.) (6–7 average)	10–18 years	13–14 years	12–18 years
Inter-birth interval	10 years	4 years	4–8 years (6 years average)	3–7 years	3–4 years
Home range size	Small (M larger)	Small	Large (F smaller)	Large?	Very large
Territory	Exclusive (?)	Overlap	Exclusive	Overlap	Exclusive?

Pan troglodytes

The focus of social life is the strong relationship between (related) males. They form the core of a social group, and patrol and defend a territory, excluding males from neighbouring communities. Territorial disputes between groups of males relate to competition for an area rather than females directly, and can be very intense. Females associate loosely with these male cliques, and reside within the defended community range in smaller core areas. Relationships between females, while often friendly, are unstable through time. Females often move from their natal group on reaching sexual maturity. They show very marked sexual swellings and consort promiscuously with males. Females have close and prolonged relationships with their offspring, maintaining these until puberty is reached at about 8 years of age. Chimpanzees are largely frugivorous, but are among the most omnivorous of primates, with insects, birds and mammals forming an important component of their diet.

Pan paniscus

The pygmy chimpanzee or bonobo is the least well known hominoid, being geographically restricted to a bend in the Congo river. It is only slightly (if at all) smaller than the common chimpanzee, but differs in several ways in its socioecology. The principal differences appear to be less consistent or intense male–male relationships, and a strengthening of those between females. Female–female relationships are characterized by consistent associations and genito–genital rubbing, which occurs when entering a feeding patch (see White, this volume). Differences between common and pygmy chimpanzee socioecology may result from the larger patches of food available in the latter's more heavily forested habitat.

Homo sapiens

Modern humans display an extraordinarily wide range of social form and organization, and it is difficult either to characterize it fully or to be sure as to what represents the fundamental form. What follows is an attempt to describe the socioecology of the basal populations of anatomically modern *Homo sapiens*.

The earliest known representatives of this species are found in sub-Saharan Africa from the early Upper Pleistocene (Smith & Spencer 1984). Their basic habitat, therefore, appears to be tropical woodland and more open savannah. Relative to earlier hominids and living hominoids they appear to have been living in larger groups, and incorporating quite large quantities (relative to other primates) of meat in their diet. Their home ranges were probably much larger than those of non-human primates of similar body size. Large group sizes, omnivory and a relatively dry tropical environment would seem to be a useful starting point, then, for describing this species (Foley 1987).

Ethnographic evidence can be used to supplement the information available for prehistoric social organization. As stated above, humans are remarkably variable in their social organization, but using sources such as the Human Relations Area Files (HRAF) certain quantifiable statements can be made that may indicate general tendencies within the species. In relation to the comparative approach adopted here, two such generalizations are of interest. First, that most human societies are characterized by at least the potential for polygynous males, rather than strict monogamy; and second, that patrilocality (i.e. female transfer) appears to be more common than matrilocality. Certain inferences can be drawn from this given the general background of hominoid socioecology discussed above.

Patrilocality suggests a predominance of male–male affinal relatedness over that of females with each other (i.e. a pattern similar to that found in common chimpanzees). However, with modern humans the lateral extent of these affinal relationships is far greater, and very frequently they are extended through several generations (descent groups). This results in characteristic institutions of alliances and patrilines. Within these kin groups, individual males will have durable relationships with one or more female. Females are often 'exchanged' between patrilines. Despite the range of variations in this pattern, it may be argued that if there is a basic human social system, one involving a combination of overall male-bonded alliances, with strong male–female bonds creating a significant set of sub-groups (families) is an appropriate candidate. Another, no doubt related, characteristic is that human young have a very prolonged period of parental dependence, requiring considerable maternal and paternal care.

A cladistic approach to the socioecology of the hominoids

The approach adopted here is to combine the pattern of socioecological diversity of living hominoids with their phylogeny. This approach leans heavily on cladistics. This is a system of phylogenetic classification employed widely in palaeontology (see Eldredge & Cracraft 1980). At its heart lies the distinction between ancestral or primitive features and derived ones. The former, known as plesiomorphies, are characteristics inherited from an ancestral species. Apomorphies are derived features which are unique in a clade relative to other clades. According to the principles of phylogenetic analysis, only the derived features are of any use in constructing a clade, since primitive features can say little about degrees of relatedness.

Cladistics can be usefully employed to investigate the evolution of social behaviour of animals (Ridley 1986), although there are certain problems in applying this formally. It is not clear at what level social 'features' may be treated as traits analogous to the morphological ones used in classic cladistic analysis, and runs the danger of assuming little intra-specific variation in behaviour. However, if the social systems are treated as baseline models from which further variability is developed, then some inferences about the evolution of social behaviour may be made. Emphasis will be placed here on the extent to which the various hominoid social systems may have evolved from one state to another.

Polarities

The essential problem, then, is to determine how any of the observed social systems can be transformed from one to another. Cladistics approaches this problem by assigning evolutionary trajectories or polarities to characters or adaptive complexes. The first task here, therefore, is to assign polarities to the various hominoid social systems. This means making assumptions about the relative antiquity of the social systems observable today, or more specifically stating which of them is more likely to be derived or primitive. However, unlike other cladistic analyses, this one is constrained (or helped) by the fact that the basic phylogenetic relationships can be established by other means (see Fig. 2).

A series of alternative polarities are possible (Fig. 3). These indicate that one of several alternative evolutionary sequences might have occurred during the course of hominoid evolution. For example, at its most simple (Fig. 3b), each hominoid might represent an evolutionary stage — from solitary (orang-utan) to harem (gorilla) to male alliances (common chimpanzees) to bilateral relationships (pygmy chimpanzee) to the modern human system. At another extreme, all the modern systems may be recent developments that have evolved independently from social systems that are no longer extant (Fig. 3a). However, as a device for exposing what transitional and/or extinct states may have existed, the close genetic relatedness of African hominoids is used as a basis for the evolution of the social systems. The evolutionary ecological basis underpinning this is discussed below.

In establishing which of the various social systems are the most primitive or derived, the presence of two vectors of change provide a guide (Fig. 4). One is the increasing links developed between males and females, the other the bonds between related males. The extent to which either or both of these are developed provides the best indication of how derived a social system may be. On this criterion, the orang-utan is clearly the most primitive, followed by the gorilla. Humans are the most derived, with the chimpanzees lying in an intermediate position (Fig. 3e). However, given the peculiar asociality of the orang-utan relative to all other living catarrhines, they themselves may be derived (a specialized south-east Asian ape grappling with large body size and habitat loss: MacKinnon 1974), and a harem structure may indeed be ancestral to all large hominoid species. Whichever is correct, they both show a pattern where each adaptive step appears to be the shortest possible. For example, the transition from orang-utan to gorilla is one of females becoming more permanently attached to the male within whose home range they forage. The transition from gorillas to common chimpanzee is one where males permanently associate to defend a

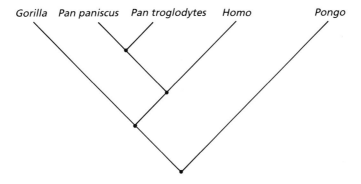

FIG. 2. Evolutionary relationships of the hominoids.

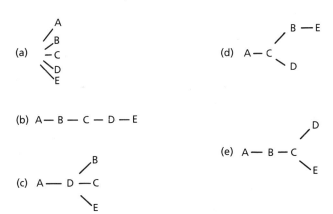

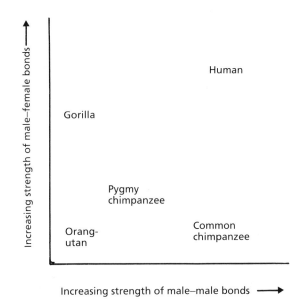

FIG. 3. Possible polarities of hominoid social systems. These diagrams show the possible routes of evolution of the various hominoid social systems (see text for a description of them), from primitive on the left of each diagram to derived on the right. For example, (b) suggests that each social system is a stage in hominoid evolution. The preferred polarity employed in this analysis is (e). A: *Pongo*; B: *Gorilla*; C: *Pan troglodytes*; D: *Pan paniscus*; E: *Homo sapiens*.

FIG. 4. Proposed vectors of change in hominoid socioecology. Development of male−male links and male−female links varies between species among the Hominoidea, providing a basis for determining polarities.

common range, while that from chimpanzees to humans is one of strengthening the male−male bonding across time and generations. This conforms to the principle of parsimony that is central to cladistic methodology.

This model of polarities provides the basis for assigning social systems at the principal branching points during hominoid evolution (Fig. 5). This can be used to suggest the nature of ancestral hominid social systems, and the phylogenetic background with which the hominids faced the selective pressures of their new environment.

Hominoid cladistics and social evolution

Figure 5 shows the basic hominoid cladogram and the social systems associated with extant species. Using the previously proposed models, the social systems at the various branching points can be established. In the figure, the last common ancestor of all the African hominoids is shown as a harem holder. If the chimpanzees and humans represent a clade relative to the gorilla, their last common ancestor was probably similar in social organization to the modern common chimpanzee. This evolutionary sequence can now be placed in an appropriate socioecological context, and what is basically an estimate based on phylogeny alone can be assessed in terms of its socioecological consistency and logic.

THE EVOLUTIONARY ECOLOGY OF HOMINID SOCIAL BEHAVIOUR

The framework developed here is that an adaptation, social or anatomical, derives from the interaction of phylogenetic characteristics and direct environmental (*sensu lato*) selective pressures. The considerations of both the palaeoenvironments of the hominids and hominoid social behaviour in a phylogenetic

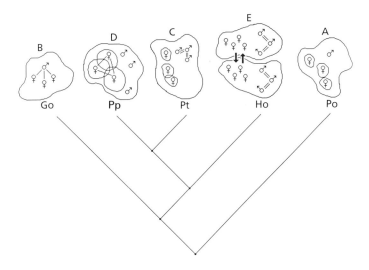

FIG. 5. Cladogram of hominoids showing their social systems graphically. Po: *Pongo*; Go: *Gorilla*; Pt: *Pan troglodytes*; Pp: *Pan paniscus*; Ho: *Homo sapiens*.

context provides us with the framework for examining this proposition in relation to the evolution of hominid social behaviour.

African hominoids

The hominoids of the early and middle Miocene were principally arboreal and inhabitated relatively moist, equatorial, forested environments (Bernor 1983). According to the argument presented above the Middle Miocene (*c.* 10 Ma) hominoids of Africa would have lived in closed, single male groups. This would be consistent with the habitat. Resources would have been relatively abundant and distributed in large uniform patches. Females would have been able to subsist within a relatively small home range, and individual males able to defend such a group of females. This pattern is maintained in *Gorilla*, full forest specialists, foraging for the abundant low quality browse of the increasingly refuge areas of central Africa, despite changes in body size (a result of increased dependence upon low quality foods?).

The late Miocene saw a shift to a more seasonal and drier climate. The main adaptive problem faced by the hominoids that colonized these new habitats that were emerging was the dispersed and patchy distribution of food. With resources more constrained, males would need to defend the resources required by females and offspring, not just the females themselves. The females in turn would be under considerable energetic and reproductive stress, requiring more widespread foraging behaviour. Overall the nuclear grouping of the single male systems of the ancestral African hominoids would have broken down, with two consequences. The first is females becoming distanced from a single male as they forage more widely and independently; the second is that as the female home range became larger several males would have to coexist in order to patrol and defend that area. This latter shift would have been achieved by male siblings remaining with their natal groups past puberty, and so forming alliances of males based on kinship. This more chimpanzee-like behaviour would be the predicted response to a more patchy resource base. It is also a response that provided one of the key elements on which human social organization is built — the kin-based male alliance. The general trends of hominoid socioecology are shown in Fig. 6.

Early African hominids

The proposition made at the beginning of this paper — that the direction of evolutionary change is the product of the interaction between phylogenetic heritage and immediate selective pressures — can now be examined in relation to the early hominids. Their reconstructed environment was yet drier and more seasonal, and so their resource base even more patchy and unpredictable in time and space. Their phylogenetically inherited socioecology (i.e. from their common ancestor with the chimpanzee) consisted of associations of males and females, in which the strongest links were between related males. The critical question is how the

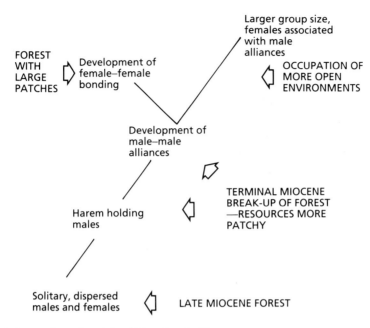

FIG. 6. Evolutionary dynamics of the African hominoids.

evolutionary dynamics of the hominids developed from this starting point in response to the occupation of full savannah conditions.

Open, seasonal and terrestrial environments appear to select, among other things, for larger group size among primates (Clutton-Brock & Harvey 1977; Foley 1987). This is partly a response to the greater threat of predation, and partly an expected outcome of the utilization of a more patchy resource base. This can be seen most clearly among the Papionini (see Dunbar 1988). There is, however, a critical difference between most of these cercopithecoids and the hominoids. Among the former, group size has grown up around a matrilineal focus (an exception is *Papio hamadryas*: Dunbar 1988; see discussion below about the robust australopithecines); in most cercopithecoid species the males transfer between groups, leaving related females as the central core. As we have seen, though, among the hominoids, and especially the chimpanzee, it is the males that form the core to any associations that develop, and I have argued that this is also the ancestral hominid condition. If larger group size was selected for among early hominids, it is the males that are more likely to form the basis for any social structure that emerges. The 'cercopithecoid option' — that is, a switch to male transfer and female-centred social groups is unlikely to be viable. The transition from one form to another would either involve both or neither sex transferring, resulting either in a breakdown of sociality or high levels of inbreeding. In other words, such a transition is evolutionarily improbable as it involves a maladaptive 'trough' between the two adaptive 'peaks' of either matrifocal or patrifocal group

Evolution of hominid social behaviour

structure. The switch the other way (such as happened with *Papio hamadryas*) may have been less problematic, occurring in the context of much larger groups where there is less of a practical difference between transferring and not transferring, resulting in opportunities for related males to stay together.

The socioecological model for the earliest hominids, therefore, would involve larger groups in which male alliances and relationships are even more developed than those found in the common chimpanzee. The proposed social system of the earliest hominids (cf. *Australopithecus afarensis*) (Fig. 7) probably consisted of groups of twenty or more individuals, with males linked by a network of kinship. In this model females are shown in a closer relationship with males as a whole than is the case with chimpanzees. This would be the result of increased selective pressures (due to food patchiness and predator threat) for females to stay closer to males (and hence increase visible group size) than occurs in the common chimpanzee. This pattern would form the basis for later hominid social evolution. Table 3 summarizes the socioecology of the earliest hominids.

The adaptive radiation of the hominids

It is now clear that the evolution of the hominids has involved considerable radiation of forms, not just a simple unilinear pattern of change within a single lineage. It follows from this that hominid social evolution is likely to have taken divergent paths in the same way that morphological evolution did. The number of species involved is still a matter of debate, but among the African hominids of the late Pliocene and early Pleistocene (2·5−1·0 Ma) two distinct trends can be recognized. One of these is towards megadonty and markedly robust cranial

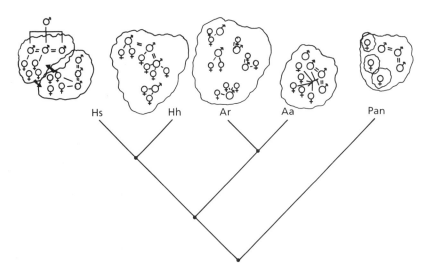

FIG. 7. Cladogram of early African hominids showing proposed social systems graphically. Aa: *Australopithecus afarensis*; Ar: the robust australopithecines; Hh: early *Homo*; Hs: *Homo sapiens*.

TABLE 3. Comparative socioecology of early African hominids

	Australopithecus afarensis	Australopithecus robustus	Homo habilis
Body size	25–30	60+	50
Sexual dimorphism	Marked	Marked	Moderate
Diet	Frugivorous	Low quality foliage (tubers?)	Omnivorous (meat)
Habitat	Woodland– savannah	More open	Woodland/bush/ grassland
Male–male relationships	Strong/kin alliances	Weakened	Strengthened
Male–female relationships	Association with male alliance	Bond to single male	Bond to males within alliance
Female–female relationships	Weak	Weak	Weak (getting stronger?)
Group size	Larger than *Pan*	Very large, but sub-structured	Larger than *Pan*
Transfer	F	F	F
Parental care	F	F, with M protection	F, with M provisioning
Home range size	cf. *Pan*	Larger	Larger

musculature (the robust australopithecines, sometimes referred to as parapithecines). The other is towards encephalization (the genus *Homo*). It is widely recognized that these two trends represent rather different adaptive strategies. It has been argued elsewhere (Foley 1987) that the robust australopithecines, as a response to more seasonal conditions, were exploiting coarse, fibrous, and generally low quality or beneath-ground plant foods, while *Homo* was more involved in omnivorous foraging, including the consumption of greater quantities of vertebrate meat than any living non-human primate. These two evolutionary trends are likely to have had divergent socioecological implications.

Table 3 and Fig. 7 show the way in which these two groups changed from the ancestral state suggested for *A. afarensis*. The robust australopithecines were considerably larger than earlier hominids, and probably lived in more open environments. If their principal food source was low-quality plants, as is indicated by their teeth, then in the savannah regions in which they were living these foods are likely to have occurred in large, quite dispersed patches. This is similar to the resource base of the gelada baboons. In these species group size has become considerably larger still. Were this to occur among the hominids, the effect would be to weaken male–male ties, as there would be a lower degree of overall relatedness. In these circumstances sub-groups are likely to develop, with a female or females becoming attached to a single male in a way similar to that of the gorilla, but in the context of a larger group. Males would have been under considerable pressure to provide some sort of protection to their offspring in this increasingly multi-male environment, but this may have been short-term only. There is some evidence (Dean, personal communication) that growth rates among immature robust australopithecines were particularly rapid, suggesting selection for the independence of the young at an early age. This may reflect their vulnerability to competitive interference within a larger group, and consequently there may have been some protection of young by their fathers.

The rapid growth, and possibly reproductive, rates of the australopithecines contrast with those of modern humans and probably the genus *Homo* as a whole. Slow growth rates among humans are most probably related to their brain enlargement. Selection for larger brain size was probably a two-fold function. On the one hand, increased diet breadth and home range size would select for greater information processing and behavioural flexibility; on the other, larger and more complexly related social groups would select for considerable mental sophistication and a potential for flexible and rapid responses to social interactions (Humphrey 1976). Encephalization therefore brings with it many benefits — increased learning capacity, increased behavioural flexibility, etc.—but it also has several costs (Martin 1983). Principal among these is the fact that most brain growth occurs while an infant is entirely dependent upon its mother for food, and the larger the brain, the greater the energetic stress upon the mother. From a socioecological perspective the central question to be asked is: what social and ecological conditions would buffer females from the extra costs of raising large-brained offspring (Martin 1983)?

In the model of early hominid social behaviour presented above, males play relatively little part in parental care. When maternal costs of child care are high, paternal investment can increase when male care can substitute for that of the female. This is usually associated with an increase in paternity certainty. It might be expected, therefore, that the level of paternal investment, possibly through provisioning, increased bonding between specific males and females. This would be occurring within the specific context of a social group in which males are also bonded by lineage relationships. Selection for the maintenance of these male–male links would occur if foraging and provisioning of mother and/or young was enhanced by cooperation or foraging in large groups. This may be the case in patchy environments in general, but more specifically, meat eating, whether hunting or scavenging, may be expected to increase the benefits derived from male–male cooperation. What is critical here is that this feeds directly back to the higher maternal costs involved in encephalization, as meat comes in large packages which can be carried easily (and so is ideal for provisioning, hence its high frequency among carnivores) and is also of very high quality (and so of great benefit to a lactating mother by providing protein to enhance brain growth). Ultimately a change in foraging behaviour provides the conditions in which the costs of brain enlargement can be sustained, while the social context of male alliances may provide the basis for that foraging. The critical development at this stage is the change in male—female relationships from loose associations within the groups to close dyadic partnerships. It should be stressed, though, that there is no reason for these partnerships to be monogamous. A male will be constrained only by the number of partners he is able to provision efficiently, and polygynous structures would almost certainly be expected.

Later hominid evolution

The change from the social system proposed for early *Homo* to that of modern humans (see Tables 2 and 3, Figs 5 and 7) is one of intensification — both of kin networks and lineages, and of male—female relationships. Both of these aspects of social organization result in greater levels of social complexity, and thus are likely to promote selection for the capacity to deal with increasingly complex social situations. This presents the conditions that Humphrey (1976) has argued are necessary for rapid increases in intelligence — selection for the ability to process social information.

It would, however, probably be wrong to assume that all later hominid social evolution is simply a process of unilinear intensification. Major transformations in ecological conditions occurred during the course of human evolution during the Pleistocene (development of hunting, colonization of temperate latitudes, control of fire, etc.) that may have had a major impact on hominid socioecology (Fig. 8). Indeed there is case to be made that the transition to both modern humans and the development of food production had a major impact on human socioecology (Foley 1988). What is important to establish here is that subsequent changes in socioecology during the course of later hominid evolution occurred in the context of the phylogenetic heritage of the socioecology of the early hominid and homi-nines, and these will have played an important part in determining the direction that evolution took.

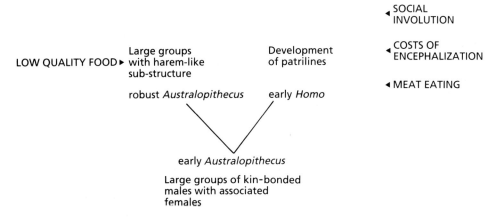

FIG. 8. Evolutionary dynamics of the early hominids.

CONCLUSIONS

This paper has attempted to map out the course of hominid social evolution, not in terms of a narrative, but through applying the general comparative principles of evolutionary ecology to the conditions in which hominid evolution took place. Central to this project has been the close relationship between hominids and living hominoids, and various methods, specifically cladistics, have been developed to use this close affinity to help construct evolutionary history.

To some extent this may run the risk of imposing the present on the past, and of placing living apes in the unenviable position of substituting for our ancestors, rather than being treated as unique species in their own right. This is undoubtedly a danger with this approach, but two points may be made in its defence. The first is that the living hominoids are being used only indirectly — they provide examples of how selection shapes social behaviour, but the basis for placing certain social systems at the branching points of hominoid evolution is derived from the link between social form and ecological conditions. This is fully in accord with the principle of uniformitarianism that underlies all reconstructions of the past. That hominids may have undergone a greater degree of evolutionary change subsequently than the other African hominoids is not a function of their inherent characteristics, but rather of the ecological circumstances in which the different species found themselves. Indeed the very presence of hominids is likely to have acted as a major constraint on the evolution of the other hominoids.

The second line of defence relates back to the point established at the outset of this paper — that the diversity of the extinct fauna is greater than that of the extant one. The variability in hominoid socioecology is likely to have been far greater than that observable today, and it would be important and exciting to model some alternative ways of being an ape. However, what the work of evolutionary ecology is gradually establishing is that the way in which selection operates can be predicted if the ecological conditions can be ascertained. What this paper has tried to add to this theoretical framework is that adaptations should be even more predictable if we know not only the ecological pressures, but also the phylogenetic heritage on which those pressures will be exerted.

REFERENCES

Aiello, L.C. (**1981**). Locomotion in the Miocene Hominoidea. *Aspects of Human Evolution* (Ed. by C. Stringer), pp. 63–98. Symposia for the Study of Human Biology, 12, Taylor & Francis, London.

Andrews, P.J. (**1986**). Molecular evidence for catarrhine evolution. *Major Topics in Primate and Human Evolution* (Ed. by B.A. Wood, L. Martin & P.J. Andrews), pp. 107–129. Cambridge University Press, Cambridge.

Andrews, P.J. & Aiello, L. (**1984**). An evolutionary model for feeding and positional behaviour. *Food Acquisition and Processing in Primates* (Ed. by D.J. Chivers, B.A. Wood, & A. Bilsborough), pp. 429–66. Plenum Press, New York.

Bernor, R.L. (**1983**). Geochronology and zoogeograhic relationships of Miocene Hominoidea. *New Interpretations of Ape and Human Ancestry* (Ed. by R.L. Ciochon, & R.S. Corruccini), pp. 149–64, Plenum Press, New York.

Bilsborough, A. (1986). Diversity, evolution and adaptation in early hominids. *Stone Age Prehistory* (Ed. by G. Bailey & P. Callow). Cambridge University Press, Cambridge.

Clutton-Brock, T. & Harvey, P. (1977). Species differences in feeding and ranging behaviour in primates. *Primate Ecology* (Ed. by T. Clutton-Brock), pp. 557–84. London.

Darwin, C. (1871). *Descent of Man and Selection in Relation to Sex.* Murray, London.

Dixson, A.F. (1981). *The Natural History of the Gorilla.* Weidenfield & Nicholson, London.

Dunbar, R.I.M. (1988). *Primate Social Systems.* Croom Helm, Beckenham and Cornell University Press, Ithaca.

Eldredge, N. & Cracraft, J. (1980). *Phylogenetic Patterns and the Evolutionary Process.* Columbia University Press, New York.

Foley, R. (1987). *Another Unique Species: Patterns in Human Evolutionary Ecology.* Longman, London.

Foley, R. (1988). Hominids, humans and hunter-gatherers: an evolutionary perspective. *History, Evolution, and Social Change in Hunting and Gathering Societies* (Ed. by T. Ingold, D. Riches & J. Woodburn), pp. 207–21. Berg, Oxford.

Fossey, D. & Harcourt, A.H. (1977). Feeding ecology of free ranging mountain gorilla. *Primate Ecology* (Ed. by T. Clutton-Brock), pp. 415–49. Academic Press, New York.

Galdikas, B.M.F. & Teleki, G. (1981). Variations in subsistence activities of female and male pongids: new perspectives on the origins of hominid labor division. *Current Anthropology*, **22**, 241–7.

Ghiglieri, M.P. (1984). Feeding ecology and sociality of chimpanzees in Kibale forest, Uganda. *Adaptations for Foraging in Nonhuman Primates* (Ed. by P.S. Rodman & J.G.H. Cant), pp. 161–94. Columbia University Press, New York.

Goodall, J. (1986). *The Chimpanzees of Gombe.* Harvard/Belknap Press, Cambridge, Massachusetts.

Gould, S.J. (1977). Eternal metaphors of palaeontology. *Patterns of Evolution, as Illustrated by the Fossil Record* (Ed. by A. Hallam), pp. 1–26. *Developments in Palaeontology and Stratigraphy.* Elsevier, Amsterdam.

Harcourt, A.H. (1977). Strategies of emigration and transfer by primates with particular reference to gorillas. *Zentralblatt für Tierpsychologie*, **48**, 401–20.

Humphrey, N.K. (1976). The social function of intellect. *Growing Points in Ethology* (Ed. by P.P.G. Bateson & R.A. Hinde), pp. 303–17. Cambridge University Press, Cambridge.

Ingold, T. (1986). *Evolution and Social Life.* Cambridge University Press, Cambridge.

Isaac, G. (1984). The archaeology of human origins: studies of the Lower Pleistocene in East Africa 1971–1981. *Advances in World Archaeology*, **3**, 1–79.

Kimbel, W.H., White, T.D. & Johanson, D.C. (1985). Craniodental morphology of the hominids from Hadar and Laetoli: evidence of *Paranthropus* and *Homo* in the mid-Pliocene of eastern Africa? *Ancestors: the Hard Evidence* (Ed. by E. Delson), pp. 120–37. Alan R. Liss, New York.

Krebs, J. & Davies, N.B. (1981). *Introduction to Behavioural Ecology.* Blackwell Scientific Publications, Oxford.

Krebs, J. & Davies, N.B. (Eds) (1984). *Behavioural Ecology*, 2nd edn., Blackwell Scientific Publications, Oxford.

MacKinnon, J.R. (1974). The ecology and behaviour of wild orang utans (*Pongo pygmaeus*). *Animal Behaviour*, **22**, 3–74.

Martin, R.D. (1983). *Human Brain Evolution in an Ecological Context.* 52nd James Arthur Lecture on the Evolution of the Brain. American Museum of Natural History, New York.

Martin, R.D. (1984). Primates: a definition. *Major Topics in Primate and Human Evolution.* (Ed. by B.A. Wood, L. Martin & P. Andrews), pp. 1–31. Cambridge University Press, Cambridge.

Maynard Smith J. (1978). Optimization theory in evolution. *Annual Review of Ecology and Systematics*, **9**, 31–56.

Nishida, T. (1979). The social structure of chimpanzees of the Mahale mountains. *The Great Apes* (Ed. by D.A. Homburg & E.R. McGown), pp. 73–122. Benjamin Cummings, Menlo Park, California.

Nishida, T. & Hiraiwa-Hasegawa, M. (1987). Chimpanzee and bonobos: co-operative relationships among males. *Primate Societies* (Ed. by W.B. Smuts, D.L. Cheney, R.M. Seyfarth, R.W. Wrangham & T.T. Struhsaker), pp. 165–77. University of Chicago Press, Chicago, Illinois.

Raup, G. & Vondra, C.F. (Eds) (1981). *Hominid Sites: Their Geologic Setting.* AAAS Selected Symposium, 63, pp. 25–86. Westview Press, Boulder, Colorado.

Ridley, M. (1986). The number of males in a primate troop. *Animal Behaviour*, **34**, 1848–58.

Rodman, P.S. (1984). Foraging and social systems of orang utans and chimpanzees. *Adaptations for Foraging in Nonhuman Primates* (Ed. by P.S. Rodman & J.G.H. Cant), pp. 134–60. Columbia University Press, New York.

Rodman, P.S. & Mitani, J.C. (1987). Orang utans: sexual dimorphism in solitary species. *Primate Societies* (Ed. by W.B. Smuts, D.L. Cheney, R.M. Seyfarth, R.W. Wrangham & T.T. Struhsaker), pp. 146–54. University of Chicago Press, Chicago, Illinois.

Sarich, V.M. (1983). Appendix: retrospective on hominoid macromolecular systematics. *New Interpretations of Ape and Human Ancestry* (Ed. By R. Ciocchon, & R.S. Corruccini), pp. 137–50. Plenum Press, New York.

Sibley, C. & Ahlquist, J. (1984). The phylogeny of hominoid primates as indicated by DNA–DNA hybridization. *Journal of Molecular Evolution*, **20**, 2–15.

Smith, F. & Spencer, F. (Eds) **(1984).** *The Origins of Modern Humans*. Alan Liss, New York.

Stanley, S.M. (1979). *Macroevolution: Pattern and Process*. Freeman, San Francisco.

Stern, J.T. & Susman, R.L. (1983). The locomotor anatomy of *Australopithecus afarensis*. *American Journal of Physical Anthropology*, **60**, 279–317.

Stewart, K.J. & Harcourt, A.H. (1987). Gorillas: variation in female relationships. *Primate Societies* (Ed. by W.B. Smuts, D.L. Cheney, R.M. Seyfarth, R.W. Wrangham & T.T. Struhsaker), pp. 155–64. University of Chicago Press, Chicago, Illinois.

Susman, R.L. (Ed.) **(1984).** *The Pygmy Chimpanzee*. Plenum Press, New York.

Szalay, F.S. & Delson, E. (1979). *Evolutionary History of the Primates*. Academic Press, London.

Tutin, C.E.G. & Fernandez, M. (1985). Foods consumed by sympatric populations of *Gorilla gorilla* and *Pan troglodytes* in Gabon: some preliminary data. *International Journal of Primatology* **6**, 27–43.

Vrba, E. (1976). The fossil Bovidae of Sterkfontein, Swartkrans, and Kromdraai. *Memoirs of the Transvaal Museum*, **21**, 1–166.

White, F.J. (1986). *The behavioural ecology of the pygmy chimpanzee*. Ph.d. thesis, SUNY, Stonybrook.

Wrangham, R.W. (1977). Feeding behaviour of chimpanzees in Gombe National Park, Tanzania. *Primate Ecology* (Ed. by T.H. Clutton-Brock), pp. 503–38. Academic Press, New York.

Wrangham, R.W. (1980). An ecological model of female-bonded groups. *Behaviour* **75**, 262–300.

Wrangham, R.W. (1986). Ecology and social relationships in two species of chimpanzee. *Ecology and Social Relationships: Birds and Mammals* (Ed. by D.I Rubenstein & R.W. Wrangham), pp. 352–78. Princeton University Press, Princeton, New Jersey.

Wrangham, R.W. & Smuts, B.B. (1980). Sex differences in the behavioural ecology of chimpanzees in the Gombe National Park, Tanzania. *Journal of Reproduction and Fertility*, Suppl. **28**, 13–21.

The social and environmental relations
of human beings and other animals

T. INGOLD

Department of Social Anthropology, University of Manchester, Roscoe Building,
Brunswick Street, Manchester, M13 9PL, UK

SUMMARY

1 In order to arrive at generalizations about the nature and causes of sociality on the basis of comparisons between humans and non-human animals, it is essential to compare like with like. Biological definitions of society, which purport to be applicable right across the animal kingdom, emphasize the criterion of *interaction* between individual conspecifics. However, for anthropologists, the essence of sociality lies in *rules* and *relationships*.

2 Institutional rules, presupposing language, are uniquely embodied in human culture, yet language and culture undoubtedly evolved within the context or interpersonal relationships of a kind that are by no means limited to humanity, and which depend on an animal's awareness of others in its milieu as fellow subjects or consociates.

3 The paper examines to what extent human social relations can be comprehended within an ecological framework of organism−environment interaction. It is first necessary to distinguish between the human 'making' of the artificial environment and the way in which any organism 'constructs' its surroundings. The pattern of sociality, in its interactive sense, derives from the organism's 'construction' of the conspecific component of its environment. But as an ordering of rules, 'society' is a conceptual framework that is imposed upon the 'raw material' of humanity. Thus the human social environment is, in a sense, 'man-made'.

4 If, however, sociality is taken to inhere in neither interactions nor rules but in relationships, there are grounds for separating the domains of social and ecological relations. Within these two domains, human beings exist as bearers of personal and natural powers respectively.

5 The link between these two domains, mediated by the technical and institutional rules of culture, lies in the purposive activity of production. The elaboration of such rules should be viewed in the context of developing power differentials in the relationships, activated in production, between human subjects.

INTRODUCTION

The aim of comparative socioecology, as I understand it, is to comprehend the various forms and manifestations of sociality among both humans and non-human animals within a single explanatory framework provided by the modern synthesis of population ecology and neo-Darwinian evolutionary theory. The considerable power of this synthesis, at least with regard to the sociality of non-human animals, is indisputable. Yet social anthropologists have been remarkably reluctant to take it on board in their comparative study of human societies. To biologists this reaction must seem puzzling, if predictable. For surely human beings are animals, and like other animals they must interact with the resources of their environments, and with each other, in order to live and reproduce. If the social behaviour of ants, wolves, or baboons can be accounted for in terms of its positive contribution to the efficiency of foraging or to reproductive fitness, surely there is no reason in principle why the same should not apply to human social organization. Most biologists are apparently unimpressed by the commonplace anthropological objection that humans are instructed in social behaviour through a learning process, rather than compelled by their genetic inheritance. For if culture is defined merely as learning-transmitted tradition, it certainly is not uniquely human; moreover the biological theory of evolutionary adaptation does not necessarily stand or fall on the stipulation that the units of behavioural encoding should be genes rather than analogous cultural instructions. Biological evolution is not the same as genetic determinism (Borgerhoff Mulder 1987).

In any comparative enterprise, however, it is essential that we should compare like with like. As an anthropologist, my reaction to the explanatory claims of evolutionary ecology is not that it fails to account for patterns of behavioural interaction (albeit generated from a cultural template) analogous to what is usually called 'social organization' in reference to non-human animals, but rather that when we anthropologists speak of 'society' and 'social relationships' we are really referring to something quite different. Indeed from a certain point of view, biological and anthropological understandings of sociality are systematically opposed: what is essentially social for the former is essentially *non*-social for the latter (Ingold 1986b).

My purpose in the first part of this paper is to indicate the kinds of answers that anthropologists might give to the question: what *is* a social relationship? A further question, of fundamental importance, is whether the relationships of which we speak are uniquely human. Should it turn out that they are not, we must be prepared to admit that a so-called 'humanistic' approach may be just as appropriate for the understanding of animal social life as is the approach from biological science for the understanding of human organization. This is not to condone anthropomorphism: it is one thing to attribute subjective awareness to animals, quite another to assume that they think and feel the same way as we do in comparable situations.

In the second part of the paper, I turn to a comparison of the ecological relations of humans and other animals. According to a widely accepted definition, ecology is the study of the interrelations between organisms and their environments (Odum 1975). We are inclined, however, to regard the environment as a set of constraints to which organisms must adapt. I argue, to the contrary, that in an important sense every animal *constructs* its environment. But if our concern is with manifestly self-conscious organisms, such as human beings, the picture is complicated by the fact that their environment may be regarded — at least to some extent — as *artificial*. This observation leads us to ask: in what respects is the environment of human beings man-made? And how does this 'making' differ from the way in which any organism 'constructs' its surroundings?

Having attempted to demarcate the sphere of human ecological relations, I go on in the third part of the paper to ask whether it includes or excludes the sphere of social relations. To what extent can social relations be comprehended within an ecological framework of organism–environment interaction? Naturally the answer will depend on which of the different senses of sociality we adopt. Granted that the environment of any free-ranging animal normally includes conspecifics, it could be argued that its social relations make up a particular subset of the total set of its ecological relations: namely, the set comprising relations between the organism and its *conspecific* environment. All of social science would then comprise but a specialized branch of the general science of ecology. Yet if the human environment is man-made, so also, in some respects, is the environment of social institutions.

We are finally forced to return to the point that humans at least, and possibly other animals too, are not just individual organisms but also intentional agents, the bearers of personal as well as natural powers, subjects as well as objects (Shotter 1974). Whilst constituted as individuals within systems of ecological relations, they are also constituted as persons by their involvement in systems of social relations of a kind that cannot simply be encompassed within the ecological domain.

INTERACTIONS, RELATIONSHIPS AND CULTURE

There are many definitions of sociality in the literature on animal behaviour, though they differ little in substance (see Ingold 1986a for a review). The following definition, by Wilson, may be considered as one example: 'Society [is] a group of individuals belonging to the same species and organized in a cooperative manner. The principal criterion for applying the term 'society' is the existence of reciprocal communication of a cooperative nature that extends beyond mere sexual activity' (Wilson 1978, p. 222). The crucial terms here are cooperation and communication. And in order that the definition should apply — as Wilson intends — not only to mammals or, more widely, to vertebrates, but right across the animal kingdom, these terms have to be understood in a rather special sense. For nothing is implied, in either, about intentional agency or conscious awareness.

Consider first the meaning of cooperation. When we speak of people cooperating, we usually mean their purposive coming together for the performance of some task — for example, to shift a heavy object that one person could not handle alone. This 'coming together' involves a coalescence of personal powers, which in unison constitute the joint agency that puts to use the combined natural powers of the participants. We could say that people *cooperate* in co*operating* the capacities of their several bodies to perform work. Now if we bracket the intentional component of joint action, we are left with only the latter of these two senses of cooperation, yielding a purely mechanical account which describes how several bodies, suitably deployed and coordinated in their movements, can deliver a force sufficient to shift the heavy object. In this restricted sense, there is no difference in principle between the cooperation of living organisms and that of the working parts of a machine, or of a number of machines in a larger industrial process. In both cases, the relations involved are between objects (mechanical or organic) rather than subjects (persons), and the 'work' of cooperation lies not in purposive labour but in the exertion of physical force.

The coordination of bodily movements which is a prerequisite for cooperation in this mechanical sense is achieved through the exchange of signs, that is, through communication. But again, if the concept of communication is to have general applicability across the animal kingdom, we must exclude from it the component of intentionality that normally inheres in the conversation of human beings (and perhaps of some other animals). What remains, having subtracted this component, are sequences of behavioural events involving the emission of signals. Thus an organism may be said to communicate when it emits signals which influence the behaviour of one or more other organisms, which may emit further signals in their turn. Every event of communication is therefore an *interaction*, where the 'interactive' is defined as individuals' 'mutual influence on one another's behaviour' (Baylis & Halpin 1982, p. 257; see also Bastian 1968). Consequently, we may infer that intra-specific interaction (excluding that between sexual partners) is the minimal condition for sociality in the usual biological sense of the term. But this is to say nothing of *relationships*. It is a fact of our own experience that we commonly interact and communicate with people with whom we do not recognize any relationship at all. At what point, then, do we deem a social relationship to exist between interacting parties?

Addressing this question, Hinde (1979) admits that although the distinction between 'interaction' and 'relationship' cannot be absolutely drawn, 'a series of interactions totally independent of each other would not constitute a relationship. An essential character of a relationship is that each interaction is influenced by other interactions in that relationship' (1979, pp. 15–16). It is important to note that this condition is not built into biological definitions of society and social organization which rest on the criteria of cooperation and communication. That is to say, such definitions can countenance societies from which relationships are totally absent. It may be that most, perhaps all insect 'societies' are of this kind, and it is certainly no accident that a number of biological conceptions of sociality (including Wilson's) have their origins in the field of entomology (Baylis & Halpin 1982). I am in no position to speculate on the point — presumably corresponding to some level of complexity in the nervous system — at which relationships emerge in the animal kingdom. Be that as it may, I believe that the idea of a society without relationships would strike most anthropologists, and probably also many primatologists, as absurd. For them, sociality is the definitive quality of relationships, and corresponds precisely to that increment by which any relationship *exceeds* the sum of its constituent interactions.

If the stuff of interactions lies in quanta of energy expended or information emitted, the essence of relationships lies in an enduring connection among the interacting parties, or a binding together of their respective life-histories. Where a relationship exists, every successive interaction between those involved appears as but a moment in its creative unfolding. Like a line in a conversation, which seems both to grow out of what was said before and yet to project it forward in directions that could not have been anticipated on the basis of past conditions, every interaction in a relationship is embedded in a previous history of mutual involvement, and will in turn have a bearing on how the participants react to one another in the future. In that sense, whereas every unit interaction (like every elementary act of speech) is in principle compressible into a single instant, being virtually over in the moment it is begun, a relationship (like a conversation) is a continuous process borne on the current of real time consubstantial with the lives of the participants. That is why to decompose relationships into their constituent interactions is to eliminate their social content, which cannot be recovered simply through a reverse process of statistical reaggregation.

I would suggest that the lack of a 'feel' for relationships, which to an anthropological eye seems to characterize much work on the social life of non-human primates and other mammals, may derive from a difference in the time-scales of observation. Focusing, as it were, on the 'small print' of behavioural interaction — on individual gestures or expressions, or on events of (say) feeding, grooming, copulation or aggression — it is easy to lose sight of the matrix of relationships in which they are embedded, relationships which are not static but which evolve over a time-scale commensurate with an animal's total lifespan. It is remarkable that many fundamental anthropological concepts denoting aspects of sociality — such as reciprocity and exchange, hierarchy and dominance — are rooted in the temporal perspective of human 'lifetime', referring to processes that run over years rather than hours or minutes. Yet when carried over into accounts of animal behaviour, the same concepts seem to catch only fleeting moments in the current of social life. We are not, of course, to suppose that human social life is somehow conducted in slow motion. It is rather that anthropologists tend to focus on the more enduring qualities of inter-personal relationships, often at the expense of the fine detail of dyadic interaction recorded by ethologists.

Returning now to the concepts of cooperation and communication, we could of course use the former term to refer to the way in which parties to a relationship may be practically immersed as fellow subjects or consociates in joint action; indeed as I have already remarked, this is how it is commonly understood in reference to everyday human affairs. However, this sense of cooperation goes far beyond the special, restricted and (as we may now add) short-term sense entailed in biological definitions of sociality such as Wilson's. For example, the cooperation of members of a human hunter-gatherer band exceeds that of bees in the hive, since although both involve a coordination of bodily powers, the hunter-gatherers additionally cooperate *in person* whereas the bees do not. The same distinction applies with regard to communication. Both bees and hunter-gatherers exchange information about the location of food resources. But the communication of band members has an intentional component which is grounded in their mutual 'face-to-face' involvement — that is in a direct *communion* of responsibility and interests. And indeed much of their conversation serves not to communicate information of strategic value for foraging, but to perpetuate the current of mutuality underwriting cooperative action. This corresponds, of course, to what Malinowski (1923) called 'phatic communion'. Thus social relationships among hunters and gatherers, or — to use Price's (1975) expression — in 'intimate economies', are based on *sharing*: not only of food but also of words and — in cooperation — of activities and company. In short, hunters and gatherers, whose communities are 'small in scale and personal in quality' (Price 1975), may be said to share *one another* (Ingold 1986b).

I should add that this characterization of human hunter-gatherer bands could apply with equal force to the intimate social groups of non-human primates such as chimpanzees. Primatologists should therefore have as much cause as anthropologists to protest the inadequacy of entomological concepts, which might work well enough for bees, when it comes to comprehending the much more complex, subtle and long-term layers of relationship with which they have to deal. Yet for the most part students of animal behaviour, whether concerned with bees or chimpanzees, seem to have been content to equate sociality with the homespun notion of 'group living' (Alexander 1974), without questioning what this might entail beyond the mere fact of association.

So far I have outlined two alternative views of the essence of sociality, according to which it inheres in interactions and relationships respectively. There is, however, a third view, very commonly encountered in a certain tradition of anthropology, which holds that social life is founded upon *rules*. Comparing animal societies and human societies, Durkheim classically observed that the latter 'present a new phenomenon of a special nature, which consists in the fact that certain ways of acting are imposed, or at least suggested *from outside* the individual and are added on to his own nature: such is the character of "institutions"' (Durkheim 1982 [1917]). Society, then, is identified with the framework of regulative institutions here reserved for humanity. In the later anthropology of Radcliffe-Brown, this framework came to be known as the *social structure*, an 'ordered arrangement of parts or components'. These components Radcliffe-Brown called *persons*, as distinct from (organic) individuals. Each person is a position in a structure, and the relations between these positions — embodied in institutions — regulate the conduct of their individual incumbents. That is to say, behaviour is governed by norms (Radcliffe-Brown 1952).

Now this view presupposes quite a different conception of the person, and of social relations, from what we have encountered up to now. For Radcliffe-Brown was invoking the classical sense of the Latin *persona*, connoting an artificial role or a part in a performance, and the relations involved connect the parts and not their players. That is to say, they represent the linkages between components of a regular programme of practical conduct, rather than a binding together of the lives being conducted (Ingold 1986a). But once we conceive of society, in these terms, as a system of relations between persons, we have to admit that its existence depends on the ability of subjects to project and externalize their common experience on the level of ideas, that is to formulate collective codes and standards against which actual performance can be monitored. This kind of accounting, the matching of behaviour to rules, is essential to what Hallowell (1960) calls the *normative orientation* of conduct. It requires of the subject a 'capacity for self-objectification', in other words for the conceptualization of the self as a focus of attention, as something that one can, reflexively, be conscious *of*. I think it may be agreed that a condition for such self-objectification is the possession of a symbolic faculty rooted in the special properties of language (above all, in the use of personal pronouns), and therefore that it is unique to humankind.

This conclusion has a critical bearing on the question of how, if at all, society may be distinguished from culture. So long as sociality is understood in purely interactive terms, it is clear that it may occur in the absence of even the most rudimentary aspects of culture: witness, for example, the apparently cultureless 'societies' of ants and bees. But though there can be social life, in this sense, without culture, the reverse does not hold. For clearly, the association and interaction of individuals is a prerequisite for the intergenerational transmission of tradition by learning. However, once we conceive social relations to be grounded in institutionalized systems of rules, the dependency of culture on society is inverted. For the rules are themselves products of the symbolic imagination that is the hallmark of *human* culture.

Arguing along these lines, Fortes has concluded that 'without the cultural equipment that distinguishes man from the rest of the animal world', there can be no rules, hence no social relationships, nor even such a thing as society (Fortes 1983). Indeed there would scarcely remain any grounds at all for distinguishing the social and the cultural; rather they appear to merge into a single domain of 'sociocultural' phenomena, which have in common that they are composed and organized by symbolic meaning (Sahlins 1976). For this reason, anthropologists habitually catch themselves using the terms 'society' and 'culture' virtually interchangeably.

Supposing that we adopt this latter view of sociality, as founded neither in interactions nor in relationships but in culture, how does this affect our understanding of the nature of sharing and cooperation? According to Peter Wilson, a fundamental feature of sharing is that it is *obligatory* rather than simply spontaneous: 'sharing has the status of a rule in [hunter-gatherer] societies' (Wilson 1975). Indeed he asserts that both sharing and cooperation critically distinguish human hunter-gatherers from non-human primate foragers. If cooperation were defined in the interactive sense implied in biological definitions of sociality, this assertion would of course be absurd, for the evidence for non-human cooperative behaviour is clear and unequivocal. It also seems rather unlikely that cooperative *relationships*, of an inter-subjective kind, should be restricted to humanity. Though Crook may well be right in supposing that 'the development of a capacity for objective self-awareness and description marks the boundary between the animal and the human' (1980, p. 267), inter-subjectivity requires that neither ego nor alter be posited as objects of attention, but only that they join in a shared act of attending, or of purposively doing things together. In other words it requires subjective awareness, but not necessarily objective self-awareness.

Thus the assertion that cooperation is distinctively human can only be sustained if we take cooperation to mean not just the conjoint engagement of personal powers but the regulation of activity in terms of a mutually agreed schedule for the division of labour. The same goes for sharing. In some ethological accounts (for example, Galdikas & Teleki 1981, on non-human primates), any distributive interaction whose consequence is the consumption of food by one or more individuals other than the procurer is taken to be sharing, even though the behaviour need involve no component of intentionality, nor any relationships enduring beyond the particular distributive situation. In that sense, sharing is rife throughout the animal kingdom. Should we, on the other hand, adopt the sense of sharing introduced above, as the experience of companionship in intimate social groups, we would expect it to be conditional upon the formation of relationships, but not necessarily of human relationships. Only if we follow Peter Wilson in regarding sharing as the observance of obligation, that is as the subjection of mutual conduct to the governance of a cultural regime in terms of which persons make themselves *accountable* for their actions, can it be regarded as a hallmark of humanity.

To sum up: I have introduced three different notions of sociality, according to which it inheres respectively in interactions, in relationships and in culture. A condition for the first is *communication* regarded as the exchange of coded 'bits' of information between discrete individuals (Cherry 1957). A condition for the second is *communion*, a mutual empathy or unity of feeling and action which, 'having no propositional contents' (Langer 1972), may be expressed just as well by non-verbal as by verbal means. A condition for the third is *language*. I agree with Sebeok (1986) that human language did not originally evolve as a means of communication, and that its communicative function — associated specifically with speech — was an entirely secondary development. For speech is no substitute for non-verbal communication, and the scope and complexity of the latter has, if anything, expanded with the transition to humanity. The primary function of language was rather as a 'modelling device', enabling its bearers to design and articulate standards of conduct and plans for action, and to monitor their performance accordingly. We may conclude that whilst this kind of regulative sociality, founded in language, is uniquely human, and whilst interactive sociality can be generalized across the entire animal kingdom, the scope or relationships — and the sociality inherent in them — is both wider than the former and more narrow than the latter, having emerged alongside the evolution of consciousness long before the appearance of the specifically human consciousness of self.

ORGANISM AND ENVIRONMENT

The environment is, literally, that which surrounds, and therefore presupposes something to be surrounded. 'Animal' and 'environment' form an inseparable dyad (Gibson 1979), or as Lewontin puts it, 'there is no organism without an environment, but no environment without an organism' (Lewontin 1982, p. 160). However, we are inclined to think of the environment as a world of nature filled with objects both living and non-living, both mobile and stationary, like a huge room cluttered with furniture and decorations. From this analogy comes the classic ecological concept of the *niche*, signifying a little corner of nature that an organism occupies, and to which it has fitted itself through a process of adaptation. If I remove a vase from an alcove in the wall, a niche remains for a small object that might appropriately fill the vacant space; by analogy it is implied that the ecological niche of an organism is independently specified by the essential properties of the environment, which imposes the conditions of functioning to which any occupant must conform. The trouble with this analogy is that it ignores the most fundamental property of living things, that with each there comes into being a particular teleonomic project, which underwrites its subsequent development. The effective environment of an organism consists not of objects as such, but of the opportunities or hindrances they offer for the realization of its project. Gibson (1979) expresses this idea nicely by designating the constituents of the niche as a set of *affordances*, but he goes on to argue that each object 'offers what it does because of what it is', or in other words that affordances exist as the invariant properties of environmental objects (1979, p. 139). Though it would appear from this that every niche is constituted 'out there' in the environment, regardless of whether any organism exists to fill it, Gibson's argument is surely fallacious (Shotter 1984). For the same essential objects, whether living or non-living, will afford quite different things to different organisms, depending on the nature of their respective behavioural projects: thus a single stone may, at one time or another, serve as a shelter for the mite concealed beneath it, as an anvil for the thrush which uses it to break open snail-shells, and as a missile for an angry human to hurl at his adversary. Indeed a catalogue of all the possible affordances of the stone would be as long as that of all the activities of all the organisms that make some use of it.

This observation is hardly novel. It was made long ago by von Uexküll, who used the German term *Umwelt* — translated as 'subjective universe' — to describe the environment as constituted within the life-project of an animal (von Uexküll 1982 [1940]). In our example, the stone figures as shelter within the *Umwelt* of the mite, as an anvil within the *Umwelt* of the thrush, and as a missile within the *Umwelt* of the man. Unlike Gibson, von Uexküll insisted that the 'quality' (*Ton*) of an object, corresponding to the Gibsonian affordance, is not given but is *acquired* by virtue of the object's having entered into a relationship with a subject organism. When the man picked up the stone to throw it, 'the stone became a missile — a new meaning became imprinted upon it. It had acquired a "throw quality" (*Wurf-Ton*)' (von Uexküll 1982 [1940], p. 27). From this it follows that, far from fitting into a given corner of the world (a niche), it is the organism that fits the world to itself, by ascribing functions to the objects it encounters, and thereby integrating them into a coherent system of its own. Hence the environment in which it lives and moves (its *Umwelt*) is the projection or 'mapping out' of its internal organization onto the world outside its body, or 'nature organized by an organism' (Lewontin 1982). Take away the organism, and the environment, in this sense, disappears with it.

Yet surely, something remains. There is still a landscape: a contoured surface sculpted by geological forces whose actions — barring quakes and eruptions — are normally so slow as to give it the impression of permanence relative to the life-cycles of even the most long-lived animals. And this landscape is cluttered with what von Uexküll called 'neutral objects'. One example of a neutral object would be a thing consisting of 'two long poles and several short poles, which are connected to the two long poles at regular intervals'. Propped up vertically, it could be a ladder; fixed horizontally to the ground it could be a fence; but as an object in the landscape it is neither. Likewise the stone, in the eyes of an indifferent observer, is neither shelter, nor anvil, nor missile (nor paperweight, pendulum bob or hammer), but an object of a certain shape and size, or a certain chemical composition, and with certain properties of hardness and durability, that could in principle be put to any number of possible uses. Or consider the cave, which as a dwelling was part of an *Umwelt* for ancient humans; but as part of the landscape is merely a semi-enclosed cavity in the rock. Both stone and cave were there in the landscape long before the arrival of the thrush and the man, but there was no anvil before the thrush began smashing shells, and no dwelling before man took up residence.

How should we distinguish between the *Umwelt*, with its array of qualities, and the landscape, with its array of neutral objects? One way would be to say that the former is an environment of *affordances*, and the latter an environment of *essences*, whose constituents are characterized in terms of form and composition rather than potential function, in terms, that is, of what they are like rather than of what can be done with them (Ingold 1986b). Von Uexküll was adamant that since every animal is inevitably a participant in its world, since it never plays the role of an indifferent observer, it cannot enter into relationships with neutral objects. It follows that for the animal, the environment of essences cannot exist. What it perceives are not objects as such, but the affordances of objects, already organized (like the body of which they form an extension) for a life-project coextensive with its own existence. By and large, animals do not deliberately choose among the different possible uses of the objects they encounter, such choices are rather prefigured in the normal course of ontogenetic development. Thus the thrush does not wonder what to do with the stone or the snail, any more than it wonders what to do with its beak.

For this reason I am unhappy about the rendering of the (virtually untranslatable) notion of *Umwelt* as 'subjective universe', and feel that Lewontin — in an otherwise admirable discussion of the ways in which organisms construct their environments — goes too far in inferring that they are 'active *subjects* transform-ing nature according to its laws' (Lewontin 1982). For what is nature if not an environment of essences, a world of objects standing apart from, and opposing, the observing subject? Animals, we are told, do not thus gaze upon the world, but are from the start immersed within it. But for a certain species of animal, the situation is apparently different. Comparing the ways in which mosquitoes and humans perceive mammalian prey, von Uexküll notes that whereas any mammal presents to the mosquito a vivid sense of warmth and juiciness, both cues for blood-sucking, 'in our human *Umwelt* a mammal does not in itself appear as a vivid object, but as a mental abstraction, a concept to be used to classify, not as an object we ever encounter' (1982 [1940], p. 57). The human *Umwelt* is there-fore, in reality, an *Innenwelt*, literally a 'subjective universe': an organization of *concepts* internal to the subject, which, is then *inscribed* upon the external world of nature. Hence the environment appears to us to consist primordially of objects, substances and raw materials that have *yet to be organized* according to a project of our own design. It is a landscape with neutral objects, an environment of essences.

Moreover we perceive an elementary disjunction between ourselves as organ-izers and this inherently resistant raw material. The environment, we feel, is like our very nature something that — initially — we *confront*, and that has first to be rendered intelligible within the context of current needs and purposes. This indicates a quite radical difference between the 'ways of worldmaking' (Goodman 1978) of human beings, and the manner in which other animals, unaided by the primary modelling device of language, 'organize' and 'construct' their environ-ments. For we are, to a degree unmatched elsewhere in the animal kingdom, the authors of our projects of construction. And in this sense Lewontin's assertion, which I cited above, that organisms are 'active subjects transforming nature ac-cording to its laws', does indeed apply in the case (and only in the case) of human beings. From our external, subjective viewpoint, the non-human animal's con-struction of its environment appears not as a 'transforming reaction on nature' but as a pattern of interactions *within* the natural world (Ingold 1986b).

However, neither the construction of the environment of affordances, nor the transformation of the environment of essences, need involve the actual physical modification of objects. The thrush does not modify the stone, it merely takes advantage of properties it already happens to have, and which render it suitable as an anvil. I refer to this harnessing of the useful properties of objects to the project of an organism as *co-option*. Sometimes animals modify objects so as to render them more efficient in use, an activity that is often, and very mislead-ingly, known as tool making. It is misleading because the essence of 'making' lies not in physical modification *per se*, but in the element of *self-conscious design* (Alexander 1964). Compare, for example, the work of the human engineer with that of the beaver. We have no doubt that the former, in building a dam for the purposes of irrigation, is the creator of an artificial environment. But does not the beaver perform an equally impressive feat in constructing its dam and lodge?

Certainly both the engineer and the beaver modify their respective environments. They differ, however, in that the engineer begins with an idea already formed of the nature of the end result and with a knowledge of the procedures required to achieve it. Lacking such knowledge, the beaver is a builder of its environment, but not a designer. Indeed the design it implements, and that is 'written out' in the course of its activity, *has no designer*, being the product of an evolutionary process of variation under natural selection. For this reason, the beaver's dam is not 'beaver-made' in the way that the engineer's dam is 'man-made'.

But by the same token, the human making of the environment need involve nothing more than the *self-conscious co-option* of the properties of neutral objects. The stone is *made*, without modification, into an anvil, or the cave into a dwelling, when it is matched to an idea or concept regarding its future use, thereby converting it into a use-value. And since as objects in a landscape, stone and cave are parts of nature, whilst as concepts, anvil and dwelling are parts of culture, we can conclude that making is equivalent to the cultural ordering of nature — the inscription of an ideal design upon the material world of things.

SOCIAL AND ECOLOGICAL RELATIONS

It is now time to link our discussion of the construction of the environment to the issues addressed in the first part of this paper concerning the nature of sociality. To do so it is necessary to recognize that for any organism, the environment normally comprises three components: the non-living or abiotic world, the world of other species, and the world of conspecifics. In principle, therefore, what has been said about the environment in general should apply with equal force to its conspecific component. Our first question must then be: what are the affordances of the conspecific environment? What qualities does an animal acquire by virtue of its entering into a relation with another animal of the same species, and thereby becoming incorporated into the latter's Umwelt? The whole field of social behaviour, Gibson surmised, could be supposed to rest on the perception (or misperception) by the individual of what other individuals afford (Gibson 1979; see Ingold 1986b). Social life, from this perspective, involves the co-optation of the capacities of conspecifics to the project of an organism, whose capacities will of course be co-opted by others in their turn. Interactions occur because for each partner, the presence of one or more others provides opportunities for fulfilling a need or disposition, hence the organization of society is a projection, onto the surrounding world of conspecifics, of a project internal to the individual.

This is all perfectly consistent with the interactive view of sociality and the purely instrumental sense of cooperation it entails. Indeed there exists a certain affinity between social cooperation and tool use, since both involve the harnessing, to an organic project, of the mechanical properties of environmental objects, though in one case the object happens to be a living conspecific which can 'act back' or reciprocate, whereas in the other case it is an inanimate thing which cannot. A chimpanzee can use a stick to reach food, but as Hall (1963) notes, it could equally 'use' another chimpanzee to bring it to him, whether through positive inducement or threat. In cooperative predation, one individual can 'manipulate' others to bring prey within range for capture. Tool use and cooperation literally merge in circus acrobatics, where human beings use each others' bodies for a host of purposes including climbing, swinging and supporting; likewise weaver ants, by linking their bodies head to abdomen, can form a living chain to bridge a gap between leaves (von Frisch 1974). In these instances, as in cooperative interaction generally, the natural powers of individuals are reciprocally augmented through their precise coordination, yielding results exceeding the sum of the effects of each individual working independently.

However if my argument in the previous section is correct, for human beings the encounter with a conspecific does not act directly as a cue for the emission of a certain type of behaviour (as the encounter with a mammal is a cue for the mosquito's blood sucking). Our response is rather mediated by a conceptual design of which we are ourselves the authors, and which we seek to impose upon the external world. Within this design, the 'neutral objects' of our conspecific environment — namely human beings *qua* individual organisms — are fitted into categories and given parts to play. That is to say, they are provided with the lineaments of personhood. In this sense, just as interactive sociality has its parallel in tool use, so also its regulation in terms of a cultural system is comparable to tool making, where 'making' refers specifically to the element of self-conscious design rather than the physical transformation of environmental objects. Thus in tool making one holds before the mind a picture of the object, or the projected act, against which one continually monitors actual performance, making appropriate adjustments until the desired end is achieved. The same is true in the normative orientation of social action. In both cases, human beings recursively subject their conduct to 'home-made' rules and standards, which precede and govern the subsequent execution. The designs we impose on the conspecific environment are 'institutions', their integration is 'social structure', and the links between their component parts ('persons') correspond to what I have called regulative relations. Or as Bidney (1953) has put it, where tools are artefacts, institutions are 'socifacts'. A precondition for the construction of these or any other kinds of 'facts' (all of which involve a matching of concepts and objects) is objective self-awareness: thus a toolmaking being, as Hallowell (1960) pointed out, must also be a moral being, since both capacities attest to the same psychological functions.

However, the analogies between society and artefact, and between social interaction and tool use, have their limitations, and these are also the limitations of an ecological approach to sociality. For if society is a man-made instrument of cooperation, we must take into account relations between the cooperators as well as the cooperated. In the latter capacity, human beings are but organisms that can be made to perform work. But as fellow workers they are mutually involved in the exercise of personal rather than natural powers, relating to one another as intentional subjects rather than natural objects. This mutual involvement cannot

be comprehended within the conventional ecological framework of organism—
environment interaction. If we adopt the view that sociality essentially inheres in
relationships, rather than in interactions or in culture, we have a basis for *separating* the domains of social and ecological relations, instead of treating the former
as a subset of the latter. Among human hunter-gatherers, for example, cooperative
'behaviour patterns' in resource extraction may be comprehended in exclusively
ecological terms (Steward 1955), but the mutuality of sharing that underwrites the
hunter-gatherer's responsibility to his or her group should be seen as not
ecological but social.

As this example indicates, it would be wrong to draw the interface between
social and ecological domains around the boundary of the species, putting all
intra-specific relations into the first domain and all inter-specific relations into
the second. With regard to human beings, intra-specific relations have both a
social and an ecological aspect: between persons (intentional subjects) they are
social, between individuals (organic objects) they are ecological. The same would
hold in the case of non-human animals, to the extent that they are similarly
involved in relationships of an inter-subjective kind. Moreover, turning from
intra- to inter-specific relations, an equivalent dualism applies. When a man goes
out to hunt, he does so as a responsible agent: his hunting is an exercise of
personal powers. And since this responsibility is founded in relations of sharing,
hunting qualifies not just as action but as *social* action. Viewed ecologically,
however, hunting reduces to predation, which we define as an interaction between
an organism of a certain species (in this case, human) and one or more organisms
of another species, resulting in the death of the latter and their conversion into
food for consumption by the former. In other words, hunting activity issues from
the person, predatory behaviour from the individual. If sharing and cooperation
are likewise distinguished as the conjunction of personal and natural powers
respectively, it follows that hunting is to predation as sharing is to cooperation. In
each pair the first term refers to the social aspect, and the second to the ecological
aspect, of inter-specific relations on the one hand, and of intra-specific relations
on the other (Ingold 1986b). Or to put the same point in another way, through
hunting and sharing environmental resources are engaged in a common nexus of
social relations; through predation and cooperation individuals are engaged in a
common nexus of ecological relations. I have shown elsewhere (Ingold 1986b)
that the same distinction separates the principles of tenure and territoriality in
human communities.

Having established a basis for distinguishing the domains of social and ecological relations, our final task is to characterize the link between them. This link,
I believe, lies in purposive activity that affects the state of the physical world, that
is, in *production*. Following a precedent set by Engels (1934), it is conventionally
supposed that humans alone produce, and that all other animals are mere 'collectors'. Engels originally based this distinction on the grounds that human
procurement activity is uniquely governed by plans, and oriented 'towards definite,
preconceived ends'. But confusing this imposition of self-conscious design on

the environment with the criterion of its physical modification (a confusion which, as we have seen, also plagues definitions of tool making), subsequent generations of anthropologists and prehistorians have come to regard production not as the hallmark of humanity *vis-à-vis* other animals, but as the particular achievement of human cultivators and pastoralists *vis-à-vis* hunters and gatherers. The fact that certain species of ants engage, apparently quite instinctively, in activities formally analogous to those of plant and animal husbandry, has been conveniently disregarded. In truth, many kinds of animals, in constructing their environments, also modify them physically to a greater or lesser extent — I mentioned earlier the example of the beaver. I concur with Engels that only humans deliberately *design* their environments, but I would regard the incorporation of environmental objects into this design as a matter of *appropriation* rather than production. It is through their appropriation that objects come to be valorized as resources, and acquire significance within a symbolically constituted, cultural system.

In my view the essence of production lies neither in its results (in terms of environmental modification), nor in its conformity to preconceived design, but in the intentional component of the activity itself. If non-human animals can act as intentional agents, they can also produce, even if that production is not preceded by a conscious model. Human production is distinctive only to the extent that it is guided by a symbolic projection of the end to be achieved and of the constituent steps towards its realization. The structures through which humans appropriate the resources of their environment include both 'technical' and 'social' rules, the former regulating interactions with animals, plants and inanimate objects, the latter regulating interactions (whether cooperative or distributive) with other human beings. For example, the rules of gardening instruct people who derive their subsistence from horticulture in their concrete dealings with particular categories of plants; likewise the rules of kinship instruct them in their behaviour towards certain categories of humans. Culture, comprising the totality of such rules, therefore has a basically instrumental function, translating the intentionality of subjects, constituted by their mutual involvement in inter-personal relationships, into material interactive effects. In short, culture mediates the relation, activated in production, between the social world of persons and the ecological world of organism—environment interactions.

CONCLUSION

A difficulty with generalized biological definitions of sociality, note Baylis & Halpin, 'arises as a result of the discrepancy in meaning given to the words 'social' and 'society' when used in common parlance and when used by entomologists. Any definition of sociality that is applicable to the vertebrates as well as to the social insects must emphasize the shared attributes instead of the dissimilarities between the two cases' (1982, pp. 257–8). Sociobiologists have been somewhat reluctant to recognize the inevitable implication: that any account of the nature and causes of sociality that purports to be equally applicable to insects, mammals and humankind can hardly be expected to throw much light on what we commonly understand by society and social relationships. This is not to reject the socio-biological approach, it is merely to point out that its explanatory remit is really very limited, and that it does not at all capture the essence of social life as it is experienced by our human selves, and which social anthropologists consider it their task to interpret and explain. It is entirely specious to claim that comparative socioecology can offer a generalized account of social *rules* and *relationships*, when a condition for comparative generalization is the adoption of a narrow definition of sociality according to which it is limited to intra-specific *interaction*.

I realize, of course, that many investigations of sociality among non-human primates, or among mammals in general, are centrally concerned with the emergent properties of relationships. However, whilst vociferously denying that the discontinuities between non-human and human sociality are as wide as anthropologists are inclined to make out, primatologists and mammalogists have not — to my knowledge — drawn explicit attention to the points first, that there is a lot more to social life than 'living in groups', and secondly, that if sociality inheres in relationships, then the so-called 'social insects' are not really social at all. Be that as it may, rather than arguing over whether sociobiology or socioecology can explain everything or nothing about social life, I think it is far more profitable to show how accounts of cooperative interaction, inspired by theoretical models from evolutionary biology and by the empirical comparison of human and non-human behaviour, may be integrated with the kinds of understandings that anthropologists can provide about social institutions and relationships, through an approach more attuned to the subjective meanings and intentions of the participants themselves.

Explanations of cooperative behaviour in terms of its adaptive functions do not, in themselves, tell us anything about the evolution of consciousness, nor about the evolution of that form of sociality that inheres in the relationships among cooper*ators* as intentional agents. For some of the most complex and large-scale forms of cooperative organization are to be found among animals, such as insects, which can hardly be said to be subjectively 'aware'. Indeed the lack of such awareness may be a *condition* for complex cooperation. In theory, the more that consciousness develops, the better equipped is the organism to overcome the determinations of its conspecific environment, and the greater are its powers of autonomous action. By and large the scale of cooperation has *decreased* with the development of personal powers (compare the insect colony with the human hunter-gatherer band), a trend that has only been reversed in a relatively late stage of human history by virtue of their systematic *repression*.

Through such repression the individual components of what Mumford (1967) has called the 'great labour machine', though human beings, 'were reduced to their bare mechanical elements' of bone, nerve and muscle. Likewise Marx (1930 [1867]) described the detail worker of capitalist industry as one cog in a 'living mechanism of manufacture', his entire body converted into a specialized instrument for performing just one type of operation (see Ingold 1986a). It is on the factory floor that we find the true human analogue of the cooperation of the insect colony, for having placed their labour-power at the disposal of an employer, workers no longer cooperate in person but are cooperated in the latter's interest. Yet of course capitalist cooperation differs from that of the insects in that it follows the dictates not of individual natures but of an imposed blueprint. More generally, I believe that it is *within the context* of emerging power differentials in the relations between social subjects that we can account for the establishment, elaboration, and eventual hypertrophy of those artificially imposed systems of rules and regulations, constituted within culture, that go by the name of human societies. For this reason a sociobiology that can countenance interactions but not relationships, and that is blind to the inter-subjective domain of social being, will never be able to provide more than a partial and lopsided view of the peculiar sociality of our species.

REFERENCES

Alexander, C. (1964). *Notes on the Synthesis of Form*. Harvard University Press, Cambridge, Massachusetts.

Alexander, R.D. (1974). The evolution of social behaviour. *Annual Review of Ecology and Systematics*, **5**, 325–83.

Bastian, J. (1968). Psychological perspectives. *Animal Communication* (Ed. by T.A. Sebeok), pp. 572–91. Indiana University Press, Bloomington.

Baylis, J.R. & Halpin, Z.T. (1982). Behavioural antecedents of sociality. *Learning, Development and Culture* (Ed. by H.C. Plotkin). Wiley, Chichester.

Bidney, D. (1953). *Theoretical Anthropology*. Columbia University Press, New York.

Borgerhoff Mulder, M. (1987). Adaptation and evolutionary approaches to anthropology. *Man* (N.S.), **22**, 25–41.

Cherry, C. (1957). *On Human Communication*. MIT Press, Cambridge, Massachusetts.

Crook, J. (1980). *The Evolution of Human Consciousness*. Clarendon Press, Oxford.

Durkheim, E. (1982 [1917]). Society. *The Rules of Sociological Method and Selected Texts on Sociology and its Method* (Ed. by S. Lukes, translated W.D. Halls), p. 248. Macmillan, Basingstoke.

Engels, F. (1934). *Dialectics of Nature*. Progress, Moscow.

Fortes, M. (1983). *Rules and the Emergence of Society*. Royal Anthropological Institute Occasional Paper, 39, London.

Frisch, K. von (1974). *Animal Architecture*. Hutchinson, London.

Galdikas, B.M.F. & Teleki, G. (1981). Variations in subsistence activities of female and male pongids: new perspectives on the origins of hominid labor division. *Current Anthropology*, **22**, 241–56.

Gibson, J.J. (1979). *The Ecological Approach to Visual Perception*. Houghton Mifflin, Boston.

Goodman, N. (1978). *Ways of Worldmaking*. Harvester Press, Brighton.

Hall, K. (1963). Tool-using performances as indicators of behavioral adaptability. *Current Anthropology*, **4**, 479–94.

Hallowell, A.I. (1960). Self, society and culture in phylogenetic perspective. *Evolution after Darwin. Vol. II. The Evolution of Man* (Ed. by S. Tax), pp. 309–71. University of Chicago Press, Chicago, Illinois.

Hinde, R.A. (1979). *Towards Understanding Relationships*. Academic Press, London.

Ingold, T. (1986a). *Evolution and Social Life*. Cambridge University Press, Cambridge.

Ingold, T. (1986b). *The Appropriation of Nature: Essays on Human Ecology and Social Relations*. Manchester University Press, Manchester.

Langer, S.K. (1972). *Mind: an Essay on Human Feeling, Vol. 2*. Johns Hopkins University Press, Baltimore, Maryland.

Lewontin, R.C. (1982). Organism and environment. *Learning, Development and Culture* (Ed. by H.C. Plotkin). Wiley, Chichester.

Malinowski, B. (1923). The problem of meaning in primitive languages. *The Meaning of Meaning* (Ed. by C.K. Ogden & I.A. Richards), Suppl. I, pp. 296–336. Routledge & Kegan Paul, London.

Marx, K. (1930 [1867]). *Capital, Vol. I.* Translated by E. and C. Paul from 4th German edition of *Das Kapital* (1890). Dent, London.

Mumford, L. (1967). *The Myth of the Machine: Technics and Human Development.* Secker & Warburg, London.

Odum, E.P. (1975). *Ecology* (2nd edn.). Holt, Rinehart & Winston, New York.

Price, J.A. (1975). Sharing: the integration of intimate economies. *Anthropologica*, **17**, 3–27.

Radcliffe-Brown, A.R. (1952). *Structure and Function in Primitive Society.* Cohen & West, London.

Sahlins, M.D. (1976). *Culture and Practical Reason.* University of Chicago Press, Chicago, Illinois.

Sebeok, T.A. (1986). *I Think I am a Verb: More Contributions to the Doctrine of Signs.* Plenum Press, New York.

Shotter, J. (1974). The development of personal powers. *The Integration of the Child into a Social World* (Ed. by M.P.M. Richards). Cambridge University Press, Cambridge.

Shotter, J. (1984). *Social Accountability and Selfhood.* Basil Blackwell, Oxford.

Steward, J.H. (1955). *Theory of Culture Change.* University of Illinois Press, Urbana.

Uexküll, J. von (1982 [1940]). The theory of meaning. Translated by B. Stone & H. Weiner from *Bedeutungslehre* (Ed. by T. von Uexküll). *Semiotica*, **42**, 25–82.

Wilson, E.O. (1978). *On Human Nature.* Harvard University Press, Cambridge, Massachusetts.

Wilson, P.J. (1975). The promising primate. *Man* (N.S.), **10**, 5–20.

Index

References to figures appear in *italic type* and to tables in **bold type**.